A8

ANIMAL PHYSIOLOGY
Adaptation and Environment

ANIMAL PHYSIOLOGY
Adaptation and Environment

Knut Schmidt-Nielsen

James B. Duke Professor of Physiology
Department of Zoology, Duke University

CAMBRIDGE UNIVERSITY PRESS

Published by the Syndics of the Cambridge University Press
Bentley House, 200 Euston Road, London NW1 2DB
American Branch: 32 East 57th Street, New York, N.Y. 10022

© Cambridge University Press 1975

Library of Congress Catalogue Card Number: 74-12983

ISBN: 0 521 20551 4 hard covers
       0 521 29075 9 paperback

First published 1975
Reprinted with corrections 1975
Reprinted 1976, 1977

Printed in the United States of America
by Vail-Ballou Press, Inc., Binghamton, N.Y.

Contents

To my students

This book was written in anger and frustration. Frustration because I was unable to give my students a book that in simple words says what I find exciting and important in animal physiology, that deals with problems and their solutions, that tells how things work.

For some 20 years I have thought about writing such a book, well knowing that physiology is a field too vast for one person to know, understand, and handle. I looked and failed to find a co-author on whom I could unload some responsibility. One day my procrastination made me very angry at myself and I decided now or never. Perhaps the disadvantages of a job too big for the author might be offset by some coherence in style and viewpoint. A sabbatical leave gave me the needed time. Returning to my students I found myself knowing more physiology than I did when I set out on the agonizing and rewarding task of writing a textbook out of my own heart.

About this book

This book is about animals and their problems. It is not only about how things are, it is about the problems and their solutions. It is also about aspects of physiology that I happen to find particularly fascinating or interesting. It is written for the student who wants to know how things work, who wants to know what animals do and how they do it.

The book deals with the familiar subjects of physiology, respiration, circulation, digestion, and so on. These subjects are treated in 13 chapters, arranged according to major environmental features, oxygen, food and energy, temperature, and water. I consider this arrangement important, for there is no way to be a good physiologist, or a good biologist for that matter, without having an understanding of how living organisms function in their environment.

The book is elementary and the needed background is minimal. I have assumed that the student is familiar with a few simple concepts, such as obtained in a good high school course, or in Introductory Biology at the college level. Otherwise, there will be few demands on prerequisite knowledge. I have included in the text sufficient background information to make physiological principles understandable in terms of simple physics and chemistry. In some cases, a more rigorous treatment has been placed in an appendix (e.g. solutions and osmosis). This makes it available to the student who wants to acquire a better understanding, and to the teacher who wants to make such information required knowledge.

The quantity and complexity of scientific information is steadily increasing, and students are already overburdened with material to remember. Furthermore, the mere recital of factual information does not increase one's understanding of general principles. I have therefore tried to present information that can provide a reasoned background for my statements or conclusions. The student will find that many problems can be understood once a few fundamental principles are familiar to him. I also feel that clear concepts are more important than the learning of terms, but since concepts cannot be conveyed without the use of words, terms are necessary. However, terms should clarify and help, and must therefore be clearly and consistently defined.

To avoid overburdening the student with information, a textbook must necessarily be selective, and many of the omissions are intentional. For example, most students will be familiar with vitamins and the physiology of reproduction, described in terms that have become household words, and there is no need to repeat these endlessly. The mere familiarity with common household words does not automatically confer an understanding of how living organisms work. It is more important to acquire coherent concepts, consistent with available information, consistent with the rules of chemistry and physics, and consistent with what the living organism needs in order to live and function in its environment.

Much of this book explores how animals can live where the environment seems to place insurmountable obstacles in their way. The book tries to compare the possible approaches, and the solutions found by different animals. The study of animals with anatomical or physiological specializations can contribute much to our understanding of general principles. However, unless we look for these general principles comparative physiology is apt to become a description of functions peculiar to uncommon animals, not uncommon because they are rare, but because they are outside our daily experience with ourselves and with well-known laboratory animals, dogs, cats, rats, frogs, etc. What we want to achieve is to put information together into general concepts which help us understand how all animals function.

The text contains literature references. These are arranged at the end of each chapter, not only to tell where I obtained some of the information, but also to help the student satisfy his curiosity without having to search for information that often is hard to come by. The vast quantity of scientific information made it necessary to be highly selective, and opinions about the proper selection will differ.

To bring more specialized and advanced information within the reach of the reader, I have arranged short lists of Useful Reference Material at the end of each of the five major parts of the book. These references can serve as a key to further reading and information. To avoid a feeling of helplessness I have made these lists short. They list titles which vary from brief and simple essays to large, comprehensive treatises. Except for a few older works, I have restricted these lists to reasonably recent and up-to-date material.

Like most authors, I hope that friendly, and perhaps not so friendly, readers will let me know about errors I have made and what they think I should have done better.

What is physiology?

Physiology is about the functions of living organisms, how they eat, breathe, move about, and what they do just to keep alive. To use more technical words, physiology is about food and feeding, digestion, respiration, transport of gases in the blood, circulation and function of the heart, excretion and kidney function, muscle and movements, and so on. The dead animal has the structures that carry out these functions, in the living animal they work. Physiology is also about how the living organism adjusts to the adversities of the environment, obtains enough water to live, or avoids too much water, escapes freezing to death or dying from excessive heat, moves about to find suitable surroundings, food, and mates, and how it obtains information about the environment through its senses. Finally, physiology is about the regulation of all of these functions, how they are correlated and integrated into a smooth-functioning organism.

Physiology is not only a description of function, it also asks the questions 'why?' and 'how?' To understand how an animal functions, it is necessary to be familiar both with its structure and with some elementary physics and chemistry. For example, we cannot understand respiration unless we know about oxygen. Since ancient times breathing movements have been known as a sign of life or death, but the true meaning of respiration could not be understood until chemists had discovered oxygen.

The understanding of how living organisms function is helped enormously by using a comparative approach. By comparing different animals and examining how they each have solved their problems of living within the constraints of the available environment, we gain insight into general principles which otherwise may remain obscure. No animal exists, or can exist, independently of an environment, and the animal which utilizes the resources of the environment must also be able to cope with the difficulties it presents. Thus, a comparative and environmental approach is a fruitful way of gaining insight into physiology.

In examining how an animal copes with its environment the answer will often tend to show what is good for the animal. This may bring us uncomfortably close to explanations that suggest evidence of purpose, or teleology, and many biologists consider this scientifically improper. However, we all do tend to ask 'why?' or 'what good is it for the animal?' Anyway, the animal has to survive, and there is nothing improper or unscientific in finding out how and why animals succeed – if they did not arrive at solutions to the problem of survival, they would no longer be around to be studied.

This book follows an environmental approach to comparative physiology. It begins with a description of how animals obtain oxygen from the environment, be it from water or air. Next it describes the role of blood in transport of oxygen to the tissues, and how the blood is pumped around in the organism. The energy supply (food) is dealt with in a chapter on feeding and digestion, followed by a discussion of energy metabolism in general. An important environmental factor, temperature and its effects, is discussed in two chapters, followed by the equally important role of water to the organism. One chapter deals with movements and locomotion, another with the ways an animal obtains information about its environment (senses). The last chapter of the book discusses how all of these functions with the aid of the hormonal and nervous systems are controlled, correlated, and integrated into a smoothly functioning whole organism.

I

OXYGEN

1
Respiration in water

All living organisms use energy which they must obtain from outside sources. Most plants capture the energy of sunlight and use carbon dioxide from the atmosphere to synthesize sugars and subsequently all the other complex compounds that make up a plant. Animals, on the other hand, use energy from chemical compounds which they obtain from plants, either directly by eating them or indirectly by eating other animals that in turn depend on plants. The chemical energy that animals use is therefore in the end derived from the energy of solar radiation.

Most animals satisfy their energy requirements by oxidation of food materials. A small number of animals can, in the absence of oxygen, utilize chemical energy from organic compounds, but the complete oxidation of these compounds makes available roughly ten or twenty times as much energy. The bulk of most animal food consists of three major groups of compounds, carbohydrates, fats, and proteins. The oxidation of carbohydrates and fats yields carbon dioxide and water as the only end products, protein oxidation yields small amounts of other end products in addition to carbon dioxide and water.

The uptake of oxygen and release of carbon dioxide is called *respiration,* a word that applies both to the whole organism and to the processes in the cells. Animals take up oxygen from the medium they live in, and give off carbon dioxide to it. Aquatic animals take up oxygen from the small amounts of this gas dissolved in water, terrestrial animals from the oxygen in atmospheric air. Many small animals can take up sufficient oxygen through the general body surface, but most animals have special respiratory organs for oxygen uptake. As the cells utilize oxygen for oxidation of foodstuffs, carbon dioxide is formed and follows the opposite path, being released through the general body surface or the respiratory organs. The water formed in the oxidation processes merely enters the general pool of water in the body and presents no special problems.

The most important and sometimes the only physical process in the movement of oxygen from the external medium to the cell is that of diffusion, i.e. the gas moves as a dissolved substance from a higher to a lower concentration. The movement of carbon dioxide in the opposite direction also follows concentration gradients. The diffusion is often aided by a bulk movement (such as the circulation of blood), but this does not change the basic fact that concentration gradients provide the fundamental driving force in the movement of the respiratory gases. To understand respiration, it is therefore necessary to have a basic knowledge of the respiratory gases, their solubility, and the physics of diffusion processes.

Life presumably originated in the sea, and it is convenient to discuss aquatic respiration first and afterwards deal with respiration in the air. The air-breathing animals (primarily vertebrates and insects) are among the most complex organisms; however, the largest number of animals, especially many of the less highly organized invertebrates, are aquatic. After a brief review of the respiratory gases and a bit of basic physics, this chapter deals with the problems of aquatic respiration.

GASES IN AIR AND WATER

Composition of atmospheric air

The physiologically most important gases are oxygen, carbon dioxide, and nitrogen. They are present in atmospheric air in the proportions shown in table 1.1. In addition, the atmosphere contains water vapor in highly variable amounts.

What physiologists usually call nitrogen is actually a mixture of nitrogen with about 1% of the noble gases, and for accuracy these should be listed as well. However, in physiology it is customary to

Table 1.1. *Composition of dry atmospheric air* (Otis, 1964)

	(%)	
Oxygen	20.95	All atmospheric air contains water vapor in highly vari-
Carbon dioxide	0.03	able amounts. The less common noble gases, helium,
Nitrogen	78.09	neon, krypton, and xenon make up a total of only
Argon	0.93	0.002% of the total.
	100.00	

lump these gases with nitrogen. The main reason that we do not separate nitrogen and the noble gases is that in most physiological processes, nitrogen and the noble gases are equally inert to the organism. Another reason is that the analysis of respiratory gases is usually carried out by the determination of oxygen and carbon dioxide, the remainder being called 'nitrogen'. To the physiologist the amount of 'nitrogen' in air is therefore $78.09 + 0.93\%$, or 79.02%. The nearly 1% argon is of physiological interest only in some quite special circumstances, for example in connection with the secretion of gases into the swimbladder of fish. The complete analysis of all the gases in an air sample can be carried out with the aid of a mass spectrometer, an expensive and rather elaborate instrument which is mostly unavailable to physiologists.

The composition of the atmosphere remains extremely constant. Convection currents cause extensive mixing to a height of at least 100 km, and no discernible changes in the percentage composition have been demonstrated, although the pressure of the air is greatly reduced at altitude. The statement that the lighter gases, notably hydrogen and helium, are enriched in the outer reaches of the atmosphere applies to the very outermost layers, which are of no physiological interest whatsoever. For our purposes, the open atmosphere has a constant gas composition, except for its water vapor (Spitzer, 1949).

The composition of the air is maintained as a balance between use of oxygen in oxidation processes (primarily oxidation of organic compounds to carbon dioxide) and the assimilation of carbon dioxide by plants which in the process release oxygen.

The fear that our use of fossil fuels, oils, coal, and natural gas, may deplete the atmosphere of oxygen and add large amounts of carbon dioxide is probably unfounded. In the year 1910 an extremely accurate oxygen analysis showed the value of 20.948%, and during the years 1967 to 1970 repeated measurements gave a value of $20.946\% \pm 0.006$. The investigators who made these very accurate analyses then calculated that if all known recoverable fossil fuel reserves were depleted, there would still be 20.8% oxygen left in the atmosphere (Machia and Hughes, 1970). Physiologically this change would be of no consequence. The slight increase in carbon dioxide caused by the combustion of all the fuel would likewise have negligible physiological effects, but this is not to say that it would be harmless. Even a slight change in carbon dioxide changes the absorption of solar radiation in the atmosphere and may have an unpredictable 'greenhouse' effect which ultimately may drastically change present climatic conditions on the earth's surface.* Another matter is that if carbon dioxide were released in fairly large amounts in the higher atmosphere, for example by high-flying aircraft, the effects may be much more drastic because photosynthetic removal will not take place until convection currents bring this air into contact with the surface vegetation of the earth.

After having stressed the constancy of the atmospheric composition, a few words about special cases are in place. For example, micro-environments such as burrows occupied by animals have more variable air composition, with the oxygen as low as 15% or even less (Darden, 1972). The carbon dioxide content is increased, but not necessarily to the same extent. However, carbon dioxide may rise to above 5%, an amount that has considerable physiological effects.

The air contained in soil, in open spaces between the soil particles, is often low in oxygen. The reason is that the soil often contains oxidizable material that can deplete the oxygen severely. Not only organic matter but also substances such as iron sulfide can in fact consume oxygen until practically all free oxygen has been removed. These oxidation processes depend on temperature, humidity, and other factors, and also how much exchange there is with the atmosphere. Rain, for example, may block the surface porosity of the soil, and at the same time provide humidity for increased oxidation, and the micro-atmosphere may then change drastically.

Water vapor in air. The preceding information about the percentage composition of the atmosphere referred to dry air, and we must now turn to its water content. The pressure of water vapor over a free water surface increases with temperature (table 1.2). At the freezing point the vapor pressure is 4.6 mm Hg, it increases with increasing temperature, and reaches 760 mm Hg at 100 °C. For this reason water boils at 100 °C if the atmospheric pressure is 760 mm Hg. However, if the atmospheric pressure is lower, water will boil at a

---

* The atmosphere is more transparent to incoming short wave radiation than to the long wave radiation emitted by the earth. The outgoing long wave radiation is absorbed in the atmosphere mainly by carbon dioxide and water vapor. It is estimated that a doubling of the atmospheric carbon dioxide content would increase world temperature by 1.3 °C if atmospheric water remained constant, but at higher temperature the atmosphere can hold more water vapor, which increases the blanketing effect and causes further temperature increase. However, increased water vapor in the atmosphere may cause increased formation of clouds, which in turn reflect more of the incoming solar radiation, thus having the opposite effect. The complexity of these relationships makes predictions of the 'greenhouse' effect of carbon dioxide increase highly uncertain (Sawyer, 1972).

Table 1.2. *Water vapor over a free water surface at various temperatures*

Temperature (°C)	Water vapor Pressure (mm Hg)	Pressure (% of 1 atm)	(mg H₂O per liter)
0	4.6	0.6	4.8
10	9.2	1.2	9.4
20	17.5	2.3	17.3
30	31.7	4.2	30.3
40	55.1	7.3	51.1
50	92.3	12.2	83.2
100	760.0	100.0	598.0
37	46.9	6.2	43.9

lower temperature. For example water boils at 20 °C if the pressure is reduced to 17.5 mm Hg.

Any mixture of gases, such as atmospheric air, which is in equilibrium with free water contains water vapor at a pressure corresponding to the temperature, and the fraction of the air sample which is made up of water vapor therefore increases with the temperature (third column of table 1.2). At 37 °C, the usual body temperature of mammals, the water vapor pressure is about 47 mm Hg, and water vapor then makes up 6.2% of the air volume.

The lung air of man and of other air-breathing vertebrates is always saturated with water vapor at body temperature, but the outside atmospheric air is usually not. When air is saturated with water vapor we say that the relative humidity is 100%. If the air contains less water vapor, the humidity can be expressed in per cent of the amount required for saturation at that temperature; for example, 50% relative humidity means that the air contains half of the water it would contain if saturated with water vapor at that temperature.

For some purposes the relative humidity is a convenient expression, but at times we may want to know the total *amount* of water vapor in the air, and this can be expressed as mg H₂O per liter air. Since cold air has a very low water vapor content, even at 100% r.h., the absolute amount of water in cold air is small. Therefore, if saturated outside air in winter enters our houses and is heated, the indoor relative humidity will be extremely low (although the absolute humidity of the air is unchanged), and we say that 'the air is very dry'. This causes moist surfaces and mucous membranes to dry out, often to great discomfort for sensitive persons.

Altitude. Climbing in high mountains or ascent to high altitude in non-pressurized aircraft has serious physiological effects. At an altitude of 3000 meters most men begin to feel the effects of altitude as a reduction in physical performance, and at 6000 meters (about 20 000 feet) most men can just barely survive. This is due to lack of oxygen, although the air still contains the usual 20.95% O_2.

At sea level, where atmospheric pressure is 760 mm Hg and 20.95% of this is oxygen, the *partial pressure* of oxygen in dry air is 159 mm Hg.* At 6000 meters the atmospheric pressure is half that at sea level, or about 380 mm Hg. The partial pressure of oxygen is also half that at sea level, or about 80 mm Hg (20.95% of 380 mm Hg). It is this decrease in partial pressure of oxygen that has such severe effects.

The relation between altitude and atmospheric pressure is shown in fig. 1.1. The abscissa has a lower scale which shows the partial pressure of oxygen of inhaled air. The zero of this scale does not coincide with zero atmospheric pressure, its zero is where atmospheric pressure is slightly less than 50 mm Hg. The reason is simple. At the body temperature of man (37 °C) the water vapor pressure is 47 mm Hg. Therefore, if a man were placed at an atmospheric pressure of 47 mm Hg (19 000 meters or 63 000 feet altitude), his lungs would be filled with water vapor and no air or oxygen could enter his lungs.

We shall later return to some of the effects that the low oxygen partial pressure has on animals at high altitude.

Solubility of gases in water

Gases are soluble in water. If a sample of pure water is brought into contact with a gas, some gas molecules will enter the water and go into solution. This continues until an equilibrium has been established and an equal number of gas molecules enter and escape from the water per unit time. The amount of gas which is then dis-

---

* In a mixture of gases, the total pressure is the sum of the pressure each gas would exert if it were present alone. Thus, in dry atmospheric air at standard barometric pressure (760 mm Hg), the partial pressure of oxygen (P_{O_2}) is 159.2 mm Hg (20.95% of 760 mm Hg), of nitrogen (P_{N_2}) 600.6 mm Hg (79.02% of 760 mm Hg), and of carbon dioxide (P_{CO_2}) 0.2 mm Hg (0.03% of 760 mm Hg).

Atmospheric air is never completely dry, and its water vapor exerts a partial pressure (P_{H_2O}) corresponding to the water vapor content in the air. The partial pressure of the other gases will then be reduced in exact proportion. If the air at 760 mm Hg contains 5% water vapor ($P_{H_2O} = 38$ mm Hg), the total pressure of the remaining gases is 722 mm Hg, and their individual partial pressures would be in the proportion of their relative concentrations to make up the total of 722 mm Hg.

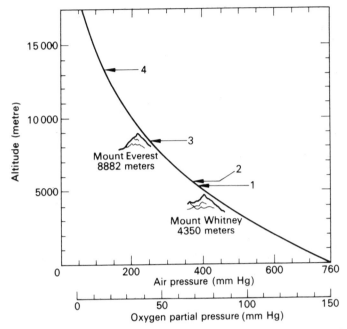

Fig. 1.1. The relationship between altitude and atmospheric pressure. Arrow 1: Altitude where most unacclimatized persons will lose consciousness due to oxygen lack. Arrow 2: Highest permanent human habitation. Arrow 3: Highest altitude where acclimatized humans can survive for a few hours when breathing air. Arrow 4: Highest altitude possible when breathing pure oxygen (from DeJours, P. 1966. *Respiration:* Oxford University Press, p. 187, Fig. 46. © Flammarion et Cie, Editeurs, Paris).

solved in the water depends on (1) the nature of the gas, for the solubility is not the same for all gases, (2) the pressure of the gas in the gas phase, (3) the temperature, and (4) the presence of other solutes. For an understanding of physiology it is necessary to be familiar with the fundamentals of these relationships.

The solubilities of different gases in water are very different, and the physiologically most important ones are listed in table 1.3. Since the solubility depends on both temperature and gas pressure, the conditions of these two variables must be specified.

We immediately see that nitrogen is only about half as soluble as oxygen, but carbon dioxide, on the other hand, is roughly 30 times as soluble as oxygen or 60 times as soluble as nitrogen. It is this high solubility of carbon dioxide in water which makes it possible to make soda water or champagne. In an unopened soda water bottle, which usually is at between two and three atmospheres pressure, the small amount of gas at the top of the neck is under this pressure, the remainder of the carbon dioxide is dissolved in the liquid. As we

Table 1.3. *Solubilities of gases in water at 15 °C when the gas is at 1 atmosphere pressure* *

Oxygen:	34.1 ml O₂ per liter water
Nitrogen:	16.9 ml N₂ per liter water
Carbon dioxide:	1019.0 ml CO₂ per liter water

* The solubility coefficient, α, is defined as the volume of gas (in ml STPD) dissolved in one liter of water, when the pressure of the gas itself, without the water vapor, is 760 mm Hg. The amount of gas is expressed as the volume this gas would occupy if the dry gas were at 0 °C and 1 atm pressure; this is designated as 'Standard Temperature and Pressure Dry', or STPD.

know, this must be a substantial amount, for a large number of bubbles will continue to rise to the surface for a long time. If nitrogen were dissolved at three atmospheres pressure, only a few bubbles would be formed for the amount of dissolved nitrogen would only be about 1/60 of the carbon dioxide.

Pressure and temperature. The amount of gas dissolved in a given volume of water depends on the pressure of the gas. If the gas pressure is doubled, twice as much gas will be dissolved. The proportionality between gas pressure and the amount dissolved is known as Henry's law, and can be expressed as follows:

$$V_g = \alpha \frac{P_g}{760} \cdot V_{H_2O}$$

α is the solubility coefficient (see table 1.3), and the equation tells us the number of milliliters of the gas, V_g, (at STPD) which is dissolved in the water at the pressure P_g, given in mm Hg.

The solubility of a mixture of gases depends on the partial pressure of each gas present in the gas phase. Any one of the gases will be dissolved according to its own partial pressure in the gas phase, independently of the presence of other gases.

The solubility of gases decreases with increasing temperature. Most of us know this from our own experience. When we open a warm bottle of soda-water or beer (God forbid!), the liquid has a much greater tendency to foam and overflow than in a cold bottle. Also, if we watch a pot of water being heated on a stove, small bubbles begin to form on its walls long before boiling begins; these bubbles consist of gas driven out of solution as temperature rises. The solubility of gases in water is thus exactly the reverse of what we know about solids, which for the most part are more soluble in hot water than in cold (sugar, for example).

We can now examine the solubility of a gas in greater detail. As an example, the solubility for oxygen is given in table 1.4. First of all, we must note that this table refers to the amount of oxygen dissolved in water in equilibrium with atmospheric air (not with 1 atm pure oxygen as in table 1.3). We can now see that the solubility for oxygen decreases to about half as the temperature is raised from the freezing point to 30 °C. This decreased solubility is quite important for many aquatic animals.

Table 1.4. *Amount of oxygen dissolved in fresh water and in sea water in equilibrium with atmospheric air* (data from Krogh, 1941)

Temperature (°C)	Fresh water (ml O_2 per liter water)	Sea water (ml O_2 per liter water)
0	10.29	7.97
10	8.02	6.35
15	7.22	5.79
20	6.57	5.31
30	5.57	4.46

Table 1.4 also gives the solubility of atmospheric oxygen in sea water, which on the whole is some 20% lower than in fresh water. This is because salts reduce the solubility for gases. This effect is characteristic for dissolved solids, but does not apply to the presence of dissolved gases that do not affect the solubility of other gases (under conditions with which we deal in physiology).

Partial pressure and tension

We have now discussed the solubility of gases in terms of the amount of a gas that enters into solution when the gas has a certain pressure. Let us look at it the other way. Take a sample of water that has a certain gas dissolved in it; the amount of gas in the water sample must correspond to one specific gas pressure in the gas phase. This pressure is called the *tension* of this gas in the water sample.* If a water sample has several gases dissolved in it, the ten-

---

* Physiologists often refer to the *tension* of a gas in a liquid, rather than its 'partial pressure', tension of a gas in solution being defined as the pressure of this gas in an atmosphere with which that particular liquid sample is in equilibrium. The major reason for saying tension is that a dissolved gas as such exerts no measurable pressure, and the term partial pressure is therefore conceptually somewhat misleading. In common usage, however, we frequently find the two terms used interchangeably.

sion of each gas corresponds to the partial pressure of that particular gas in the atmosphere with which the water is equilibrated. The tension of a gas in solution is thus defined as the partial pressure of that gas in an atmosphere in equilibrium with the solution.

When the gas pressure over a water sample is reduced, gases tend to leave the solution. If we reduce the gas pressure to about half of the original value, gas will leave the solution until equilibrium is reached when the amount of dissolved gas has reached half of its original value. If the gas pressure is reduced to zero, which is the same as exposing the water sample to a vacuum, all the gas will leave or be extracted from the water. Such vacuum extraction is one way of removing all dissolved gas from a liquid, in fact it is a commonly used method in the analysis of the gas content of blood samples.

Since the gas which is dissolved in a liquid is in equilibrium with a given partial pressure in the gas phase, we could say that the gas in the liquid is under that particular 'partial pressure'. If we have a container of water which has been equilibrated with atmospheric air, and we introduce a tiny bubble of e.g. pure nitrogen into the water, oxygen (as well as carbon dioxide) will diffuse from the water into the bubble, equilibrium being reached when the bubble contains 20.95% O_2. Some nitrogen will initially dissolve into the water, for the initial nitrogen pressure in the bubble is 1 atm and the tension in the water only 0.79 atm. However, as oxygen enters, the loss of nitrogen subsides, and the final concentrations within the bubble will be those of the initial equilibration atmosphere. (This argument depends on the bubble being so small, relative to the volume of water, that it does not materially influence the gas concentrations in the water.)

Solubility of carbon dioxide and its rate of diffusion

The solubility for carbon dioxide in water is some 30 times as high as that for oxygen (see table 1.3, p. 10). However, since the amount of carbon dioxide in the atmosphere is very small (0.03%) the total quantity dissolved in water will be very small. The amount can be calculated from the solubility coefficient and the fractional concentration of carbon dioxide in the atmosphere as follows:

(at 15 °C)

$$\text{Volume dissolved } CO_2 = \frac{1019 \times 0.03}{100} = 0.3 \text{ ml } CO_2 \text{ per liter water}$$

In addition to the dissolved carbon dioxide, natural water contains a variable amount of carbon dioxide found in bicarbonates and carbonates. The total amount of carbon dioxide present in water in nature therefore can be quite high, and will vary with the cations present. Hard water, for example, contains large amounts of dissolved calcium bicarbonate ($CaHCO_3$), which adds to the total amount of carbon dioxide. In sea water, which is slightly alkaline (pH = *c.* 8.2), the total amount of carbon dioxide may range between 34 and 56 ml CO_2 per liter sea water (Nicol, 1960), yet the amount of carbon dioxide present as dissolved gas is still that which is in equilibrium with the atmosphere, about 0.3 ml CO_2 per liter, the remainder being primarily in the form of bicarbonate ion. Thus, in spite of the high carbon dioxide *content* of sea water (and many other natural waters), sea water still has a carbon dioxide tension close to that in the atmosphere, 0.23 mm Hg (photosynthesis and respiration may cause minor changes).

In the respiratory organs of animals, gills, lungs, and so on, gases diffuse between the environment and the organism, oxygen enters and carbon dioxide leaves the animal. It is therefore of interest to know how fast the gases diffuse, the *rate of diffusion*. This subject needs special attention, for many biologists have been led to believe that carbon dioxide diffuses much faster than oxygen. This is not so. The rate of diffusion of a gas is inversely proportional to the square root of its molecular weight; carbon dioxide is heavier than oxygen, and therefore diffuses more slowly. The molecular weight of carbon dioxide is 44 and of oxygen 32, and the square roots of these numbers are 6.6 and 5.7. Since the diffusion rates are inversely proportional to the square roots, the diffusing rates of the carbon dioxide and oxygen will be in the proportion of 5.7/6.6, or 0.86. In other words, carbon dioxide diffuses at a rate which is 0.86 of that of oxygen.

When carbon dioxide diffuses between air and water the high solubility of carbon dioxide in water makes it appear that carbon dioxide diffuses faster. The situation can be explained by reference to the diagram in fig. 1.2. Let us assume that we have an atmosphere which contains oxygen at 100 mm Hg and carbon dioxide at 100 mm Hg pressure, i.e., the partial pressures, or concentrations, of the two gases are equal. Each one will dissolve in the surface water independently of the other, in proportion to its concentration in the gas phase and the solubility coefficient. The amount of the two gases dissolved at the surface will therefore be 4.5 ml O_2 liter^{-1} and 134 ml CO_2 liter^{-1} (at 15 °C). In other words, at the surface the concentration of carbon dioxide in solution will be 30 times as high as that of oxygen (29.8 times, to be exact), although in the gas phase their concentrations are equal. Carbon dioxide and oxygen will now dif-

GAS

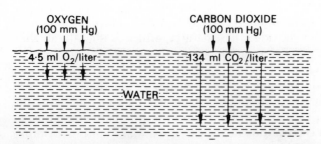

Fig. 1.2. Diffusion between air and water of oxygen and carbon dioxide. The amounts of these two gases dissolved at the surface of water (at 15 ° C) will be 29.8 times as high for carbon dioxide as for oxygen when the two gases are at the same partial pressure in the gas phase. Carbon dioxide molecules diffuse more slowly (at a rate 0.86 times the rate for oxygen), but because of their higher concentration in the surface layer, the amount of carbon dioxide diffusing from the same pressure in the gas phase into the water will be 25.6 times as high for carbon dioxide as for oxygen ($29.8 \times 0.86 = 25.6$).

fuse from the surface into the body of water. The amount of carbon dioxide dissolved at the surface is much higher, and more carbon dioxide therefore diffuses into the water. The carbon dioxide molecules as such diffuse at a rate of 0.86 times that of the oxygen molecules, and the amount of carbon dioxide diffusing, relative to oxygen, will be $29.8 \times 0.86 = 25.6$.

We now understand that, when carbon dioxide and oxygen diffuse from equal concentrations in air into water, the total amount of carbon dioxide diffusing will be about 25 times higher. Likewise, if the carbon dioxide and oxygen tensions in the water are equal (in our case, 100 mm Hg at the surface of the water), the diffusion into the atmosphere will also be higher for carbon dioxide. It is important to note that this apparent faster diffusion of carbon dioxide is only applicable when diffusion takes place in water or between gas and water and we refer to the tension or partial pressure of the gases rather than to their molar concentrations. This fast diffusion is due to the higher solubility of carbon dioxide, and as a molecular species carbon dioxide still diffuses according to the laws of physics, i.e. somewhat slower than oxygen.*

With a knowledge of the physical basis of gas diffusion we can now move on to the question of how animals live and function within these bounds.

* In a gas the molar concentrations are directly proportional to the pressures and no confusion can arise as to the rate of diffusion.

AQUATIC RESPIRATION

Some of the simplest arrangements for respiratory gas exchange are found in aquatic animals. In fact, many small organisms obtain their oxygen by diffusion through the surface without having any special respiratory organs, and without circulating blood. Larger and more complex animals often have specialized surfaces for gas exchange and a blood system that transports oxygen more rapidly than diffusion alone can provide.

Animals without specialized respiratory organs

The simplest geometrical shape of an organism is that of a sphere. For our considerations it is important that a sphere has the smallest possible surface corresponding to a given volume, and any deviation from the spherical shape therefore gives a relative enlargement of the surface area. If we assume that a spherical organism shall be supplied with oxygen by diffusion through the surface and into every part of the body, the longest diffusion distance is from that surface to the center. For oxygen to reach the center, the oxygen concentration at the surface must be of a certain magnitude, for as oxygen diffuses inward it is consumed by the metabolism of the organism. The necessary oxygen tension at the surface, sufficient to supply the entire organism with oxygen by diffusion, can be calculated from an equation developed by E. Newton Harvey (1928), who was well known for his studies of luminescent organisms in the sea.

$$F_{O_2} = \frac{\dot{V}_{O_2}\, r^2}{6\,K}$$

In this equation F_{O_2} is the concentration of oxygen at the surface expressed in fractions of an atmosphere, $\dot{V}_{O_2}$ is the rate of oxygen consumption as ml O_2 per gram per minute (ml O_2 g^{-1} min^{-1}), r is the radius of the sphere in cm, and K the diffusion constant in cm^2 atm^{-1} min^{-1}. (K signifies the number of ml of oxygen which will diffuse per minute through an area of 1 cm^2 when the gradient is one atmosphere per cm. For a further discussion of diffusion, see appendix 2.)

Taking as a hypothetical example a spherical organism with a radius of 1 cm, an oxygen consumption of 0.001 ml O_2 g^{-1} min^{-1}, and a diffusion constant of 11×10^{-6} cm^2 atm^{-1} min^{-1} (the same as for connective tissue and many other animal tissues), we find that the required oxygen concentration at the surface, necessary to supply the entire organism to the center by diffusion, would be 15 atm.

This shows clearly that the organism cannot be supplied by diffusion alone if it has the postulated metabolic rate, which, incidentally, is rather low even for an invertebrate. The conclusion is therefore that the organism, to be supplied with oxygen, must either be much smaller, or have a much lower metabolic rate. If we take a smaller organism, choosing a radius of 1 mm, we find that the required oxygen concentration at the surface is 0.15 atm. Since well-aerated water is in equilibrium with the atmosphere, which contains 0.21 atm O_2, such an organism could obtain enough oxygen by diffusion only and would be quite feasible.

If we consider real animals we find that these calculations have given quite reasonable orders of magnitude. Organisms that are supplied with oxygen by diffusion only, protozoans, flatworms, etc. are mostly quite small, less than a mm or so, or have very low metabolic rates, as jellyfish do. Although some jellyfish can be very large, they may contain less than 1% organic matter, the rest is water and salts. They have a very low average rate of oxygen consumption, and the actively metabolizing cells are located along the surfaces where the diffusion distances are relatively short.

Organisms that deviate from the spherical shape have larger relative surface and shorter diffusion distances than a sphere. This holds for a variety of relatively simple organisms, they are flattened, thread-like, have pseudopodia, or have very large and complex surfaces such as corals and sponges, and can thus obtain enough oxygen by diffusion, although some may be much larger than the sizes used in the preceding calculations.

Animals with respiratory organs

While small organisms can get enough oxygen by diffusion through the surface, this is usually not true for larger organisms. Of course, any shape deviating from the sphere has a larger surface, and the diffusion distances are also reduced. However, in most cases this does not suffice at all, and we find specialized respiratory organs with greatly enlarged surfaces. Often these organs also have a thinner cuticle than other parts of the body, thus facilitating gas exchange.

If the respiratory surface is turned out, forming an evagination, the resulting organ is usually called a *gill*. Secondarily, the gill may be enclosed in a cavity, such as in fish, but this does not change the fact that gills fundamentally are evaginations.

If the general body surface is turned in, or invaginated, the resulting hollow is called a *lung*. Our own lungs are a good example, although secondarily they are finely subdivided and have a quite complex structure. Simpler lungs exist; pulmonate land snails, for example, have a lung that is little more than a simple sac-like invagination in which gas exchange takes place. The term lung is used whether the respiratory medium is water or air.

Insects have a special form of respiratory system. Small openings on their body surface connect to a system of tubes (*tracheae*) which branch and lead to all parts of the body. In this case the respiratory organ combines a distribution system (the tubes) with the gas exchange system, for most of the gas passes through the walls of the finest branches of this system and diffuses directly to the cells.

In general, gills mostly serve for aquatic breathing, and lungs for breathing in air. There are exceptions, sea cucumbers have a water-lung in which most of the gas exchange seems to take place. Gills may also be modified for use in air, but on the whole they are rather unsuited for atmospheric respiration. For example, most fish when taken out of water rapidly become asphyxiated, although there is far more oxygen in air than in water. The reason is that in water the weight of the gills is well supported, but they do not have the mechanical strength and rigidity for support of their own weight in air.

A requirement for an effective respiratory organ is that it should have (1) a large surface, and (2) a thin cuticle. Both these demands are contrary to providing the mechanical rigidity for support of a gill in air. Furthermore, in air the surfaces of the fish gill tend to stick together due to surface adhesion. Therefore, the surface area exposed to air is reduced to a minute fraction of what it is in water, thus severely impeding oxygen uptake.

Ventilation of gills. If a gill removes oxygen from completely still water, the immediately adjacent boundary layer of water will soon be depleted of oxygen. Renewal of this water is therefore important in supplying oxygen, and various mechanical devices serve to increase the flow of water over the gills. Increased flow can be achieved in two ways, either by moving the gill through the water, or by moving water over the gill.

Moving the gill through the water is practical only for small organisms. Some aquatic insect larvae achieve the necessary ventilation in this way. Mayfly larvae (Ephemeridae), for example, ventilate their gills in this way. The difficulty in moving a gill through the water is that the force needed to overcome the resistance to the movement is too great. This resistance increases with the square of the linear velocity of the organ, and therefore the energy needed to move the gill also increases in the same proportion. The mechanical strength of the gill would also need to be increased, again with the square of the linear velocity, as would the force applied to the base of the gill to make it move through the water. The large aquatic salamander known under the name 'mud-puppy' (*Necturus*) does move its gills, but the movements are very slow.

Moving water over the respiratory surface is a much more feasible solution. The movement may be achieved by ciliary action, as in protozoans and in the gills of mussels and clams. Sponges move water through their ostia by the action of flagella, and so on.

Moving the water with a mechanical pump-like device is more common. Fish and crabs, for example, move water over their gills in this way. As a matter of principle, it is less expensive to move water slowly over a large surface, than to move water fast over a smaller surface.

For some animals their own locomotion may contribute to the movement of water. This is true of many pelagic fish; the large and fast-swimming tunas have practically immobile gill covers and obtain the required high water flow over the gills through their rapid swimming through the water. They probably cannot survive if kept from swimming forward, and when these fish are maintained in aquaria it is common to keep them in large circular tanks so that they can keep moving without meeting obstacles.

Also in squid and octopus there is a close correlation between locomotion and water flow over the gills, but in these animals the relationship is in a way reversed. They ventilate their gills by taking water into the mantle cavity, and by ejecting the water through the siphon, they propel themselves through the water by jet propulsion. In this case the ventilatory system has been modified for locomotion, but like fish, as increased swimming increases the call for oxygen, oxygen is automatically provided in greater amounts.

Other gill functions. Gills may have functions other than the respiratory gas exchange, and sometimes it is difficult to decide what is the primary or only function of a gill. Some so-called 'anal gills' of mosquito larvae function in osmotic regulation, they serve in absorbing ions from the surrounding water, and it is doubtful that they play any major role in respiration. It is well known that the gills of both fish and crabs serve in osmotic regulation, but in these cases it is quite clear that the gills also serve a primary function in respiration.

To evaluate the role of a gill or other suspected 'respiratory organ' in gas exchange, it is necessary to have information about the amount of oxygen taken up, and preferably also about the carbon dioxide given off, through this organ. By comparing this with the oxygen consumption of the entire organism, we can see whether the organ is responsible for virtually all oxygen uptake, a large part, a small part, or just a trivial amount.

If an organism has definite and well developed respiratory organs, there is usually a necessity for a circulatory system as well, which can carry oxygen to the various parts of the body. (This, of course, does not hold for the tracheal system of insects, which in principle is independent of a circulatory system.)

The gills of mussels and clams were mentioned above as establishing water currents by ciliary action. In the filter-feeding bivalves the gills are arranged so that they act as a sieve, retaining particles suspended in the water, which afterwards are carried to the mouth and ingested. The gills therefore have a primary function in food uptake; whether they also are of importance in gas exchange is less certain. Bivalves on the whole have quite low metabolic rates, and it is possible that the surface of the mantle is sufficient to provide the required gas exchange.

Gas exchange and water flow. The fact that gill surface area must be large enough to provide adequate gas exchange is well expressed in fish. Highly active fish have the largest relative gill areas (fig. 1.3). The fast-swimming mackerel has a gill surface area, expressed per unit body weight, which is some 50 times as high as the sluggish, bottom-living goose fish.

For the gas exchange to be adequate, it is necessary to have a high rate of water flow and also a close contact between the water and the gill. This is achieved by the anatomical structure of the gill apparatus. The gills are enclosed in a gill cavity which provides protection of these rather fragile organs, but equally important, it permits water to be perfused over the gills in a most effective way. A special advantage is gained by an arrangement which makes the stream of water flowing over the gill and the stream of blood running within the gill flow in opposite directions to each other. This we call a *countercurrent flow*. To understand the importance of this arrangement, it is necessary to know the structure of the gill (fig. 1.4).

The fish gills consist of several major *gill arches* on each side. From each gill arch extend two rows of *gill filaments*. The tips of these filaments from adjoining arches meet, forcing water to flow between the filaments. Each filament carries densely packed, flat *lamellae* in rows. It is in these lamellae that gas exchange takes place as the blood flows through them in the opposite direction to the water flowing between them.

This countercurrent type of flow has an important consequence.

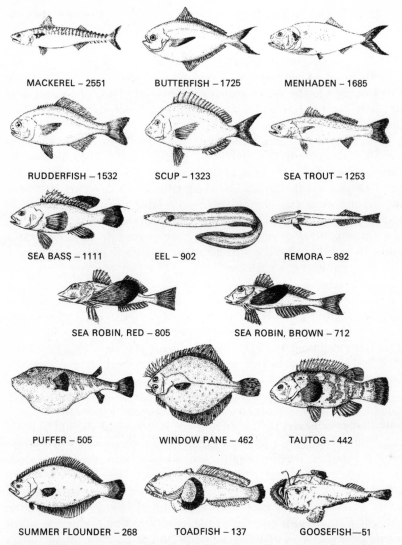

MACKEREL – 2551   BUTTERFISH – 1725   MENHADEN – 1685

RUDDERFISH – 1532   SCUP – 1323   SEA TROUT – 1253

SEA BASS – 1111   EEL – 902   REMORA – 892

SEA ROBIN, RED – 805   SEA ROBIN, BROWN – 712

PUFFER – 505   WINDOW PANE – 462   TAUTOG – 442

SUMMER FLOUNDER – 268   TOADFISH – 137   GOOSEFISH—51

Fig. 1.3. Highly active, fast-swimming fish have larger gill areas than sluggish bottom-living fish. The number after each name indicates the total surface area of the gills expressed in arbitrary units per gram body weight of the fish. (Drawing courtesy of I.E. Gray, based on Gray, 1954.)

Just as the blood is about to leave the gill lamella, it encounters water which has not yet had any oxygen removed from it. Thus, this blood takes up oxygen from water which still has the full oxygen content of inhaled water, and this permits the oxygen content of the blood to reach the highest possible level. As the water runs further between

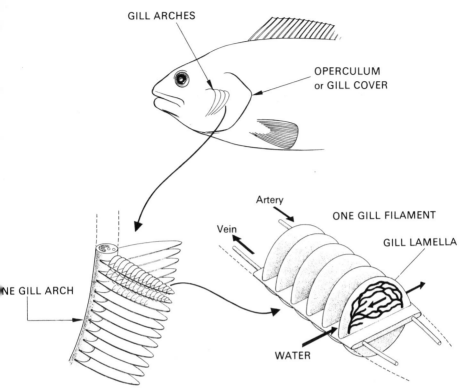

Fig. 1.4. The gills of fish consist of several gill arches on each side. Each arch carries two rows of gill filaments. Each filament carries thin, parallel, plate-like lamellae. In these lamellae the blood flows in a direction opposite to that of the water which flows between the lamellae (from Randall, D. J. 1968. Fish Physiology. *Am. Zool.,* **8,** 179–189. Fig. 1. American Society of Zoologists).

the lamellae, it meets blood with a lower and lower oxygen content, and it therefore continues to give up more oxygen. Thus, the lamella, along its entire length, serves in taking up oxygen from the water, and the water may leave the gill having lost as much as 80 or 90% of its initial oxygen content (Hazelhoff and Evenhuis, 1952). This is considered a very high oxygen extraction; compare it, for example, with mammals, which remove only about one-quarter of the oxygen present in the lung air before it is exhaled.

We can express the effect of the countercurrent type of gas exchange in a diagram (fig. 1.5). This figure shows how blood, as it flows through the gill lamellae takes up more and more oxygen and approaches the oxygen tension of the incoming water. The outflowing water has had most of its oxygen removed, and has a tension far lower than that of the blood leaving the gill. If the flows of water

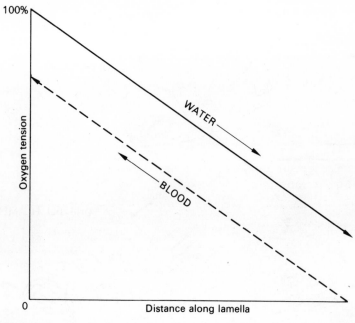

Fig. 1.5. Diagram to show that the countercurrent flow in the fish gill permits the blood (dashed line) to leave the gill with an oxygen tension almost as high as that in the incoming water. The water (solid line) gives up oxygen all along the lamella, and when it leaves the gill it may have lost nearly all its oxygen.

and blood were in the same direction, this would be impossible, for the blood could at best reach the oxygen tension of the outflowing water. Since pumping water over the gills requires energy, the countercurrent flow through the increased oxygen extraction also reduces energetic cost of pumping.

Until recently it was believed that sharks have no countercurrent flow in their gills, although they are excellent swimmers which we might expect to have superbly well-functioning respiratory organs. Recent investigations, however, have shown that sharks do indeed have countercurrent flow (Grigg, 1970). A similar countercurrent type of flow is found also in the gills of some crabs, but in these the efficiency of oxygen removal from the water is far less than in fish (Hughes, Knights and Scammell, 1969). This may in part be because the water flow is less effective, but probably it is mainly because the gill–blood diffusion barrier is greater. This seems to be the case in the European shore crab (*Carcinus*), in which the oxygen extraction, in spite of countercurrent flow, ranges between no more than 7 and 23%.

Because the lamellae in the fish gill are very close, it has been suggested that there would be a high resistance and virtually no flow of
water between them, for the space may be no more than 0.02 mm
wide. The pressure that drives the water through the gills is often
less than 10 mm H_2O, and such a low pressure appears insufficient
to drive water through the narrow space. A careful analysis of this
problem has been carried out by Hughes (1966), based on the width
of the passageway between the lamellae, its height, and its length.
Using a modified form of Poiseuille's equation (see page 144),
Hughes calculated that the flow through the gills of a 150 g tench
(*Tinca*), for a pressure of 5 mm H_2O, would be 10.1 ml s^{-1}. The normal volume of water pumped through the gills of this fish is about 1
to 2 ml s^{-1}; the obvious conclusion is that the gill lamellae do not
offer much resistance to flow, compared to the flow rates that actually occur in the living fish. Similar calculations for more than a
dozen other species of fish in all cases showed that the calculated
ventilation volumes could easily be greater than those measured in
the living fish.

It is interesting that the highest water flows are found in some antarctic fish, the so-called ice-fish, which are very unusual in the sense
that there is no hemoglobin in their blood. We shall return later to
these interesting fish and how they can live without the normal
oxygen-carrying mechanism of the blood.

To move water over the gills, teleost fish use a combined pumping
action of the mouth and the opercular covers, aided by suitable
valves to control the flow. A model of this pump is shown in fig. 1.6.
The system actually consists of a double set of pumps. The first, the
oral cavity, can be enlarged by lowering the jaw and especially the
floor of the mouth. The second consists of the opercular cavity; its
volume can be increased by movements of the opercular covers
while backflow of water around the edges is prevented by a skin flap
acting as a passive valve. The diagram shows only one opercular
pump, in reality there are two opercular chambers, one on each side.

The action of the two pumps is such that a flow of water through
the gills is maintained nearly throughout the entire respiratory cycle.
This flow continues although the pressure in the mouth during part
of the cycle may be less than that in the surrounding water; the
reason is simply that the pressure in the opercular cavity is maintained even lower than in the mouth.

This is evident from pressure recordings made in the mouth and
the opercular cavities during the respiratory cycle (fig. 1.7). The
graph shows that the pressure changes are synchronized with the
movements of the mouth and the operculum. Furthermore, the dif-

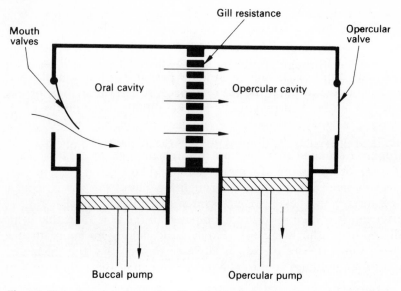

Fig. 1.6. Water is pumped over the gills of fish by a dual pumping system. With the aid of suitable valves the pumps provide a unidirectional flow of water over the gill surface (Hughes, 1960).

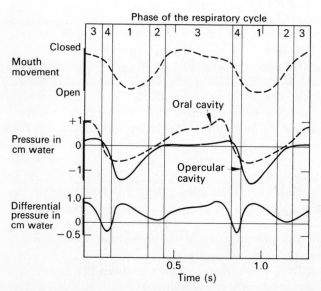

Fig. 1.7. Record of pressure changes in the respiratory pump of a carp-like fish, the roach (*Rutilus*). The lower curve shows the difference between the pressures in the oral and the opercular cavities (Hughes and Shelton, 1958).

ference between the pressures in the mouth and the opercular cavities (as shown by the curve at the bottom of the graph) remains positive almost throughout the cycle and provides the pressure that drives water through the gill. Only during a brief moment is there a slight pressure reversal. The details of the pressure curves differ from fish to fish, but all those that have been studied so far are, in principle, the same; during almost the entire respiratory cycle the pressure in the oral cavity remains higher than in the opercular cavity, thus providing for a virtually continuous flow of water over the gills.

With a continuous flow of water, solid particles suspended in the water will tend to get caught in the gills. Such material can be dislodged by a sudden reversal of the flow, which is brought about by enlarging the oral cavity with closed lips, thus causing a sudden lowering of the pressure in the mouth. This maneuver is analogous to the use of coughing to remove material from the respiratory passageways of mammals.

Crabs keep their gills clean in a somewhat similar way. The flow of water over the gills is unidirectional practically all the time, but at intervals the pumping is stopped and there is a sudden reversal of the flow for a few seconds. The frequency of these abrupt reversals varies greatly. It may happen once a minute or once in ten minutes, or even less frequently (Hughes *et al.*, 1969). We assume that such abrupt reversals of water flow serve to remove particles that have become lodged in the gills.

REFERENCES

DARDEN, T. R. (1972). Respiratory adaptations of a fossorial mammal, the pocket gopher (*Thomomys bottae*). *J. comp. Physiol.*, **78**, 121–37.

DEJOURS, P. (1966). *Respiration*, New York: Oxford University Press, 244 pp.

GRAY, I. E. (1954). Comparative study of the gill area of marine fishes. *Biol. Bull.*, **107**, 219–25.

GRIGG, G. C. (1970). Use of the first gill slits for water intake in a shark. *J. Exp. Biol.*, **52**, 569–74.

HARVEY, E. N. (1928). The oxygen consumption of luminous bacteria. *J. Gen. Physiol.*, **11**, 469–75.

HAZELHOFF, E. H. and EVENHUIS, H. H. (1952). Importance of the 'countercurrent principle' for the oxygen uptake in fishes. *Nature, Lond.*, **169**, 77.

HUGHES, G. M. (1960). A comparative study of gill ventilation in marine teleosts. *J. Exp. Biol.*, **37**, 28–45.

HUGHES, G. M. (1966). The dimensions of fish gills in relation to their function. *J. Exp. Biol.*, **45**, 177–95.

HUGHES, G. M., KNIGHTS, B. and SCAMMELL, C. A. (1969). The distribution of P_{O_2} and hydrostatic pressure changes within the branchial chambers in relation to gill ventilation of the shore crab *Carcinus maenas* L. *J. Exp. Biol.*, **51**, 203–20.

HUGHES, G. M. and SHELTON, G. (1958). The mechanism of gill ventilation in three freshwater teleosts. *J. Exp. Biol.*, **35**, 807–23.

KROGH, A. (1941). *The Comparative Physiology of Respiratory Mechanisms*, Philadelphia, Pennsylvania: Univ. of Pennsylvania Press, 172 pp.

MACHIA, L. and HUGHES, E. (1970). Atmospheric oxygen in 1967 to 1970. *Science*, **168**, 1582–4.

NICOL, J. A. C. (1960). *The Biology of Marine Animals*, New York: Interscience Publishers, Inc., 707 pp.

OTIS, A. B. (1964). Quantitative relationships in steady-state gas exchange. In *Handbook of Physiology*, sect. 3, *Respiration*, vol. i (W. O. Fenn and H. Rahn, eds), pp. 681–98, Washington, D.C.: American Physiological Society.

SAWYER, J. S. (1972). Man-made carbon dioxide and the 'greenhouse' effect. *Nature, Lond.*, **239**, 23–6.

SPITZER, L. (1949). The terrestrial atmosphere above 300 km. In *The Atmospheres of the Earth and Planets* (G. P. Kuiper, ed.), pp. 211–47, Chicago, Ill.: University of Chicago.

2
Respiration in air

Successful, large-scale evolutionary adaptation to air breathing and terrestrial life has occurred only in arthropods and vertebrates. In addition, some snails are well adapted to terrestrial life, some even live in deserts, and a small number of other invertebrates live in various terrestrial microhabitats.

The atmosphere provides a high and constant oxygen concentration, available practically everywhere. The greatest drawback of breathing in air is the evaporation of water.

The easy and troublefree access to oxygen permits a high rate of metabolism, and a high degree of organizational development, both structurally and physiologically. Another consequence of life in air is that the low heat conductivity and heat capacity of air, as compared to water, permits an animal to maintain a considerable difference between its own body temperature and that of the surroundings. This is nearly impossible for aquatic animals. In other words, life in air permits the evolution of warm-blooded animals, such as birds and mammals. [There are schemes through which some aquatic animals, especially large and highly active fish, maintain their bodies at temperatures substantially above that of the water (see p. 354).]

Let us compare how much oxygen is available in water and in air. Water in equilibrium with atmospheric air at 15 °C, contains 7 ml O_2 per liter (1000 ml). 7 ml O_2 weighs 0.01 g, and this amount is found in a weight of water 100 000 times as great. To obtain a given amount of oxygen, we must therefore move 100 000 times its weight * of water over the respiratory organs.

One liter of air, in contrast, contains 210 ml O_2 which weigh 280 mg. The remainder of the air is 790 ml N_2, which weigh 910 mg. In other words, to obtain the oxygen, we must move only 3.5

---

* More correctly, mass. Since in daily speech we say that we weigh an object when we determine its mass, it is very common to say weight when we should say mass. Cf. footnote on p. 37.

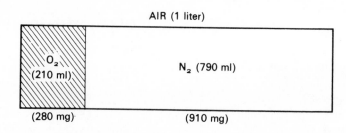

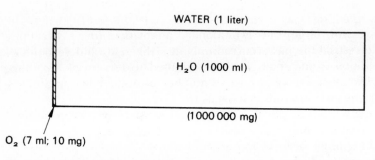

O_2 (7 ml; 10 mg)

Fig. 2.1. Air has a high oxygen content (21%) and contains inert gas (nitrogen) in an amount weighing only 3.5 times as much as the weight of the oxygen. Water, in contrast, contains only 0.7% of its volume of dissolved oxygen (at 15 °C, P_{O_2} 159 mm Hg). In this case the weight of inert medium is 100 000 times as great as the weight of oxygen.

times its mass of inert gas. This difference is illustrated in fig. 2.1.

One consequence of this tremendous difference in the mass of the inert medium is that in aquatic respiratory systems the movement of water is almost universally unidirectional. If the flow of water were to be in-and-out or back-and-forth, a large mass of water would have to be accelerated, then stopped, and again accelerated in the opposite direction. This would require a high expenditure of energy for the continuous changes in kinetic energy of the water. In respiratory organs that use air, such as the lung, an in-and-out flow is not very expensive, for it need move only a few times more inert mass than the mass of oxygen used. We do find, however, that air-breathing animals may use a unidirectional flow of air, as we shall see later when we discuss bird and insect respiration.

In addition to the mass of the respiratory medium, another factor contributes to the amount of work required to move it. Water has a higher viscosity than air, and this increases the work required to pump the fluid, for the driving pressure must be increased in proportion to the higher viscosity. Since the energy required for moving the fluid increases in direct proportion to the pressure, it also increases in proportion to the viscosity.

The viscosity of water at 20 °C is 1 cP (centipoise), and the viscosity of air is 0.02 cP, i.e. water is about 50 times as viscous as air, and the work required for pumping increases accordingly.

Another advantage of air respiration is the high rate of diffusion of oxygen in air, which, at the same partial pressure or tension, is some 300 000 times as rapid as in water (see appendix 2, p. 671). This fast diffusion in air permits very different dimensions in the respiratory organs. The distances over which a gas can diffuse, for example in a lung, may be several millimeters, while the diffusion distances in the fish gill is a small fraction of a millimeter.

To prevent undue evaporation from the respiratory surfaces, these cannot be freely accessible to the outside air. The gas exchange surfaces are usually located in specialized respiratory cavities (lungs), and this greatly limits the access of air. The renewal of air in the cavity is often very carefully regulated, being no greater than dictated by the requirement for oxygen.

Gas exchange across the general body surface is usually possible only in a moist habitat. Earthworms, for example, in which the entire respiratory gas exchange takes place through the body surface, are very susceptible to water loss. They live in moist habitats, and if they remain exposed on the surface of the earth they rapidly dry out and succumb.

With regard to water and the need for gas exchange, plants are really in a much worse situation than animals. Plants need carbon dioxide for photosynthesis, and they must obtain this gas from the air, which, as we have seen, contains only 0.03% CO_2 (0.23 mm Hg). The diffusion gradient for carbon dioxide into the plant is therefore extremely small. Even if the plant can maintain a zero concentration of carbon dioxide inside its tissues, the driving force for carbon dioxide diffusion will be no more than 0.23 mm Hg. The plant tissues will have a water vapor pressure which varies with temperature; at 25 °C it will be 24 mm Hg. The driving force for diffusion of water to dry air will therefore be 24 mm; in moist air it will be less, at 50% r.h., for example, it would be half as much. In any event, the magnitude of the vapor pressure that drives the outward diffusion of water will be of a magnitude 100 times as high as the inward driving pressure for carbon dioxide diffusion. As a consequence of this unfavorable situation, plants have a very high requirement for water. (One scheme for alleviating this situation is used by the pineapple. These plants have open stomata during the night when the air hu-

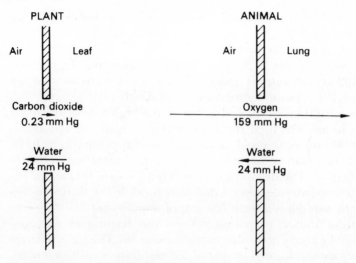

Fig. 2.2. For plants it is difficult to obtain by diffusion the necessary carbon dioxide from the atmosphere without incurring a relatively tremendous water loss. For animals to obtain oxygen, the water loss presents less of a problem because of the high partial pressure of oxygen in air (*c.* 700 times as high as the carbon dioxide pressure).

midity is high, carbon dioxide diffuses in and is stored in the plant (it cannot be photosynthesized in the dark, of course). In the daytime the stomata are kept closed, thus reducing water loss, while the now available sunlight is used for photosynthesis of the stored carbon dioxide (Joshi, Boyer and Kramer, 1965).)

The situation for plants is compared to that for animals in fig. 2.2. In animals the inward driving force for oxygen is the 21% O_2 in the air, or 159 mm Hg. At 25 °C the outward driving force for loss of water to dry air is 24 mm Hg, or only a small fraction of the inward driving pressure for oxygen. Even at high temperatures, say 37 °C, the water vapor pressure is small compared to the pressure for inward movement of oxygen. There is thus a great contrast between the unfavorable water balance that plants must cope with, and the favorably high oxygen concentration that animals enjoy (at least, hopefully so).

Respiratory organs

Air-breathing animals have three major types of respiratory organs, gills, lungs, and tracheae.

Gills. On the whole, gills are rather poorly suited for respiration in air and are used only by few animals. These are mostly animals that relatively recently have invaded the terrestrial habitat, taking with them the remnants of their previous mode of aquatic respiration. Land crabs are a good example. The coconut crab (*Birgus latro*), which has adopted an almost completely terrestrial existence (including climbing coconut palms) has gills which are sufficiently rigid to remain useful for respiration in air. Another land crab, *Cardisoma*, is particularly interesting because it can survive indefinitely either in air *or* in water; this is not common for animals that live in the transition between the two media. Another crustacean group with air-breathing gills is the terrestrial isopods (commonly known as sow bugs, pill bugs, or wood lice); these animals usually prefer to live in moist surroundings, and those that are most successful in the terrestrial habitat have their gills in cavities that can be regarded as functional lungs.

Among fish that can breathe air, functional gills have been maintained in some, but not in all. The common eel (*Anguilla vulgaris*) survives quite well in air if it is kept reasonably cool and moist. Much of the oxygen is then taken up through the skin and less through the gills, for the filaments tend to stick together and expose only a small surface to the air in the gill chamber. As a consequence, the eel does not obtain its normal oxygen requirement, and the oxygen uptake in air is reduced to only about one-half of that in water (Berg and Steen, 1965).

Lungs. We can separate two types of lungs, *diffusion lungs,* and *ventilation lungs.* (*a*) Diffusion lungs are characterized by the fact that air exchange with the surrounding atmosphere takes place by diffusion only. Such lungs are found in relatively small animals, pulmonate snails, scorpions, and some isopods. (*b*) Ventilation lungs are found only among vertebrates. Substantial and regular renewal of the air in the lung is necessary for a large body size combined with a high metabolic rate. The vertebrate respiratory systems are ventilated by an in-and-out, or tidal, flow of air. The respiratory system of birds, however, which is far more complex than that of mammals, is arranged so that the air can flow unidirectionally through the lung, both during inspiration and expiration (see p. 52ff).

Tracheae. This type of respiratory organ is characteristic of insects. It is defined as a system of tubes which supply oxygen directly to the tissues, thus obviating the need for circulation of blood for the purpose of gas transport. Exchange of gas in the tracheal system may take place by diffusion only, but in many large, and especially in highly active insects, there is active unidirectional pumping of air through parts of the tracheal system. The advantage of unidirectional flow is that it permits a far better gas exchange than obtained by pumping air in and out.

Respiratory movements

Vertebrate lungs are ventilated by active pumping of air. (Tracheal ventilation will be discussed later, for insects differ from other animals in so many respects that it is convenient to treat them separately.) The ventilation of the vertebrate lung can be achieved in two different ways, filling of the lung can take place with the use of a pressure pump, as in amphibians, or by a suction pump, as in most reptiles, birds, and mammals.

Frogs fill the lungs by taking air into the mouth cavity, closing the

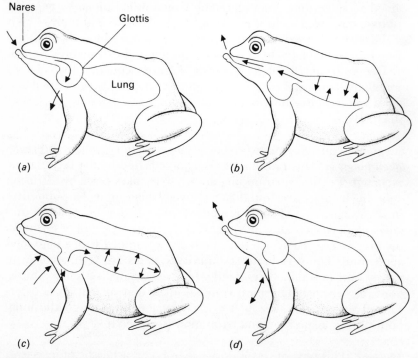

Fig. 2.3. The breathing cycle in frogs (*Rana*) has several successive stages. In (a) air is taken into the buccal cavity by lowering the floor of the mouth, in (b) air is permitted to escape from the lungs, passing over the buccal cavity, in (c) the external nares are closed and air forced into the lungs, and in (d), while air is retained in the lungs by the closed glottis, the cycle can be repeated by again taking air into the mouth (from Gans, C. 1969. Ventilation: How Does the Bullfrog Breathe? *Science,* **163,** 1223–5, Fig. 1. © American Association for the Advancement of Science).

mouth and the nostrils, and pressing air into the lungs by elevating the floor of the mouth. The detailed sequence is somewhat more complex, as explained in fig. 2.3. As a result of this filling mechanism, a frog can continue to take in repeated volumes of air several times in sequence without letting air out, and thus blow itself up to a considerable size.

Bentley and Shield (1973) have recently shown that, contrary to commonly accepted opinion, at least some amphibians can also breathe with the aid of a suction-type pumping. Whether this is universal for amphibians is not known.

Positive pumping, similar to the amphibian mechanism, is found in some reptiles. The chuckawalla (*Sauromalus*) is a desert lizard from southwestern North America which often hides in a rock crevice where, by inflating its lungs, it lodges itself so firmly that it cannot be pulled out. One way to dislodge the animal, used by Indians who want it for food, is to puncture it with a pointed stick.

The normal mechanism for filling the lungs in reptiles is the same as in birds and mammals, the lungs are filled by suction. Exhalation can either be passive, following inhalation by elastic recoil, or it can be actively aided by muscular contraction. A suction-type pump requires a closed thoracic cavity where the pressure during inhalation is less than the surrounding atmosphere. In mammals the inhalation is aided by contraction of the muscular diaphragm. Birds have a membranous diaphragm attached to the body wall by muscles, but its function differs from the mammalian diaphragm. The common statement that birds have no diaphragm is incorrect.

Role of the skin in respiration

Gas exchange through the skin is normal and important for amphibians, which have a moist and well vascularized skin. In fact, some small salamanders (plethodont salamanders) have no lungs at all, and all gas exchange takes place through the skin surface, except for a small contribution by the oral mucosa.

The relative roles of skin and lungs in frog respiration changes through the year (fig. 2.4). In winter, when the oxygen uptake is quite low, the skin takes up more oxygen than the lungs. In summer, however, when oxygen consumption is high, the uptake through the lung increases several-fold and now far exceeds the cutaneous uptake. The fact that oxygen uptake through the skin remains nearly constant throughout the year is related to the constant oxygen concentration in the atmosphere, thus providing a constant diffusion head. If the oxygen concentration in the blood remains uniformly

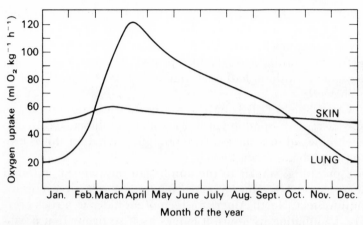

Fig. 2.4. Oxygen uptake through the skin of frogs is nearly constant throughout the year, and the increased oxygen consumption during summer is covered by a greatly increased oxygen uptake through the lung (Dolk and Postma, 1927).

low through the year, the diffusion through the skin should not change much, for diffusion rates change very little with temperature. Since the need for oxygen increases greatly in summer, the increase could not be handled by the skin and must be covered by additional uptake in the lung, as is indeed the case.

Is the change with the seasons a result of the changing temperature? This question has been studied with toads (*Bufo americanus*), which were moved between three different temperatures, 5, 15, and 25 °C (see fig. 2.5). At the two higher temperatures, the pulmonary oxygen uptake exceeds that through the skin, but at the lowest temperature the cutaneous oxygen uptake is greater. This is similar to the situation in the frog. For carbon dioxide exchange the skin is more important at all temperatures. At the lowest temperature the skin is therefore more important than the lungs for gas exchange, both for carbon dioxide and for oxygen.

In mammals the gas exchange through the skin is trivial. There is an oft-repeated legend about some children who for a religious procession in Italy were painted with gold paint; the story goes that they all died of asphyxiation because the skin could not 'breathe.' Death from asphyxiation is out of the question, for oxygen uptake through the skin is barely measureable and the carbon dioxide loss from the skin is less than 1% of that from the lung (Alkalay, Suetsugu, Constantine and Stein, 1971). The gold-painted children must have died from other causes, a plausible explanation is that the gold paint was made by amalgamating gold and mercury and suspending the

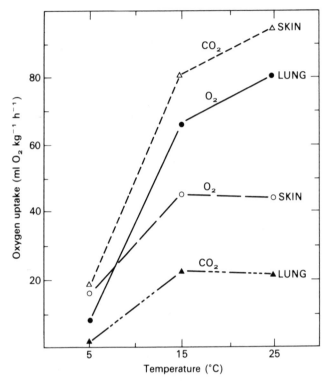

Fig. 2.5. Pulmonary and cutaneous gas exchange in the toad *Bufo americanus* at different temperatures (Hutchison, Whitford and Kohl, 1968, *Physiol. Zool.*, **41**, 65–85, © 1968. By the University of Chicago. All rights reserved).

amalgam in oil, a common paint base. Mercury emulsifies readily in oil and is then rapidly absorbed through the skin, and the children may well have died from acute mercury poisoning.

Bats have much larger relative skin surface than other mammals; the large, thin, hairless wing membranes are highly vascularized and might contribute to gas exchange. There is, in fact, some carbon dioxide loss from the wing membrane. In the bat *Eptesicus fuscus* at 18 °C 0.4% of the total carbon dioxide production is lost from the wing skin. The amount increases with temperature, and at an air temperature of 27.5 °C as much as 11.5% of the total carbon dioxide is lost this way (Herreid, Bretz and Schmidt-Nielsen, 1968). The uptake of oxygen through the wing membranes, however, is not sufficiently great to be of any significance; as was discussed earlier, the diffusion of oxygen between water and air is some 25-fold slower than for carbon dioxide.

Mammalian lungs

As we move up through the vertebrate classes, the lungs become increasingly complex. In amphibians the lung is a single sac, sub-divided by a few ridges which give an increased surface. The mammalian lung is much more finely divided into small sacs, the *alveoli*, which vastly increase the surface area available for gas exchange. Measurements of the surface of the frog lung indicate that 1 ml lung tissue has a total gas exchange surface of 20 cm²; the corresponding figure for the human lung is 300 cm² surface per ml lung tissue. The large surface area is essential for the high rate of oxygen uptake required by the high metabolic rate of warm-blooded animals.

The lung volume of mammals constitutes about 6% of the body volume, irrespective of the body weight (fig. 2.6). If the lung volume were to remain the exact same proportion of body size, the slope of

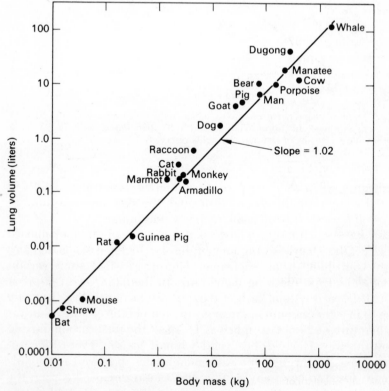

Fig. 2.6. The lung volume of various mammals plotted against the body weight shows that the lung size is a regular function of body size (Tenney and Remmers, 1963).

the regression line in fig. 2.6 should be exactly 1.0. The best fitting regression line has a slope of 1.02, i.e. the deviation from strict proportionality is insignificant. As expected, the individual points do not fall exactly on the line, but there are no large characteristic deviations. This means that small mammals, which have high metabolic rates, obtain sufficient oxygen with a lung of the same relative size as in large animals.

It is worth noting that diving animals, such as porpoise, manatee, and whale, follow the common mammalian pattern in regard to lung size. One could expect that these animals, in order to stay under longer, should have very large lungs which they could fill with air before a dive. This is not the case, and as we shall see later, diving animals do not depend for diving on the oxygen reserves in their lungs.

The regression line in fig. 2.6 can be described by the equation

$$V_1 = 0.0567 \cdot M_b{}^{1.02},$$

where V_1 is the lung volume in liters and M_b the body mass in kg.* For a mammal of 1 kg body mass the expected lung volume would therefore be 0.0567 liters, or 56.7 ml. Assuming that the body volume in liters equals the body mass in kg (this is for practical purposes correct, we know that man is close to neutral buoyancy in water and therefore has a density near 1.0), the lung volume is 5.67% of the body volume. (For those unfamiliar with the arithmetic manipulation of exponential equations, a brief discussion is found in appendix 3.)

Gas exchange in the lung takes place in the alveoli, and the trachea, the bronchi, and their branches are to be considered only as connecting tubes. At the end of an exhalation these tubes are filled with 'used' air from the lung, and when inhalation follows this air is pulled back into the lung before fresh outside air enters. The volume of air in the passageways thus reduces the amount of fresh air that enters the lung, and is therefore called the *dead space*. The volume of air inhaled in a single breath is the *tidal volume*. A normal man at rest has a tidal volume of about 500 ml. Since the dead space

---

* Mass is a fundamental property of matter, and weight refers to the force exerted on a given mass by a specified gravitational field. In the gravitational field of the earth mass and weight will, if both are expressed in kg units, numerically have the same value. In a different gravitational field the mass is unchanged but the weight will be different. Thus, on the moon a 70 kg man still has a mass of 70 kg, but his weight will be only one-sixth of that on the earth, or about 12 kg (more correctly, kilogram force or kgf). Because of the similarity between mass and weight on earth, it is very common not to distinguish between them.

is about 150 ml, only 350 ml of fresh air will reach the lungs. The dead space thus constitutes about one-third of the tidal volume at rest. In exercise, the relative role of the dead space is less. Let us say that a man who breathes heavily inhales 3000 ml of air in a single breath, a dead space of 150 ml is now only about $^1/_{20}$ of the tidal volume. While the dead space is a substantial fraction of the tidal volume at rest, in exercise it is relatively insignificant.

An important aspect of respiration is that the lungs are never completely emptied of air. Even if a man exhales as much as possible, there is still about 1000 ml of air left in his lungs. It is therefore impossible for a man to fill the lungs completely with 'fresh' air, for the inhaled air is always mixed with air that remained in the lungs and the dead space.

In respiration at rest a man may have, say, 1650 ml of air in the lungs when inhalation begins. During inhalation, 350 ml fresh air reach the lung and are mixed with the 1650 ml already there. The renewal of air is therefore only about one part in five. The result is that the composition of the alveolar gas remains quite constant at about 15% O_2 and 5% CO_2. This composition of alveolar air remains the same during exercise, in other words, the increased ventilation during exercise is adjusted to match accurately the increased use of oxygen.

Surface tension. Every person who has blown soap bubbles knows that, when the connection to the atmosphere is open, the bubble tends to contract and expel air until it collapses. The vertebrate lung is somewhat similar; the bubble-like shape and the high curvature of the alveoli means that the surface tension of the moist inner surface will tend to make the 'bubbles' contract and disappear. The surface tension should make the lung collapse, but the presence of substances which reduce the surface tension greatly reduces this tendency.

These substances are phospholipids, and their effect on surface tension has given them the name *surfactants*. Surfactants are found in the lungs of all vertebrates, mammals, birds, reptiles, and amphibians. The amount of surfactant present in the vertebrate lung seems to be in a constant excess above the minimum which is required to cover the pulmonary surface with a mono-molecular layer (see fig. 2.7).

Mechanical work of breathing. The movement of air in and out of the lung requires work, and it is of interest to compare the cost of pumping with the amount of oxygen that is provided. The simplest

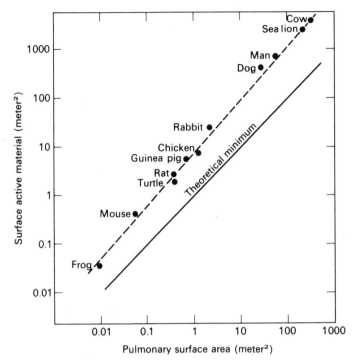

Fig. 2.7. The amount of surfactant that can be extracted from the lungs of various vertebrates plotted against the surface area of the lung. The extracted amount is uniformly greater than the theoretical minimum amount needed to cover the lung surface with a mono-molecular layer (Clements, Nellenbogen and Trahan, 1970, copyright 1970 by the American Association for the Advancement of Science).

way is to compare the amount of oxygen it takes to run the pump to the amount of oxygen the organism obtains in the same period of time.

Determinations of the work of breathing are rather difficult and have a substantial margin of error. Most determinations indicate that the cost of breathing in man at rest is 1.2% of the total resting oxygen consumption. At rest the lung ventilation is about 5 liters min^{-1} and the cost of respiration about 0.5 ml O_2 per liter ventilated air. With increasing ventilation, however, the cost of breathing increases. If the ventilation increases to 10 liters min^{-1}, the cost increases to 1 ml O_2 per liter air, and in very heavy respiration, 50 liter min^{-1}, the cost increases to 2 ml O_2 per liter ventilated air (Otis, 1954). Other determinations indicate that the maximum work of breathing during exercise is no more than 3% of the total oxygen consumed, or slightly less than the figure just mentioned (Margaria,

Milic-Emili, Petit and Cavagna, 1960). Cost of respiration has also been determined with some success in dogs; the figures are of the same magnitude as those for man.

What is the cost of pumping for fish, which must move a much heavier and more viscous medium? Accurate determinations are difficult to make, and the question remains somewhat controversial. It has been suggested that the cost of breathing in fish may be as high as 30 or even 50% of the oxygen obtained (Schumann and Piiper, 1966). These figures seem unrealistic, and recent determinations indicate that the cost of breathing is much lower, a few per cent of the total oxygen uptake. If the mechanical work of breathing is calculated from the pressure drop across the gill and the volume of water flow, the calculated work comes out to less than 1%, a much more reasonable figure. Without a unidirectional flow of water, such a low cost of ventilation would probably be impossible.

Regulation of respiration

If the need for oxygen increases, the ventilation of the respiratory organs must be increased accordingly. Likewise, if the oxygen concentration in the medium falls, there must be a compensation, either by increasing the ventilation or by increasing the oxygen extraction, or both.

In warm-blooded vertebrates, mammals and birds, the ventilation of the lungs is very precisely adjusted to the need for oxygen, but interestingly, the primary agent responsible for the regulation is the carbon dioxide concentration in the lung air. This is readily demonstrated by adding carbon dioxide to the inhaled air. This causes a rapid increase in pulmonary ventilation (fig. 2.8). Normal atmospheric air contains virtually no carbon dioxide (0.03%), and if 2.5% CO_2 is added to the inhaled air, the ventilation volume is approximately doubled. The effect on mammals and birds is similar. 2.5% CO_2 is really not very much, for the lung air of mammals already contains about 5% CO_2. If the carbon dioxide concentration in inhaled air is increased to what is normally found in the lung, the increase in respiratory ventilation volume is sevenfold. In much higher concentrations carbon dioxide becomes narcotic and therefore gives abnormal responses.

Oxygen has a much smaller effect on ventilation. If we reduce the oxygen concentration in inhaled air by 2.5%, from 21% to 18.5% O_2, there is virtually no change in respiration.

The sensitivity of respiration to the carbon dioxide concentration is often used by swimmers who wish to stay under water for long

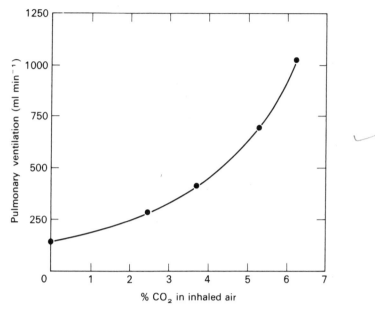

Fig. 2.8. The effect on pulmonary ventilation volume of increased carbon dioxide content of the inhaled air of chickens (Johnston and Jukes, 1966).

periods of time. By breathing deeply for some time, a person can increase the loss of carbon dioxide from the lungs and from the blood. This removes the usual stimulus to respiration, and as a consequence, a person who has hyperventilated can remain under water longer before he is forced to the surface by the urge to breathe. This practice is extremely dangerous. As a person swims under water, the oxygen in his blood is gradually depleted, but in the absence of the usual concentration of carbon dioxide the urge to breathe is not very strong. He therefore remains submerged, and as the blood oxygen falls, he may lose consciousness without even being aware of the danger. If in this state he is not immediately discovered and rescued, he will drown. This sequence of events has in fact been the cause of many drowning accidents, in particular in swimming pools, when good swimmers competitively attempt long underwater swimming feats (Craig, 1961*a*; 1961*b*).

It has been stated that seals and whales are less sensitive to carbon dioxide than other animals, and therefore are able to stay under water longer. Merely remaining under water for a longer period will, of course, not increase the amount of oxygen available, and oxygen is probably the limiting factor in the duration of a dive.

If we examine the response to inhaled carbon dioxide in various

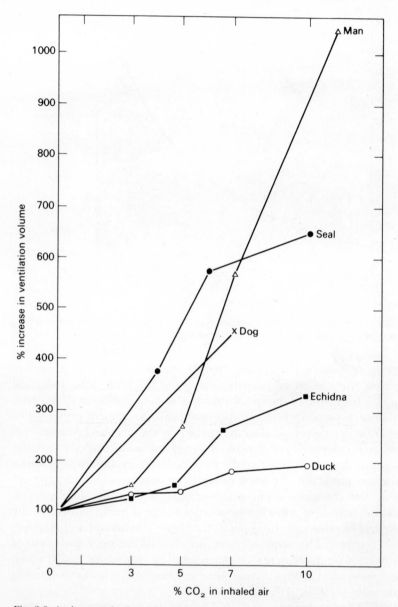

Fig. 2.9. An increase in the carbon dioxide content of inhaled air causes a several-fold increase in respiration volume. Note that the seal is more sensitive to carbon dioxide than the dog (Bentley, Herreid and Schmidt-Nielsen, 1967).

animals, it is not enough to measure the respiration frequency. In some animals the respiration frequency increases considerably in response to carbon dioxide, but in others there may be little or no change in frequency and yet a considerable increase in tidal volume. This has, for example, been found in the spiny anteater (the echidna, *Tachyglossus*) (Bentley, Herreid and Schmidt-Nielsen, 1967). The information we need is the ventilation volume, that is, the product of respiration frequency and tidal volume.

If we determine the increase in ventilation volume in response to carbon dioxide, it turns out that the seal, at carbon dioxide concentrations up to 6%, is even more sensitive to carbon dioxide than non-diving animals such as man and dog (see fig. 2.9). This contradicts the statements about diving seals being insensitive to carbon dioxide.

Comparison of air and aquatic respiration

In many aquatic invertebrates the regulation of respiration is poor or even absent. This is true especially in marine species which normally live in well-aerated water with a relatively constant oxygen supply. Some lower animals are quite tolerant to lack of oxygen; bivalve molluscs can keep the shells closed for long periods and, in the absence of ventilation, utilize anaerobic metabolic processes (see p. 218–9).

For most aquatic animals the primary stimulus to respiration is a lack of oxygen. This is usual in crustaceans, octopus, fishes, and so on. The effect of carbon dioxide on aquatic invertebrate animals is never very pronounced, and may be completely absent. The carbon dioxide tension in natural water is almost always low, and as we shall see later (p. 109), because of the high solubility of carbon dioxide in water, aquatic animals cannot build up a high tension of this gas. If they were to depend on an increase in carbon dioxide tension for stimulation of respiration, an adequate supply of oxygen could not be assured.

If we compare two closely related forms, the marine lobster (*Homarus*) and the fresh-water crayfish (*Astacus*), we notice a characteristic difference. The lobster, a marine animal, does not show much change in ventilation with a decrease of oxygen in the water (Thomas, 1954). The fresh-water crayfish, in contrast, responds to a decrease in oxygen with an increase in ventilation. The difference is readily understood, the lobster lives in cold waters where the oxygen is always high, and an elaborate mechanism for regulation of ventilation would be somewhat superfluous. The crayfish, on the other hand, may readily encounter fresh-water environments where oxygen is depleted, and a ventilatory response is needed.

Fishes in general respond to a decrease in oxygen, and their response to changes in carbon dioxide is minimal. In this regard they resemble other aquatic animals, rather than the air-breathing vertebrates. Insects, incidentally, are for the most part highly sensitive to carbon dioxide.

Can we generalize these differences? We have already mentioned that aquatic animals cannot readily base their respiratory regulation on carbon dioxide, not only because the carbon dioxide tension in natural waters usually is low, but also because it is an unreliable measure of the oxygen content. Sea water is highly buffered, so that the carbon dioxide tension never builds up to any appreciable extent. Stagnant fresh water, on the other hand, may have high carbon dioxide concentrations, usually associated with low oxygen. Aquatic animals could not possibly depend on something as unreliable as the carbon dioxide concentration, and whenever we find their respiration to be regulated, it is in response to oxygen concentration.

The question now becomes: 'why have terrestrial air-breathing animals abandoned regulation on oxygen and gone over to carbon dioxide as the primary stimulus?' The answer is probably that, as oxygen becomes easily available in the atmosphere, the carbon dioxide concentration tends to build up in the respiratory organs. If, for example, a mammal reduces the oxygen in the inhaled air from 21 to 16%, this entails a simultaneous build-up of carbon dioxide to about 5%, an amount which greatly affects the acid–base balance of the organism. A decrease in oxygen to 16% does not have any profound physiological effect, but a change in 1% of the carbon dioxide concentration, say from 4 to 5%, constitutes a 25% increase in carbonic acid. Also, it may be easier to design, physiologically speaking, a precise control system based on carbon dioxide and the detection of small changes in hydrogen ion concentration, than a system sensitive to small changes in oxygen concentration. Be this as it may, we universally find that air-breathing animals are far more sensitive to changes in carbon dioxide than to changes in oxygen.

Air-breathing fish

Everybody has heard about lung-fish, but many other fish also can breathe air. Many of these resort to air only when the oxygen content in the water is low; relatively few depend on air to such an extent that they drown if kept submerged. There are two main ecological reasons for using accessory or exclusive air breathing, one is depletion of oxygen in the water, the other is the occurrence of periodic droughts. Lung-fish, for example, during dry periods burrow deep into the mud and encase themselves in a cocoon and remain inactive until the next flood.

Most of the air-breathing fish are tropical fresh-water or estuarine species, few if any are truly marine. Oxygen-deficient fresh water is much more common in the tropics than in temperate climates. This is because there is much decaying organic matter in the water, the temperature is high and speeds up bacterial action, small bodies of water are often heavily shaded by over-hanging jungle (which reduces photosynthesis and oxygen production in the water), and there is little temperature change between day and night and therefore minimal thermal convection to bring oxygen-rich surface water to deeper layers. Not all air-breathing fish belong in the tropics, however, the well-known bowfin (*Amia calva*) is found in the northern United States where the lakes are frozen over in winter. (During such periods they manage well without air breathing because the low temperature reduces their oxygen consumption.)

It was explained before (p. 17) that fish gills are not well suited for respiration in air, they lack the necessary rigidity and tend to stick together, but even so, some oxygen can be taken up through gills in air. Any other moist surface will supplement the gas exchange, provided it is a surface that has access to air and is supplied with blood. Some gas exchange can always take place through the skin and the surface of the mouth cavity, but in addition there may be other anatomically more specialized organs that aid in gas exchange. Organs commonly utilized in air breathing are:

Gills; skin; mouth and opercular cavities; stomach; intestine; swimbladder; lung.

Some air-breathing fish are listed in table 2.1, with a few comments that should help to place them in a familiar frame of reference. It is worth noting that all of these fish, except the last five, are more or less ordinary higher bony fishes (Actinopterygii). The bowfin (*Amia*) is a representative of the primitive group Holostei and the bichir (*Polypterus*) is peculiar because it has a lung, but both are considered primitive representatives of the Actinopterygii. Only the last three fishes in table 2.1 are true lung-fishes (Dipnoi), which presumably are closely related to the Crossopterygii (the sub-class which contains the famous coelacanth, *Latimeria*). There is one genus of true lung-fish in each of the three continents, Australia, Africa, and South America.

The need for air breathing depends on two factors, the amount of oxygen in the water and the temperature, for the rate of oxygen

Table 2.1. *Fish which can use accessory air breathing or depend completely on air. Most of these fish are modern teleosts; only the last three are true lung-fishes. The first column gives the major organs used for respiration in air*

Gill	*Symbranchus,* an eel-shaped South American fresh-water fish without any common English name.
Skin	*Anguilla,* common eel. North America and Europe: breeds in the sea, larva migrates to fresh water.
	Periophthalmus, mud-skipper. Common on tropical estuarine beaches.
Mouth and associated cavities	*Electrophorus,* electric eel. South America: fresh-water.
	Anabas, 'climbing perch'. Not a perch, related to Betta, the Siamese fighting fish.
	Clarias, a catfish, the 'walking catfish' in Florida.
	Gillichtys, mudsucker. Pacific coast of North America.
Stomach	*Plecostomus,* a small catfish. Common in home aquaria.
	Anicistrus, an armored catfish. Protected by heavy spines and bone plates; South America.
Intestine	*Hoplosternum,* armored catfish. South America: fresh-water.
Swimbladder	*Arapaima,* South American rivers. The world's largest fresh-water fish.
	Amia, bowfin. North America, fresh-water, extends north to areas where lakes remain ice-covered throughout winter. Amia belongs to the primitive group Holostei.
Lung	*Polypterus,* the bichir. Africa: fresh-water. Has a lung, although it is not a true lung-fish (cf. text).
	Lepidosiren. South America: fresh-water.
	Protopterus. Africa: fresh-water.
	Neoceratodus. Australia: fresh-water, rivers.

consumption increases with temperature. For this reason it is not always possible to say whether a given fish does, or does not, depend on air breathing, it will depend on circumstances. However, some fish are so dependent on air breathing that they cannot survive even in well-aerated water, they are obligatory air breathers and will 'drown' if they are deprived of access to air. The following fish are such obligatory air breathers; interestingly, the Australian lung-fish, *Neoceratodus,* is not included in the list.

Protopterus	Lung-fish	Africa
Lepidosiren	Lung-fish	South America
Arapaima	Swimbladder	South America
Hoplosternum	Intestine	South America
Ophiocephalus	Pharyngeal cavities	South Asia and Africa
Electrophorus	Mouth	South America

We shall now discuss how some of the various mechanisms function and how effective they are by describing some fish that have been studied by physiologists.

The common eel. The common European and North American eel (*Anguilla vulgaris*) supposedly is able to crawl considerable distances over land and cross from one water-course to another, especially at night and through moist grass. Fishermen often keep live eel in a box for days on end, merely covered by a wet sack to keep them from drying out.

When an eel is kept out of water, it keeps its gill cavity inflated with air. About once a minute it renews the air in the gill cavity; in contrast, in water at 20 °C an eel respires at a frequency of about 20 per minute.

The oxygen consumption of an eel in air is about one-half of what it is in water at the same temperature. When an eel is moved to air, oxygen taken from the air is initially supplemented with oxygen removed from the swimbladder; this amount may during the first hour nearly equal that taken from the air. However, when eels are kept in air for longer periods of time they gradually accumulate a

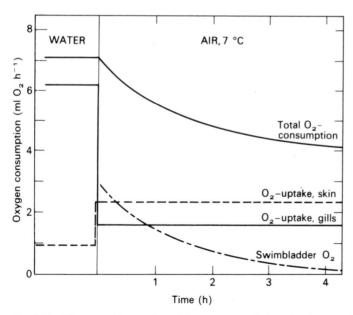

Fig. 2.10. When an eel is transferred from water to air (at 7 °C), its rate of oxygen consumption gradually decreases and stabilizes at a lower level (top curve). Initially, oxygen is taken from the swimbladder, but when this is used up, oxygen uptake from skin and gills suffices to sustain the lowered rate of oxygen consumption (Berg and Steen, 1965).

substantial oxygen debt, and the lactic acid concentration in the blood increases. When returned to water, recovery is rapid, and in two hours the blood lactic acid is normal again. At a lower temperature, 7 °C, the rate of oxygen consumption is much lower and there is no rise in lactic acid in the blood and apparently no oxygen debt. The permeability of the skin to oxygen should be nearly independent of temperature, and at the low temperature the combined oxygen uptake through skin and gills seems sufficient to cover the metabolic rate (fig. 2.10).

The relative importance to the eel of gills and skin differs in water and in air. In water (at 20 °C) about 90% of the total oxygen uptake is via the gills, the remainder through the skin. In air, only about one-third of the total oxygen uptake is via the gills, the remaining two-thirds via the skin. Although the gills are still of some use in air, they are far from adequate, and even with the aid of the skin the total available oxygen is insufficient to avoid an oxygen debt, except at the lowest temperatures.

The mud-skipper. Another fish that, like the eel, depends primarily on cutaneous respiration is the well-known mud-skipper (*Periophthalmus*), which is common on tropical beaches, especially in muddy estuarine areas. These fish are up to about 20 cm long; they prefer to remain in air for extended periods of time and move about supported on their pectoral fins. When disturbed they rapidly disappear into a mud hole which usually has some water at the bottom. When kept submerged in water, however, they display circulatory reactions which indicate that they suffer from partial asphyxia.

Symbranchus. This South American swamp fish has no common English name; it is notable because it seems able to breathe equally well in water and air. It uses primarily the gills for respiration, but the walls of the gill chamber are highly vascularized and contribute to gas exchange. When the fish is in air, the gill chamber is periodically inflated, and the air is retained for some ten to fifteen minutes while about half of its oxygen is extracted.

When the fish is in water, the arterial oxygen saturation is always rather low, never more than 50 or 60%. Inflation of the gill chamber with air is invariably followed by a substantial increase in the arterial oxygen, which then approaches saturation (Johansen, 1966).

When *Symbranchus* breathes air, the arterial carbon dioxide concentration increases, a change characteristic of all air breathers, and when the fish is returned to water, there is a prompt fall in the carbon dioxide again. If the carbon dioxide concentration in the water is increased, the respiration rate decreases; thus carbon dioxide inhibits rather than stimulates respiration, as is universal for air breathers. The decreased respiration caused by a high carbon dioxide level in the water leads to hypoxia, which in turn stimulates air breathing. The primary stimulus in the respiratory regulation of *Symbranchus* is therefore a decrease in oxygen, which makes the fish turn to aerial respiration.

Electric eel. The South American electric eel (*Electrophorus electricus*) is known for its powerful electric discharges which have been measured at 550 volts, enough to stun or kill other fish and perhaps even a man. The electric eel is a mouth-breather, and will drown if denied access to air. The oral cavity, which is immensely vascularized, has multiple foldings and papillations which greatly enlarge its surface area.

The electric eel takes air into the mouth at intervals from a few seconds to several minutes. The oxygen in the mouth air gradually decreases while the carbon dioxide increases, although not in the same proportion (see fig. 2.11). When the oxygen has fallen by about

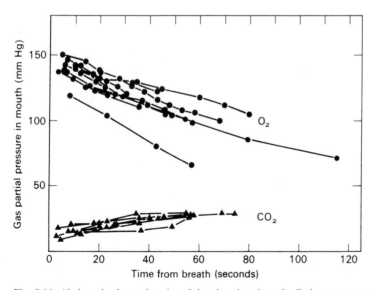

Fig. 2.11. Air kept in the oral cavity of the electric eel gradually loses oxygen as it is used by the fish. The carbon dioxide in the mouth air gradually increases, but not to the same degree. When about one-third of the oxygen has been removed, the fish seeks to surface for renewal of the air (Johansen, Lenfant, Schmidt-Nielsen and Petersen, 1968).

one-third or so, the fish slowly moves to the surface for renewal of the air.

In an air breather in which the entire gas exchange takes place between the blood and a body of air, the build-up of carbon dioxide should correspond roughly to the oxygen depletion. Since this is not the case for the electric eel, carbon dioxide must be lost elsewhere, perhaps through the rather rudimentary gills, or more likely through the skin.

The regulation of respiration in the electric eel can be examined by changing the carbon dioxide and oxygen concentrations in the atmosphere it is allowed to breathe. It turns out that a substantial increase in the carbon dioxide content in the air gives only a moderate increase in the rate of breathing, but a drop in the oxygen content causes a several-fold increase. Likewise, if the oxygen tension is increased above that in normal atmospheric air, it has a somewhat depressing action on the respiration (Johansen, 1968). It is interesting, then, that we have a fish which is an obligatory air breather, but has retained the aquatic mode of regulating the respiration on oxygen. Evidently, this is associated with the fact that carbon dioxide escapes through the relatively permeable skin, thus not building up to sufficient concentrations to be physiologically of importance.

Lung-fish. The African and South American lung-fishes (*Protopterus* and *Lepidosiren*) were listed on page 46 as obligatory air breathers. They live in stagnant bodies of water and in lakes where long droughts may cause complete drying out of their habitat. They will estivate (p. 405) until the next wet period, when they come out from their cocoons in the mud and resume normal life. The Australian lung-fish (*Neoceratodus*) lives in rivers and slow streams, it will also estivate in dry periods, but it depends much less on the lung and is primarily a gill-breather. If it is kept in a tank from which the water is drained slowly, it becomes frantic and seeks to keep itself submerged in what little water remains. It does fill its lung with air, however, which contributes substantially to the gas exchange.

When the three kinds of lung-fish are kept in water with access to air, the relative roles of gills and lungs are as shown in fig. 2.12. In the Australian lung-fish the gills are the primary organ of gas exchange. In the African lung-fish most of the oxygen is taken up by the lung and only a small fraction through the gills. In the South American lung-fish, virtually all the oxygen uptake is via the lung and only a minute amount via the gills. With carbon dioxide the situation is different; both the African and the South American lung-

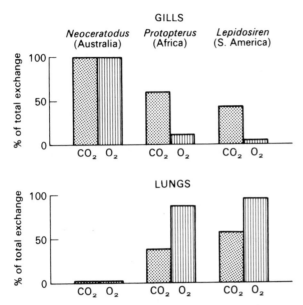

Fig. 2.12. The relative roles of gills and lungs to respiratory gas exchange in three kinds of lung-fish when kept in water and with access to air (Lenfant, Johansen and Hanson, 1970, reprinted from *Federation Proceedings*, **29**, 1124–1129, 1970).

fish exchange about half through the lung, the remainder through the gills. Hence, gills are important in carbon dioxide exchange, while the lung plays the dominant role in oxygen uptake.

The changes that take place as the lung-fish are removed from water to air show a similar relationship between the three species (see fig. 2.13). The oxygen consumption of the African and South American lung-fish remains nearly unchanged in air. In contrast, the Australian lung-fish shows a precipitous drop in oxygen consumption, because normally it is an aquatic breather and in air it cannot obtain sufficient oxygen.

The oxygen saturation of the arterial blood (lower part of fig. 2.13) shows that the two obligatory air breathers maintain or slightly increase the oxygen saturation of the blood when they are removed to air, while the Australian lung-fish drops to nearly zero arterial oxygen content, a clear indication that the fish is becoming asphyxiated.

The lung-fishes are among the best studied of the air-breathing fish. The information on many others is inadequate or virtually nil. The study of these could be very rewarding, although often technically difficult, for many of them are quite small.

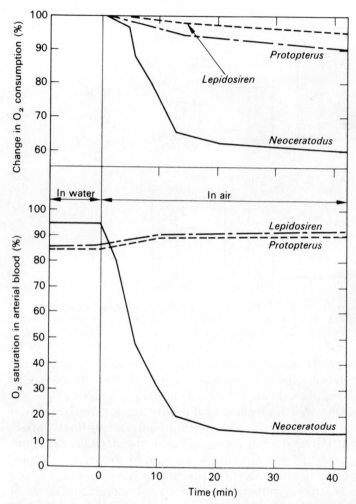

Fig. 2.13. When African and South American lung-fish are kept out of water their rates of oxygen consumption remain nearly unchanged. The Australian lung-fish shows a precipitous decrease in oxygen consumption when kept out of water. In this species the oxygen saturation in the arterial blood shows a corresponding drop (Lenfant *et al.*, 1970, reprinted from *Federation Proceedings*, **29,** 1124–1129, 1970).

Bird respiration

Structure of the respiratory system. The respiratory organs of birds are very different from their counterparts in mammals. The small, compact bird lungs communicate with voluminous, thin-walled <u>air sacs</u> and air spaces which extend between the internal organs and even ramify into the bones of the skull and the extremities. This extensive and intricate respiratory system has been considered as an adaptation to flight. We can immediately say, however, that it is not *necessary* for flight, because bats (which have typical mammalian lungs) are good fliers and at times even migrate over long distances.

We may imagine that flying birds should have a very high oxygen consumption, and that the respiratory system should be seen in this light. However, the rates of oxygen consumption of resting birds and mammals of equal body sizes are very similar, and although normal flight requires an eight or ten-fold increase in the oxygen consumption, many mammals are capable of similar increases. Finally, during flight bats have an oxygen consumption very similar to that of birds of the same body mass. On the other hand, although the avian-type respiratory system is not a prerequisite for flight, it may still have considerable advantages.

The presence of air spaces in the body of a bird can be said to make the bird lighter, but only in a very limited sense. The bird needs a digestive system, a liver, kidneys, and so on, and if we merely add large sacs of air to the abdominal cavity, it does not make the bird any lighter. If we were to remove this air, or double its volume, the bird would still have exactly the same weight to carry during flight. On the other hand, if the marrow of a bone is replaced by an equal volume of air, of course, the bone weighs less. The air-filled bones therefore do contribute to making a bird lighter, but the other large air spaces do not.

If we compare the volumes of the respiratory systems of birds and mammals we find some conspicuous differences (see table 2.2). A typical bird has a lung volume which is only a little more than half of that in a mammal of the same body size. In contrast, the tracheal vol-

Table 2.2. *Volumes of the respiratory systems of typical birds and mammals of 1 kg body size.* Estimated from data collected by Lasiewski and Calder (1971)

	Bird (1 kg)	Mammal (1 kg)
Lung volume (ml)	29.6*	53.5
Tracheal volume (ml)	3.7	0.9
Air sacs (ml)	127.5	—
Total resp. system (ml)	160.8	54.4
Tidal volume (ml)	13.2	7.7
Resp. frequency (min^{-1})	17.2	53.5

* This is the total volume of the bird lung, which contains only 9.9 ml air.

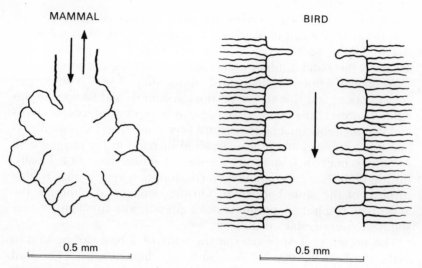

MAMMAL

BIRD

0.5 mm

0.5 mm

Fig. 2.14. The smallest units of the mammalian lung are the sac-like alveoli. In the bird lung the finest branches are tubes which are open at both ends and permit through-flow of air (Schmidt-Nielsen, 1972).

ume of birds is much larger than that of mammals. This can easily be understood in view of their long necks, but we shall later see that it has other implications as well. The air sacs of birds are large, several times as large as the lung, and mammals have no air sacs at all. Therefore the total volume of the respiratory system of birds is some three times as large as that of mammals.

The difference between birds and mammals is not restricted to the air sacs; the avian lungs have a structure which differs radically from that of mammals. In mammals the finest branches of the bronchi terminate in sac-like alveoli (see fig. 2.14). In birds the finest branches of the bronchial system (known as tertiary bronchi) permit through-passage of air. This permits air to flow *through* the bird lung and continuously past the exchange surface, while in mammals air must flow in and out. This is the most important difference in the respiratory system of the two kinds of animals, and it has profound physiological consequences.

To understand how the avian respiratory system works, we must know a few additional facts about its complex anatomy, and a diagram of the lungs and the major air sacs will help (fig. 2.15). It is not necessary to learn the names of all the sacs, in particular since different investigators often use different names that merely confuse the non-specialist. It is important, however, to know that the air sacs anatomically and functionally form two groups, a *posterior group* that

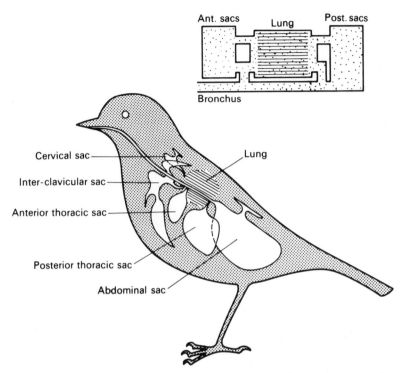

Fig. 2.15. The body of birds is filled with several large, thin-walled air sacs. The paired lungs are small and located along the vertebral column. The main bronchus which runs through the lung has connections to the air sacs as well as to the lung. In the upper right hand corner is a diagram of the system, simplified by combining all anterior sacs into one single space, and all the posterior sacs into another (Schmidt-Nielsen, 1972).

includes the large abdominal sacs, and an *anterior group* which consists of several somewhat smaller sacs.

The trachea divides into two bronchi, and each bronchus runs to and then actually through the lung, and terminates in the abdominal sac. The anterior sacs connect to this main bronchus in the anterior part of the lung, the posterior sacs connect to the posterior part of the main bronchus. The main bronchus also connects to the lung, and furthermore, some of the air sacs connect directly to the lung tissue.

Function. One of the first questions to be asked is: 'what is the function of the air sacs?' Do they function in gas exchange, or do they serve another function, such as working as bellows to move air in and out?

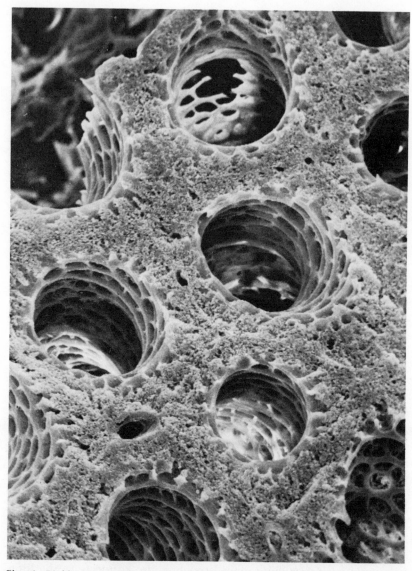

Plate 1. *Bird lung.* In cross-section a bird lung shows cylindrical tubes (parabronchi) which permit the unidirectional through-flow of air characteristic of bird lungs. The diameter of each tube in this picture, which is from a chick of domestic fowl, is slightly less than 0.5 mm (H.-R. Duncker, *Zentrum für Anatomie und Cytobiologie,* Giessen, Germany).

To begin with, the morphology of the air sacs does not indicate that they have any major role in gas exchange between air and blood. Their walls are thin and flimsy and poorly vascularized, and there are no foldings or ridges to increase the surface area. A simple experiment that excluded a direct role in gas exchange was made in the last century by a French investigator, who plugged the openings from the large abdominal air sacs to the rest of the respiratory system, and then introduced carbon monoxide into the sacs (Soum, 1896). Birds are quite sensitive to carbon monoxide, but these birds showed no signs of carbon monoxide poisoning. Therefore, carbon monoxide was not taken up by the blood from the air sacs. This conclusion can be generalized; it is valid for other gases that have similar diffusion properties, oxygen, for example.

A more plausible hypothesis for the function of the air sacs is that they serve as bellows to move air in and out. As an inspiration begins, there is a simultaneous pressure fall in both the anterior and posterior sacs as they expand (see diagram in fig. 2.15). This means that during inspiration air will flow into all sacs (but, as we shall see later, not all sacs fill with outside air). During exhalation the pressure in the sacs increases, and air flows out again.

If we want to follow the flow of gas in greater detail, we could use a gas mixture of a composition different from the usual air, and follow the movements of this gas through the respiratory system. A convenient gas is pure oxygen, for it is harmless and can easily be measured with an oxygen electrode. For such experiments the ostrich has the advantage that it breathes quite slowly (a single respiration lasts about 10 s), and changes in gas composition can therefore be followed readily. When an ostrich is permitted to inhale one single breath of pure oxygen instead of air, the oxygen shows up in the posterior air sacs towards the end of this breath (fig. 2.16). This must mean that the inhaled oxygen reaches these sacs directly through the main bronchus. In the anterior sacs, on the other hand, the oxygen concentration never increases during the inhalation of oxygen; however, these sacs do expand during inhalation, and this can only mean that they receive air from somewhere else. Towards the end of the second inspiration, however, when the bird again breathes ordinary air, the oxygen in the anterior sacs begins to increase. This can only mean that the oxygen which now appears in the intervening time has been located elsewhere in the respiratory system, presumably in the posterior sacs, and has passed through the lung to reach the anterior sacs.

The gas concentrations found in the air sacs are interesting. The posterior sacs contain some 4% CO_2 (see fig. 2.16), and relative to at-

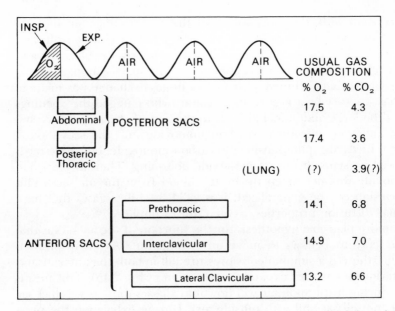

Fig. 2.16. When an ostrich is given a single breath of pure oxygen to inhale, and then immediately returns to breathing ordinary air, the oxygen functions as a tracer gas which first is discovered in the posterior air sacs. Only later, during the second or third respiratory cycle (indicated at the top) does the tracer gas show up in the anterior sacs. Respiratory rate about 6 cycle min⁻¹ i.e. each cycle about 10 seconds. The ordinary gas composition in the air sacs is indicated at the right hand of the graph (Schmidt-Nielsen *et al.*, 1969).

mospheric air the oxygen is depleted by a similar amount, from 21 to about 17%. In the anterior sacs, however, the carbon dioxide concentration is higher, between 6 and 7%, and the oxygen is correspondingly reduced to some 13–14%. These differences in gas composition could indicate that the posterior sacs are better ventilated, and that the anterior sacs contain more stagnant air which therefore reaches a higher carbon dioxide concentration. This conclusion is incorrect.

A marker gas can also serve to determine the renewal of air in a sac. After introduction directly into a sac, the concentration of the marker will decrease step-wise in synchrony with each respiratory cycle, and the rate of wash-out indicates the extent of air renewal. The time required for the marker gas to be reduced to one-half of the initial level can be designated as the half-time. In the ostrich this half-time was between two and five respiratory cycles for both anterior and posterior air sacs, and this means that both sets of sacs are about equally well ventilated. Similar determinations on ducks also gave about equal half-times for anterior and posterior sacs (Bretz

CYCLE 1

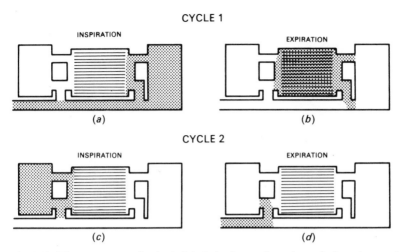

Fig. 2.17. The movement of a single inhaled volume of gas through the avian respiratory system. It takes two full respiratory cycles to move the gas through its complete path (Bretz and Schmidt-Nielsen, 1972).

and Schmidt-Nielsen, 1972). Since tracer gas experiments indicate that air flows into the anterior sacs from the lung (their carbon dioxide concentration is consistent with this conclusion), and since their air renewal is high, the evidence indicates that they serve as a holding chamber for air from the lung, to be exhaled on the next exhalation.

How air flows can also be determined by placing small probes, sensitive to air flow, in the various passageways. Studies on ducks indicate flow patterns as shown in fig. 2.17, which shows how a single bolus of air would flow. During inhalation (a) most of the air flows directly to the posterior sacs. Although the anterior sacs expand on inhalation, they do not receive any of the inhaled outside-air, instead they receive air from the lung. On exhalation (b), air from the posterior sacs flows into the lung instead of out through the main bronchus. On the following inhalation (c) air from the lung flows to the anterior sacs. Finally on the second exhalation (d) air from the anterior sacs flows directly to the outside. The movement of a single bolus of gas thus requires two full respiratory cycles to move through the respiratory system. This does not mean that the two cycles differ in any way, they are completely alike, each bolus of gas followed by another, in tandem, on the next cycle.

The most notable characteristic of this pattern is that air always flows through the lung from the posterior to the anterior, and furthermore, that air moves through the lung both during inhalation and exhalation.

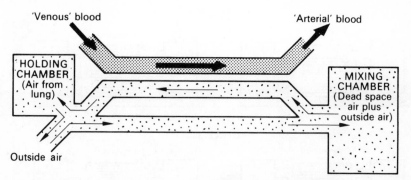

Fig. 2.18. Gas exchange in the bird lung at high altitude. This highly simplified diagram shows the blood and air flows through the lung, each represented by one single stream with opposite flow directions. This permits the oxygenated blood to leave the lung with the highest possible oxygen tension (Schmidt-Nielsen, 1972).

This flow pattern has an important consequence for gas exchange between air and blood, which in principle is similar to the counter-current flow in the fish gill (see fig. 1.4, p. 21). It means that the oxygenated blood that leaves the lung can have a higher oxygen tension than the oxygen partial pressure in exhaled air. A highly simplified diagram will explain this (fig. 2.18). Blood that is just about to leave the lung (right-hand side of the diagram) is in exchange with air that has just entered the lung as it comes directly from the posterior sacs with a high oxygen partial pressure. As the air flows through the lung (towards the left in the diagram) it loses oxygen and takes up carbon dioxide. All along, this air encounters blood with a low oxygen tension, and therefore gives up more and more oxygen to the blood. Due to this type of flow, the blood can become well saturated with oxygen, and yet is able to extract more oxygen from the pulmonary air and deliver more carbon dioxide to it, than in mammals.

Recent evidence indicates that the flow of air in the bird lung is not an ideal countercurrent exchange system, but rather a *cross-current* type of flow (fig. 2.19). The result, with regard to arterial gas tensions, would be similar to that described above, although a cross-current system is not as effective in achieving a maximum advantage as is a true countercurrent system.

The effectiveness of the unidirectional air flow in the avian lung is particularly important at high altitude. Experiments in which mice and sparrows were exposed to an atmospheric pressure of 350 mm Hg, corresponding to 6100 meters or 20 000 ft altitude (slightly less than 0.5 atm), the mice were lying on their bellies and barely able to crawl, while the sparrows were still able to fly (Tucker, 1968). Mice

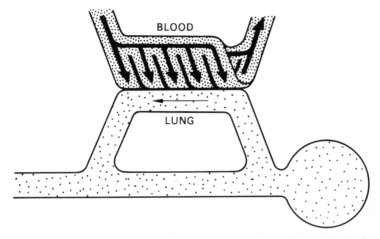

Fig. 2.19. The blood in the avian lung does not flow in parallel capillaries, but rather in an irregular, complex network. This can be represented by a simplified diagram, showing that blood which leaves the lung is a mixture of blood flowing through different parts of the lung, thus having different degrees of oxygenation. This diagramatic flow pattern can be described as cross-current flow (Scheid and Piiper, 1972).

and sparrows have the same body weight, their blood has the same affinity for oxygen, and their metabolic rates are similar, so the difference cannot be explained in terms of their rates of oxygen consumption or the chemistry of their blood. The flow pattern in the bird lung is the most plausible explanation, for it permits blood to take up oxygen from air which has a higher concentration than in the mammalian system, and in addition, because of the unidirectional and continuous flow through the lung, extract more oxygen from it. It is consistent with these experiments that birds in nature have been seen in the high Himalayas, flying overhead at altitudes where mountain climbers can barely walk without breathing oxygen.

Insect respiration

Terrestrial life poses a continual conflict between the need for oxygen and the need for water. Conditions that favor the entry of oxygen also favor the loss of water. Insects, the most successful terrestrial animals, have a hard cuticle which is highly impermeable to gases, and a covering wax layer makes it virtually impermeable to water as well. Gas exchange takes place through a system of internal air-filled tubes, the *tracheae,* which connect to the outside by openings called *spiracles* (see fig. 2.20). The spiracles usually have a closing mechanism which permits accurate control of the exchange be-

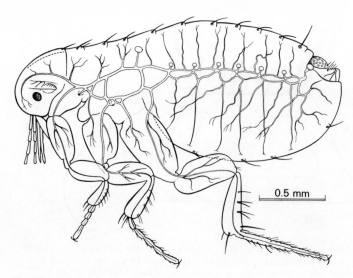

Fig. 2.20. The tracheal system of insects (e.g. the flea) consists of fine air-filled tubes, tracheae, which convey respiratory gases to and from all parts of the body. The tracheae connect to the outside air through the spiracles (Wigglesworth, 1972).

tween the air in the tracheal system and the outside atmosphere. The tubes branch and ramify, and extend to all parts of the body. The finest branches, the *tracheoles,* are about 1 μm or less in diameter, and can even extend into individual cells, such as muscle fibers.

The tracheal system conveys oxygen directly to the tissues, and carbon dioxide in the opposite direction. This makes insect respiration independent of a circulatory system. Insect blood has no direct role in oxygen transport, and this is different from the role of the blood in vertebrates. The comparison with vertebrates may at first seem beside the point, but it is by no means irrelevant, for the rate of oxygen consumption of a large moth and a hummingbird in flight are similar; the oxygen supply mechanisms, however, differ radically.

We cannot conclude, however, that the circulatory system in insects is unimportant; it has many other functions. One such role is obvious, during flight the muscles have a high power output, and while the oxygen is provided through the tracheal system, the fuel must be supplied by the blood.

The tracheal system

Tracheal systems are suitable primarily for respiration in air and must have evolved in air. Insects that secondarily have become

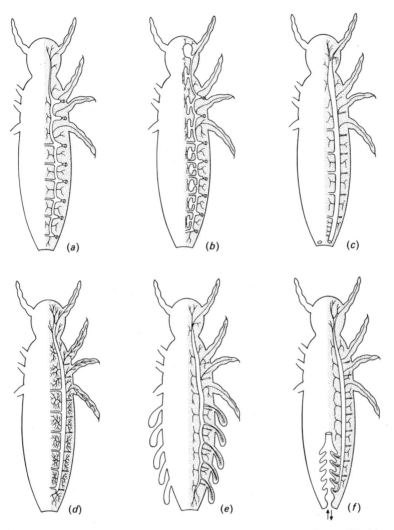

Fig. 2.21. The basic pattern of the insect tracheal system (*a*) may be modified into a variety of other patterns (*b*)–(*f*). For details, see text (Wigglesworth, 1972).

aquatic have in principle maintained air respiration, but they have many interesting modifications which make the system suitable for gas exchange in water.

The tracheal system may show various modifications from a basic pattern, in which there characteristically are 12 pairs of spiracles, three pairs on the thorax and nine on the abdomen. Often there are fewer spiracles, and there may be none. It will be helpful to be familiar with some of the possible variations.

In a typical pattern the larger tracheae connect so that the system

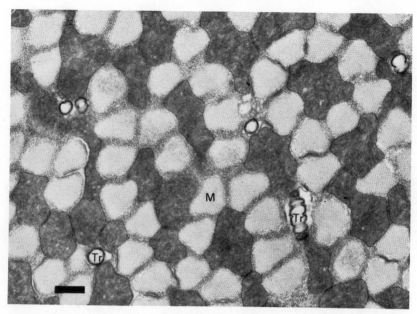

Plate 2. *Insect tracheoles.* Tracheoles in the tymbal muscle of a cicada (*Tibicen* sp.). The tracheoles (Tr) are cut in cross-section, except for one which is cut tangentially and shows the spirally ribbed wall. The diameter of these tracheoles is about 0.5 μm. The distance between the two labelled tracheoles is about 6.5 μm, i.e. about the diameter of a human red blood cell. The light areas (M) are muscle fibers in cross-section, the dark areas are mitochondria (David S. Smith, University of Miami).

consists of tubes that all interconnect (fig. 2.21*a*). This pattern is commonly modified by the addition of enlarged parts of the tracheae or air sacs which are compressible (*b*). The volume of these sacs can be altered by movements of the body, thus pumping air in and out of the tracheal system. This is important, for in large and highly active insects diffusion alone does not provide sufficient gas exchange. The tracheae themselves have internal spiral ribs and are rather incompressible, the large air sacs are therefore necessary for ventilation. It is possible to achieve a unidirectional stream of air through the larger tracheal stems by having the spiracles open and close in synchrony with the respiratory movements, but out of phase with each other.

A modification characteristic of many aquatic insects is shown in fig. 2.21*c*. Most of the spiracles are non-functional, and only the two hindmost open to the outside. They are located so that the insect, by penetrating the water surface with the tip of the abdomen, can make contact with the atmosphere and obtain gas exchange, either by diffusion, or more effectively, by respiratory movements.

The tracheal system may be completely closed, without any opening to the surface, although it is filled with air (fig. 2.21*d*). In this event gas exchange must be entirely by diffusion through the cuticle. Many small aquatic insects obtain sufficient gas exchange in this way, their cuticle is relatively thin, and they can remain submerged without the need to make contact with the atmosphere. Yet, their tracheal system is necessary for gas transport inside the animal, otherwise diffusion through the tissues would be too slow. As we saw before, diffusion in air is some 300 000 times faster than in water; this clearly points out the advantage of a tracheal system in conveying gases when the circulation of blood is unsuited for this purpose.

A further development on the same theme again represents an aquatic insect (fig. 2.21*e*). The closed tracheal system extends into abdominal appendages or 'gills', which, with their relatively large surface and thin cuticle, permit effective gas exchange between the water and the air within the tracheal system. Such tracheal gills are found, for example, in mayfly larvae (Ephemeroptera).

Another variation is shown in fig. 2.21*f*. The tracheal gills are located within the lumen of the rectum, and ventilation takes place by moving water in and out of the rectum. This system is found in some dragonfly larvae. These larvae may also use the water in the rectum for locomotion by expelling it rapidly, thus moving by jet propulsion in the same way as squid and octopus.

Diffusion and ventilation. Diffusion alone suffices for gas exchange within the tracheal system of many small insects, and also in relatively inactive large insects. As an example we will use a large beetle larva of the genus *Cossus* which was studied by Krogh (1920). This larva weighs about 3.4 g and is 60 mm long. There are nine pairs of spiracles along the body, and careful measurements of the tracheae leading from these showed that the aggregate cross-sectional area of all the tracheae combined was 6.7 mm². The average length of the tracheal system was 6 mm. Interestingly, as the tracheae branched and subdivided, the aggregate cross-sectional area of the system remained rather constant, not changing much with the distance from the spiracles. Therefore, the diffusion through the entire tracheal system can be represented by a single cylindrical tube of 6.7 mm² cross-sectional area and a length of 6 mm. The oxygen consumed by the animal, $0.3 \mu l$ O_2 s^{-1}, will diffuse through a tube of these dimensions if the partial pressure difference between the two ends is 11 mm Hg. This means that, with an oxygen pressure in the atmosphere of 155 mm Hg, the tissues could still have an oxygen tension of 144 mm Hg. Obviously, for the *Cossus* larva an adequate oxygen supply is secured by diffusion alone, even if the metabolic rate during activity should be increased several-fold.

Intuitively it may seem that the diffusion through 6 mm of air would be slow, but since the diffusion in air is 300 000 times faster than in water, the diffusion through 6 mm of air is as fast as the diffusion through a water layer of 0.02 μm. The greatest barrier to oxygen reaching the tissues is therefore likely to be between the finest branches of the tracheoles and the cells. In very active tissues, such as insect flight muscles, electron-micrographs reveal that the tracheoles are as close as 0.07 μm to a mitochondrion.

A quantitative consideration of the branched tracheal system gives us information about another aspect of gas exchange. As the tracheae divide and subdivide the aggregate wall area of the system increases. In the *Cossus* larva the largest tracheae are about 0.6 mm in diameter and the finest tracheoles 0.001 mm. The total cross-section area of the tubes remains constant, and the wall area of the tracheoles therefore increases to 600 times that of an equal length of the largest tracheae. Thus, the reduction in diameter alone increases the wall surface area 600-fold. This immediately tells us that virtually the entire area available for gas exchange is in the finest branches. Consider now that the wall thickness in the tracheoles is less than one-tenth of the wall thickness of the large tracheae, and it becomes evident that diffusion through the walls of the larger branches must be insignificant.

The spiracles. The openings of the tracheal system to the outside, the spiracles, are highly complex structures which can be opened or closed to permit a variable amount of gas exchange, and their accurate control helps impede the loss of water.

The spiracles open more frequently and more widely at high temperature and when there is increased activity, in accord with the increased need for oxygen. The spiracles do not necessarily all open simultaneously, they are under control of the central nervous system, and out-of-phase opening and closing permits the control of air flow through the tracheal system.

The ventilation of the tracheal system, and especially the function of the spiracles is influenced both by carbon dioxide and lack of oxygen. Carbon dioxide seems to be a primary stimulus for opening of the spiracles. If a tiny stream or carbon dioxide is directed towards one single spiracle, only this spiracle will open; this shows that the single spiracle may respond independently. The necessary concentration of carbon dioxide is fairly small, in the cockroach, for example, 1% carbon dioxide in the air has a perceptible effect, 2%

keeps the spiracles open, and 3% makes them remain widely open.
The effect of open spiracles on the water loss is considerable. In
the mealworm, which spends its entire life in the dry environment of
flour, the control of the spiracles is very important. If the spiracles
are made to remain open by adding carbon dioxide to the air, the
water loss immediately increases several-fold (fig. 2.22).

Carbon dioxide cannot be the only agent that controls the spiracles, however, the spiracles can be made to open up by pure nitrogen, i.e. oxygen deficiency is also a stimulus. This could be interpreted as an accumulation of acid metabolites, which are often
formed when there is a lack of oxygen, and acids (lactic acid, for example) increase the carbon dioxide tension. The anoxic effect could
therefore be interpreted as an indirect carbon dioxide effect. Even a
moderate oxygen deficiency, which also stimulates respiration, can
be interpreted similarly, but a direct effect of oxygen is more likely.

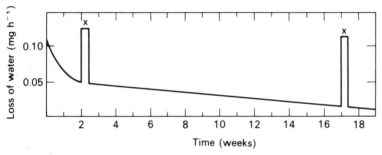

Fig. 2.22. The water loss from a mealworm when kept in dry air. During periods
when the spiracles were made to remain open by adding carbon dioxide to the air
(marked with an x), the water loss increased several-fold (Mellanby, 1934).

Ventilation. The tracheae are relatively rigid and do not easily collapse, but the spiral folding of the wall permits some accordion-like
shortening. This can serve to reduce the volume sufficiently to contribute to the active ventilation of the system. If the tracheae are oval
or flat in cross-section, they will be more compressible. However, an
effective ventilation depends on thin-walled dilations or air sacs that
are connected with the larger tracheae, for when these are compressed, a large volume of air can be expelled. Expiration is usually
the active phase, inspiration being passive.

In many insects the major respiratory movements are made by the
abdomen. During flight synchronous pressure changes and movements may increase the ventilation, and even the head may have
some active ventilation due to transmission of blood pressure. The
ventilation volume may be quite large, as much as one-third or one-

half of the total capacity of the air system being emptied in one expiration, thus giving a renewal of about half of the volume of the respiratory system. This is much greater than in a mammal at rest, where the renewal is more in the order of one-tenth of the air contained in the system, although the maximum possible renewal in mammals is closer to two-thirds of the total volume of the respiratory system.

The breathing movements are synchronized with the opening and closing of spiracles and are similarly controlled by lack of oxygen or excess carbon dioxide. Excess oxygen may cause complete arrest of respiratory movements, and since high oxygen should have no influence on the carbon dioxide production, it probably affects the nervous system directly.

The breathing movements are directed by centers in the segmented ventral nerve chain, for isolated segments may perform respiratory movements. Coordination between the segments is brought about by higher centers located in the prothorax. The head, apparently, is not involved, insects in general have few higher centers in the head, and decapitation has little effect on respiration and many other functions.

Respiration in aquatic insects

Insects that live in water have evolved from terrestrial forms and have retained many characteristics of their terrestrial ancestors. Almost all are fresh-water forms, very few live in brackish water, and virtually no insect is truly marine.

Aquatic insects may have retained air breathing through spiracles. The spiracles are then usually reduced in number and found at the hind end of the body, which is brought into contact with the atmosphere. The spiracular openings are hydrophobic and often surrounded by hydrophobic hairs, the water surface is therefore easily broken and contact established with the air.

Small insects, e.g. mosquito larvae, depend on diffusion exchange between the atmosphere and the tracheal system, but larger larvae use respiratory movements to aid in the replacement of the tracheal air. The tracheae are often large and voluminous so that they can contain enough air to serve as an oxygen store, permitting submersion for longer periods. There are strict limits to this solution, however, because an unlimited increase of air stores gives too much buoyancy and the insect will be unable to submerge.

Many insects carry air attached to the outside of the body, and the spiracles then open into this air mass. The air is held in place by non-wettable surfaces, often helped by hydrophobic hairs. The adult *Dytiscus* beetle has a large air space under the wings; when they make contact with the water surface this air space is ventilated, and on submersion it serves as an air store into which the spiracles open. A similar means is used by the back-swimmer, *Notonecta,* which carries air attached to the ventral surface of the abdomen. The rather large air bubble gives it positive buoyancy, so that it must swim vigorously to descend from the surface and must attach itself to vegetation or other solid objects in order to remain submerged.

A simple experiment can demonstrate that the air carried by *Notonecta* has a respiratory role and is not related to buoyancy problems. If a *Notonecta* is kept submerged in water that has been equilibrated with pure nitrogen instead of pure air, it will live for only five minutes. If it is kept submerged in water saturated with air, it will live for six hours. If it is submerged in water equilibrated with 100% O_2 instead of air, and is allowed to fill its air space with oxygen before submerging, it will live for only 35 minutes. This seemingly paradoxical situation is readily explained. It comes about because when the bubble contains no nitrogen, it disappears as rapidly as oxygen is used up. When nitrogen is present, the bubble helps in obtaining additional oxygen by diffusion from the water (Ege, 1915). Let us examine this further.

The air mass carried by *Notonecta* serves two purposes, firstly, it contains a store of oxygen which is gradually used up, and secondly, it serves as a 'gill' into which oxygen diffuses from the surrounding water. Say that a *Notonecta* is submerged in air-saturated water, and that the air mass at a given time contains 5% O_2, 1% CO_2, and 94% N_2. Since the oxygen partial pressure in the air mass is lower than in the water, oxygen will diffuse from the water into the air. The nitrogen in the bubble, on the other hand, is at 94%, while the water is in equilibrium with the normal 79% N_2 in the atmosphere. The partial pressure of nitrogen in the bubble is therefore higher than in the water, nitrogen diffuses out, and the volume of the air mass gradually diminishes. Therefore, the air mass will be slowly lost. With increasing depth in the water, the total pressure within the air mass increases, while the partial pressures of dissolved gases in the water remain unchanged. The bubble, therefore, is lost faster. For a one meter increase in depth, the partial pressure of nitrogen in the bubble increases by about one-tenth, thus accelerating the rate of loss.

At ten meters' depth, the total pressure within a submerged air mass is 2 atm. This high pressure causes a rapid loss of gas to the water, and an air mass at this depth could function only for a very short time. Therefore, there are limitations to the depth to which insects with air masses can dive, they would have to return to the surface very often for renewal of the air, the distance to be traversed would require additional work and use of oxygen, exposure to predators would be increased, and the total time spent at the bottom would be very limited.

For insects that carry an air mass as a diffusion gill, the total obtainable oxygen far exceeds the initial supply carried in the mass when the insect submerges. Nitrogen is lost more slowly than oxygen diffuses in, primarily because nitrogen, due to its lower solubility, diffuses only half as fast as oxygen between air and water. Because of the initial nitrogen supply, an air mass will last for sufficient time to obtain eight times as much oxygen by diffusion from the surrounding water as was originally in the bubble. This eight-fold ratio remains constant and independent of the metabolic rate of the insect, of the exposed surface area, of the initial volume of the air mass, and of the thickness of the boundary layer of the water (meaning that turbulence in the water in the form of active ventilation will not influence the amount of obtainable oxygen) (Rahn and Paganelli, 1968).

A similar type of respiratory device is found in the 'plastron' breathers. These insects have surfaces densely covered with hydrophobic hairs which provide them with a non-wettable surface where air remains permanently, thus permitting these insects to remain submerged without time limit. In the bug *Aphelocheirus,* the number of such hairs is 2 000 000 single hairs per mm^2 plastron surface, each single hair being 5 μm long (Thorpe and Crisp, 1947).

The plastron breathers have a tremendous advantage above those that carry single bubbles, for water will not penetrate between the hydrophobic hairs. As a result the plastron serves as a non-compressible gill into which oxygen diffuses from the water. Since the volume remains constant, the net diffusion of nitrogen in this gill must be zero, and the total gas pressure within the air mass is negative relative to the gas tensions in the surrounding water. This is possible because the surface film of the water is supported by the hydrophobic hairs, and to force water into the air spaces between the hairs takes a pressure of 3.5 to 5 atm.

Two hemipteran insects, *Buenoa* and *Anisops,* which are relatives of the back-swimmer *Notonecta,* are remarkable for their ability to remain poised in mid-water for several minutes, almost effortlessly. This is most unusual among insects, but it permits these predaceous animals to exploit the mid-water zone in oxygen-poor waters. In small temporary bodies of water where fish are absent, the mid-water zone is relatively free from predators. The bottom is dangerous because of voracious dragonfly and beetle larvae, and other predators hunt at the surface, but there is safety as well as little competition for food in mid-water.

Both *Buenoa* and *Anisops* have large hemoglobin-filled cells in the abdomen. These are important because they serve as an oxygen store which is consumed during the dive. The insects carry with them only a small air mass, but this external supply is depleted only slowly while oxygen from the hemoglobin is being used. When the oxygen of the hemoglobin has been exhausted, the external oxygen is rapidly consumed, and the insects lose buoyancy and become heavier than water.

The long-lasting internal oxygen store confers an advantage on these animals. After they have replenished their oxygen store at the surface, they swim actively to descend, and then remain poised in mid-water where they are in near-neutral buoyancy for an extended period of time while the oxygen stores are being used. Not until they begin to use the external oxygen store and begin to sink, do they return to the surface to replenish their oxygen supply (Miller, 1964).

If an *Anisops* is placed in air-free water, the duration of the dive is not much shorter. This shows that the relatively small air mass does not act as a gill, as it does in *Notonecta*. However, if the hemoglobin is poisoned with carbon monoxide, the dive is much shortened, for the hemoglobin now carries no oxygen. Even if the insect is permitted to take normal air with it, the situation is unchanged, the oxygen store is absent, and the external air mass does not help appreciably.

Discontinuous or cyclic respiration

Many insects show a peculiar phenomenon which is characterized by a periodic or cyclic release of large amounts of carbon dioxide. The carbon dioxide is released during brief periods or 'bursts', while oxygen is taken up at a more or less constant rate. The carbon dioxide bursts may last for a few minutes and alternate regularly with long periods of very slow carbon dioxide release, lasting several hours or even days. A record of a single carbon dioxide burst from the pupa of a silkworm is shown in fig. 2.23. Such bursts may occur from once every week to many times per hour, depending on temperature and metabolic rate.

Insect groups in which cyclic respiration has been observed include roaches, grasshoppers, beetles, larvae and pupae of butterflies and moths (Lepidoptera), and also diapausing adults of Lepidoptera. The phenomenon is rather common but seems to be characteristic of insects with low metabolic rate. In some it can be produced by

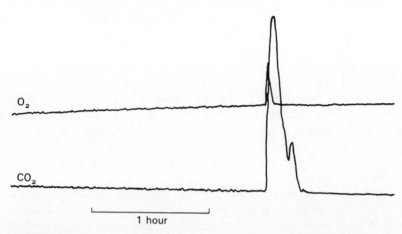

Fig. 2.23. A record of the gas exchange of the pupa of a *Cecropia* silkworm (at 20 °C) shows a steady oxygen uptake. The concurrent release of carbon dioxide is minimal until a major burst occurs after several hours (Punt, Parser and Kuchlein, 1957).

lowering the temperature, thus lowering the metabolic rate (Schneiderman, 1960).

Between the bursts, during the interburst period, a minute amount of carbon dioxide is being released, but the concurrent oxygen uptake is many times greater, in some cases 20 or even 100 times as high. It is clear that gas exchange takes place through the spiracles and not through the cuticle, for if all the spiracles are sealed with wax, there is no detectable oxygen uptake or carbon dioxide release at all, i.e. the cuticle is highly impermeable to gases. This means that, during the interburst period, the spiracles permit inward passage of oxygen while virtually no carbon dioxide escapes. How is this possible? This and many other questions need answers.

Amount of carbon dioxide released. In a *Cecropia* pupa which weighs 6 g, more than 0.5 ml CO_2 is released during a single burst. This volume exceeds the total volume of the tracheal system, so that even if the tracheal system were filled with carbon dioxide only, the carbon dioxide would still have to come from the blood or tissues during the short duration of the burst.

Accurate measurements of the carbon dioxide released from the pupa of another moth, *Agapema,* showed the tracheal volume to be $60\mu l$. Just before a burst, the tracheal air contained about 5.9% CO_2. This means that 3.5 μl CO_2 was present within the tracheal system when the burst started. However, during the burst 30 μl CO_2 was released, in other words, only 10% of the carbon dioxide could come from the tracheal air and the remainder would have to come from elsewhere.

Is metabolism a cyclic event? The record in fig. 2.23 showed that the oxygen uptake, although continuing between the bursts, was intensified during the burst. This periodic nature of both carbon dioxide release and oxygen uptake raises the question of whether the metabolism of the pupa is cyclic with short periods of sudden or cataclysmic carbon dioxide production interspersed by long quiet periods of virtual metabolic standstill.

This question can be answered by a very simple experiment. If a single spiracle is kept permanently open by inserting a fine glass tube through the opening, the cyclic phenomenon disappears completely. We can therefore discount the hypothesis of cyclic biochemical events in the metabolism. It then becomes necessary to examine the function of the spiracles in relation to the cycles.

Movements of the spiracles. Direct observation of the spiracles with a microscope shows that they are widely open during the carbon dioxide burst. Shortly after the burst, the valve at the opening of each spiracle flutters for a short while, and then closes completely and remains closed for as long as an hour. Following this period the valve begins to pulse or flutter faintly, slightly enlarging the spiracular opening. The flutter continues, often for several hours, until the next period of wide opening and burst of carbon dioxide.

The movements of the spiracles of a cecropia silkworm pupa is shown in fig. 2.24. The top record (*a*) was obtained with the insect in pure oxygen. After a long period with closed spiracles, they opened

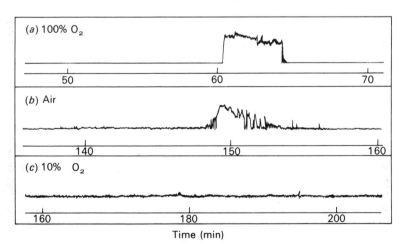

Fig. 2.24 Record of the movements of the spiracular opening of a *Cecropia* pupa. (*a*) pupa in pure oxygen (100% O_2), note absence of flutter before the burst; (*b*) pupa in atmospheric air (21% O_2), burst is preceded by flutter; (*c*) pupa in 10% O_2, spiracle flutters continuously (Schneiderman, 1960).

and remained open for about four minutes, and then closed again. In atmospheric air (*b*), after a long closed period the spiracles fluttered for about ten minutes before they opened completely, then they fluttered for some time and gradually closed down, and finally remained closed until the next fluttering period started. The third graph (*c*) shows that 10% O_2 caused the spiracles to flutter continuously.

During the long period of flutter, when the spiracles are slightly open, the release of carbon dioxide is minimal but oxygen uptake is almost continuous. We therefore are faced with the perplexing fact that, when the spiracle flutters, oxygen enters at a rate up to 100 times as high as that with which carbon dioxide leaves.

How can the spiracles, which are the only site of gas exchange, allow oxygen to enter but effectively prevent the carbon dioxide from leaving? To answer this question we need information about the gas concentrations inside the tracheal system.

The insects which we have discussed usually have no visible respiratory movements, and we might therefore assume that gas exchange between the tracheal system and outside air takes place by diffusion. However, the gas concentrations actually found in the tracheal system cannot be explained by diffusion processes only.

Schneiderman and his collaborators have succeeded in removing minute gas samples for analysis (fig. 2.25). As expected, they found the lowest carbon dioxide concentration, about 3%, immediately after a burst, and during the next six hours the carbon dioxide concentration rose very slowly to 6.5%. During the burst the oxygen concentration increased to nearly atmospheric concentrations, 18 to 20%, but after the burst it declined rapidly again and remained constant at 3.5% throughout the interburst period. These amazing figures cannot be explained by diffusion as the only driving force for the movement of gases through the spiracles. However, gases can also be moved by pressure, and it was suggested by Buck (1958) that a slightly negative pressure within the tracheal system would cause a mass flow of air in through the spiracle.

It is possible to measure the pressure within the tracheal system by sealing a small glass capillary into one spiracle while leaving the others intact, and then connecting the tube to a manometer. During periods of flutter the pressure within the tracheal system is very close to atmospheric, although slightly negative. This slightly negative pressure causes an inward mass flow of air. When the spiracles close completely, the pressure in the tracheal system drops precipitously. As the spiracles remain closed, a considerable negative pressure develops, corresponding to the rapid decrease in oxygen con-

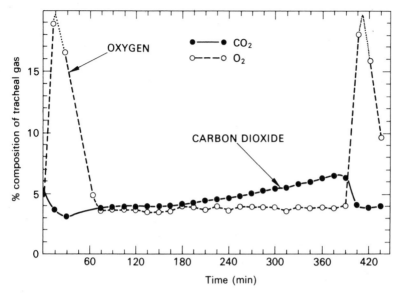

Fig. 2.25. Composition of air obtained from the tracheal system at various stages of cyclic respiration. A sudden rise in oxygen concentration occurs when the spiracles are fully open during the burst. Pupal weight 8.4 g; temperature, 25 °C (Levy and Schneiderman, 1958).

centration within the tracheal system immediately after the closing of the spiracles (see fig. 2.25). When flutter begins again, the pressure rises step-wise, and when flutter continues regularly, the pressure has again returned to near-atmospheric (Levy and Schneiderman, 1966*b*).

The rapid drop in pressure when the spiracles close is caused by the rapid consumption of oxygen present in the closed tracheal system. Most of the carbon dioxide produced remains dissolved or buffered in blood and tissues, and therefore a negative pressure must develop. If the pupa is placed in pure oxygen before the spiracles close, so that the tracheal system is filled with only oxygen, the negative pressure becomes even greater. This is precisely what could be expected when the oxygen is used and no nitrogen is present.

If we combine all the information at hand, the facts are consistent with the following interpretation (see fig. 2.26). The top graph (fig. 2.26*a*) shows flutter, followed by a period of open spiracles, and then a period of completely closed spiracles, after which the flutter is resumed. The carbon dioxide output (*b*) is low during the closed and flutter periods, most of the carbon dioxide output takes place in a short burst during the open period. The oxygen uptake is recorded as being continuous; this corresponds to the actual consumption of

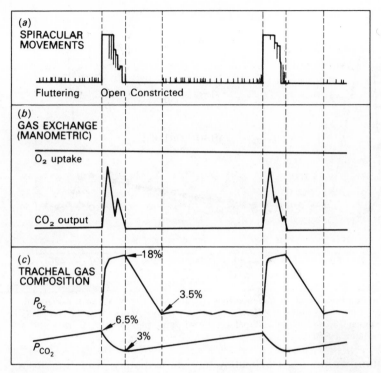

Fig. 2.26. Summary graph, showing the synchronous events in spiracular movements during cyclic respiration (Levy and Schneiderman, 1966*a*).

oxygen by the tissues and can be measured by suitable manometric methods (Schneiderman and Williams, 1955).

The composition of the tracheal air (*c*) shows that the carbon dioxide concentration falls during the burst from 6.5 to 3%, and again rises slowly during the interburst period. The changes in the oxygen concentration are more spectacular. When the spiracles open the oxygen rises rapidly to 18%, not far from atmospheric concentration. During the ensuing constricted period, the tracheal oxygen concentration falls steadily as the tissues use oxygen at a constant rate. When flutter begins, a negative pressure has already developed within the tracheal system, and inward bulk flow of air begins. This partly replenishes the oxygen, but nitrogen will also enter and gradually increase. If there is, say, 4% CO_2 and 4% O_2 in the tracheal system, the remainder must be nitrogen in a concentration of 92%.

Significance of cyclic respiration. Terrestrial animals are vulnerable to desiccation and insects often live in very dry environments. The diapausing pupa, which may remain encased for nearly a year from one summer to the next, has no possibility for replenishing lost water. It must be extremely economical in the use of water while permitting sufficient gas exchange for its metabolic processes. The spiracles must therefore be governed by a compromise between permitting oxygen uptake and preventing water loss.

We can now see a rationale for the periodic opening of the spiracles. The water loss depends on the vapor pressure within the tracheae (a function of temperature), the humidity in the outside atmosphere, and the spiracular opening and closing. If the spiracles are closed, water loss from the tracheal system is nil, when they are fully open, the water loss will be maximal. As we have seen, partial closing (flutter) permits a sufficient oxygen uptake, for in addition to the inward mass flow of air, the high oxygen in the atmosphere gives a steep inward diffusion gradient. However, as the carbon dioxide concentration in the tracheal system builds up and reaches a limit of 6% or so, the spiracles open widely and permit this gas to escape (the carbon dioxide burst). If instead the carbon dioxide were permitted to leave continuously through open spiracles, as fast as it is formed, water would also be lost continuously. The cyclic nature of the gas exchange, which at first glance appears as a puzzling deviation, can therefore be considered as a very clever device for water conservation.

REFERENCES

ALKALAY, I., SUETSUGU, S., CONSTANTINE, H. and STEIN, M. (1971). Carbon dioxide elimination across human skin. *Am. J. Physiol.*, **220**, 1434–6.

BENTLEY, P. J., HERREID, C. F. and SCHMIDT-NIELSEN, K. (1967). Respiration of a monotreme, the echidna, *Tachyglossus aculeatus*. *Am. J. Physiol.*, **212**, 957–61.

BENTLEY, P. J. and SHIELD, J. W. (1973). Ventilation of toad lungs in the absence of the buccopharyngeal pump. *Nature, Lond.*, **243**, 538–9.

BERG, T. and STEEN, J. B. (1965). Physiological mechanisms for aerial respiration in the eel. *Comp. Biochem. Physiol.*, **15**, 469–84.

BRETZ, W. L. and SCHMIDT-NIELSEN, K. (1972). Movement of gas in the respiratory system of the duck. *J. Exp. Biol.*, **56**, 57–65.

BUCK, J. (1958). Cyclic CO_2 release in insects. IV. A theory of mechanism. *Biol. Bull.*, **114**, 118–40.

CLEMENTS, J. A., NELLENBOGEN, J. and TRAHAN, H. J. (1970). Pulmonary surfactant and evolution of the lungs. *Science*, **169**, 603–4.

CRAIG, A. B., JR. (1961a). Causes of loss of consciousness during underwater swimming. *J. Appl. Physiol.*, **16**, 583–6.

CRAIG, A. B., JR. (1961b). Underwater simming and loss of consciousness. *J. Amer. Med. Assoc.*, **176**, 255–8.

DOLK, H. E. and POSTMA, N. (1927). Über die Haut- und die Lungenatmung von *Rana temporaria*. *Zeits. vergl. Physiol.*, **5**, 417–44.

EGE, R. (1915). On the respiratory function of the air stores carried by some aquatic insects (Corixidae, Dytiscidae, and *Notonecta*). *Zeits. f. allg. Physiol.* **17**, 81–124.

✕ GANS, C., DE JONGH, H. J. and FARBER, J. (1969). Bullfrog (*Rana catesbeiana*) ventilation: How does the frog breathe? *Science*, **163**, 1223–5.

HERREID, C. F. II, BRETZ, W. L. and SCHMIDT-NIELSEN, K. (1968). Cutaneous gas exchange in bats. *Am. J. Physiol.*, **215**, 506–8.

HUTCHISON, V. H., WHITFORD, W. G. and KOHL, M. (1968). Relation of body size and surface area to gas exchange in anurans. *Physiol. Zool.*, **41**, 65–85.

JOHANSEN, K. (1966). Air breathing in the teleost *Symbranchus marmoratus*. *Comp. Biochem. Physiol.*, **18**, 383–95.

JOHANSEN, K. (1968). Air-breathing fishes. *Sci. Amer.*, **219**, 102–11.

✕ JOHANSEN, K., LENFANT, C., SCHMIDT-NIELSEN, K. and PETERSEN, J. A. (1968). Gas exchange and control of breathing in the electric eel, *Electrophorus electricus*. *Zeits. vergl. Physiol.*, **61**, 137–63.

JOHNSTON, A. M. and JUKES, M.G.M. (1966). The respiratory response of the decerebrate domestic hen to inhaled carbon dixoide–air mixture. *J. Physiol.*, **184**, 38–39P.

JOSHI, M. C., BOYER, J. S. and KRAMER, P. J. (1965). Growth, carbon dioxide exchange, transpiration, and transpiration ratio of pineapple. *Botan. Gaz.*, **126**, 174–9.

KROGH, A. (1920). Studien über Tracheenrespiration. II. Über Gasdiffusion in den Tracheen. *Pflügers Archiv*, **179**, 95–112.

LASIEWSKI, R. C. and CALDER, W. A., JR. (1971). A preliminary allometric analysis of respiratory variables in resting birds. *Respir. Physiol.*, **11**, 152–66.

LENFANT, C., JOHANSEN, K. and HANSON, D. (1970). Bimodal gas exchange and ventilation–perfusion relationship in lower vertebrates. *Fed. Proc.*, **29**, 1124–9.

LEVY, R. I. and SCHNEIDERMAN, H. A. (1958). An experimental solution to the paradox of discontinuous respiration in insects. *Nature, Lond.*, **182**, 491–3.

LEVY, R. I. and SCHNEIDERMAN, H. A. (1966a). Discontinuous respiration in insects – II. The direct measurement and significance of changes in tracheal gas composition during the respiratory cycle of silkworm pupae. *J. Insect Physiol.*, **12**, 83–104.

LEVY, R. I. and SCHNEIDERMAN, H. A. (1966b). Discontinuous respiration in insects – IV. Changes in intratracheal pressure during the respiratory cycle of silkworm pupae. *J. Insect. Physiol.*, **12**, 465–92.

MARGARIA, R., MILIC-EMILI, G., PETIT, J. M. and CAVAGNA, G. (1960). Mechanical work of breathing during muscular exercise. *J. Appl. Physiol.*, **15**, 354–8.

MELLANBY, K. (1934). The site of loss of water from insects. *Proc. Roy. Soc. Lond. B,* **116,** 139–49.

MILLER, P. L. (1964). The possible role of haemoglobin in *Anisops* and *Buenoa* (Hemiptera:Notonectidae). *Proc. Roy. Entomol. Soc. Lond. A,* **39,** 166–75.

OTIS, A. B. (1954). The work of breathing. *Physiol. Rev.,* **34,** 449–58.

PUNT, A., PARSER, W. J. and KUCHLEIN, J. (1957). Oxygen uptake in insects with cyclic CO_2 release. *Biol. Bull.,* **112,** 108–19.

RAHN, H. and PAGANELLI, C. V. (1968). Gas exchange in gas gills of diving insects. *Respir. Physiol.,* **5,** 145–64.

SCHEID, P. and PIIPER, J. (1972). Cross-current gas exchange in avian lungs: effects of reversed parabronchial air flow in ducks. *Respir. Physiol.,* **16,** 304–12.

SCHMIDT-NIELSEN, K. (1972). *How Animals Work,* Cambridge, England: Cambridge University Press, 114 pp.

SCHMIDT-NIELSEN, K., KANWISHER, J., LASIEWSKI, R. C., COHN, J. E. and BRETZ, W. L. (1969). Temperature regulation and respiration in the ostrich. *Condor,* **71,** 341–52.

SCHNEIDERMAN, H. A. (1960). Discontinuous respiration in insects: role of the spiracles. *Biol. Bull.,* **119,** 494–528.

SCHNEIDERMAN, H. A. and WILLIAMS, C. M. (1955). An experimental analysis of the discontinuous respiration of the cecropia silkworm. *Biol. Bull.,* **109,** 123–43.

SCHUMANN, D. and PIIPER, J. (1966). Der Sauerstoffbedarf der Atmung bei Fischen nach Messungen an der narkotisierten Schleie (*Tinca tinca*). *Pflügers Archiv,* **288,** 15–26.

SOUM, J. M. (1896). Récherches physiologiques sur l'appareil réspiratoire des oiseaux. *Ann. Univ. Lyon,* **28,** 1–126.

TENNEY, S. M. and REMMERS, J. E. (1963). Comparative quantitative morphology of the mammalian lung: diffusing area. *Nature, Lond.,* **197,** 54–6.

THOMAS, H. J. (1954). The oxygen uptake of the lobster (*Homarus vulgaris* Edw.). *J. Exp. Biol.,* **31,** 228–51.

THORPE, W. H. and CRISP, D. J. (1947). Studies on plastron respiration. I. The biology of *Aphelocheirus* [Hemiptera, Aphelocheiridae (Naucoridae)] and the mechanim of plastron retention. *J. Exp. Biol.,* **24,** 227–69.

TUCKER, V. A. (1968). Respiratory physiology of house sparrows in relation to high-altitude flight. *J. Exp. Biol.,* **48,** 55–66.

WIGGLESWORTH, V. B. (1972). *The Principles of Insect Physiology,* 7th edn, London: Chapman and Hall, 827 pp.

3
Blood

The preceding two chapters dealt with respiratory gases and the gas exchange between the organism and the environment. It focused on the two most important gases, oxygen and carbon dioxide.

We learnt that in all but very small organisms diffusion alone does not suffice to distribute the gases within the organism, and that a mechanical transport system is necessary. Nearly all animals of large size, except some that have very low oxygen demands, have a distribution system that is designed around the mechanical movement in the organism of a fluid, the blood. Insects do not use blood for gas transport, they use air-filled tubes, and yet, insects do have blood which is pumped around in the body, for many other substances need to be transported at rates faster than diffusion alone can provide.

Blood, in fact, has many functions that may not be apparent at first thought. This becomes clear when we make a list of the major functions that are served by this important fluid (table 3.1).

We always tend to think of gas transport as the primary function of the blood, but the list contains as many as six major categories for transport of material items and a seventh for heat transport. An eighth and very important although often overlooked item on our list concerns the transmission of force. Of the two last categories, one is concerned with a highly specific inherent characteristic of blood, the ability to coagulate, and the other with the general chemical composition of the blood fluid.

Most of the functions on our list can be carried out by nearly any aqueous medium, i.e. no special functional characteristics of the blood are necessary, for example, to carry metabolites or excretory products. Exceptions must be made, however, in regard to gas transport and coagulation, both of which are associated with highly complex biochemical properties of the blood.

In this chapter we shall be concerned primarily with the function

Table 3.1. *The most important functions of blood*

(1) *Nutrients.* Transport from digestive tract to tissues; to and from storage organs, e.g., adipose tissue, liver.

(2) *Metabolites.* e.g. lactic acid from muscle to liver. Permits metabolic specializations.

(3) *Excretory products.* From tissues to excretory organs; from organ of synthesis (e.g. urea in liver) to kidney.

(4) *Gases.* Oxygen and carbon dioxide between respiratory organs and tissues. Blood may also provide oxygen storage.

(5) *Communication.* Transport of hormones, e.g., adrenalin (fast response), growth hormone (slow response).

(6) *Cells.* Many types of blood cells of non-respiratory function; e.g. vertebrate leukocytes. Insect blood lacks respiratory function, yet has numerous types of blood cells.

(7) *Transport of heat.* From deeper organs to surface for dissipation. Essential for large animals with high metabolic rates.

(8) *Transmission of force.* e.g. earthworm locomotion; crustaceans for breaking shell during molting; movements of organs such as penis, siphon of bivalve, extension of spider's legs; also force for ultrafiltration in capillaries and kidney.

(9) *Coagulation.* Inherent characteristic of many blood and hemolymph fluids. Serves to protect against blood loss.

(10) *'Milieu interieur'.* Provides internal medium suitable for cells in regard to pH, ions, nutrients, etc.

of blood in the transport of gases and the special properties of blood that serve this purpose.

GAS TRANSPORT IN BLOOD

Respiratory pigments

In many invertebrates oxygen is carried in the blood or hemolymph in simple physical solution. This aids in bringing oxygen from the surface to the various parts of the organism, for diffusion alone is too slow for any but the smallest organisms.

The amount of oxygen that can be carried in simple solution is small, however, and many highly organized animals (vertebrates almost without exception), have blood that can bind larger quantities of oxygen reversibly, thus greatly increasing the amount of oxygen that is carried. In mammalian blood, the amount of physically dissolved oxygen is about 0.2 ml O_2 per 100 ml blood, and the amount bound reversibly to hemoglobin is up to some 100 times as great, about 20 ml O_2 per 100 ml blood. Dissolved oxygen is therefore of miniscule importance as compared to the role of hemoglobin.

The substances we know as oxygen carriers in blood are proteins which contain a metal (commonly iron or copper). Usually they are colored, and therefore they are often called respiratory pigments. The commonest ones are listed in table 3.2.

Table 3.2. *Common respiratory pigments and examples of their occurrence in the animal kingdom*

Hemocyanin. Copper-containing protein, carried in solution.
　Mol. wt. = 300 000 – 9 000 000
　Molluscs: chitons, cephalopods, prosobranch and pulmonate gastropods; not in lamellibranchs.
　Arthropods: Malacostraca (sole pigment in these); Arachnomorpha: *Limulus, Euscorpius*
Hemerythrin. Iron-containing protein, always in cells, nonporphyrin structure.
　Mol. wt. = 108 000
　Sipunculids: all species examined
　Polychaetes: *Magelona*
　Priapulids: *Halicryptus, Priapulus*
　Brachiopods: *Lingula*
Chlorocruorin. Iron-porphyrin protein, carried in solution.
　Mol. wt. = 2 750 000
　Restricted to four families of Polychaetes:
　　Sabellidae; Serpulidae; Chlorhaemidae; Ampharetidae.
　Prosthetic group alone has been found in starfishes, *Luidia* and *Astropecten*.
Hemoglobin. Most extensively distributed pigment: iron-porphyrin protein: in solution or in cells.
　Mol. wt. = 17 000 – 3 000 000
　Vertebrates: almost all, except leptocephalus larvae and some Antarctic fish (*Chaenichtys*, etc.)
　Echinoderms: sea cucumbers
　Molluscs: Planorbis, Pismo clam (*Tivella*)
　Arthropods: insects *Chironomus, Gastrophilus.* Crustacea *Daphnia, Artemia*
　Annelids: *Lumbricus, Tubifex, Arenicola, Spirorbis* (some species have hemoglobin, some chlorocruorin, others no blood pigment). *Serpula,* both hemoglobin and chlorocruorin
　Nematodes: *Ascaris*
　Flatworms: Parasitic trematodes
　Protozoa: *Paramecium, Tetrahymena*
　Plants: Yeasts, *Neurospora,* root nodules of leguminous plants (clover, alfalfa)

Blood corpuscles

In some animals the respiratory pigments occur dissolved in the blood fluid, in others (such as vertebrates) it is enclosed in cells and the blood fluid contains no dissolved respiratory pigment. The distribution between cells and blood fluid for the most common respiratory pigments is listed in table 3.3, which in addition gives representative animal names and molecular weights of the pigments.

This table immediately reveals that when the pigments occur enclosed in cells, their molecular weights are relatively low and range from some 20 000 to 120 000; if the pigments occur dissolved in the plasma, the molecular weight is much higher, from 400 000 to several million (the sole exception being the insect *Chironomus*).

Table 3.3. *List of the molecular weights of respiratory pigments with indication of whether they are found within cells or dissolved in the blood fluid*

	CELLS		PLASMA	
		(Mol. wt)		(Mol. wt)
HEMOGLOBIN	Mammals ⎫		Oligochaetes	
	Birds ⎬ *c.* 68 000		*Lumbricus*	2 946 000
	Fish ⎭		Polychaetes	
	(myoglobin = 17 000		*Arenicola*	3 000 000
	Cyclostomes		*Serpula*	3 000 000
	Lampetra	19 100	Molluscs	
	Myxine	23 100	*Planorbis*	1 539 000
	Polychaetes		Insects	
	Notomastus	36 000	*Chironomus*	31 400
	Echinoderms			
	Thyone	23 600		
	Molluscs			
	Arca	33 600		
	Insects			
	Gastrophilus	34 000		
CHLOROCRUORIN			Polychaetes	
			Spirographis	3 400 000
HEMERYTHRIN	*Sipunculus*	66 000		
	Phascolosoma	120 000		
HEMOCYANIN			Molluscs	
			Helix	6 680 000
			Cephalopods	
			Rossia (squid)	3 316 000
			Octopus	2 785 000
			Eledone	2 791 000
			Arthropods	
			Limulus	1 300 000
			Crustacea	
			Pandalus	397 000
			Palinurus	447 000
			Nephrops	812 000
			Homarus	803 000

The large molecules of respiratory pigments which are dissolved in the plasma are in fact aggregates of smaller molecules. They can be regarded as a means to increase the total amount of the pigment without an increase in the number of dissolved protein molecules in the blood plasma. A large increase in the number of dissolved particles would increase the colloidal osmotic pressure of the plasma, which in turn would influence many other physiological processes such as the passage of fluid through capillary walls and the ultrafiltration which is the initial process in the formation of urine in the kidney.

It also has been said that an advantage of enclosing a high concentration of hemoglobin within cells is that it decreases the viscosity of the blood, as compared to a hemoglobin solution with the same oxygen capacity. Mammalian plasma contains about 7% protein, and the red blood cells contain about 35% hemoglobin. If all the hemoglobin from the red cells were to be carried dissolved in the blood instead of enclosed in the red blood cells, the total protein concentration would be about 20%, or three times as high as the typical mammalian plasma protein concentration. The argument goes that a 20% protein solution is highly viscous or syrupy, far more viscous than blood. However, if we first measure the viscosity of a blood sample, and then disrupt the red cells with ultrasound so that the hemoglobin is set free, the resulting hemoglobin solution has a viscosity less than half of the viscosity of blood with the hemoglobin carried in the cells (fig. 3.1).

There are, however, some other advantages to be gained by enclosing hemoglobin in cells. The most important is probably that the chemical environment within the red cell can differ from that of the blood plasma. The reaction between oxygen and hemoglobin is greatly influenced by inorganic ions as well as certain organic compounds, notably organic phosphates. Therefore, by keeping the hemoglobin within cells, the hemoglobin can be provided with a separate well-adjusted environment different from the plasma.

Size and shape of the red cell. The red blood cells of mammals are round discs which are slightly biconcave. There is one exception, all members of the camel family have oval red cells. The diameter of the red cell is characteristic for each species, the total range for mammals being from about 5 to 10 μm. There is no particular relationship between the size of the red cell and the size of the animal, the smallest mammalian red cell is found in a small Asiatic deer, but a mouse and an elephant have approximately the same red cell size.

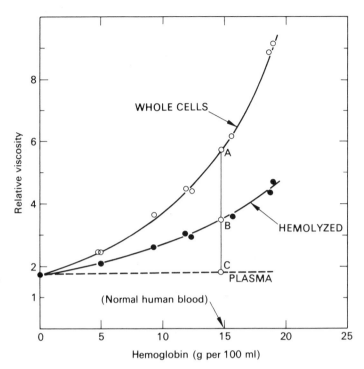

Fig. 3.1. Blood viscosity increases with increasing hemoglobin content. Blood viscosity is here expressed relative to the viscosity of pure water (water = 1.00). The viscosity of goat blood containing intact red cells is shown in the upper curve. The lower curve shows the relative viscosity of hemolyzed blood, and the observed plasma viscosity of the same blood samples is shown by the dashed line. Comparing points A, B and C shows that a hemoglobin solution has a relative viscosity of about half that of intact blood which has the same total hemoglobin content (Schmidt-Nielsen and Taylor, 1968, copyright 1968 by the American Association for the Advancement of Science).

The mammalian red cells lack nuclei, although they have a life span of about 100 days and carry out complex metabolic functions. The red cells of all other vertebrates, birds, reptiles, amphibians, and fish have nuclei, they are almost universally oval in shape, and frequently they are much larger than the mammalian cells (especially in urodele amphibians). The importance of a nucleated versus a non-nucleated red cell, if any, is not understood. Likewise, whether the size of the red cell is of functional importance is unclear.

The only vertebrates that lack hemoglobin and red cells are the larvae of eels ('leptocephalus' larvae) and a few antarctic fish of the family Chaenichtyidae.

Oxygen dissociation curves

The property of blood that is of greatest importance to oxygen transport is the reversible binding of oxygen to the hemoglobin molecule. This binding between hemoglobin and oxygen can be written as an ordinary reversible chemical reaction:

$$Hb + O_2 \rightleftarrows HbO_2.$$

At high oxygen concentration the hemoglobin (Hb) combines with oxygen to form oxyhemoglobin (HbO_2) and the reaction goes to the right, at low concentration oxygen is given up again and the reaction proceeds to the left. If the oxygen concentration is reduced to zero, the hemoglobin gives up all the oxygen it carries.

Each iron atom in the hemoglobin molecule binds one oxygen molecule, and when at high oxygen concentration (or pressure) all available binding sites are occupied, the hemoglobin cannot take up any more oxygen and is fully saturated. At any given oxygen concentration there is a definite proportion between the amount of hemoglobin and oxyhemoglobin. If we plot the amount of oxy-hemoglobin present at each oxygen concentration, we obtain an

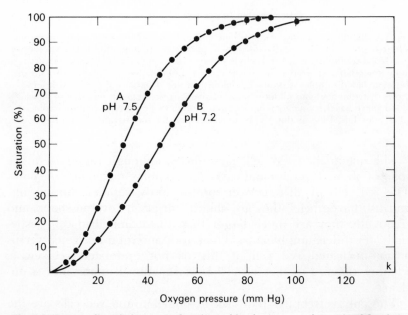

Fig. 3.2 Oxygen dissociation curve for pigeon blood. (A) curve determined for the normal temperature (41 °C), blood P_{CO_2} (35 mm Hg), and pH (7.5) for this bird. (B) curve determined after shifting the pH from 7.5 to 7.2 with unchanged P_{CO_2} (Lutz, Longmuir, Tuttle and Schmidt-Nielsen, 1973).

oxygen–hemoglobin dissociation curve, which describes how the reaction depends on the oxygen concentration or partial pressure. Such a dissociation curve is shown in fig. 3.2. This curve shows that pigeon blood is fully saturated with oxygen (100% HbO_2) at or above about 80–100 mm Hg. At still higher oxygen pressures, the hemoglobin takes on no more oxygen and we say that it is fully saturated. At lower oxygen pressure, however, it gives off oxygen, and at $P_{O_2} = 30$ mm Hg, half of the hemoglobin is present as oxyhemoglobin, the other half as hemoglobin. This particular oxygen pressure is referred to as the *half-saturation pressure*, or P_{50}. As the oxygen pressure is decreased further, more oxygen is given off, until the hemoglobin is fully deoxygenated at zero oxygen pressure.

The oxygen dissociation curves of other animals have a similar shape, but are not identical to the one shown in fig. 3.2. Some animals have blood which has a higher affinity for oxygen, it gives up oxygen less readily, and the dissociation curve is located further to the left. Others have blood that gives up oxygen more readily, the affinity for oxygen is lower, and the dissociation curve is located further to the right. In addition, a number of other parameters (pH, ions, organic phosphates, temperature) influence the binding of oxygen and thus the dissociation curve. The functional significance of these variations are of great importance.

Temperature effect. Increased temperature weakens the bond between hemoglobin and oxygen, and therefore causes an increased dissociation of the bond. As a consequence, at higher temperature the hemoglobin gives up oxygen more readily and the dissociation curve is shifted to the right. This is of physiological importance, because increased temperature is usually accompanied by an increased metabolic rate (or need for oxygen), and therefore it is an advantage that hemoglobin delivers oxygen more readily at higher temperature. Even in warm-blooded animals this may be of some importance; in fever or in exercise, for example, when there is an increase in the rate of oxygen consumption.

Carbon dioxide and pH. Another important influence on the dissociation curve is the pH of the blood plasma. Increase in carbon dioxide or other acids lowers the pH of the plasma, and this shifts the dissociation curve to the right (fig. 3.2). The significance of this shift to the right is that a high carbon dioxide concentration causes more oxygen to be given up at any given oxygen pressure. This effect is known as the *Bohr effect.* In the capillaries of the tissues, as carbon dioxide enters the blood, the hemoglobin gives up a larger amount of oxygen than would be the case if there were no effect of carbon dioxide on the binding. The Bohr effect can therefore be said to facilitate the delivery of additional oxygen to the tissues. In some fish the effect of acid on the dissociation curve is much more pronounced than in mammals; this stronger effect is known as the *Root effect*.

The Bohr effect is not of equal magnitude in all mammals. We can express the Bohr effect as the shift in the P_{50} which occurs for a unit change in pH. We then find that the Bohr effect is body size dependent, and that the shift in P_{50} is greater in a mouse than in an elephant (see fig. 3.3), in other words, mouse hemoglobin is more acid-sensitive than elephant hemoglobin. We shall later see that this is of importance to the delivery of oxygen to the tissues of small animals, which have a higher metabolic rate per gram than larger animals (Schmidt-Nielsen, 1972).

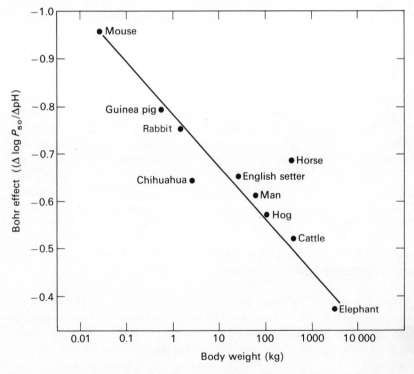

Fig. 3.3. The Bohr shift of hemoglobin in relation to body size. The hemoglobin of small mammals has a greater Bohr shift (i.e. is more acid-sensitive) than the hemoglobin of large mammals and therefore will release more oxygen at a given P_{O_2} (Riggs, 1960).

Fetal hemoglobin

For many mammals, including man, the dissociation curve of the fetus blood is located to the left of that of the mother (fig. 3.4). This is related to how the fetus obtains its oxygen by diffusion from the maternal blood. Since the fetal blood has a higher affinity for

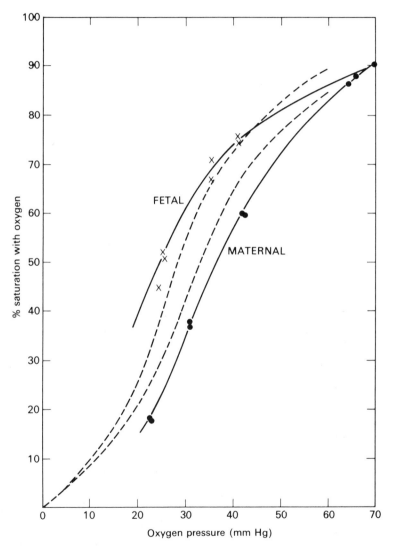

Fig. 3.4. Oxygen dissociation curves for fetal and maternal blood of goat. The higher oxygen affinity of the fetal blood helps in the transfer of oxygen to the fetal blood in the placenta. Dotted lines indicate limits of non-pregnant adult goats (Barcroft, 1935).

oxygen than the maternal blood, it can take up oxygen more readily. At a given oxygen pressure the fetal blood will contain more oxygen (have a higher per cent saturation) than the maternal blood at the same oxygen pressure, and this helps facilitate the uptake of oxygen by the fetal blood in the placenta. The difference in dissociation curve between fetal and maternal blood is in part due to the fact that fetal hemoglobin is slightly different from the maternal hemoglobin,

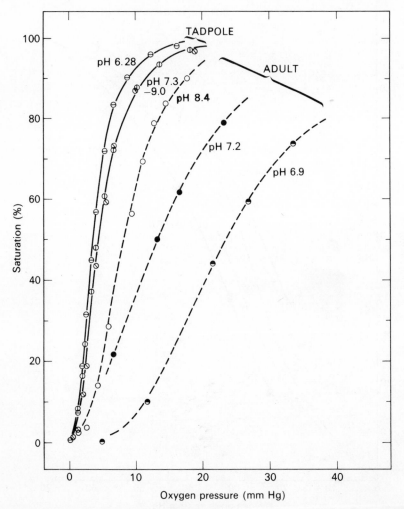

Fig. 3.5. The oxygen dissociation curve for tadpole blood is located to the left of that for adult frogs, i.e. tadpole blood has a higher oxygen affinity. Furthermore, tadpole blood is nearly acid-insensitive, while adult blood has a pronounced Bohr shift (Riggs, 1951).

and in part to a difference in the organic phosphate within the red cell (cf. p. 92). After birth the fetal hemoglobin gradually disappears and is replaced by adult-type hemoglobin.

Displacement of the dissociation curve to the left, so that the blood more readily takes up oxygen, is found also in the chick embryo and in frog tadpoles. The tadpole hemoglobin has a higher oxygen affinity than adult-type frog hemoglobin, and it is particularly interesting that it is acid-insensitive while adult-type hemoglobin has a pronounced Bohr effect and is highly acid-sensitive (fig. 3.5). If we consider the normal habitat of tadpoles, we can readily see the advantage; tadpoles often live in stagnant pools with low oxygen content, and at times also high carbon dioxide content in the water. A high oxygen affinity therefore is an advantage. Since low oxygen and high carbon dioxide are likely to occur at the same time, a pronounced Bohr effect would be undesirable, for it would decrease the oxygen affinity of the blood when a high affinity is most needed.

Altitude. The low atmospheric pressure at high altitude means that the partial pressure of oxygen is lower than at sea level. Animals that

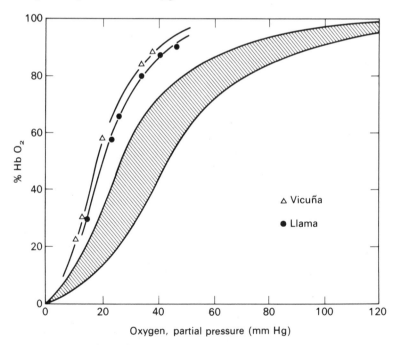

Fig. 3.6. Oxygen dissociation curves of llama and vicuña blood is located to the left of other mammals. The higher oxygen affinity of the blood of these animals aids the oxygen uptake at the low pressure of high altitude (Hall, Dill and Barron, 1936).

normally live in this environment of relative oxygen deficiency have corresponding adaptations in the oxygen dissociation curve of their blood. For example, the dissociation curve of the llama is located to the left of the usual range for mammals. The llama lives in the high Andes of South America, often above 5000 meters, and the high oxygen affinity of its hemoglobin enables its blood to take up oxygen at the low pressures. This adaptation to altitude is an inborn characteristic, for the oxygen dissociation curve of llamas that have been brought up in zoos at sea level also show the same high oxygen affinity (fig. 3.6).

If a man moves to high altitude, he gradually becomes more adapted to the low oxygen pressure, he becomes acclimated to the altitude and performs better, both physically and mentally. However, the adaptation does not include an increased oxygen affinity of the blood, paradoxically the dissociation curve moves slightly to the right as altitude tolerance is improved. This is due to other physiological adaptations which are of a more complex nature (Lenfant, Ways, Aucutt and Cruz, 1969).

Diving animals. It might be expected that diving mammals, such as seals, which are normally exposed to a relative oxygen deficiency when they dive, could gain an advantage by having blood with a high oxygen affinity. This is not so, however, and a little thought will clarify why they would gain no special advantage in this way. Since the seal breathes ordinary atmospheric air, the uptake of oxygen into the blood in the lung takes place under circumstances similar to those in other mammals, and the loading of the blood in the lung therefore does not require any particularly high oxygen affinity. During diving the unloading of oxygen in the tissues (primarily the central nervous system and the heart muscle), would not be aided by a high oxygen affinity of the blood, on the contrary, perhaps, a high oxygen affinity of the hemoglobin would tend to impede delivery of oxygen to the most vital organs. Accordingly, diving animals generally have oxygen dissociation curves which differ little from what we find in similar non-diving animals.

Effects of organic phosphates. The function of the red cell in gas transport is intimately related to the metabolic activities of the cell itself. Due to the discovery of the importance of organic phosphates we can now understand many peculiarities of the oxygen dissociation curve which recently seemed extremely confusing.

The mammalian red cell has no nucleus, and for many years this cell was considered little more than an inert sac packed full of hemoglobin. The red cell, however, has an active carbohydrate metabolism which not only is essential to the viability of the cell, but also has profound effects on its function in oxygen transport. It has a high content of adenosintriphosphate (ATP), and an even higher level of 2,3-diphosphoglycerate (DPG). The presence of these organic phosphates helps explain why a purified solution of hemoglobin has a much higher oxygen affinity than whole blood (a hemoglobin solution has a dissociation curve far to the left of whole blood). If organic phosphates, notably DPG, are added to a hemoglobin solution, its oxygen affinity is greatly decreased and approaches that of intact cells. The decrease is due to a combination of the hemoglobin with the DPG which alters the oxygen affinity (Benesch, Benesch and Enoki, 1968; Benesch, Benesch and Yu, 1968).

The discovery of the DPG effect explains many observations that previously were quite puzzling. It also explains why, if we examine whole blood, we can observe many characteristics of the dissociation curve that remain unobserved if hemoglobin solutions are examined. For example, the DPG level of the red cell is higher in people living at high altitude than in people living at sea level. As a consequence, there is a shift to the right of the oxygen dissociation curve at high altitude. This shift to the right has been well known, but the mechanism remained unclear until it was discovered that the level of the DPG increases in response to hypoxia (Aste-Salazar and Hurtado, 1944; Lenfant *et al.*, 1968).

The role of DPG and other phosphates explains the long-known fact that the oxygen affinity of blood stored in blood banks increases with time. It also helps explain why purified fetal and adult hemoglobins, when stripped of the DPG, have similar oxygen affinities while, as we saw before, fetal blood has a higher oxygen affinity. Furthermore, a shift to the right of the maternal dissociation curve relative to non-pregnant adults can now be understood, although the pregnant animal (as opposed to the fetus) has the same type of hemoglobin throughout.

The shift to the right of the dissociation curve of man at high altitude (due to increased DPG levels) is exactly the opposite of the change we find in the blood of the llama and other animals native to high altitude. This may be difficult to reconcile, but we should realize that the affinity of hemoglobin for oxygen is of importance at two different locations in the body, first, when the hemoglobin takes up oxygen in the lung, and second, when the hemoglobin again gives up oxygen to the tissues. For the uptake in the lung, a high oxygen affinity (shift to the left) increases the effectiveness of the uptake; in the tissues, on the other hand, a low oxygen affinity (shift to the right) facilitates the delivery of oxygen. It therefore appears that the left-shift of the llama dissociation curve is primarily an adaptation to low oxygen pressures at high altitude, and that the right-shift which is observed in man under similar conditions serves to augment the oxygen delivery to the tissues.

The red cells of various mammals do not have equally high DPG levels, nor are their hemoglobins equally sensitive to the DPG level. The red cells of man, horse, dog, rabbit, guinea pig, and rat all have high levels of DPG, and their hemoglobins, when stripped of DPG, have high oxygen affinities. In contrast, sheep, goat, cow, and cat have low concentrations of DPG, and their hemoglobins interact only weakly with the DPG (Bunn, 1971). At the present time these differences are difficult to understand, they neither seem to follow phylogenetic relationships nor any characteristic differences in the biology of the animals.

Dissociation curve and body size. If the dissociation curves for various mammals are plotted together, we see that mammals of large body size have curves located to the left, and small animals have curves to the right (fig. 3.7).

As stated before, the loading–unloading reactions of hemoglobin are of importance at two locations in the body, first in the lung when oxygen is taken on, and secondly in the tissues when oxygen is unloaded. Fig. 3.7 shows that the blood of all mammals is practically 100% saturated at the normal oxygen pressure in the mammalian lung, which is about 100 mm Hg. The blood therefore becomes fully loaded with oxygen in the lung, and there is no great difference between the species in this regard. Let us therefore examine the unloading of oxygen in the tissues.

The blood of a small mammal releases oxygen more readily because of the lower affinity for oxygen. Usually, blood does not give up all its oxygen in the tissues; as an approximation we can say that on the average about half of the oxygen is given up. This corresponds to half-saturation of the hemoglobin (50% oxyhemoglobin and 50% hemoglobin), which is a convenient way of describing the location of the dissociation curve. The corresponding oxygen pressure, which we can consider an average unloading pressure, is about 22.5 mm Hg for elephant blood, 28 mm Hg for man, and about 45 mm Hg for the mouse.

These different unloading pressures of the blood can be related to the metabolic need for oxygen. Small animals have a higher rate of oxygen consumption per gram body weight than large animals. For

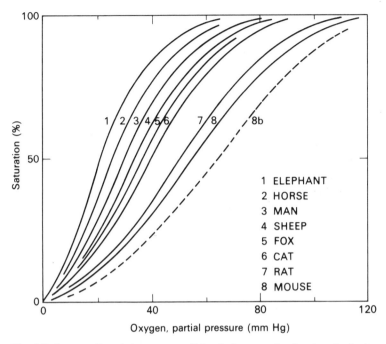

Fig. 3.7. Oxygen dissociation curves of blood of mammals of various body sizes. Mammals of small size have a lower oxygen affinity. This helps in the delivery of oxygen in the tissues to sustain the high metabolic rate of a small animal. Dashed curve (8B) indicates effect of acid (Bohr effect) on mouse blood (8) (Schmidt-Nielsen, 1972, reprinted from *Federation Proceedings*, **29**, 1524–1532, 1970).

example, the oxygen consumption of a horse weighing 700 kg is 1.7 μl O_2 g^{-1} min^{-1}, and of a 20 g mouse is 28 μl O_2 g^{-1} min^{-1}, i.e. the mean oxygen consumption of one gram of mouse tissue is about 15 times as high as that of one gram of horse tissue.

It follows that mouse tissue must be supplied with oxygen at a rate 15 times higher than horse tissue, and this can only be achieved by a diffusion gradient from capillary to cells 15 times as steep. This is shown diagrammatically in fig. 3.8. Each capillary supplies the surrounding tissue with oxygen, and the distance over which oxygen must diffuse is determined by the distance to the next capillary, the half-way point being the most distant point to which oxygen must diffuse. Horses and mice have approximately the same size red cells, therefore their capillaries are also of approximately the same size, and this simplifies the analysis. To achieve a 15 times as steep diffusion gradient in the mouse, two variables can be adjusted. Firstly, the diffusion distance can be reduced by reducing the inter-capillary

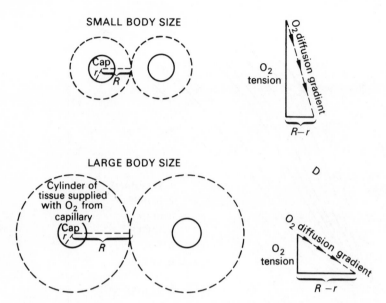

Fig. 3.8. Each capillary is surrounded by an approximately cylindrical mass of tissue which is supplied with oxygen by diffusion from the capillary. The steepness of the diffusion gradient depends on the radius of the tissue cylinder (R), the radius of the capillary (r), and the unloading pressure (tension) for oxygen.

distance (i.e. increase the capillary density in the tissue), and secondly, the unloading pressure for oxygen can be increased (i.e. a dissociation curve located to the right).

The capillary density is indeed larger in the small animal. In a mouse, for example, a 1 mm² cross-section of a muscle will contain, say, 2000 capillaries, while a similar cross-section of a horse muscle will contain less than 1000 capillaries per mm² (Schmidt-Nielsen and Pennycuik, 1961). A reference to fig. 3.7 shows that the unloading pressure (or P_{50}) for mouse blood is about 2.5 times higher than for horse blood. We thus find that the higher metabolic rate in the mouse, and the necessary steep diffusion gradient for oxygen from capillary to tissue cells, is sustained by a combination of a reduced diffusion path and a higher unloading pressure for oxygen.

Respiratory function of fish blood. When we compare the oxygen affinities of the blood of various fishes, we immediately meet some problems which are difficult to resolve. For example, at what temperature should the comparison be made? If we compare blood from two different fishes, should they be examined at the same temperature, or should they be examined at the temperature at which each fish normally lives? The question is not easy to answer if we compare a

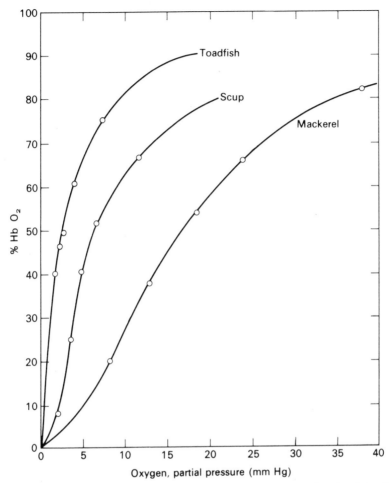

Fig. 3.9. Highly active fish have oxygen dissociation curves located to the right of those of more sluggish fish. Their blood therefore gives up oxygen to the tissues more readily: pH 7.38, 25.0 °C (Hall and McCutcheon, 1938).

tropical and an arctic fish, but with fish from similar habitats we can use the same temperature.

The dissociation curves for fish hemoglobin show a wide range in oxygen affinity with characteristic differences which are related to the activity of the fish (see fig. 3.9). The mackerel, an active fast-swimming fish which normally moves in well-oxygenated water, has a dissociation curve to the right of the other fish. This is in accord with what we have just seen, a low oxygen affinity facilitates the delivery of oxygen to the tissues at a high rate, and in well-aerated water

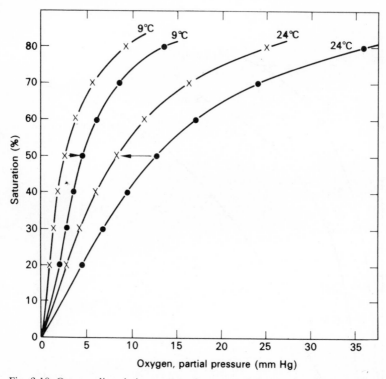

Fig. 3.10. Oxygen dissociation curves of warm-acclimated (x) and cold-acclimated (•) bullhead measured at the temperatures marked on each curve: pH 7.8 (Grigg, 1969).

the arterial blood will still be 100% saturated. The toadfish, with a curve to the extreme left, is a slow-moving and relatively sluggish bottom-living fish (Randall, 1970). It is often found in less well oxygenated water, and it is highly tolerant to oxygen deprivation. The hemoglobin of the toadfish has a particularly high oxygen affinity, and this is in accord both with the environment in which it normally lives and with the relatively low metabolic rate of this fish, the demands on the loading mechanism apparently being more important than the unloading in the tissues.

It might seem to be a disadvantage to fish that the oxygen affinity of their blood changes with temperature, for they are exposed to considerable temperature changes from season to season, especially if they live in fresh water. The change in dissociation curve as fish become acclimated to different temperatures is therefore interesting. The dissociation curve for blood of the brown bullhead, *Ictalurus nebulosus* (a catfish), depends on the acclimation temperature (see fig. 3.10). The blood of the warm-acclimated fish has higher oxygen affinity than blood of cold-acclimated fish, when measured at the same temperature. If instead we compare the curves for the blood measured at the temperature to which the fish has been acclimated, we should examine the two curves in the middle of the group of four. The fish acclimated at 9 °C has a curve (when measured at 9 °C) which is located to the left of the curve of the blood from fish acclimated at 24 °C (and measured at 24 °C). Acclimation therefore brings the two sets of curves closer together, but the warmer fish, which has a higher metabolic rate, is still shifted to the right relative to the cold-acclimated fish.

The differences in the blood and the seasonal changes depend on the characteristics of the red cell, and not on the hemoglobin. If solutions of hemoglobin from warm- and cold-acclimated fish are examined, there is no measurable difference in the oxygen affinities of the two hemoglobins.

Fish blood is sensitive to a lowered pH, even more so than mammalian blood. In the blood of some fish the effect of acid is far more pronounced than the Bohr effect of mammalian blood, so much, in fact, that at low pH the blood cannot be saturated with oxygen at any partial pressure of oxygen. This exaggerated Bohr effect on fish blood is known as the *Root effect*, it is of importance in the secretion of oxygen in the swimbladder of fish, and will be discussed on page 558.

Invertebrate respiratory pigments

We saw before that invertebrates have a variety of respiratory pigments (table 3.2, p. 82). In many, the oxygen affinity is very high, and the dissociation curve is located far to the left in the customary coordinate system. The hemoglobin from a bivalve mollusc, *Phacoides,* for example, has a P_{50} for oxygen which is about 0.2 mm Hg (Read, 1962).

The hemoglobin of the *Chironomus* larva also has an extremely high oxygen affinity. The dissociation curve is influenced by both temperature and pH, and within a temperature range of 5 to 24 °C, the P_{50} varies from 0.1 to 0.6 mm Hg. By mammalian standards this is an extremely high oxygen affinity. Characteristically, these animals live in environments that may periodically have extremely low oxygen concentrations, and it seems that their hemoglobins are primarily of importance under circumstances of oxygen lack.

The suggestion that hemoglobin is of particular value at low environmental oxygen is further supported by the fact that those invertebrates that do have hemoglobin, often develop much higher hemo-

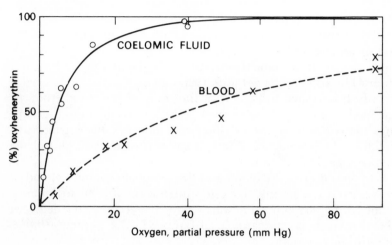

Fig. 3.11. Oxygen dissociation curves of blood and coelomic fluid of the sipunculid worm *Dendrostomum zostericolum* (Manwell, 1960).

globin concentrations when they are kept in water low in oxygen. This holds for *Daphnia,* the brine shrimp (*Artemia*), *Chironomus* larvae, and many others.

Hemerythrin differs chemically from hemoglobin in that it contains no porphyrin group. It is found only in a few marine forms (see table 3.2), which show an interesting relationship between oxygen affinity and the ecological conditions under which the animals normally live. The two sipunculid worms, *Dendrostomum* and *Siphonosoma* have hemerythrin both in their blood and in their coelomic fluid. In *Dendrostomum* the tentacles around the mouth are richly supplied with blood, they are kept above the surface of the sand in which the animal lives, and serve respiratory purposes. Characteristically, in *Dendrostomum* the oxygen affinity of the coelomic fluid is higher than that of the blood (see fig. 3.11). This means that oxygen taken up by the blood in the respiratory organ is readily transferred to the pigment of the coelomic fluid. In the other animal, *Siphonosoma,* the blood and coelomic fluid do not differ much, in fact, the blood has a slightly higher oxygen affinity. This is in accord with the habits of the worm; the oral tentacles are used for feeding but not for respiration, the worm remains in its burrow deep below the surface of the mud, and oxygen is taken up through the general body surface. Hence, both blood and coelomic fluid have a similar high oxygen affinity.

A similar relationship in the oxygen affinities of two different pigments is found in the amphineuran mollusc, *Cryptochiton.* This ani-

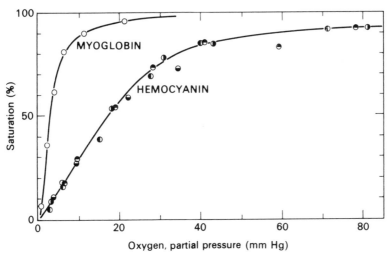

Fig. 3.12. Oxygen dissociation curves of hemocyanin (from blood) and myoglobin (from radula muscle) of the amphineuran mollusc *Cryptochiton* (Manwell, 1958).

mal has hemocyanin in its blood, while its radula contains myoglobin. The blood has a relatively low oxygen affinity and the myoglobin a high oxygen affinity (fig. 3.12). This situation is completely analogous to what we find in vertebrates. In these, oxygen is transferred from the hemoglobin of the blood to the muscle hemoglobin (myoglobin), which has a far greater oxygen affinity and a dissociation curve displaced to the far left with a P_{50} of about 0.5 mm Hg.

Oxygen diffusion in hemoglobin solutions

It is easy to see that hemoglobin carried in the blood vastly improves the transport of oxygen to the tissues. But hemoglobin occurs not only in blood, it is also found in various tissues, in higher animals especially in muscle. Can this hemoglobin, lodged inside the cells and removed from circulation, serve any function?

An increased oxygen capacity of tissues can help smooth out fluctuations when there are rapid increases in the demand for oxygen. For example, during muscle contraction the blood flow to a muscle may be reduced or momentarily stopped due to mechanical compression of the blood vessels by the contracting muscle. In this situation the myoglobin gives up its oxygen and thus acts as a short-time oxygen store (Millikan, 1937). It is unlikely, however, that muscle contractions regularly cause complete cessation of blood flow, on the contrary, the circulation in heavily working muscles is usually greatly increased.

The discovery that hemoglobin greatly accelerates the diffusion rate of oxygen throws a different light on the presence of hemoglobin in muscle, as well as the wide occurrence of hemoglobin in animals that live in environments that at least periodically subject them to low oxygen tensions. Some such animals were mentioned in table 3.2 (p. 82), and it is probable that these also enjoy the advantages of accelerated or *facilitated diffusion* of oxygen.

It was discovered by Scholander (1960) that while nitrogen diffuses through a hemoglobin solution a little more slowly than through water, the rate of oxygen transfer through the same solution is greatly enhanced.

Scholander made these gases diffuse through a layer of liquid in the following way. The liquid was absorbed in a highly porous membrane, air was present on one side, and a vacuum was applied on the other side of the membrane. Oxygen and nitrogen would diffuse from air through a layer of water, or of blood plasma, in a constant ratio of about 0.5. (This particular ratio comes about because the air contains about four times as much nitrogen as oxygen, but nitrogen is only one-half as soluble as oxygen.) If the air pressure above the membrane was lowered, say to 0.5 atm, only half as much air would diffuse across to the vacuum per unit time, but the ratio between the two gases remained unchanged at 0.5 volumes oxygen per volume of nitrogen (see fig. 3.13). If the plasma is replaced by a hemoglobin solution containing as much hemoglobin as normal blood (obtained by hemolyzing blood), the ratio of oxygen to nitrogen diffusion changes to nearly 1. This means that relative to the nitrogen the diffusion of oxygen is speeded up by the presence of hemoglobin. With further lowering of the pressure in the air chamber the ratio increasingly favors the oxygen above the nitrogen, until the ratio indicates that the diffusion of oxygen is speeded up about eight-fold relative to its diffusion in the absence of hemoglobin.

Facilitated diffusion is associated with the ability of hemoglobin to bind oxygen reversibly. Hemoglobin which has been changed to methemoglobin (achieved by oxidizing the divalent iron to trivalent iron) can no longer bind oxygen, and a solution of methemoglobin causes no facilitation of oxygen diffusion. However, myoglobin and various hemoglobins from vertebrates as well as invertebrates all show facilitated diffusion.

One important aspect of facilitated diffusion is that the enhanced diffusion occurs only when the oxygen pressure at the down-hill end

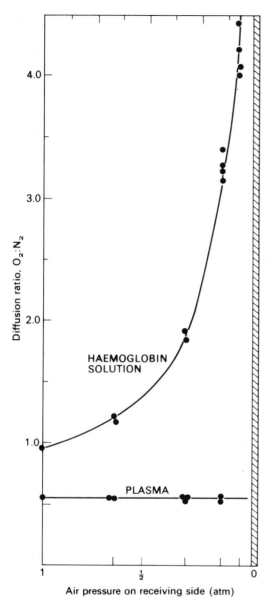

Air pressure on receiving side (atm)

Fig. 3.13. The diffusion of oxygen and nitrogen from air to a vacuum through a layer of blood plasma takes place at the constant ratio of 0.5. Through a hemoglobin solution the relative rate of oxygen diffusion is higher, and increases with decreasing pressure. For further explanation, see text (after Scholander, 1960).

of the diffusion path is close to zero. In fact, maximal facilitation is found when one side of the hemoglobin layer is fully saturated and the other is fully deoxygenated.

The question has been raised as to whether non-hemoglobin respiratory pigments, such as hemocyanin, also facilitate diffusion. This, however, is not the case. This seems related to the large molecular size of hemocyanin (several millions) which greatly reduces the motility of the hemocyanin molecule in solution. Although the exact mechanism of oxygen transfer is inadequately understood, it seems that the normal thermal motility of the hemoglobin molecules in solution play a role. If this motility is slowed down, for example, by adding gelatin to the hemoglobin solution, the facilitation is reduced. Myoglobin, which has a lower molecular weight (17 000) than hemoglobin (68 000) shows a greater facilitation, roughly in inverse proportion to the square root of the molecular weights of the two pigments. This indicates that the difference results from the thermal motion of the pigment in solution (Hemmingsen, 1965).

The main points of facilitated diffusion of oxygen in hemoglobin solutions can be summarized as follows:

(1) The net diffusion of oxygen is accelerated when the exit pressure is low, (2) the transfer equals the sum of simple diffusion and the enhancement due to the pigment, (3) the facilitation increases with increasing hemoglobin concentration, (4) the transfer rate is inversely related to the square root of the molecular weight of the pigment, so that a small hemoglobin molecule gives greater facilitation.

Physiological implications. In vertebrates hemoglobin is present in high concentration primarily in the red blood cells, and secondly in red muscles. It is doubtful that facilitated diffusion has much importance in the erythrocytes, for facilitation requires a near-zero oxygen pressure at the receiving end of the diffusion path. The red blood cells, which already have a very short diffusion distance, will rarely if ever be exposed to near-zero oxygen pressures, except in the capillaries of heavily working muscles.

For myoglobin the situation is different. At some distance from a capillary the oxygen tension in a working muscle may readily approach zero, and this establishes an ideal situation for facilitated diffusion, the oxygen tension in the capillary being somewhere between arterial and venous tension, and in the tissue (the receiving end of the diffusion path) nearly zero. A several-fold diffusion facilitation, as was observed in the laboratory experiments, therefore would indeed be important in improving the oxygen supply to the muscle.

It now appears particularly noteworthy that those invertebrate organisms that have hemoglobin frequently live in environments that, at least at times, have low oxygen tensions. When the oxygen supply is poor, the oxygen tensions in the tissue will be particularly low, and facilitation will be most important. Hemoglobin is especially important for these animals, and many of them adapt to prolonged exposure to oxygen deficiency by increasing the amount of hemoglobin in their bodies. It therefore seems that the role of hemoglobin in these animals is to facilitate oxygen diffusion.

Evolutionary implications. It seems curious that a molecule as complex as hemoblogin occurs so widely among animals, yet in such a scattered or sporadic way. In addition to being nearly universally present in vertebrates, hemoglobin occurs here and there in a number of animal phyla, including protozoans, and even in plants and bacteria. Often, it is found in only one or a few species within a large phylum, at times in a parasite and not in its free-living relative, and so on. Once the advantage of hemoglobin in facilitated diffusion has been established, however, the peculiar distribution of hemoglobin is easier to understand.

The basic structure of the heme part of the hemoglobin molecule is a porphyrin nucleus, which is universally present in plants and animals as part of cellular enzymes that belong to the cytochrome system. Relatively minor biochemical alterations can change this raw material into a hemoglobin-type pigment which binds oxygen reversibly; for such mutations to be permanently established requires some suitable evolutionary 'advantage' that favors their maintenance.

The most obvious and immediate advantage of hemoglobin is the improvement in oxygen delivery due to facilitated diffusion. Once hemoglobin synthesis is established, further improvement in oxygen transport can be achieved by adding convective transport, in other words, blood and a circulatory system.

The establishment of a complete circulation system is a complex change which would require not only relatively minor biochemical changes in porphyrin, but also complex morphological and physiological alterations of the entire organism. Such drastic changes are, of course, possible but can hardly be achieved as a single or few-step event. In contrast, an initial occurrence of hemoglobin in any tissue immediately endows the organism with the advantage of facilitated diffusion, thus giving a selective advantage to its maintenance. A later step would then be to permit this hemoglobin to enter an already existing blood system, thus further augmenting oxygen delivery by convection.

It follows from this hypothetical sequence of events that the scattered occurrence of hemoglobin is due to the universal presence of iron-porphyrin compounds, that facilitated diffusion gives the primary impetus to maintaining the biochemical mechanism for hemoglobin synthesis, and that adding hemoglobin to a pre-existing convective system then becomes a logical later step in the evolutionary history.

Carbon dioxide transport in blood

As blood passes through the tissues and gives off oxygen, it concurrently takes up carbon dioxide; as it passes through the lungs the reverse takes place. Let us examine the events in the lung as carbon dioxide is given off.

The venous blood of mammals, as it reaches the lungs, contains about 55 ml CO_2 per 100 ml blood and has a carbon dioxide tension of about 46 mm Hg. As the blood leaves the lungs it contains about 50 ml CO_2 per 100 ml blood at a tension of about 40 mm Hg. Thus, the blood has given up only a small amount of the total carbon dioxide it contains, the arterio–venous difference being about 5 ml CO_2 per 100 ml blood. In order to understand the carbon dioxide transport, we need to know more about carbon dioxide and its behavior in solution.

The total amount of carbon dioxide carried in the blood far exceeds the amount that would be dissolved in water at the carbon dioxide tension of the blood. This is because the largest part of the carbon dioxide is carried in chemical combination, rather than as free dissolved carbon dioxide gas. The two most important such 'compounds' are (1) bicarbonate ions, and (2) a combination of carbon dioxide with hemoglobin. (Note that the carbon dioxide binds to terminal amino groups of the protein molecule and not to the binding site for oxygen.)

When carbon dioxide is dissolved in water, it combines with water to form carbonic acid (H_2CO_3). This can be described as follows:

$$CO_2 + H_2O \rightleftarrows H_2CO_3 \rightleftarrows H^+ + HCO_3^- \rightleftarrows (H^+ + CO_3^{--})$$

The H_2CO_3 dissociates as an acid into hydrogen ion and bicarbonate ion (HCO_3^-). Carbonic acid (H_2CO_3) acts as a very weak acid with an apparent dissociation constant of 8.0×10^{-7} at 37 °C (pK = 6.1).* (The further dissociation of bicarbonate ion into one ad-

---

* Carbonic acid is actually a stronger acid with a true pK value of 3.8. The reason for the discrepancy in pK values is that only about 0.5% of the dissolved carbon dioxide combines with water to form the acid. Thus the bulk of the carbon dioxide remains

ditional hydrogen ion and the carbonate ion (CO_3^{--}) is included for completeness, but at the pH of living organisms the amount of carbonate ion present is minuscule and can usually be disregarded.)

At the normal pH of mammalian blood, most of the total carbon dioxide present will be in the form of bicarbonate ion. This can readily be calculated from the Henderson–Hasselbalch equation in the following way:

$$\frac{[H^+][A^-]}{[HA]} = K$$

$$[H^+] = K \cdot \frac{[HA]}{[A^-]}$$

$$pH = pK - \log\frac{[HA]}{[A^-]}$$

for $pK = 6.1$ and at $pH = 7.4$

$$-\log\frac{[HA]}{[A^-]} = pH - pK = 7.4 - 6.1 = 1.3$$

$$\log\frac{[HA]}{[A^-]} = -1.3 = 0.7 - 2$$

$$\frac{[HA]}{[A^-]} = \frac{[H_2CO_3]}{[HCO_3^-]} = 0.05 = \frac{1}{20}$$

This means that at $pH = 7.4$ the ratio of carbonic acid to bicarbonate ion is 1 : 20, i.e. for each one part of carbon dioxide present as acid, twenty times as much carbon dioxide is present as bicarbonate ion.

The hydrogen ion formed as a result of the ionic dissociation of the carbonic acid is buffered by various buffering substances in the blood, and the effect of carbon dioxide on the pH of the blood is therefore only moderate. Arterial blood may have a normal pH of, say, 7.45 and venous blood 7.42, the difference in pH caused by the uptake of carbon dioxide in the tissues thus is no more than 0.03.

as dissolved gas, and only the fraction 0.005 is hydrated to form H_2CO_3. In physiology it has been customary to consider that all carbon dioxide is in the form of carbonic acid, hence the carbonic acid appears to be weaker than the molecular species, H_2CO_3, in fact is. If to the true pK for carbonic acid (3.8) we add the negative log of the hydration constant (log $0.005 = -2.3$), the sum is 6.1, which is the apparent pK for carbon dioxide as used in physiology. In the further discussions we will follow physiological convention and assume that all dissolved carbon dioxide forms H_2CO_3.

The most important buffering substances in the blood are the carbonic acid–bicarbonate system, the phosphates, and the proteins in the blood. Proteins act as excellent buffers because they contain groups that can dissociate both as acids and as bases, the result being that proteins can either take up or give off hydrogen ions. Hemoglobin is the protein present in the largest amount in blood, and it has the greatest role in buffering; the plasma proteins are second in importance in this regard.

Carbon dioxide dissociation curve. The amount of carbon dioxide taken up by blood varies with its partial pressure. It is therefore possible to construct a carbon dioxide dissociation curve for blood, analogous to the oxygen dissociation curve. This is done by plotting on the ordinate the total amount of carbon dioxide in the blood and on the abscissa the various carbon dioxide partial pressures. The result is a curve as in fig. 3.14. This graph shows a slightly different dissociation curve for oxygenated and deoxygenated blood. This is because oxyhemoglobin is slightly more acid than hemoglobin; hence, oxygenated blood will bind slightly less carbon dioxide. This phenomenon is closely related to the Bohr effect, which merely shows the other side of the coin. When carbon dioxide (acid) is added to the blood, it pushes the equilibrium between oxyhemoglobin and hemoglobin in the direction of the weaker acid (hemoglobin). This therefore tends to release more oxygen, which, in fact, is the Bohr effect.

In fig. 3.14 the point marked A stands for normal arterial blood, and the point marked V for mixed venous blood (venous blood is never completely deoxygenated). In the body the actual dissociation curve for carbon dioxide will be the fully drawn line A–V, and this curve therefore is said to be the *functional dissociation curve*.

Fig. 3.14 also shows the 'dissociation curve' for a sodium bicarbonate solution with the same total available base as in blood. The slope of the upper part of this curve represents the solubility for carbon dioxide in an aqueous medium. As carbon dioxide pressure is reduced, this dissolved carbon dioxide is given up. When the carbon dioxide pressure falls to about 10 mm Hg, the bicarbonate solution begins to give up carbon dioxide from the bicarbonate ion, leaving carbonate behind. By exposing a solution of sodium bicarbonate to a vacuum, half of the total carbon dioxide of the bicarbonate can be removed, leaving a solution of sodium carbonate (Na_2CO_3) behind. Sodium carbonate, however, does not give off carbon dioxide to a vacuum, and the final solution therefore retains half of the total carbon dioxide that was originally present as bicarbonate. In contrast, whole blood exposed to a vacuum gives up all the carbon dioxide

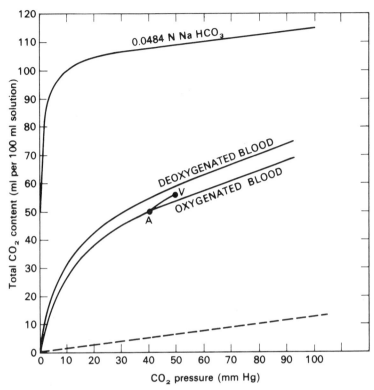

Fig. 3.14. Carbon dioxide dissociation curves of mammalian blood. The curves for oxygenated and deoxygenated blood differ somewhat. The line A–V is the physiological dissociation curve in the body, describing the difference between arterial and venous blood. The dashed line indicates the amount of carbon dioxide held by physical solution in water (Winton and Bayliss, 1955).

present. This is because other ions, primarily proteins, furnish the necessary anions. If separated plasma is subjected to a vacuum, it behaves more like a solution of sodium bicarbonate; the presence of red cells is necessary to obtain the normal dissociation curve shown in fig. 3.14.

Carbon dioxide in aquatic respiration

The amount of carbon dioxide produced in metabolism is closely related to the amount of oxygen consumed. The ratio between the respiratory exchange of carbon dioxide and oxygen, the *respiratory exchange ratio* (also called the respiratory quotient) normally remains between 0.7 and 1.0 (see p. 211). For the following discussion we shall consider an animal in which the exchange ratio is 1.0, i.e.

for each molecule oxygen taken up, exactly one molecule of carbon dioxide is given off.

An air-breathing animal, a mammal for example, with a gas exchange ratio of one, will have an easily predicted carbon dioxide concentration in exhaled air. If the 21% O_2 in inhaled air is reduced to 16% in the exhaled air, the exhaled air will of necessity contain 5% CO_2. If instead we consider the change in partial pressure of the two gases, we can say that if oxygen in the respired air is reduced by 50 mm Hg, the carbon dioxide is increased by the same amount, 50 mm Hg. Each millimeter change in the partial pressure of oxygen in the respiratory air results in an equal and opposite change in the carbon dioxide.

If the exchange ratio of unity is applied to gas exchange in water, the amounts of oxygen and carbon dioxide will, of course, be equal, but the changes in the partial pressures of the two gases will be very different because the solubility of the two gases are so different. For example, if a fish removes two-thirds of the oxygen present in the water, thus reducing the P_{O_2} from 150 to 50 mm Hg, the increase in P_{CO_2} will be about 3 mm Hg (provided that the water is unbuffered, a buffered medium may give an even smaller increase). Depending on the temperature, the 100 mm Hg reduction in P_{O_2} may mean that, say, 5 ml O_2 was removed from one liter of water. Returning 5 ml CO_2 will increase the P_{CO_2} by only about 3 mm Hg because the solubility of carbon dioxide is very high, roughly 30 times as high as the solubility for oxygen.

This relationship in the carbon dioxide tensions resulting from respiration in air and in water is expressed in fig. 3.15. This graph shows the P_{CO_2} that will result from any given decrease in oxygen in the respired medium. Even if a fish were to remove all oxygen in the water, the carbon dioxide in the water would increase to only 5 mm Hg partial pressure. Since carbon dioxide diffuses across the gill epithelium nearly as readily as oxygen, it means that the blood P_{CO_2} of the fish can never exceed 5 mm Hg (in air-breathing vertebrates it is about ten times as high). As a consequence, in regard to the effect on acid–base regulation, air and aquatic respiration are very different. This is particularly important for animals that shift between air and water respiration, such as, for example, amphibians and air-breathing fish. For these animals a successful shift between the two media requires substantial physiological adjustments in their acid–base regulation.

Blood carbon dioxide and pH. We have seen that the carbon dioxide tension in the arterial blood is very different in aquatic and terrestrial animals. The carbon dioxide content also varies greatly, it is

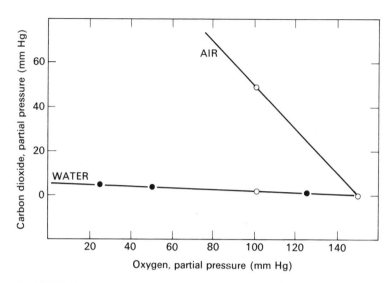

Fig. 3.15. Relation between P_{O_2} and P_{CO_2} in air and in water, when the respiratory exchange ratio (*RQ*) between oxygen and carbon dioxide is unity (*RQ* = 1). For further explanation, see text (Rahn, 1966).

usually low in fishes (often less than 10 ml CO_2 per 100 ml blood), and may be very high in turtles (in excess of 100 ml CO_2 per 100 ml blood). It is therefore evident that there is neither uniformity in the carbon dioxide tension nor in the total carbon dioxide content of vertebrate blood. It is therefore surprising to find a much greater uniformity in the pH of the blood of various vertebrates than would be expected from the large variation in carbon dioxide. This results from corresponding adjustments in the acid–base balance, achieved primarily through the amount of sodium (often called 'free bicarbonate') present in the blood.

The pH in the blood of a variety of vertebrates, both warm- and cold-blooded, usually varies around values between about 7.4 and 8.2. If the data are plotted against the temperature of the animals, however, we obtain a more regular distribution than indicated by this rather wide pH range (fig. 3.16). At low temperatures, the pH often reaches values well over 8.0 as the temperature approaches 0 °C, and at 40 °C it is closer to 7.5. At any given temperature, however, the blood pH remains within a narrow range of about 0.2 pH units. It is important to remember that the dissociation of water, and therefore the pH of neutrality, changes with temperature. The neutrality point is at 7.0 only at room temperature (25 °C), and changes with temperature according to the line in the center of fig. 3.16, marked 'water.'

It now appears that the blood pH of all vertebrates deviates from

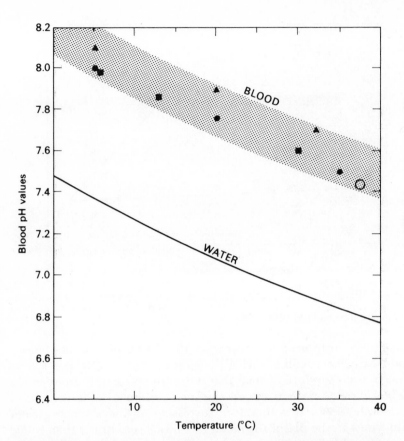

Fig. 3.16. Blood pH values for vertebrates of all classes, determined at the normal body temperature for each animal. Upper shaded band, range of observed values; central fully drawn line, pH of water at neutrality as a function of temperature (after Rahn, 1966).

neutrality by about 0.6 pH units, always in the alkaline direction. Consequently, although the blood pH of various vertebrates may differ considerably, they are amazingly uniform in the extent of deviation from neutrality when measured at the 'normal' temperature for the animal.

The very different conditions for achieving an appropriate acid–base balance in water and in air causes great physiological difficulties for animals that live in a transition between the two media. A fish that normally lives in water has its acid–base balance adjusted to a carbon dioxide tension of, say, 2 or 3 mm Hg. If it changes to air respiration the blood carbon dioxide tension rises sharply, and this requires an adjustment in the blood buffering system, primarily of the sodium ion concentration, or some auxiliary mechanism for the elimination of carbon dioxide. Amphibious animals are, to some extent, able to eliminate carbon dioxide through the skin, and this seems to be of help to both lung-fish and aquatic amphibians. If an animal emerges upon land and becomes completely terrestrial, it cannot retain a highly permeable skin due to the danger of desiccation. If it were to keep the P_{CO_2} low by lung ventilation, the ventilation would be vastly increased (relative to that needed for the oxygen supply), and evaporation would become a major problem. Thus, a radical readjustment in the acid–base regulatory system is a primary prerequisite for moving onto land.

Carbonic anhydrase and velocity of carbon dioxide–water interaction

The events which take place when carbon dioxide is dissolved in water (see equations on p. 106) are not instantaneous. When carbon dioxide enters water and becomes dissolved, the initial hydration of the carbon dioxide molecule to form H_2CO_3 is a relatively slow process. Similarly, the reverse process, the release of carbon dioxide from H_2CO_3 is also a slow process which requires several seconds to a fraction of a minute. (The dissociation of H_2CO_3 into H^+ and HCO_3^- and the reverse process are, for physiological considerations, instantaneous.) If we now consider that the time that blood remains in a capillary is usually only a fraction of a second, how is it possible for the blood to take up carbon dioxide in the tissues as rapidly as it does, and again give off carbon dioxide during the short time the blood is in the lung capillary?

The problem appeared to be solved with the discovery of an enzyme which accelerates the formation of carbon dioxide from H_2CO_3 so that the process becomes very fast. The enzyme was given the name *carbonic anhydrase*, which is somewhat of a misnomer because both the combination of carbon dioxide with water and the release of carbon dioxide from carbonic acid are speeded up, i.e. the process is catalyzed in both directions. The enzyme is not present in blood plasma, but is found in high concentration inside the red blood cells. It also occurs in other organs, notably in glands such as kidneys, the secretory epithelium of the stomach, pancreas, salivary glands, etc., but this is of no concern to us here. It should be noted, however, that carbonic anhydrase is highly specific for the hydration reaction of carbon dioxide and that it catalyzes no other known biochemical events in the organism.

Several years after the discovery of carbonic anhydrase, a series of highly potent carbonic anhydrase inhibitors were synthesized and became important as drugs, especially in the treatment of certain types of kidney malfunction. Much to the surprise of physiologists, it

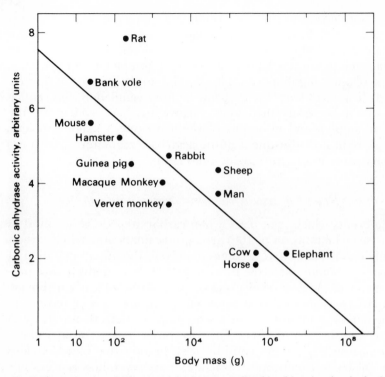

Fig. 3.17. Carbonic anhydrase activity in mammalian blood is related to body size, and is higher in small than large mammals (Magid, 1967).

turned out that these enzyme inhibitors have a very low toxicity, and that the complete inhibition of the blood carbonic anhydrase has only minor effects on the carbon dioxide transport. In other words, the enzyme is not a critical factor in the carbon dioxide transport as such, although its inhibition causes some changes in the acid–base balance.

An examination of the amount of carbonic anhydrase found in various animals might be helpful in examining its possible role. We know that small animals have higher metabolic rates, per unit body weight, than large animals. They use oxygen and produce carbon dioxide more rapidly, and carbon dioxide must therefore be released in a shorter time in the lungs of a small animal. An examination of the carbonic anhydrase of the blood in relation to the body size shows that small animals indeed have significantly higher concentrations of the enzyme in their blood (fig. 3.17). This seems to be in accord with the need for carbon dioxide transport, but the enzyme is present in amounts far in excess of that which is needed.

There is, however, another process where rapid hydration of carbon dioxide is of importance. As carbon dioxide in the tissue capillary enters the blood, the presence of carbonic anhydrase permits an immediate formation of carbonic acid, which in turn immediately affects the dissociation of the acid-sensitive oxyhemoglobin. This acid sensitivity, the Bohr effect, obviously would not have time to take effect while the blood still remains in the capillary unless the hydration of the carbon dioxide molecule were catalyzed. In the absence of carbonic anhydrase, therefore, the Bohr effect could not be of much importance.

It seems, then, that the carbonic anhydrase, an enzyme which specifically catalyzes the hydration of the carbon dioxide molecule, perhaps is of minor importance in carbon dioxide transport, and that its role is to contribute to the efficient delivery of oxygen to the tissues. Since inhibition of the enzyme in the living animal does not cause great difficulties to the animal, the enzyme is probably of importance primarily under conditions of extreme demands on the gas transport mechanism.

In other organs where carbonic anhydrase occurs, it always seems to be related to functions where ion transport processes are important, such as in glands (kidney, pancreas, salivary gland) and in the ciliary body of the eye. The effect of carbonic anhydrase inhibition on these processes is profound and is the basis for the widespread and important use of inhibitor drugs in clinical medicine as well as in physiological research.

REFERENCES

ASTE-SALAZAR, H. and HURTADO, A. (1944). The affinity of hemoglobin for oxygen at sea level and at high altitudes. *Am. J. Physiol.,* **142,** 733–43.

BARCROFT, J. (1935). Foetal respiration. *Proc. Roy. Soc. Lond. B,* **118,** 242–63.

BENESCH, R., BENESCH, R. E. and ENOKI, Y. (1968). The interaction of hemoglobin and its subunits with 2,3-diphosphoglycerate. *Proc. Nat. Acad. Sci.,* **61,** 1102–6.

BENESCH, R., BENESCH, R. E. and YU, C. I. (1968). Reciprocal binding of oxygen and diphosphoglycerate by human hemoglobin. *Proc. Nat. Acad. Sci.,* **59,** 526–32.

BUNN, H. F. (1971). Differences in the interaction of 2,3-diphosphoglycerate with certain mammalian hemoglobins. *Science,* **172,** 1049–50.

GRIGG, G. C. (1969). Temperature-induced changes in the oxygen equilibrium curve of the blood of the brown bullhead, *Ictalurus nebulosus. Comp. Biochem. Physiol.,* **28,** 1203–23.

HALL, F. G., DILL, D. B. and BARRON, E. S. G. (1936). Comparative physiology in high altitudes. *J. Cell. Comp. Physiol.,* **8,** 301–13.

◆HALL, F. G. and MCCUTCHEON, F. H. (1938). The affinity of hemoglobin for oxygen in marine fishes. *J. Cell. Comp. Physiol.,* **11,** 205–12.

HEMMINGSEN, E. A. (1965). Accelerated transfer of oxygen through solutions of heme pigments. *Acta Physiol. Scand., Suppl.,* **246,** 1–53.

LENFANT, C., TORRANCE, J., ENGLISH, E., FINCH, C. A., REYNAFARJE, C., RAMOS. J. and FAURA, J. (1968). Effect of altitude on oxygen binding by hemoglobin and on organic phosphate levels. *J. Clin. Invest.,* **47,** 2652–6.

◆LENFANT, C., WAYS, P., AUCUTT, C. and CRUZ, J. (1969). Effect of chronic hypoxic hypoxia on the O_2–Hb dissociation curve and respiratory gas transport in man. *Respir. Physiol.,* **7,** 7–29.

LUTZ, P. L., LONGMUIR, I. S., TUTTLE, J. V. and SCHMIDT-NIELSEN, K. (1973). Dissociation curve of bird blood and effect of red cell oxygen consumption. *Respir. Physiol.,* **17,** 269–75.

MAGID, E. (1967). Activity of carbonic anhydrase in mammalian blood in relation to body size. *Comp. Biochem. Physiol.,* **21,** 357–60.

MANWELL, C. (1958). The oxygen-respiratory pigment equilibrium of the hemocyanin and myoglobin of the amphineuran mollusc *Cryptochiton stelleri. J. Cell. Comp. Physiol.,* **52,** 341–52.

MANWELL, C. (1960). Histological specificity of respiratory pigments – II. Oxygen transfer systems involving hemerythrins in sipunculid worms of different ecologies. *Comp. Biochem. Physiol.,* **1,** 277–85.

MILLIKAN, G. A. (1937). Experiments on muscle haemoglobin *in vivo;* the instantaneous measurement of muscle metabolism. *Proc. Roy. Soc. Lond. B,* **123,** 218–41.

RAHN, H. (1966). Gas transport from the external environment to the cell. In *Development of the Lung* (A.V.S. de Reuck and R. Porter, eds), pp. 3–23, London: The CIBA Foundation.

RANDALL, D. J. (1970). Gas exchange in fish. *Fish Physiology,* **4** (W. S. Hoar and D. J. Randall, eds), pp. 253–92, New York: Academic Press.

READ, K.R.H. (1962). The hemoglobin of the bivalved mollusc, *Phacoides pectinatus* Gmelin. *Biol. Bull.,* **123,** 605–17.

RIGGS, A. (1951). The metamorphosis of hemoglobin in the bullfrog. *J. Gen. Physiol.* **35,** 23–44.

RIGGS, A. (1960). The nature and significance of the Bohr effect in mammalian hemoglobins. *J. Gen. Physiol.,* **43,** 737–52.

SCHMIDT-NIELSEN, K. (1972). *How Animals Work.* Cambridge, England: Cambridge University Press, 114 pp.

SCHMIDT-NIELSEN, K. and PENNYCUIK, P. (1961). Capillary density in mammals in relation to body size and oxygen consumption. *Am. J. Physiol.,* **200,** 746–50.

SCHMIDT-NIELSEN, K. and TAYLOR, C. R. (1968). Red blood cells: why or why not? *Science,* **162,** 274–5.

SCHOLANDER, P. F. (1960). Oxygen transport through hemoglobin solutions. *Science,* **131,** 585–90.

WINTON, F. R. and BAYLISS, L. E. (1955). Blood and the transport of oxygen and carbon dioxide. In *Human Physiology* (F. R. Winton and L. E. Bayliss, eds), pp. 78–118, Boston: Little, Brown & Co.

4
Circulation

The main role of circulating fluids in the body is to provide rapid mass transport over distances where diffusion is inadequate or too slow. Circulation is therefore important in virtually all animals of more than a few millimeters size, and for all large-size animals with high metabolic activity circulation is a necessity. The transport of gases, oxygen and carbon dioxide, is very important, but far from the only function of circulation.

The main functions of blood were listed in table 3.1 (p. 81). They fall into three major categories, (1) mass transport of solutes and cells, (2) transport of heat, and (3) transmission of force. Functions that depend on the transmission of force are related mostly to the movement of organs or to the locomotion of the entire animal, and to providing pressure for ultrafiltration in the kidney; these functions will be discussed elsewhere. The transport of heat and solutes (including gases) are intimately related and will be discussed in this chapter. A system that is adequate for the convective transport of oxygen will invariably suffice for transport of other solutes as well. Therefore, our main emphasis will be on the role of circulation in gas transport, notably the need for oxygen.

Circulatory systems of vertebrates are well known and will be discussed first. The function of most invertebrate circulatory systems is less well known; in many cases the morphology has been adequately worked out but information about function is often less than satisfactory. As a result, there is little possibility for generalizations, and the main emphasis in this chapter will be on the highly developed and better understood circulatory systems of vertebrates.

General principles

An adequate circulatory system depends on one or more pumps and on channels or conduits in which the blood can flow. The pump, or heart, is based on the ability of muscle to contract or shorten. By wrapping muscle around a tube or chamber, it is possible to achieve a reduction of volume, and two different types of pump can be designed this way, either (a) a peristaltic pump, or (b) a chamber pump with valves (see fig. 4.1). A chamber pump may be of two types (1) it may have contractile walls, or (2) the reduction in volume may be achieved by external pressure from other body parts.

Peristaltic hearts are found mostly in invertebrates, and the vertebrate heart is almost without exception a chamber pump with contractile walls. A typical pump for which the force is provided by outside pressure is found in the larger veins in the legs of man. The walls of these veins are relatively thin, they are provided with valves

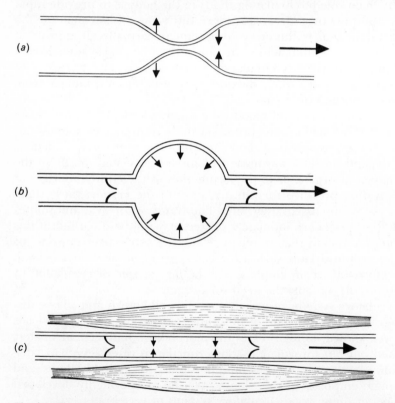

Fig. 4.1. The pumps that move blood in the circulatory systems are of three major types. (*a*) is a peristaltic pump. A constriction in a tube moves along the tube and pushes blood ahead of it. (*b*) is a chamber pump. Rhythmical contractions of the walls force blood out. Valves prevent backflow, and the blood is therefore expelled in one direction only. (*c*) Blood is pushed out of a collapsible tube by pressure from surrounding tissues. In this diagram contracting muscles provide the pressure and valves prevent backflow. The venous 'pump' in the legs of man is of this type.

which prevent the backflow of blood, and when the leg muscles contract the veins are collapsed and the valves insure that blood is forced in the direction of the heart. This pumping action aids greatly in moving the blood out of the leg veins against the force of gravity, which tends to make blood accumulate in the legs.

Since muscle can only provide force by shortening, the heart cannot pull blood, unless special arrangements are made. The elasmobranch heart is an example, it is enclosed in a stiff capsule, and when one chamber contracts, the other chamber is filled with blood by 'suction' while valves prevent backflow from the arterial side. This is described in further detail on page 127–8.

The circulating fluid or blood is carried from the heart in channels or conduits, and eventually returns to the heart again. Vertebrate blood is carried in a system of elastic tubes or pipes (arteries, capillaries, veins). Blood returns to the heart without leaving this system of tubes, and since the blood remains within this closed system, we refer to the vertebrate circulation as a '*closed circulation*'. Many invertebrates (insects, most crustaceans, many molluscs, and so on) pump the blood from the heart into the blood vessels, but these terminate and blood flows more or less freely between the tissues before it eventually returns to the heart. Such a system is called an '*open circulation*'. Some of the characteristics of open and closed circulatory systems are listed in table 4.1.

Table 4.1. *Major differences between 'closed' and 'open' circulation*

CLOSED CIRCULATORY SYSTEMS	OPEN CIRCULATORY SYSTEMS
Usually high pressure systems	Usually low pressure systems
Development of high pressure requires closed system and resistance	High pressure not possible
Sustained high pressure between heart beats requires elastic walls	Sustained pressure not possible
Blood conveyed directly to organs	Similar to closed systems
Distribution to different organs can be regulated	Distribution of blood not readily regulated
Blood returns rapidly to heart	Blood return to heart slow

Closed circulatory systems are found in vertebrates and in cephalopod molluscs and echinoderms, while open systems are found in most arthropods, non-cephalopod molluscs, and in tunicates. The circulatory system of tunicates is interesting in several ways. It is an open system, and the tube-shaped heart pumps by means of a peristaltic wave passing along from one end to the other. The heart has no valves. After a series of contractions, say several dozen or a hundred, the heart gradually slows and stops. After a pause the beat reverses direction, and blood is now forced in the opposite direction (Krijgsman, 1956). This is an unusual arrangement, and in most circulatory systems the heart pumps, and the blood always flows, in one direction only.

Vertebrate circulation

Body water compartments and blood volume

Roughly two-thirds of the vertebrate body consists of water. The exact figure varies a great deal, in particular as a result of variations in the fat content. Adipose tissue contains only some 10% water, and a very fat individual, who may have half or more of the body weight as fat, will therefore have a low overall water content. If the water content is expressed as a percentage of the fat-free body weight, the figure shows much less variation. Most of the water is located inside the cells (*intracellular water*), and a smaller fraction outside the cell membranes (*extracellular water*). The extracellular water is partly located in tissue spaces (*interstitial fluid*) and partly in the blood (*plasma water*). We can therefore consider the body water as being

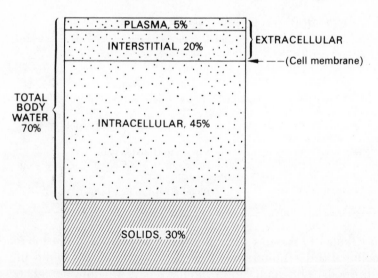

Fig. 4.2. The approximate distribution of water and solids in the vertebrate body, given in per cent of the fat-free body weight. More detailed information is given in table 4.2.

distributed among three compartments: intracellular, interstitial, and plasma water (see fig. 4.2).

The volumes of the various body fluid compartments can be measured in various ways, often without harm to the organism. One of the most convenient ways consists in introducing some suitable substance which is distributed uniformly within the compartment to be measured. The dilution of this substance can then be used for a calculation of the volume into which it is distributed. The compartments which are most readily measured in this way are plasma volume, extracellular volume, and total body water.

Total blood volume cannot be measured accurately by draining the blood from an animal, for a substantial amount of blood will remain in the vascular system. The dilution technique, however, gives reliable and reproducible results. One of the commonly used substances for this purpose is a blue dye, Evans blue (also known as T-1824). A measured small amount of this dye is injected as an aqueous solution into the blood stream, and the dye binds to the plasma proteins and therefore remains within the vascular system. After complete mixing has taken place (after several minutes), a blood sample is withdrawn and the dye concentration in the plasma is measured. The *plasma volume* is then readily calculated from the dilution of the dye. The *total blood volume* can then be calculated from a determination of the per cent red cells in the blood (the *hematocrit*) which is obtained by centrifugation of a blood sample.

The extracellular fluid volume is determined by injecting a substance which penetrates through the capillary walls and becomes distributed in blood and in interstitial spaces, but does not penetrate into the cells. One such substance is the polysaccharide inulin (also used in renal function studies, see page 461). Inulin, which has a molecular weight of about 5000, will distribute itself uniformly in extracellular space, but it is not metabolized, and it will not penetrate into the cells. Its dilution therefore will indicate the total extracellular volume. The volume of interstitial fluid is then calculated as the difference between total extracellular volume and plasma volume.

Total body water can, of course, be determined by killing the organism and determining the water content by drying. A dilution technique is, however, less dramatic and more convenient. Presently the commonest substance used is water labelled with the radioactive hydrogen isotope tritium. A small amount of tritiated water is injected, it diffuses through capillary walls and cell membranes without hindrance, and it rapidly becomes evenly distributed throughout all body water. A determination of the radioactivity of any sample of a body fluid can therefore be used for a calculation of the volume into which the injected tritiated water was diluted. In fact, it is not even necessary to take a blood sample, for if the bladder has been emptied before the experiment, a freshly voided urine sample will be as representative of the body water as a water sample obtained from any other source.

Water compartments in vertebrates. The total body water and its distribution in intracellular and extracellular volumes of some selected vertebrates are given in table 4.2. The total body water in aquatic and terrestrial vertebrates is of the same magnitude, about 70%, but there are variations in both directions. The distribution between intracellular and extracellular volume again shows a rather consistent pattern, although there are some variations.

The blood volume of vertebrates is of particular interest. The low values for teleost fish (carp and red snapper) given in table 4.2, are

Table 4.2. *Body fluid compartments in representative vertebrates. All values in per cent of body weight. The minor differences in this table are of uncertain significance, and they may in some cases be due to use of different techniques, to differences in the fat content of the animal body, etc. The blood volume of teleost fish is, however, consistently much lower than in other vertebrates*

	Total body water (%)	Intracellular water (%)	Extracellular water (%)	Blood volume (%)
Lamprey (a)	76	52	24	8.5
Dogfish (b)	71	58	13	6.8
Carp (c) (fresh water)	71	56	15	3.0
Red Snapper (c) (marine)	71	57	14	2.2
Bullfrog (d)	79	57	22	5.3
Alligator (e)	73	58	15	5.1
Gopher snake (e)	70	52	17	6.0
Pigeon (f)	—	—	—	9.2
Great Horned owl (V)	—	—	—	6.4
Dog (g)	63	37	19	9.1

(a) Thorson (1959)
(b) Thorson (1958)
(c) Thorson (1961)
(d) Thorson (1964)

(e) Thorson (1968)
(f) Bond and Gilbert (1958)
(g) Hopper, Tabor and Winkler (1944)

characteristic of a large number of fish on which determinations have been made. While teleost fish in general have blood volumes of 2 to 3%, elasmobranchs and cyclostomes have blood volumes of 6 to 10%. Air-breathing vertebrates (amphibians, reptiles, birds, and mammals) have blood volumes between 5 and 10%. We can therefore say that vertebrates on the whole have blood volumes between 5 and 10%, with the exception of teleost fish, which have substantially smaller blood volumes. At the present time it is uncertain what functional significance this difference may have.

Circulation patterns

Each class of vertebrates shows a quite uniform type of circulation, but differences between the classes are substantial. As vertebrate life changes from aquatic to terrestrial, the circulation becomes more complex.

There are two important consequences of the arrangement of the mammalian circulation, both immediately apparent from fig. 4.3. First, the blood flow through the lungs must equal the blood flow through the entire remaining part of the body (except for minute transient variations that can be caused by slight changes in heart volume during one or a few strokes). The other important consequence is due to the fact that the two halves of the heart contract simulta-

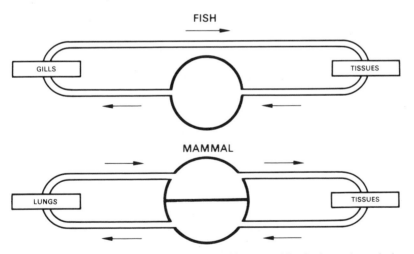

Fig. 4.3. In the circulatory system of fish, blood is pumped by the heart through the gills, from where it flows on to the tissues. In the mammalian system, blood is pumped through the lungs and then returns to the heart before being passed on to the tissues by a second pump. The mammalian arrangement permits a higher blood pressure in the tissues than in the lungs.

neously. This means that, as blood is ejected from the heart, the entire ejected volume must be taken up by changes in the volume of the elastic blood vessels.

Fish and mammals represent two extremes in vertebrate circulation. The gradual separation of the heart into two separate pumps as the vertebrates progress from aquatic life to fully terrestrial respiration is shown in fig. 4.4. Each vertebrate class has certain characteristics, and a few comments will help to explain these differences.

Cyclostomes. The hagfish have a circulatory system which differs from that of all other vertebrates. It is partly an open system with large blood sinuses, rather than a closed system as in other vertebrates. The other characteristic of hagfish circulation is that in addition to the regular heart (the *branchial heart*), they have several accessory hearts, especially in the venous system (see fig. 4.4*a*). There are three sets of such accessory hearts, the *portal heart,* which receives venous blood from the cardinal vein and from the intestine and pumps this blood to the liver, the *cardinal hearts* which are located in the cardinal veins and help to propel the blood, and the *caudal hearts* which are paired expansions of the caudal veins. In addition to these accessory hearts, all located in the venous system, the gills take active part in the forward propulsion of the arterialized blood. This is accomplished by contraction of striated muscular elements in the gills and gill ducts which help to propel the blood in the arterial system (Johansen, 1960).

The caudal hearts of the hagfish are particularly interesting because they differ in design from all other hearts (see fig. 4.5). A longitudinal rod of cartilage separates two chambers, and alternate contractions of muscles on the two sides causes the rod to be flexed. As the muscles on one side contract, those on the opposite side provide pressure for expulsion of the blood on that side. Simultaneously, the volume on the contracting side increases, so that this chamber becomes filled with blood. By alternate contractions, the two chambers fill and empty in opposing phase, while appropriate valving assures a unidirectional flow.

Fishes. The circulation in fishes, both teleosts and elasmobranchs, is as shown in fig. 4.4*b*. The heart consists of two chambers in series, an atrium and a ventricle. On the venous side the heart is preceded by an enlarged chamber or sinus on the vein, the *sinus venosus,* which helps assure a continuous flow of blood to the heart. On the arterial side the teleost heart is immediately followed by a thickened muscular part of the ventral aorta, called the *bulbus arteriosus* (fig. 4.6). The

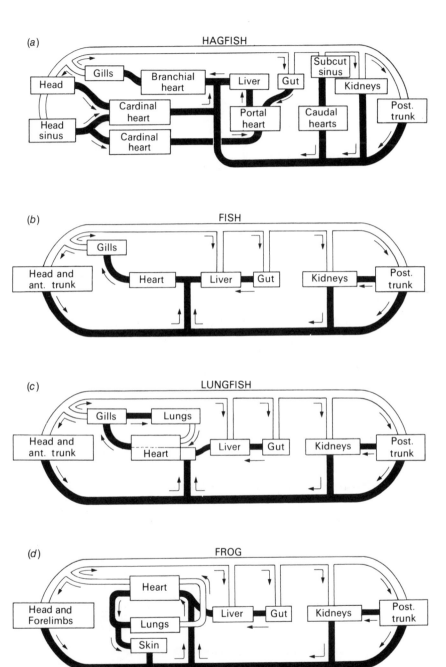

Fig. 4.4. The main circulatory patterns of the vertebrate classes. (a) Hagfish; (b) fish; (c) lung-fish; (d) frog; (e) non-crocodilian reptile; (f) crocodilian reptile; (g) mammal; (h) bird.

(e)
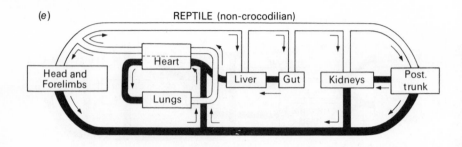

(f)
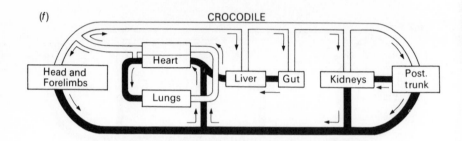

(g)
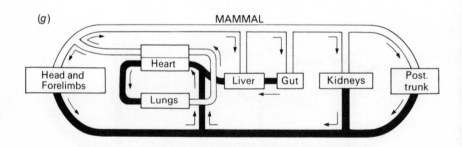

(h)
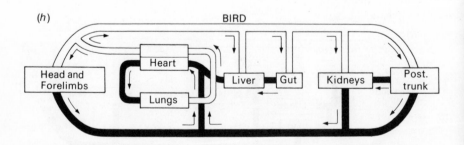

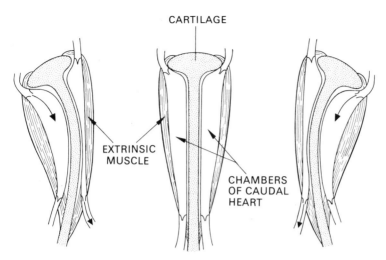

CARTILAGE

EXTRINSIC
MUSCLE

CHAMBERS
OF CAUDAL
HEART

Fig. 4.5. The caudal heart of hagfish differs from other vertebrate hearts. A central cartilage is flexed by alternate contractions of muscles on the two sides, which also serve as walls for the two chambers. As the muscles on one side contract, the chamber on this side is filled with blood, while blood is expelled from the other side. Valves help to obtain a unidirectional flow of blood (from Gordon, M. S. (ed). 1972. *Animal Physiology: Principles and Adaptations,* 2nd ed. © Macmillan, New York, p. 172, Fig. 6-6).

elasmobranch heart has a similarly located thickened part, the *conus arteriosus,* developed from the heart muscle. It is fibrous and is equipped with valves which prevent backflow of blood into the ventricle. This is particularly important because the heart, due to its location in a rigid chamber, can produce negative pressures. A negative pressure in the heart facilitates the filling by 'suction' of the atrium from the large sinus venosus.

The sequence of events in a shark heart during one contraction cycle is evident from the pressure tracing shown in fig. 4.7. During contraction of the ventricle the blood pressure rises and this pressure is transmitted to the conus arteriosus. As the ventricle relaxes, backflow from the conus is prevented by valves, and the high pressure therefore persists in the conus after the ventricle begins to relax. During contraction of the ventricle, the heart decreases in volume, and being located in a rigid chamber, it causes a negative pressure to develop in this chamber. Unless the large sinus and the adjoining veins could provide an equal volume of blood to flow into the atrium as the ventricle contracts, the negative pressure would only impede the contraction. With the inflow of blood, however, the negative pressure does not become excessive, but merely serves to fill the atrium. As the atrium afterwards contracts, the now relaxed ven-

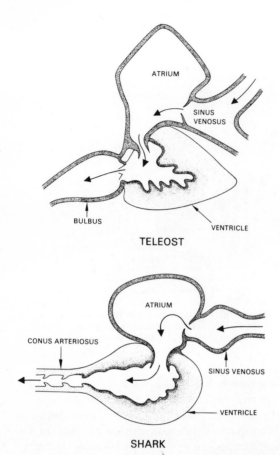

TELEOST

SHARK

Fig. 4.6. The two-chambered heart of fish is preceded by an enlarged venous sinus which supplies blood for filling of the atrium. The heart ejects blood into a thickened part of the artery, the bulbus arteriosus (teleosts) or conus arteriosus (elasmobranchs) (Randall, 1970).

tricle becomes filled with blood from the atrium, backflow into the sinus being prevented by valves. Thus, in this part of the cycle blood is merely shifted from atrium to ventricle, while the volume of the pericardial contents remains unchanged (Satchell, 1970; Randall, 1970).

Because of its elasticity, the bulbus or conus serves as an initial damping device on the pressure developed by the heart beat, thus aiding in producing a uniform and continuous flow of blood through the gills.

In the air-breathing vertebrates, the sinus venosus and bulbus arteriosus decrease in importance and in the mammalian heart they are not present as separate structures.

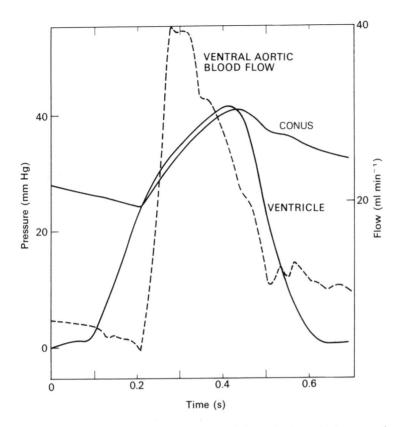

Fig. 4.7. Pressure changes in the elasmobranch heart during a single contraction cycle in relation to ventral aortic blood flow (Randall, 1970).

Lung-fish. The major evolutionary change in lung-fish is that, in addition to gills, they have lungs as respiratory organs (fig. 4.4c). The gills in part receive blood that has already passed through the lung. If the gills were similar to the gills of ordinary fishes, this might be a disadvantage, for a lung-fish swimming in oxygen-depleted water would then lose oxygen from the blood to the water which flows over the gills. The lung-fish gills, however, have degenerated, and some of the gill arches permit a direct through-flow of blood (Johansen, 1968; 1970).

The atrium of the heart is divided into two chambers by a septum, and the ventricle is partially divided. In this way the lung-fish heart somewhat resembles the completely divided heart of mammals, birds, and crocodiles. The lung-fish heart, in fact, shows an amount of structural division greater than that of any amphibian (Foxon, 1955). Blood from the lung returns to the left atrium, and the right atrium receives blood from the general circulation. The partial division of the ventricle tends to keep the two bloodstreams separated, so that oxygenated blood tends to flow into the first two gill arches and supply the head with relatively oxygen-rich blood. The less well-oxygenated blood from the right side of the heart flows through the posterior gill arches and passes on to the dorsal aorta, and in part to the lungs. Thus, the lung-fish represents the beginning of a complete separation between circulation to the lungs and to the remaining parts of the body (Johansen, Lenfant and Hanson, 1968).

The anatomy of lung-fish does not explain how blood actually flows. Studies of live African lung-fish (*Protopterus*) show that circulation to the lungs and to the systemic vascular circuit have a high degree of functional separation with preferential passage of oxygen-poor blood to the lungs and oxygen-rich blood to the systemic circulation. It is particularly significant that the functional separation is highest immediately after a breath of air when the oxygen in the lung is at its highest, whereas later in the interval between breaths, the degree of separation diminishes. This is of obvious importance for the efficiency of gas exchange in a lung which is filled with fresh air only at intervals (Johansen *et al.*, 1968).

Amphibia. The heart of modern amphibians (frogs, toads, and salamanders) has two completely separate atria, but there is only one undivided ventricle (fig. 4.4*d*). The left atrium receives oxygenated blood from the lungs, and the right atrium receives venous blood from the general systemic circulation. Although the ventricle is undivided, the two kinds of blood tend to remain unmixed, so that oxygenated blood enters the general circulation and oxygen-poor blood flows separately into the pulmonary circulation. The pulmonary artery also sends branches off to the skin, and this is of importance because the moist amphibian skin is a major site of oxygen uptake. The anatomical arrangement of the heart includes a longitudinal ridge-like baffle in the bulbus arteriosus (known as the spiral valve), which seems important in keeping the blood streams separate.

Reptiles. Non-crocodilian reptiles have a complete separation of the atria, but the ventricle is only partially divided (fig. 4.4*e*). Even so, the streams of oxygenated and non-oxygenated blood are kept well separated so that there is very little mixing of the blood, and there is in effect a fully developed double circulation. Thus, the incomplete division of the ventricle in amphibians and non-crocodilian reptiles cannot be interpreted simply on the basis of anatomical appearance. As more information is obtained it becomes increasingly clear that the blood streams remain much more separated than anatomical considerations would indicate, both in amphibians and in reptiles. The method used to evaluate the mixing problem is to determine the oxygen content in blood entering the heart, and again in the vessles leaving the heart. Such studies have shown nearly complete separation of the blood streams (Johansen, 1962; White, 1959).

Crocodiles have both chambers of the heart completely divided (fig. 4.4*f*). This complete separation is confounded by the peculiar fact that the left aortic arch originates from the right ventricle and thus should receive venous blood. There is, however, a hole or foramen connecting the two aortic arches. On an anatomical basis this has been used to argue that crocodilian circulation allows mixing of oxygenated and non-oxygenated blood, but a more careful study shows that both types of blood remain separate, and that both aortic arches carry unmixed oxygenated blood (White, 1956). During diving, however, the circulation changes, blood flow to the lungs is decreased, and a major part of the output of the right ventricle is now ejected into the left aortic arch. This shunt thus permits a rerouting of the blood during diving and a partial or complete bypass of the lungs.

Birds and mammals. The division of the heart and separation of pulmonary and systemic circulation is complete in birds and mammals (fig. 4.4*g, h*). This has one important consequence, the pressure can be different in the pulmonary and the systemic circulation. The resistance to flow in the pulmonary system is much lower than in the systemic circulation, and the blood pressure in the pulmonary circulation is therefore only a small fraction of the pressure in the systemic part. Such a difference is, of course, not possible if separation of the heart is incomplete. Since an incomplete separation has been retained in both amphibians and reptiles, it seems that this arrangement may have other advantages that are not well understood.

There are some differences between the circulation in birds and mammals that are of great significance in comparative anatomy, for example, mammals retain the left aortic arch whereas birds retain the right. A difference of physiological importance is that the kidneys of all non-mammalian vertebrates receive venous blood from the posterior part of the body (the *renal portal circulation*). Birds have retained this renal portal circulation, but it is completely absent in mammals. This difference is of importance to the understanding of renal function (p. 465) (Johansen and Martin, 1965).

Heart and cardiac output

We know that, as a rule, small animals have a higher rate of oxygen consumption per unit body weight than large animals, and thus the heart of the small animal must supply oxygen at a higher rate. We have already seen that the oxygen capacity of the blood of small and large mammals is similar; as a consequence the heart of the small mammal must pump blood at a higher rate. Is the necessary increase achieved by a larger pump, by a larger stroke volume, or by a higher frequency?

Size of vertebrate hearts. The heart size of a number of mammals is plotted in fig. 4.8. As expected, the heart size increases with body size, but it may be more surprising that, relative to their body size, small and large mammals have about the same heart size. If the heart mass were exactly proportional to body mass, the slope of the regression line in fig. 4.8 would be 1.0; the actual slope in the figure is 0.98, which is statistically indistinguishable from exact propor-

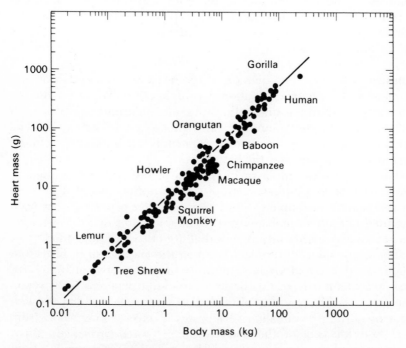

Fig. 4.8. The heart size of mammals in relation to body size. The size of the heart is nearly proportional to body size and makes up approximately 0.6% of the body mass in small and large mammals alike. The heart size of a wide variety of primates falls within the range of other mammals. (Stahl, 1965, copyright 1965 by the American Association for the Advancement of Science.)

tionality. The equation for the line is $M_h = 0.0059 M_b^{0.98}$, which says that the average mass (M_h) of the mammalian heart is 0.59% of the body mass (M_b in kg), irrespective of the size of the mammal.

The heart size of birds can be described by a similar equation to that of mammals, $M_h = 0.0082 M_b^{0.91}$ (Lasiewski and Calder, 1971). This equation says that the heart weight is not strictly proportional to body weight, the weight exponent is significantly less than 1.0, which means that, relative to their body size, larger birds tend to have slightly smaller hearts than birds of small body size. A 1 kg bird can be expected to have a heart weight of 8.2 g, while for a mammal of the same size the expectation is a heart weight of 5.9 g.

The heart size of reptiles and amphibians is less well studied, but available data indicate that reptiles have a heart size of about 0.51% of body weight and amphibians 0.46%. This is only slightly lower than mammals, although the metabolic rates of these animals is in the order of one-tenth of the mammalian rate. Fishes have smaller hearts again, about 0.2% of their body weight.

To summarize, whether we compare the different classes, or animals of different body size within one class, the large differences in metabolic rates of vertebrates are not conspicuously reflected in the size of the heart. Differences in the need for oxygen must therefore be reflected primarily in pumping frequency, for the stroke volume depends on heart size, and the amount of oxygen contained in each volume of blood is not size-dependent.

Heart frequency. The heart frequency, or pulse rate, is usually given as the number of heart beats per minute. The pulse rate for an adult human at rest is about 70 per minute; in exercise the rate increases several-fold.

The heart frequency is clearly inversely related to body size. An elephant which weighs 3000 kg may have a resting pulse rate of 25 per minute and a 3 g shrew, the smallest living mammal, has a resting pulse rate of over 600 (Morrison, Ryser and Dawe, 1959). This means that in the shrew the heart goes through ten complete contraction cycles per second, an almost unbelievable rate. During activity the heart frequency increases further, and as much as 1200 beats per minute have been measured in hummingbirds (Lasiewski, Weathers and Bernstein, 1967) and in small bats in flight (Studier and Howell, 1969).

If heart frequency is plotted relative to body mass on logarithmic coordinates, the points are located close to a straight line (see fig. 4.9). The equation for the line is $f_h = 241 M_b^{-0.25}$. The slope of the regression line is negative, that is, the larger the body mass (M_b), the lower is the heart frequency (f_h).

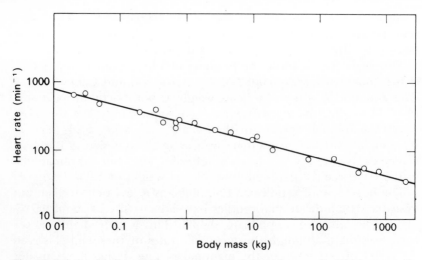

Fig. 4.9. Relationship between heart rate and body size of mammals. The heart rate is higher in small than in large animals. If the observed data are plotted on logarithmic coordinates, they are grouped around a straight line, which can be represented by a logarithmic equation (see text, data from Stahl, 1967).

In addition to the negative slope, the most significant information contained in this equation is the numerical value of the slope, which is 0.25. This is exactly the same slope as that of a regression line between body mass and resting oxygen consumption of mammals (see p. 237–8).

We now have two important pieces of information, (1) the size of the pump, the heart, remains a constant percentage of the body mass, and (2) the increase in the pumping rate (the heart rate) in the smaller mammal increases in exact proportion to the need for oxygen. Let us turn to the volume of blood pumped by the heart.

Cardiac output. The volume of blood pumped by the heart per unit time is usually called the *cardiac output* or the cardiac minute volume. (In hearts with a complete separation of the ventricles, the expression refers to the output of one side of the heart only.) The cardiac output can be determined in a number of different ways. One commonly used method is known as *Fick's principle*. It depends on the simple fact that all the oxygen consumed ($\dot{V}_{O_2}$) * by the animal is carried by the blood ejected from the heart. Therefore, if we know

* In equations related to gas metabolism V usually stands for volume, and $\dot{V}$-dot for the time derivative of V, i.e. the volume per unit time. The subscript (in this case O_2) is chosen to aid in easy recognition. Q stands for quantity of blood, $\dot{Q}$ for blood flow per unit time, and $\dot{Q}_h$ for blood flow from the heart, i.e. cardiac output. C stands for gas concentration in the blood, C_a for arterial and C_v for venous blood.

the difference in the oxygen content of arterial and venous blood, we can calculate cardiac output ($\dot{Q}_h$) from the following equation:

$$\dot{V}_{O_2} = \dot{Q}_h (C_{aO_2} - C_{vO_2}).$$

In order to determine the cardiac output, we can determine (1) the oxygen consumption, (2) the oxygen content in arterial blood, and (3) oxygen content in venous blood. An arterial blood sample can be obtained from any artery, but the venous blood must be obtained as mixed venous blood, i.e. from the right side of the heart, or from the pulmonary artery. Venous blood from some other place will not suffice, for different organs remove different amounts of oxygen from the blood passing through them. A sample of mixed venous blood is most easily obtained from the right side of the heart by inserting a catheter in a vein and threading it in until it reaches the heart. This is a widely used and relatively harmless procedure.

During activity, when the oxygen consumption increases, the cardiac output also increases, but not in proportion to the increased oxygen consumption. Part of the increase is covered by an increase in the arterio–venous oxygen difference, i.e. the amount of oxygen extracted from the blood increases during activity.

Determinations of cardiac output according to Fick's principle can only be carried out in animals that have a complete separation between arterial and venous blood in the heart, that is, in mammals and birds. In other animals, where there is a possibility for mixing arterial and venous blood in the heart, Fick's principle does not apply.

Other methods for determining cardiac output are available. These are based on a variety of principles. One method consists of injecting an easily measured dye at a known moment in time in a vein leading to the heart. The dye will shortly afterwards appear on the arterial side, and by integrating the curve representing the concentration of the dye as it rapidly increases and again disappears in the artery, it is possible to calculate the cardiac output. A method that in principle is similar is the thermodilution method, in which a known volume of saline is injected in the vein; if the saline is colder than the blood, there will be a transient decrease in arterial blood temperature; again, the integrated curve gives information about the blood flow, i.e. the cardiac output.

Regulation of heart beat

The heart has the inherent capacity to contract rhythmically without any external stimulus. Even if the heart is completely removed from the body, it may continue to beat for a considerable period of time, and an isolated frog or turtle heart beats rhythmically for hours. Mammalian hearts will not continue for as long, for they are more 'sentitive to temperature and an adequate oxygen supply. Since the isolated heart continues beating by itself, its ability to contract rhythmically must be inherent. An excellent proof of the independent contractility of the heart is the fact that the heart in a developing chick embryo begins to beat before any nerves have reached it, and that heart muscle cells grown in tissue culture also contract rhythmically without any external stimulus.

The contraction in the mammalian heart begins in a small piece of embryonic-type muscle located where the vena cava enters the right atrium. It is the vestige of the sinus venosus, and is known as the *sinus node*. The contraction spreads rapidly through the muscle of the two atria, and, after a slight delay, to the muscle of the ventricles. When the wave of contraction reaches the partition between the atria and the ventricles, a patch of tissue known as the atrioventricular bundle conducts the impulse to the ventricles, which then, after the minute delay due to conduction, contract simultaneously.

The sinus node, where the heart contraction originates, is also known as the *pacemaker* of the heart. It starts the beat and sets the rate by its inherent rhythmicity. As we know, the heart rate can vary by several-fold. Two different mechanisms are involved in the control of the rate, either nerve impulses to the pacemaker, or hormonal influences. A branch of the vagus nerve, when stimulated, slows down the rate of heart beat. The nerve is of parasympathetic origin, and it acts by release of acetylcholine. Another nerve, the accelerans nerve, is of sympathetic origin. When it is stimulated, the heart accelerates due to release of adrenaline from the nerve endings at the pacemaker.

Adrenaline is also known as a hormone, and when it is released from the adrenal medulla into the blood, it accelerates the heart. This effect on the heart is one of the major effects of adrenaline, well known as part of the 'fight-or-flight' syndrome. There is no corresponding release of the decelerating substance, acetylcholine, except from the endings of the vagus nerve. If acetylcholine were released elsewhere in the body, it would not affect the heart for the simple reason that the blood contains an enzyme which rapidly splits acetylcholine by hydrolysis.

The amount of blood pumped by the heart can be changed, not only by changing the frequency, but also by changing the volume of blood ejected in a single beat, the stroke volume. When the heart contracts, it does not eject all the blood contained in the ventricles, a substantial but variable volume of blood remains in the ventricle at the end of contraction. Two major factors influence the stroke volume, one is the hormone adrenaline, which augments the force of the cardiac muscle contraction, thus forcing a larger amount of blood out of the heart in a single stroke. The other important effect on stroke volume is the amount of blood in the ventricle when contraction begins. This effect is due to an inherent characteristic of heart muscle, which will contract with greater force if it is more distended at the moment contraction begins. If the return of venous blood to the heart is increased for some reason, the ventricle will be filled with more blood, the muscle is stretched more than before, and the following contraction will eject a larger volume of blood. This relationship between increased venous return and increased cardiac output was discovered by the famous English physiologist Starling, and is known as *Starling's Law of the Heart*. In exercise the venous return to the heart increases, partly due to the increased pumping effect of the contracting muscles on the veins, and partly due to contraction in the venous system of the abdominal region, thus causing an increase in cardiac output.

In summary, the heart beat originates in the sinus node or pacemaker. Cardiac output is under the influence of three major controlling systems, the nervous system, hormonal control, and the autoregulation due to the effect of the venous return on the heart muscle.

The conduits – blood vessels

The blood vessels are not merely tubes of different sizes, they have elastic walls and a layer of smooth muscle within their walls enables them to change diameter.

There are characteristic differences between arteries, capillaries, and veins. The arteries have relatively thick walls which consist of heavy and strong layers of elastic fibers and smooth muscle. As the arteries branch and become smaller, the relative amount of muscle gradually increases in proportion to the elastic tissue in the wall. The capillaries, the smallest units of the system, consist of a single cell layer, and virtually all exchange of substances between blood and tissues takes place through the walls of the capillaries. On the venous side the vessels have thinner walls than the arteries, but both elastic fibers and smooth muscle are found throughout the venous system.

Some measurements made on the vascular system of dogs are given in table 4.3 (Burton, 1965). Every time a larger vessel branches, the number of branches increases and the diameter de-

Table 4.3. *Geometry of the blood vessels in the mesentery of the dog*

Kind of vessel	Diameter (mm)	Number	Total cross-sectional area (cm²)	Length, approx. (cm)	Total volume * (cm³)
Aorta	10	1	0.8	40	
Large arteries	3	40	3	20	
Arterial branches	1	2 400	5	5	190
Arterioles	0.02	40 000 000	125	0.2	
Capillaries	0.008	1 200 000 000	600	0.1	60
Venules	0.03	80 000 000	570	0.2	
Veins	2	2 400	30	5	
Large veins	6	40	11	20	680
Vena cava	12.5	1	1.2	40	

* The estimates of total volume are based on a more detailed analysis than indicated by the figures for length given in the preceding column.

creases. As the number of branches increases, their total, combined cross-sectional area increases, and at the level of the capillaries the aggregate cross-section of all the capillaries in the mesentery alone is some 800 times that of the aorta. From that point on, the vessels converge into larger and larger veins, their number decreases, and so does the total cross-sectional area, until in the vena cava the cross-section slightly exceeds that of the aorta. Another interesting fact is evident from the table, the amount of blood located in the venous system exceeds by several-fold the amount of blood in the arterial system.

In addition to the volume of the vascular system, we are interested in the pressure and the flow velocity in the various parts. Such information is given in fig. 4.10, which refers to measurements on man. The first column shows that most of the blood at any given moment is located on the venous side. The last column gives the velocity of the blood in the various vessels, showing the highest velocity in the aorta. As the total cross-sectional area increases, the velocity decreases drastically, until in the capillaries it is between one hundredth and one thousandth of the velocity in the aorta. On the venous side, the velocity increases again, but does not reach the high velocity of the arterial system.

The center column in fig. 4.10 shows the pressure throughout the

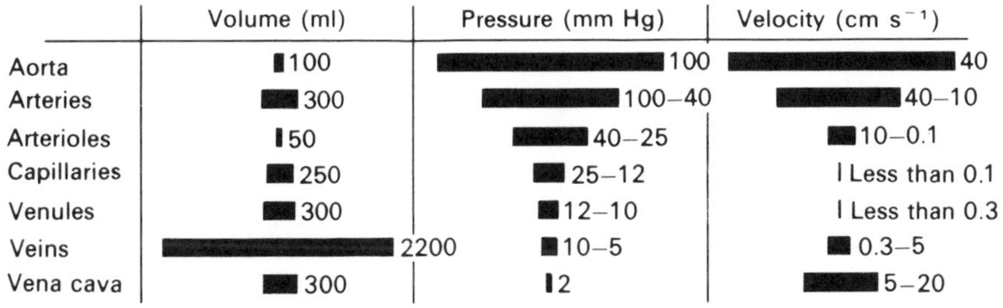

	Volume (ml)	Pressure (mm Hg)	Velocity (cm s^{-1})
Aorta	▌100	▰▰▰▰▰▰▰▰▰▰100	▰▰▰▰▰▰▰▰40
Arteries	▰▰300	▰▰▰▰▰▰▰100–40	▰▰▰▰▰40–10
Arterioles	▌50	▰▰▰▰40–25	▰▰10–0.1
Capillaries	▰▰250	▰▰25–12	I Less than 0.1
Venules	▰▰300	▰12–10	I Less than 0.3
Veins	▰▰▰▰▰▰▰▰2200	▰10–5	▰0.3–5
Vena cava	▰▰300	I 2	▰▰▰5–20

Fig. 4.10. Distribution of blood volume, blood pressure, and velocity, in the various parts of the vascular system of man.

system. The pressure must, of course, gradually decrease throughout, from the aorta to the larger veins. The greatest pressure drop takes place in the smallest arteries, called arterioles, which have a high proportion of smooth muscle in their walls and, by changing their diameter, are the most important factor in changing the resistance to flow and thus in regulating the distribution of blood flow to various organs.

The pulmonary circulation is different from the systemic in that the total resistance is less, the blood pressure can therefore be lower and the arteries have thinner walls. One interesting characteristic of the pulmonary artery is that its cross-sectional area is not circular but oval. The significance of this is that during the contraction of the heart, as blood is injected into the pulmonary artery, the flattened vessel absorbs a considerable increase in volume as the cross-section becomes circular, and only then does the wall become stretched. In this way much of the ejected blood is taken up in the increased volume of the artery, which thus serves as a damping device so that the blood does not flow through the lung capillaries in spurts with each heart beat (Melbin and Noordergraaf, 1971).

The physics of flow in tubes

The flow of a fluid (whether gas or liquid) may be smooth and regular so that in a straight tube each particle of fluid moves in a straight line. This is called *laminar flow*. If in a curved tube the direction of flow at any given point remains constant, the flow is also said to be laminar. The path followed by a fluid particle in laminar flow is called a streamline. If the fluid particles flow irregularly and swirl about, however, the flow is *turbulent*. In a given tube the flow will change from laminar to turbulent if the velocity increases above a critical point. The following discussion will refer specifically to la-

minar flow; the fluid dynamics of turbulent flow is more complex
and has little application to the flow in blood vessels.

Bernoulli's theorem. The steady flow in an ideal fluid along a streamline is described by *Bernoulli's theorem,* named after Bernoulli
(1700–82), a Swiss M.D. who at the age of 25 became professor of
mathematics. First we must note that flow in a tube is called steady if
the velocity at every point of the tube remains constant, even if the
pipe gets narrower or wider and the fluid therefore moves faster or
slower as it flows along the tube.

Bernoulli's theorem states that the total fluid energy (E) is the sum
of (1) the potential energy due to internal pressure, (2) the potential
energy due to gravity, and (3) the kinetic energy of the moving fluid.

$$E = (p\ v) + (m\ g\ h) + (\tfrac{1}{2}\ m\ u^2). \tag{1}$$

On the right side of the equation the first term is the pressure
energy of the fluid or the product of pressure (p) and volume (v).
The second term is the gravitational potential energy or the product
of the mass (m), gravitational constant (g), and the height (h). The
third term is the kinetic energy due to the velocity (u) of the fluid.
The sum of these three is the total fluid energy. The energy content
per unit volume (E') can be obtained by dividing both sides of the
equation by the volume, thus obtaining the following equation:

$$E' = p + \rho\ g\ h + \rho\ \frac{u^2}{2}, \tag{2}$$

in which ρ (rho) is the density of the fluid.

Flow of a fluid in a horizontal tube permits a considerable simplification of the equation. Since there is no change in the gravitational
potential energy from one end of the tube to the other, this term
remains constant and can be dropped. Likewise, since the density of
the fluid remains constant, this constant can be removed so that we
obtain:

$$E' \propto p + \frac{u^2}{2}. \tag{3}$$

If the flow is in a frictionless tube, the energy content of the fluid
(E') remains constant, and if the velocity (u) changes due to a change
in diameter, the pressure must change in the opposite direction, or:

$$\Delta \frac{u^2}{2} = -\Delta p. \tag{4}$$

Let us next look at a tube of uniform diameter in which there is friction. With constant diameter, the velocity u remains constant. Due to

resistance to flow, energy is used to drive the fluid through the tube, and the energy loss is therefore expressed as a decrease in pressure, p, as indicated in fig. 4.11. The energy loss due to frictional forces is degraded as heat, and this shows up as a temperature increase in the fluid, but in the bloodstream this is of no significance.

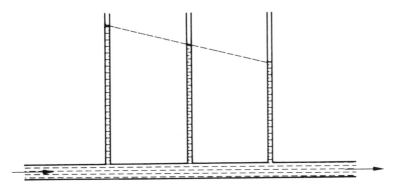

Fig. 4.11. When a fluid flows in a horizontal tube, the frictional resistance leads to a drop in pressure along the tube.

If the diameter of the tube changes, there must be a change in the velocity (u) as well. Again, in a horizontal tube the gravity term of equation (2) can be disregarded. Let us, for the moment, assume that no energy is dissipated by frictional forces and the energy content remains constant. We now see that velocity and pressure must change inversely according to expression (4), which states that if velocity increases, pressure in the fluid must decrease. This is illustrated in fig. 4.12 for a tube with frictional resistance.

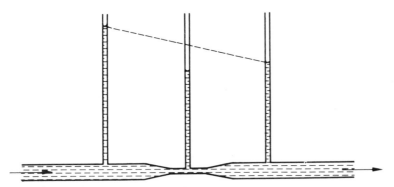

Fig. 4.12. When a fluid flows through a narrow part of a tube, the velocity of the fluid is increased at this point. The increased velocity is accompanied by a decrease in pressure. For further explanation, see text.

It is often stated that fluid always moves from higher pressure to lower pressure. Fig. 4.12 shows that this is not necessarily so, for the pressure in the narrow tube is lower than further downstream. The correct statement is that fluid always moves from a point where the total fluid energy is higher, to a point where it is lower.

In the preceding discussion the gravitational term in equation (2) was disregarded. If we have a U-tube, as in fig. 4.13, a liquid at rest has the same energy content per unit volume throughout the tube. Although the pressure at the bottom of the U is higher than in the arms, the fluid remains still. Reference to equation (2) shows that, as the height h decreases, the gravitational potential energy per unit volume $(\rho g h)$ decreases, and the pressure energy (p) must increase accordingly. This increase in pressure energy is what we frequently refer to as the *hydrostatic pressure* of a fluid, as we describe the increase in pressure with depth, relative to the pressure at the surface.

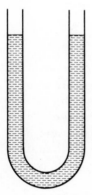

Fig. 4.13. A fluid in a U-shaped tube remains at rest. At the bottom of the U the pressure is higher than in the arms, due to the effect of gravity on the fluid.

The venous pressure in the foot of a man is analogous to the U-tube. In the veins of the foot, the venous pressure is increased due to the height of the column of blood, so that the pressure at the feet actually exceeds the arterial pressure. This is shown in fig. 4.14. The high pressure in the venous system of the legs distends the veins, and contractions of the leg muscles and valves in the veins are main aids in returning the venous blood to the heart.

Another consequence of the gravitational term is that a certain arterial pressure is necessary to drive the blood to the head. Due to the gravity effect the arterial blood pressure in the head of man is reduced to about 50 mm Hg. For a giraffe, which carries its head

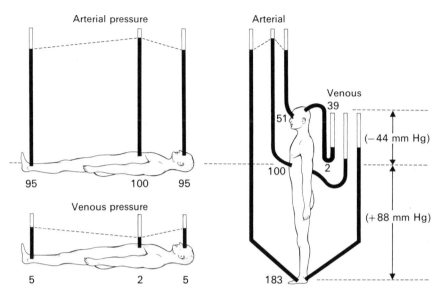

Fig. 4.14. Arterial and venous pressures in a man as he assumes different postures. The figures indicate the pressures at various points in relation to the pressure in the right atrium of the heart. (From *Physiology and Biophysics of the Circulation* by A. C. Burton, second edition. Copyright © 1972 by Year Book Medical Publishers. Used by permission.)

about 2 meters above the level of the heart, sufficient pressure to supply the brain with blood requires a much higher blood pressure than in man. Measurements of blood pressures in the giraffe have shown a systolic blood pressure as high as 260 mm Hg (Van Citters, Kemper and Franklin, 1968). (The normal systolic blood pressure in man is about 100 mm Hg.) To withstand the high pressures the giraffe has exceptionally thick walls in the arterial system, and the venous system is equipped throughout with valves that facilitate the return of blood from the limbs. When a giraffe that stands 4.5 meters tall lowers its head to drink, valves in the neck veins help to impede backflow of blood to the head and prevent the increase in hydrostatic pressure in the brain that otherwise would be caused by swinging the head down.

Wall thickness and tube diameter. In a hollow cylinder the tension in the wall (T) equals the product of the pressure across the wall (p) and the radius (r):

$$T = p \times r.$$

This relationship was derived by Bernoulli, but is usually known as Laplace's Law. It tells us that, for a given pressure, the tension in the wall increases in direct proportion with increasing radius. For the wall to withstand the tension, the thickness must therefore be increased accordingly. This is the reason that a large artery must have a thicker wall than a small artery.

The veins, which have much lower blood pressures, have thinner walls. Again, smaller veins can withstand the venous pressure with a thinner wall than a larger vein. Capillaries, of course, have a higher blood pressure than the veins, but nevertheless, because of their very small radius, a wall consisting of a single layer of cells has sufficient strength. Thus, the smallness of the capillary is a prerequisite for making the capillary wall thin enough to permit rapid exchange of material between the blood and the tissues.

Viscosity. The resistance to flow in a tube is due to inner friction in the fluid, the *viscosity* of the fluid. Everybody has some experience with viscosity – water and syrup don't flow equally fast out of a bottle. We say that water has a low viscosity and that syrup has a high viscosity.

The flow through a tube, $\dot{Q}$, is proportional to the pressure drop through the tube and inversely proportional to the resistance:

$$\dot{Q} = \frac{p}{R}$$

The resistance term, R, is a function of the dimensions of the tube and the nature of the fluid (viscosity). This relationship was clarified by the French physician Poiseuille (1799–1869) who became interested in the flow of blood in capillaries. Poiseuille worked with tubes of different diameters and found that the flow through a tube is proportional to the pressure and proportional to the fourth power of the radius of the tubes, but inversely proportional to the length of the tube and to the viscosity of the fluid. This is formally expressed in *Poiseuille's equation,* which can be written as follows:

$$\dot{Q} = \Delta p \, \frac{r^4}{l\eta} \times \frac{\pi}{8},$$

in which $\dot{Q}$ = rate of blood flow, Δp = pressure drop, r = radius, l = length of tube, and η (eta) = viscosity.*

The most striking aspect of this equation is the tremendous effect which the radius of the tube has on the flow. For our purposes here,

---

* The viscosity of a fluid changes with temperature. A reduction in temperature from 37 to 0 °C causes the viscosity of water to increase about 2.6-fold.

however, we shall be more concerned with the viscosity. The equation shows that as viscosity increases, the fluid flow decreases proportionately.

For many purposes it is convenient to express the viscosity of a fluid relative to that of water, assigning unity to the viscosity of water. Blood plasma is more viscous than water and has a relative viscosity of about 1.8, mostly as a result of the 7% dissolved protein. Assume that we measure the flow of water and of plasma through a given tube; if we assign the value of 1.0 to the flow of water, the flow of plasma will be 1/1.8. Let us now reduce the radius of the tube by one-half, keeping pressure and length of the tube constant. According to Poiseuille's equation the flow will be reduced to $(\frac{1}{2})^4$ or $1/16$. So will the flow of plasma, and the ratio of the flows will still be 1.8, i.e. the relative viscosity of plasma is independent of the size of the tube.

If instead of plasma we use whole blood, we find that its viscosity is higher than that of plasma, and that its viscosity increases with the concentration of red cells (see fig. 3.1, p. 85). Blood, however, does not behave quite as expected. Its relative viscosity changes with the radius of the tube through which it flows. With decreasing radius, the relative viscosity of blood decreases; in other words, it becomes more similar to water and flows relatively more easily. Such an anomalous fluid is called a non-Newtonian fluid, and its behavior makes it difficult to describe the flow of blood in the body, particularly in view of the fact that blood vessels are elastic and by no means maintain a constant radius. The end result, however, is important, flow of blood through the capillary bed is easier, and the apparent viscosity of blood is lower than expected merely from the dimensions of the blood vessels.

Another peculiar aspect of blood flow through capillaries is that the capillary diameter is frequently considerably less than the size of a red blood cell. Unexpectedly, this seems to be no impediment to flow in the capillary, the red cell is easily deformed and passes readily through the capillary (Skalak and Branemark, 1969). This gives rise to a very different type of flow in the capillary, known as bolus flow, in which the red cell acts as a 'plug' which causes a rapid renewal of liquid in the boundary layer along the capillary wall, thus facilitating the renewal of diffusible substances in this layer.

The capillary system

The total number of capillaries in the body is enormous. Merely in the mesentery of a dog the number of capillaries exceeds one thousand million, as was shown in table 4.3 (p. 138). Muscle is particularly well suited for an exact count of the number of capillaries, for here the capillaries run between the muscle fibers and parallel to these. In a cross-section of a muscle it is therefore relatively easy to count the number of capillaries per unit cross-sectional area. The number is very great, but the capillaries usually are not all open and filled with blood. In the resting muscle of a guinea pig, a cross-section of 1 mm² contains about 100 open capillaries through which blood flows. If the muscle is performing work, however, the constricted arterioles open up, more capillaries carry blood, and in maximal exercise a 1 mm² cross-section may have more than 3000 open capillaries. In an ordinary pencil the cross-section of the lead is about 3 mm² – imagine nearly 10 000 parallel tiny tubes running down the pencil lead!

Mammals of small body size have a higher capillary density than mammals of large body size. This is consistent with the higher metabolic rate and need for oxygen in the small animal, as discussed on page 96.

Two major processes account for the exchange of material between the fluid within the capillary on the one hand and the interstitial spaces of the tissues on the other. The capillary walls are quite thin, about 1 μm thick, and consist of a single layer of cells. Water and dissolved substances of small molecular weight (gases, salts, sugars, amino acids, etc.) can diffuse relatively unhindered across the capillary wall, but in addition, pressure within the capillary forces fluid out through the wall by bulk filtration. Molecules as large as most proteins (molecular weight about 70 000 or larger), however, do not pass through the capillary wall. Proteins are therefore withheld within the capillary, and a relatively protein-free filtrate is forced out through the capillary wall. The concentrations of the small-size solutes in this filtrate are similar to those in the blood plasma, but not quite identical because the proteins have some influence on the distribution of ions (the Gibbs–Donnan effect).

If we regard the capillary wall as a semi-permeable membrane, we can see that since the non-protein solutes penetrate freely, they have no osmotic effect. The proteins, however, are retained inside the capillary, and therefore exert a certain osmotic effect, called the colloidal osmotic pressure. In mammalian plasma it is about 25 mm Hg, and tends to draw water back into the capillary from the surrounding tissue fluid. Whenever the hydrostatic pressure (blood pressure) within the capillary exceeds the colloidal osmotic pressure, fluid is forced out through the capillary wall; whenever the hydrostatic pressure within falls below the colloidal osmotic pressure, fluid is drawn in again. The blood pressure within the capillary is variable, but in the arterial end it is often higher and in the venous

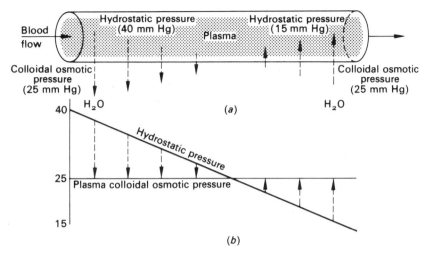

Fig. 4.15. The capillary wall is semi-permeable, and the blood pressue forces fluid out (ultrafiltration (*a*)). The plasma proteins remain within the capillary and oppose the ultrafiltration process. The blood pressure decreases along the length of the capillary, and when the blood pressure falls below the colloidal osmotic pressure exerted by the proteins, these cause an osmotic flow of fluid back into the capillary (*b*). The wall is freely permeable to salts and other small-molecular size solutes, and the fluid movements therefore take place as if these were not present at all.

end lower than the colloidal osmotic pressure. As a result, fluid is filtered out in the arterial end, and re-enters in the venous end of the capillary (see fig. 4.15). This theory for bulk filtration and reentry of fluid through the capillary wall was originally proposed by Starling (1896).

The amount of fluid filtered through the capillary wall, and the amount withdrawn again due to the colloidal osmotic pressure, fluctuate greatly. Usually outflow exceeds inflow, and the excess fluid remains in the interstitial spaces between the tissue cells. This fluid slowly drains into a system of lymph vessels, which gradually join into larger ducts. These in turn open into the larger veins, thus returning the fluid to the blood system. The lymphatic system can be considered part of the vascular system, but there is no direct circulation of the lymph as such; it is rather to be regarded as a drainage system which returns to the blood excess fluid lost from the capillary system.

If plasma proteins were absent, any hydrostatic pressure in the capillary in excess of that in the surrounding tissue spaces would cause fluid to filter out, and no force would be available to cause fluid to re-enter. The plasma proteins are therefore essential to retain the blood fluid within the vascular system.

Mammals have higher colloidal osmotic pressure in their plasma than other vertebrates (see table 4.4). The most striking fact is that birds have a rather low colloidal pressure, although their arterial blood pressure usually far exceeds the common values for mammals. Some birds have blood pressures above 200 or even 250 mm Hg, against about 100 mm Hg commonly found in mammals. These differences are not well understood. It seems that birds either must have a greater resistance to blood flow at the level of the arterioles so that the capillary pressure is not very high, or that the capillary wall has properties that differ from those of mammalian capillaries. A higher resistance in the arterioles would not seem to make much sense, for why should birds then have the high arterial pressure to begin with?

Table 4.4. *Colloidal osmotic pressure of vertebrate blood plasma* (Data from Altman and Dittmer, 1971)

	(mm Hg)		(mm Hg)
Cow	21	Alligator	9.9
Sheep	22	Turtle	6.4
Dog	20	Frog	5.1
Cat	24	Toad	9.8
Chicken	11	Codfish	8.3
Dove	8.1	Flounder	8.5

Other vertebrates, reptiles, amphibians, and fish, also have relatively low colloidal osmotic pressure. This is consistent with the fact that these groups in general have lower arterial blood pressures than mammals.

Circulation during exercise

During muscular activity the demand for oxygen is increased and thus the amount of oxygen that the heart must deliver to the tissues. Two avenues are open to meet this increased demand, increasing the volume of blood pumped by the heart (the cardiac output), or increasing the amount of oxygen delivered by each volume of blood. Arterial blood is already fully saturated and cannot take up more oxygen, but venous blood normally still contains more than half the oxygen content of arterial blood. Increasing the oxygen extraction from the blood is therefore an obvious way of obtaining more oxygen from each volume of blood.

Let us first examine the oxygen extraction. The total muscle mass of a lean man, which makes up about half of his weight, uses about 50 ml O_2 min^{-1}. This oxygen is supplied by a blood flow of about 1 liter, i.e. the oxygen content of arterial blood, 200 ml O_2 per liter blood, is reduced to 150 ml O_2 per liter in venous blood. Since a quarter of the oxygen in arterial blood is removed we say that the oxygen extraction is 25%. In a normal man during heavy exercise the blood flow to the muscles may be 20 liters min^{-1} (in well-trained athletes it may be even higher), and the oxygen extraction in the muscles now increases to 80 or 90%; in other words, the venous blood coming from heavily working muscles has very little oxygen left in it (Folkow and Neil, 1971).

The second avenue available for increasing the delivery of oxygen is, of course, an increased cardiac output. This can be achieved by increasing both heart rate and stroke volume. Because of the interest in man, both for medical reasons and in relation to athletic performance, much more information is available for man than for other animals. At rest the human heart beats at a rate of about 70 per minute, with a stroke volume of about 70 ml (from each side), giving

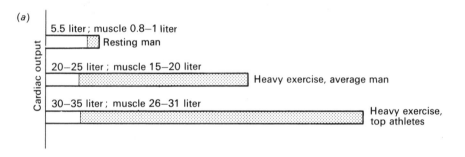

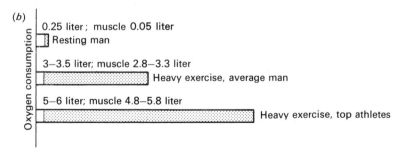

Fig. 4.16. Diagram of the distribution of total blood flow (cardiac minute volume) (*a*) and of oxygen consumption (*b*) between the muscles (shaded bars) and all other parts of the body (unshaded bars). Data for resting man, heavy exercise in a normal man; and heavy excercise in a top athlete (after Folkow and Neil, 1971).

a total cardiac output of about 5 liters per minute. In heavy exercise, the cardiac output can readily increase fivefold or more (if the oxygen extraction is doubled, this corresponds to a tenfold increase in oxygen delivery). Most of the increase in cardiac output is due to increased pulse rate, which may rise to 200 strokes per minute, but there is also an increase in stroke volume, which may exceed 100 ml.

The distribution of the increased cardiac output to the muscles and to the remainder of the body in rest and exercise is shown in fig. 4.16. The circulation to the muscles of an athlete in top condition may increase as much as 25 or 30-fold, while there is a slight decrease in circulation to the remainder of the body. In the athlete the oxygen consumption of the muscles may be increased as much as 100-fold; this is possible because of an approximately threefold increase in the oxygen extraction.

Invertebrate circulation

Many invertebrates have well-developed circulatory systems, e.g. annelids, echinoderms, arthropods and molluscs. Most of these have open systems. However, cephalopods (octopus and squid) have a closed circulation although the other classes of molluscs have open systems. It is worth noting that insects, which are highly organized and complex animals and can sustain exceptionally high metabolic activity, have an open circulatory system.

Annelids. These have a well-organized circulatory system and blood that often contains respiratory pigments dissolved in the plasma. Hemoglobin is the most common pigment, but some annelids use chlorocruorin or hemerythrin.

For the earthworm, *Lumbricus,* the general body surface serves as a respiratory organ, and is supplied with a dense capillary network (see fig. 4.17). The circulatory system has two longitudinal vessels, a dorsal vessel in which the blood is pumped forward, and a ventral vessel in which it flows in the opposite direction. The nerve cord of many annelids, including the earthworm, is supplied through additional longitudinal vessels, and some of the oxygenated blood from the integument is routed directly to this important organ. The closed system of blood vessels thus has a distinctive advantage in controlling precisely where the blood will flow.

Annelids do not have one single distinct heart, although several of the blood vessels have dilations which are contractile. The dorsal blood vessel is the most important vessel, in which peristaltic waves propel blood forward. In addition, blood vessels, which on each side

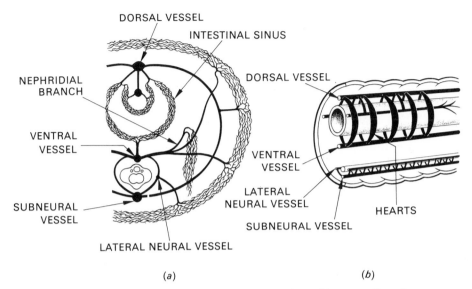

Fig. 4.17. The circulation in the earthworm. The dorsal vessel is contractile and pumps blood into the larger vessels. The integument serves in gas exchange and is heavily vascularized. Special vessels from the integument supply the ventral nerve cord with oxygen-rich blood directly from the skin (from Meglitsch, P. A. 1972. *Invertebrate Zoology*, 2nd ed. Oxford University Press, p. 395, Fig. 11.17).

connect the main dorsal with the main ventral longitudinal vessel, are contractile and serve as accessory hearts.

Echinoderms. These (e.g. starfish, sea urchins, and sea cucumbers) have three fluid systems, the *coelomic system,* the blood or *hemal system,* and the *water vascular system.* The latter functions primarily in locomotion; it is a hydraulic system used in the movement of the tube feet, and is filled with a fluid similar to sea water.

Echinoderms have a large coelom between the body wall and the digestive tract, filled with a coelomic fluid which seems to be important in transporting nutrients between the digestive tract and other parts of the body (Farmanfarmaian and Phillips, 1962).

The hemal or blood vascular system of echinoderms contains a separate fluid which, in some sea cucumbers (holothurians), contains hemoglobin. The hemoglobin from the sea cucumber *Cucumaria* has a rather high oxygen affinity, with half saturation pressures between 5 and 10 mm Hg (depending on temperature) and a striking absence of a Bohr effect between pH 7.5 and 6.5 (Manwell, 1959). These characteristics are apparently well suited to the often oxygen-deficient mud environment in which sea cucumbers live.

The relationship between the hemal system and the coelomic system is not well understood, and the role of the hemal system in respiration and oxygen supply to the animal is not clear (Martin and Johansen, 1965). The role of the water vascular system is primarily in locomotion and is better understood.

Molluscs. With the exception of octopus and squid, molluscs have open circulatory systems. The blood of many molluscs contains hemocyanin, and a few have hemoglobin. The oxygen-carrying capacity of the blood tends to be correlated with the size of the animal, and in particular with its activity.

Molluscs in general have a well-developed heart, and the heart beat is adjusted to meet the physiological demands for oxygen. An increased venous return of blood to the heart causes an increase in both amplitude and frequency of the heart beat. The mollusc heart continues to beat if it is separated from the nervous system, and thus it has an inherent rhythmicity with a pacemaker. However, the heart is also under the influence of neurosecretions which modify its beat. Acetylcholine inhibits the heart, and serotonin acts as an excitatory substance.

In clams, active movement of the foot is based on the use of blood as a hydraulic fluid. Large blood sinuses in the foot are controlled by valves which help in adjusting both the size and the movements of the foot. This is particularly important in burrowing species which use the foot in moving through the bottom material.

Cephalopods (octopus and squid) have a highly developed closed circulatory system with distinct arteries, veins, and capillary networks. This is related to the high organization and activity of these animals and the importance of the blood in respiratory exchange in the gills, as well as in the function of the kidney.

The fact that cephalopods have a closed circulatory system means that the blood remains confined within the blood vessels. Therefore the blood remains distinctly separate from the interstitial fluid. The blood makes up about 6% of the body weight, and the interstitial fluid is about 15% of the body weight. These magnitudes are strikingly similar to those of the corresponding fluid volumes in vertebrate animals. The open system of non-cephalopod molluscs does not separate blood from interstitial fluid, and blood flows freely throughout the extracellular space. In these, the blood volume is therefore quite large, and may in some make up as much as 40 or 50% of the wet body weight (without the shell) (Martin, Harrison, Huston and Stewart, 1958).

Insects. These usually have a major blood vessel running along the dorsal side. The posterior part of this vessel functions as a 'heart', and is provided with a series of valved openings through which blood can enter. The anterior part of the blood vessel, which could be called the 'aorta' is contractile and may have peristaltic waves propelling the blood forward. This main blood vessel branches and continues to the head where it ends. The branches from the aorta supply most of the body, where blood flows freely among the tissues and slowly percolates back towards the heart.

In many insects the blood that flows among the tissues is given some direction by longitudinal membranes. In the antennae and the limbs, the blood may enter on one side and leave on the other. In this way the blood is led into certain pathways, although it does not run in discrete blood vessels (see fig. 4.18).

Since the blood system is open, the blood pressure of insects can barely exceed the tissue pressure. Since insect blood plays no particular role in respiration, the main function of the blood is to carry nutrients and metabolites around in the body, and also to provide a transport system for hormones, which, in insects, are of great importance in physiological coordination (growth, molting, etc.).

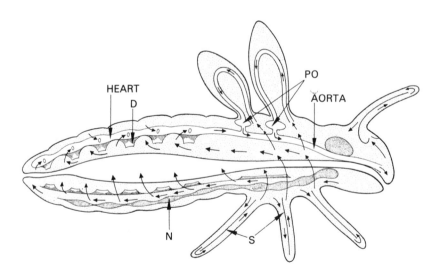

Fig. 4.18. The well-developed circulatory system in insects has a main dorsal vessel that carries blood from the heart. Although the circulation is open, the blood follows certain channels due to the presence of longitudinal membranes, especially in the legs. Arrows indicate course of circulation: D, dorsal diaphragm with muscles; N, nerve cord; PO, pulsatile organs; S, septa dividing appendages (Wigglesworth, 1972).

Although the heart and the dorsal vessel are the principal components of the circulatory system in insects, many also have accessory pumping organs or hearts. These are particularly important for maintaining circulation in the appendages, the wings, legs, and antennae. The flow of blood in the wings is aided by pulsating organs within the wings themselves, located in the channels which return the blood to the body (Thomsen, 1938). Blood flow into the antennae is aided by contractile organs located at their base. These function by aspiration of blood from blood sinuses in the head. Many insects also have leg hearts which aid in circulating blood to these important appendages. The accessory hearts of insects and their contraction are independent of the dorsal heart, they may beat at different rates, one may stop while others continue to beat, while none is dependent on the dorsal heart.

The filling of the dorsal heart of insects takes place by suction. A set of muscles, known as the aliform muscles, radiate from the heart, and when they contract, the expansion of the heart pulls blood through valved openings, the ostia, into the lumen of the heart. The subsequent contraction of the heart propels blood forward, in the direction of the head, and the valves prevent the blood from flowing out through the ostia (fig. 4.19).

Arachnids. The circulatory system of spiders and scorpions is similar to that of insects, but it may have a greater role in respiration than usual in insects. Hemocyanin has been found in the blood of some scorpions, and both scorpions and spiders have definite respiratory organs that are perfused by blood. This should not be surprising, for arachnids do not have the tracheal system which is so characteristic of insects.

The arachnid heart is located dorsally in the abdomen. It is filled through valved openings, ostia, and empties into arteries. In spiders, distinct arteries lead to the legs, where a relatively high blood pressure is of importance in locomotion. Spider legs lack extensor muscles, and blood is used as a hydraulic fluid for leg extension (see p. 535).

Crustaceans. The circulatory system of crustaceans is extremely variable. Small crustaceans have poorly developed circulatory systems, often without any heart, while large crustaceans, in particular decapods (lobster, crabs, and crayfish) have well-developed circulatory systems and blood with respiratory pigments (hemocyanin).

The crustacean circulatory system is based on the same morphological structures as in other arthropods, but it differs from insect

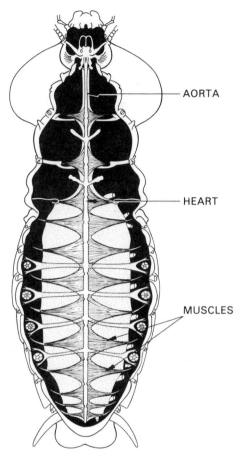

Fig. 4.19. The insect heart consists of a contractile portion of the main dorsal blood vessel. Filling of the heart with blood is achieved by suction, provided by a set of muscles which pull externally on the walls of the heart. This figure shows the dissection of a cockroach (*Blaberus*) (Nutting, 1951).

circulation in one important physiological aspect, the crustaceans have gills and therefore also a well-defined circulation to the gills.

The dorsal heart of large crustaceans lies in a pericardial sinus from which blood enters the heart through valved ostia. From the heart there is usually a main artery running in the anterior and another in the posterior direction. As the arteries branch, the blood leaves the vessels and flows between the tissues to a system of ventral sinuses. From these the blood flows into the gills, and then in discrete vessels back to the heart.

As a result of this arrangement the decapod heart is directly sup-

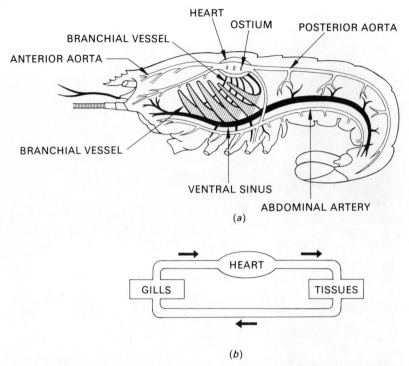

Fig. 4.20. Circulation in the lobster (*a*). In the larger crustaceans the oxygenated blood from the gills is conveyed directly to the heart, which pumps it on to the tissues (*b*). In fish the heart is located differently, before the gills (see fig. 4.3.) (from Meglitsch, P. A. 1972. *Invertebrate Zoology,* 2nd ed. Oxford University Press, p. 593, Fig. 15.38).

plied with oxygenated blood, which is then pumped to the tissues (fig. 4.20). This is the opposite arrangement of what we saw in fishes, in which the heart receives deoxygenated venous blood, which is then pumped to the gills and on to the tissues. The fish heart itself, however, is supplied with oxygenated blood through a branch of the gill circulation from which oxygenated blood reaches the heart muscle directly.

Blood coagulation and hemostasis

Several mechanisms help to prevent loss of blood from ruptured blood vessels. Severe blood loss leads to a decrease in blood pressure, and thus reduces the flow of blood from the damaged area. It also helps that damaged blood vessels constrict and thereby decrease blood flow. The most important mechanism, however, is the closing

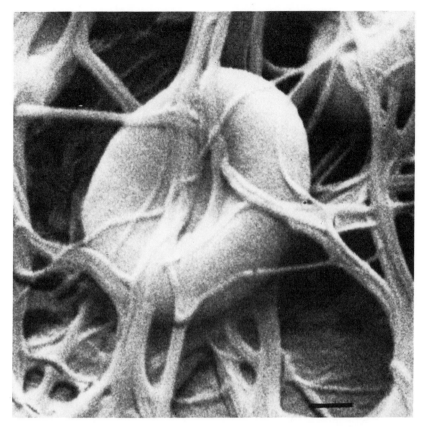

Plate 3. *Red blood cell.* A human red blood cell caught in the meshwork of fibrin in a blood clot. The cell, whose diameter is about 6 μm, has the biconcave shape characteristic of mammalian red blood cells (Emil Bernstein, the Gillette Company Research Institute; from *Science*, 1971, **173**, No. 3993, cover picture).

of blood vessels at the site of injury by a plug consisting of coagulated protein and blood cells. Such a plug or clot is important in the complete arrest of bleeding from minor injuries, but if major blood vessels have been ruptured, it will not suffice.

The clotting or *coagulation* mechanism has been well studied in mammals, and in particular in man. This is because blood clotting is of great medical importance. To be effective a clotting mechanism must act rapidly, and yet, the organism must be assured that blood does not clot within the vascular system. Blood must therefore have the inherent ability to clot, and the clotting mechanism should be ready to be turned on when needed; on the other hand, having this mechanism is like sitting on a bomb, it must not go off inadvertently.

In vertebrates the blood clot consists of the protein *fibrin,* an insoluble fibrous protein which is formed from *fibrinogen,* a soluble protein present in normal plasma in an amount of about 0.3%. The transformation of fibrinogen to fibrin is catalyzed by the enzyme *thrombin,* and the reason that blood does not clot in the vascular system is that thrombin is absent from the circulating blood. Thrombin, however, can be formed rapidly because its precursor, prothrombin, is already present in the plasma. What is necessary to initiate coagulation is therefore the formation of thrombin from prothrombin. This is only the final step in a complex sequence of biochemical events which has been slowly unravelled in studies of human patients with various deficiencies in the clotting mechanism (e.g. hemophilia). A total of 12 clotting factors have been identified, numbered I through XIII (factor VI is a term which is not used any more). Of these, only a few of the final steps of the clotting mechanism are shown in the diagram below.

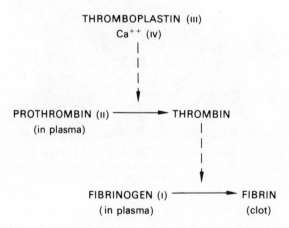

The many steps in the clotting mechanism may seem an unnecessary complication, and it has indeed been a severe hindrance in the clarification of what takes place. Biologically, the importance of the many steps seems to be explained by the fact that the clotting mechanism works as a biochemical amplifier (Macfarlane, 1964). Clotting is normally initiated when blood contacts 'foreign' surfaces or damaged tissues. This initiates a series of enzymatic steps in which the enzyme formed in the first step serves as a catalyst or activator for the next step, and so on. The series of steps thus forms an 'enzyme cascade', which ends up with the final formation of the clot in which the soluble fibrinogen is changed to the insoluble fibrin.

The enzymatic amplification has the effect that clotting can take place rapidly, and yet, there is a considerable safety factor to prevent spontaneous coagulation within the vascular system. The analogy to an electronic amplification system is obvious. If we want an amplification system with a low noise level, we use several steps with low amplification in each step, rather than a single step with a high amplification. This minimizes the chances that random noise in the system may set off the final step, and thus insures an adequate safety margin.

For most invertebrates it is as essential to have hemostatic mechanisms as it is for vertebrates. The fact that many have an open circulatory system makes the situation more difficult, for in these the contraction of blood vessels is of no help. On the other hand, open systems always have lower blood pressures, and this decreases the tendency for loss of large volumes of fluid.

The two distinct hemostatic mechanisms of vertebrates, clotting of blood and local vasoconstriction, have their counterparts in invertebrates. The simplest invertebrate mechanism is the agglutination of blood corpuscles without participation of plasma proteins (Grégoire and Tagnon, 1962). Such agglutination is followed by the formation of a cellular mesh work which shrinks and helps close a wound. This is often helped by the contraction of the muscles of the body wall, thus aiding in wound closure.

Among invertebrates, a true clotting, induced by enzymatic changes of unstable blood proteins, has been described for many arthropods, especially crustaceans. The clotting mechanism of invertebrates, where it occurs, is biochemically distinctly different from the vertebrate mechanism. For example, vertebrate clotting is inhibited by heparin, a mucopolysaccharide which can be isolated from mammalian liver. Heparin has no effect on the clotting system of the horseshoe crab *Limulus* (an arachnid, not a crab), and virtually no effect on crustacean blood (Needham, 1970).

Knowledge of the clotting mechanisms of invertebrates is very incomplete, but present information indicates that such mechanisms must have evolved independently many times in the course of evolution.

REFERENCES

ALTMAN, P. L. and DITTMER, D. S. (1971). *Respiration and Circulation,* pp. 251–3, Bethesda, Maryland: Federation of American Societies for Experimental Biology.

BOND, C. F. and GILBERT, P. W. (1958). Comparative study of blood volume in representative aquatic and nonaquatic birds. *Am. J. Physiol.*, **194**, 519–21.

BURTON, A. C. (1965). *Physiology and Biophysics of the Circulation. An Introductory Text*, Chicago: Year Book Medical Publ., Inc., 217 pp.

FARMANFARMAIAN, A. and PHILLIPS, J. H. (1962). Digestion, storage, and translocation of nutrients in the purple sea urchin (*Strongylocentrotus purpuratus*). *Biol. Bull.*, **123**, 105–20.

FOLKOW, B. and NEIL, E. (1971). *Circulation*, New York: Oxford University Press, 593 pp.

FOXON, G. E. H. (1955). Problems of the double circulation in vertebrates. *Biol. Rev.*, **30**, 196–228.

GORDON, M. S. (1972). *Animal Physiology: Principles and Adaptations*, 2nd edn, New York: Macmillan Co., 592 pp.

GRÉGOIRE, C. and TAGNON, H. J. (1962). Blood coagulation. *Comparative Biochemistry*, **4**, part *B*, pp. 435—82, London: Academic Press.

HOPPER, J., JR., TABOR, H. and WINKLER, A. W. (1944). Simultaneous measurements of the blood volume in man and dog by means of Evans blue dye, T1824, and by means of carbon monoxide. I. Normal subjects. *J. Clin. Invest.*, **23**, 628–35.

JOHANSEN, K. (1960). Circulation in the hagfish, *Myxine glutinosa* L. *Biol. Bull.*, **118**, 289–95.

JOHANSEN, K. (1962). Double circulation in the amphibian *Amphiuma tridactylum. Nature, Lond.*, **194**, 991–2.

JOHANSEN, K. (1968). Air-breathing fishes. *Sci. Am.*, **219**, 102–11.

JOHANSEN, K. (1970). Air breathing in fishes. *Fish Physiology*, **4** (W. S. Hoar and D. J. Randall, eds), pp. 361–411. New York: Academic Press.

JOHANSEN, K., LENFANT, C. and HANSON, D. (1968). Cardiovascular dynamics in the lungfishes. *Zeits. vergl. Physiol.*, **59**, 157–86.

JOHANSEN, K. and MARTIN, A. W. (1965). Comparative aspects of cardiovascular function in vertebrates. *Handbook of Physiology*, sect. 2: *Circulation*, vol. III, 2583–641.

KRIJGSMAN, B. J. (1956). Contractile and pacemaker mechanisms of the heart of tunicates. *Biol. Rev.*, **31**, 288–312.

LASIEWSKI, R. C. and CALDER, W. A., JR. (1971). A preliminary allometric analysis of respiration variables in resting birds. *Respir. Physiol.*, **11**, 152–66.

LASIEWSKI, R. C., WEATHERS, W. W. and BERNSTEIN, M. H. (1967). Physiological responses of the giant hummingbird, *Patagona gigas. Comp. Biochem. Physiol.*, **23**, 797–813.

MACFARLANE, R. G. (1964). An enzyme cascade in the blood clotting mechanism, and its function as a biochemical amplifier. *Nature, Lond.*, **202**, 498–9.

MACFARLANE, R. G. (ed.) (1970). *The Haemostatic Mechanism in Man and Other Animals. Zool. Soc. Lond. Symp.*, **27**, London: Academic Press, 248 pp.

MANWELL, C. (1959). Oxygen equilibrium of *Cucumaria miniata* hemoglobin and the absence of the Bohr effect. *J. Cell. Comp. Physiol.*, **53**, 75–84.

MARTIN, A. W., HARRISON, F. M., HUSTON, M. J. and STEWART, D. M. (1958). The blood volumes of some representative molluscs. *J. Exp. Biol.*, **35**, 260–79.

MARTIN, A. W. and JOHANSEN, K. (1965). Adaptations of the circulation in invertebrate animals. *Handbook of Physiology*, sect. 2, *Circulation*, vol. III, 2545–81.

MEGLITSCH, P. A. (1972). *Invertebrate Zoology*, 2nd edn. New York: Oxford University Press, 834 pp.

MELBIN, J. and NOORDERGRAAF, A. (1971). Elastic deformation in orthotropic vessels. Theoretical and experimental results. *Circulation Res.*, **29**, 680–92.

MORRISON, P., RYSER, F. A. and DAWE, A. A. (1959). Studies on the physiology of the masked shrew *Sorex cinereus*. *Physiol. Zool.*, **32**, 256–71.

NEEDHAM, A. E. (1970). Haemostatic mechanisms in the invertebrata. *Symposia of the Zoological Society of London*, no. **27** (R. G. MacFarlane, ed.), pp. 19–44. London: Academic Press.

NUTTING, W. L. (1951). A comparative anatomical study of the heart and accessory structures of the orthopteroid insects. *J. Morphol.*, **89**, 501–97.

RANDALL, D. J. (1970). The circulatory system. *Fish Physiology*, **4** (W. S. Hoar and D. J. Randall, eds), pp. 133–72. New York: Academic Press.

SATCHELL, G. H. (1970). A functional appraisal of the fish heart. *Fed. Proc.*, **29**, 1120–3.

SCHMIDT-NIELSEN, K. (1970). *Animal Physiology*, 3rd edn, Englewood Cliffs, N.J.: Prentice-Hall, Inc., 145 pp.

SKALAK, R. and BRANEMARK, P. I. (1969). Deformation of red blood cells in capillaries. *Science*, **164**, 717–19.

STAHL, W. R. (1965). Organ weights in primates and other mammals. *Science*, **150**, 1039–42.

STAHL, W. R. (1967). Scaling of respiratory variables in mammals. *J. Appl. Physiol.*, **22**, 453–60.

STARLING, E. H. (1896). On the absorption of fluids from the connective tissue spaces. *J. Physiol., Lond.*, **19**, 312–26.

STUDIER, E. H. and HOWELL, D. J. (1969). Heart rate of female big brown bats in flight. *J. Mammal*, **50**, 842–5.

THOMSEN, E. (1938). Über den Kreislauf im Flügel der Musciden, mit Besonderer Berücksichtigung der Akzessorischen Pulsierenden Organe. *Zeits. Morphol. Oekol. Tiere*, **34**, 416–38.

THORSON, T. B. (1958). Measurement of the fluid compartments of four species of marine Chondrichthyes. *Physiol. Zool.*, **31**, 16–23.

THORSON, T. B. (1959). Partitioning of body water in sea lamprey. *Science*, **130**, 99–100.

THORSON, T. B. (1961). The partitioning of body water in Osteichthyes: phylogenic and ecological implications in aquatic vertebrates. *Biol. Bull.*, **120**, 238–54.

THORSON, T. B. (1964). The partitioning of body water in Amphibia. *Physiol. Zool.*, **37**, 395–9.

THORSON, T. B. (1968). Body fluid partitioning in Reptilia. *Copeia, 1968*, no. 3, 592–601.

VAN CITTERS, R. L., KEMPER, W. S. and FRANKLIN, D. L. (1968). Blood flow and pressure in the giraffe carotid artery. *Comp. Biochem. Physiol.,* **24,** 1035–42.

WHITE, F. N. (1956). Circulation in the reptilian heart (*Caiman sclerops*). *Anat. Rec.,* **125,** 417–31.

WHITE, F. N. (1959). Circulation in the reptilian heart (*Squamata*). *Anat. Rec.,* **135,** 129–34.

WIGGLESWORTH, V. B. (1972). *The Principles of Insect Physiology,* 7th edn, London: Chapman and Hall, 827 pp.

USEFUL REFERENCE MATERIAL FOR PART I

ALTMAN, P. L. and DITTMER, D. S. (eds) (1971). *Biological Handbooks: Respiration and Circulation,* Bethesda, Md.: Federation of American Societies for Experimental Biology, 930 pp.

ALTMAN, P. L. and DITTMER. D. S., (eds) (1971). *Biological Handbooks: Blood and Other Body Fluids,* 3rd printing with minor corrections, Bethesda, Md.: Federation of American Societies for Experimental Biology, 540 pp.

BURTON, A. C. (1965). *Physiology and Biophysics of the Circulation. An Introuctory Text,* Chicago, Ill.: Year Book Medical Publ., Inc., 217 pp.

COMROE, J. H., JR. (1965). *Physiology of Respiration. An Introductory Text,* Chicago, Ill.: Year Book Medical Publ. Inc., 245 pp.

FENN, W. O. and RAHN, H., (eds). *Handbook of Physiology,* sect. 3, *Respiration,* vol. I, 1–926 (1964); vol. II, 927–1696 (1965).

FOLKOW, B. and NEIL, E. (1971). *Circulation,* New York: Oxford University Press, 593 pp.

HAMILTON, W. F. and DOW, P. (eds) *Handbook of Physiology.* sect. 2: *Circulation,* vol. I, 1–758 (1962); vol. II, 759–1786 (1963); vol. III, 1787–2765 (1965). Washington, D.C.: American Physiological Society.

HEATH, A. G. and MANGUM, C. (eds) (1973). Influence of the environment on respiratory function. *Amer. Zool.* **13,** 446–563.

JOHANSEN, K. (1971). Comparative physiology: gas exchange and circulation in fishes. *Ann. Rev. Physiol.,* **33,** 569–612.

JOHANSEN, K. *et al.* (1970). Symposium on cardiorespiratory adaptations in the transition from water breathing to air breathing. *Fed. Proc.,* **29**(3), 1118–53.

JONES, J. D. (1972). *Comparative Physiology of Respiration,* London: Edward Arnold Publ. Ltd, 202 pp.

KROGH, A. (1939). *The Comparative Physiology of Respiratory Mechanisms*, Philadelphia, Pa.: Univ. of Pennsylvania Press; reprinted by Dover Publ. Inc., New York, 1968, 172 pp.

MACFARLANE, R. C. (1970). The Haemostatic Mechanism in Man and Other Animals. *Zool. Soc. Lond. Symp.* **27**, London: Academic Press, 248 pp.

MARTIN, A. W. (1974). Circulation in invertebrates. *Ann. Rev. Physiol.*, **36**, 171–86.

MILLER, P. L. (1964). Respiration – aerial gas transport. In *The Physiology of Insecta*, vol. III (M. Rockstein, ed.), pp. 557–615, New York: Academic Press.

RIGGS, A. (1965). Functional properties of hemoglobins. *Physiol. Rev*, **45**, 618–73.

RØRTH, M. and ASTRUP, P., (eds) (1972). Oxygen Affinity of Hemoglobin and Red Cell Acid Base Status. *Proc. of the Alfred Benzon Symp.* vol. IV, Copenhagen, 1971, Copenhagen: Munksgaard, 832 pp.

TAYLOR, M. G. (1973). Hemodynamics. *Ann. Rev. Physiol.*, **35**, 87–116.

WEIBEL, E. R. (1973). Morphological basis of alveolar-capillary gas exchange. *Physiol. Rev.*, **53**, 419–95.

The following books are useful sources for references in all fields of animal physiology:

DILL, D. B., ADOLPH, E. F. and WILBER, C. G., (eds) (1964). *Handbook of Physiology, sect. 4, Adaptation to the Environment*, Washington, D.C.: American Physiological Society, 1056 pp.

PROSSER, C. L. (1973). *Comparative Animal Physiology*, 3rd edn, Philadelphia, Pa.: W. B. Saunders Co., 966 pp.

FOOD AND ENERGY

5
Food, fuel, and energy

Animals need food (1) for energy to keep alive and carry on their functions, and (2) for building and maintaining their cellular and metabolic machinery.

Most plants use the energy of sunlight and carbon dioxide from the atmosphere to synthesize sugars and, indirectly, all the complicated compounds that constitute a plant. All animals use chemical compounds to supply energy as well as building materials; they must obtain these from outside sources, either directly from plants by eating them, or by eating other organic material. Ultimately the chemical energy and organic compounds that animals need are derived from plants, and thus indirectly from the radiation energy of sunlight.

The acquisition and ingestion of food is referred to as *feeding*. Virtually all food, whether of plant or animal origin, consists of highly complex compounds which cannot be incorporated by the organism nor used for fuel without being broken down to simpler compounds. In fact, they usually cannot even be taken into or absorbed by the organism without such breakdown, called *digestion*.

A variety of chemical compounds can be used to provide energy, but in addition animals have very specific needs for compounds that they cannot synthesize, amino acids, vitamins, certain salts, etc. Both the need for food to provide energy and the need for specific food components belong to the subject of *nutrition*.

In this chapter we shall discuss the subjects which are related to food and food intake, beginning with the mechanism of feeding, then proceed to the processes of digestion which make the food compounds available to the organism, and finally discuss the needs with regard to the quality of the food and specific nutritional requirements.

FEEDING

Food is obtained by a diversity of mechanical means, and these very much determine the nature of the food a given animal can obtain and utilize. A listing of major feeding mechanisms and examples of animals that use the different means to obtain food is given in table 5.1.

Table 5.1. *Methods used by various animals to obtain food can be classified into a few major groups, depending mainly on the nature or character of the food*

i. Methods for dealing with small particles:
   (1) Formation of digestive vacuoles (amoeba, radiolarians)
   (2) Use of cilia (ciliates, sponges, bivalves, tadpoles)
   (3) Formation of mucus traps (gastropods, tunicates)
   (4) Use of tentacles (sea cucumbers)
   (5) Use of setae, filtering (small crustaceans, e.g. daphnia, herring, baleen whales, flamingo, petrel)
ii. Methods for dealing with large particles or masses:
   (1) Ingestion of inactive masses (detritus feeders, earthworm)
   (2) Scraping, chewing, boring (sea urchins, snails, insects, vertebrates)
   (3) Capture and swallowing of prey (coelenterates, fish, snakes, birds, bats)
iii. Methods for taking fluids or soft tissues:
   (1) Sucking plant sap, nectar (aphids, bees, hummingbirds)
   (2) Ingestion of blood (leeches, ticks, insects, vampire bats)
   (3) Sucking milk (young mammals)
   (4) External digestion (spiders)
   (5) Uptake through body surface (parasites, tapeworm)
   (6) Uptake from dilute solutions
iv. Symbiotic supply of nutrients:
   Intracellular symbiotic algae (paramecium, sponges, corals, hydra, flatworms, clams)

Small particles

Microscopic algae and bacteria can be taken directly into a cell, e.g. into the digestive vacuole of an amoeba. A variety of organisms, even some quite large ones, catch microscopic organisms with the aid of cilia. A few animals, notably tunicates and some gastropods, use a sheet of mucus that traps fine suspended particles; the mucus sheet is then ingested and the food organisms digested.

Tentacular feeding is used by sea cucumbers which live burrowed in the mud with their tentacles extended above the mud surface. They entangle fine particles, and at intervals retract the tentacles into the mouth where digestible material is removed.

Various filtering devices are common in a wide variety of small planktonic crustaceans such as copepods, and in amphipods, sponges, bivalves, etc. Quite a few vertebrates, ranging from fish to mammals, also utilize filtering to obtain their food, but mostly their filters strain off organisms larger than the microscopic-size food commonly utilized by invertebrates.

Many pelagic fish are plankton eaters. Herring and mackerel have the gill rakers structured so that they function as a sieve which catches plankton, predominantly small crustaceans. Some of the largest sharks, the basking sharks and whale sharks, feed exclusively on plankton which is strained from the water which enters the mouth and flows over the gills. It has been estimated that a large basking shark, in one hour, strains the plankton from 2000 tons of water (Matthews and Parker, 1950).

A few marine birds have also specialized on plankton feeding. One of the petrels, the whale bird or prion, has a series of lamellae extending along the edge of the upper beak. The whale bird feeds by straining crustaceans from the surface water in a way similar to what baleen whales do on a much larger scale. The flamingo, also, is a plankton eater; it has a beak which is used to strain small organisms from the water.

The baleen whales are entirely specialized for feeding on plankton. Their filtering apparatus consists of a series of horny plates attached to the upper jaw and hanging down on both sides. As the whale swims, water flows over and between the plates, and plankton is caught in the hair-like edges of the plates. The plankton-eating whales include the largest living animal, the blue whale, which may reach a weight of over 100 tons. It is interesting that both the largest whales and the largest existing sharks are plankton feeders rather than carnivores, and thus avoid any extra links in the food chain.

Large food masses

The methods for dealing with large particles or masses of food include a tremendous variety of mechanisms and structures. Some animals ingest the medium in which they live and digest the organic material it contains, whether small living organisms or dead organic matter. Such animals therefore frequently pass large amounts of relatively inert material through their digestive tracts and utilize whatever organic material it contains.

A broad range of animals use various mechanical methods, chewing, scraping, and so on, to obtain their food, which is frequently of plant origin. This includes such a variety that to list them would be pointless, we can only think of the variety of insects and other invertebrates, as well as the many vertebrates that are herbivores. Some carnivores are also included in this group, they capture prey and tear, shred, or chew it before swallowing.

Some carnivores capture and swallow their prey whole. Relatively few invertebrates feed this way, but a large number of vertebrates do, including many representatives of all classes, fish, amphibians, reptiles, birds, and mammals.

The mouth and oral cavity may have various kinds of jaws or teeth that serve in the mechanical breakdown of food into smaller particles, *mastication*, thus making it more accessible to the digestive enzymes.

In some animals the mechanical breakdown of food takes place in a special compartment with thick muscular walls, the *gizzard*, where a grinding action takes place. An earthworm, for example, has a gizzard for such grinding, called *trituration*. Birds, which have no teeth, also have a gizzard and they often swallow small stones that help in the trituration of the food. This is particularly important in seed-eating birds, for their food is hard and would, in the absence of grinding, take a long time to soften in the digestive tract. The gizzard may be preceded by a simple storage chamber, a *crop*, in which the moisturizing or softening begins, but without any important grinding action.

At the beginning of the digestive tract we usually find a special larger compartment, a *stomach*, which serves as a storage organ, and in which digestive action also takes place. In all vertebrates, with a few exceptions, the stomach has a major role in the digestion of proteins, and the most important enzyme in the stomach is the protein-digesting pepsin, which acts best in highly acidic solution. The stomach of vertebrates therefore also has special cells that secrete acid (hydrochloric acid).

The stomach is followed by a hollow tube, the intestine, in which further digestion takes place, and which is particularly important for the absorption of the products of digestion. These products are mainly simple sugars, glycerol and fatty acids, and amino acids. The length of the intestine varies a great deal, and it can usually be subdivided into several parts. The posterior portion is of little importance in digestion and serves mainly in absorption. Periodically the remaining material is expelled as feces through the anus.

Ingestion of fluids

A wide variety of liquids are of value as foods. Many of the animals feeding on liquids are highly specialized and adapted to their food source, and in some cases there is a mutual benefit to the feeding animal and the provider. For example, insects obtain nectar from flowers and in turn provide pollination by moving from flower to flower. In many cases, fluid-feeders are parasitic; this is true of aphids which suck plant sap, and of animals that suck the blood of other animals, such as leeches, mosquitoes, ticks, etc.

Incidentally, all mammals begin their lives as fluid-feeders when for a period they live exclusively on the milk produced by their mother.

Spiders provide a special case of fluid-feeding. Their prey is often as large or larger than their own body, usually covered with hard chitin, and not easily torn apart. Spiders have solved the problem of obtaining food from their insect prey by piercing the prey with their hollow jaws and pumping digestive juices into their body. After the tissues have been liquefied the spiders then suck the prey empty.

A number of parasites, e.g. tapeworms, live in a highly nutritive medium, and they have completely lost the digestive tract. Since their hosts have already provided the digestive system, they lack both a digestive tract and the digestive enzymes that are otherwise necessary before food can be absorbed, and all their nutrients are taken up directly through the body surface.

Whether organic matter dissolved in the water can be utilized by aquatic animals is a question that still remains controversial. It was suggested early in this century by the German biologist Pütter that dissolved organic matter in the sea, although present in a highly dilute state, could be used directly by aquatic animals. Pütter even suggested that such dissolved organic matter was more important than plankton, for plankton organisms are scattered so far apart in the water that often their capture is not feasible.

Whether the uptake of dissolved organic material is important has been very difficult to study experimentally because of the extremely dilute solutions to be examined. With the introduction of radioisotopes to label organic compounds this changed, and it became possible to demonstrate clearly that many aquatic invertebrates indeed can absorb both glucose and amino acids from exceedingly dilute solutions (Stephens, 1962; 1964). In such experiments it is not difficult to show that uptake has taken place, this is beyond doubt; the difficulty is to show that the amount of a compound taken up exceeds the amount given off in the same period of time. The uptake must exceed the loss in order to be of any net value to the animal. A comparison of simultaneous uptake and release of free amino acids in sea water has been made by Johannes, Coward and Webb (1969). In experiments with one single species, a turbellarian, they showed that the release of free amino acids to the water exceeded by about four-fold the simultaneous uptake of ^{14}C-labelled amino acid. Another question that remains open is whether bacteria may play a role in the uptake, the bacteria could absorb organic solutes and later be ingested by the animal. Bacteria can effectively remove glucose or acetate from solutions containing no more than 1 to 10 μg per liter, while for common fresh-water algae this is impossible (Wright and Hobbie, 1966).

Although present evidence is uncertain, there is no strong indication that any aquatic invertebrate depends primarily on the uptake of organic materials from the highly dilute solutions in which they live. The elaborate mechanical feeding mechanisms which animals possess also speak against any major role of dilute dissolved organic substances.

Symbiotic supply of nutrients

A number of invertebrate animals contain within their cells symbiotic algae. It has been suggested that these algae are of importance in supplying nutrients to the host, thus providing a shorter pathway between photosynthesis and animal than possible in any other way. Such a symbiotic relationship between animal and plant occurs in protozoans, sponges, corals, hydra, flatworms, and clams.

In this symbiotic relationship the animal obtains organic compounds (primarily carbohydrates) from the living cells of the plant, and not from the destruction of the plant cells and their digestion. In this respect the symbiotic supply of nutrients differs in principle from all the feeding methods discussed above, which always involve destruction of part or the whole of the organism which provides the food.

Two major types of algae occur in the host animals, either green algae (Zoochlorellae) or brown algae (Zooxanthellae) (see table 5.2). If we include in the term 'carbohydrate' not only sugars, but also derivatives such as sugar alcohols and glycerol, we can say that the bulk of material transferred is of carbohydrate nature. The soluble carbohydrate given off to the host animal is invariably different from the major intracellular carbohydrate of the algae. Zoochlorellae, for example, synthesize and retain in their cells sucrose, but they excrete maltose or glucose to the host animal.

The amount of organic material obtained in a symbiotic relationship is substantial. The algae found in green hydra release as much as 45 to 50% of the carbon fixed in photosynthesis to the host animal. Isolated algae will release carbohydrate, mainly maltose, to the medium, and the amount depends on the pH of the solution. At

Table 5.2. *Release of carbohydrate from symbiotic algae to host animal*
(Smith, Muscatine and Lewis, 1969)

TYPE OF ALGAE	HOST ANIMALS	CARBOHYDRATE RELEASED TO HOST
Zoochlorellae	(*a*) Protozoa, *Paramecium bursaria*	maltose
	(*b*) Porifera, *Spongilla lacustris*	glucose
	(*c*) Coelenterata	
	Chlorohydra viridissima (wild type)	maltose
	C. viridissima (mutant)	glucose
	(*d*) Platyhelminthes	unknown
	Convoluta roscoffensis	
	(*e*) Mollusca	
	Placobranchus ianthobapsus	unknown
	Tridacna crispata	
Zooxanthellae	(*a*) Coelenterata, *Pocillopora damicornis,*	glycerol
	Anthopleura elegantissima,	glycerol
	Zoanthus confertus,	glycerol
	Fungia scutaria	glycerol
	(*b*) Mollusca, *Tridacna crocea*	glycerol

pH 4.5 these algae release as much as 85% of their photosynthesized carbon, but in alkaline medium the release declines to a few per cent. This opens the possibility that the host animal may be able to control the amount of carbohydrate release by the algae by adjusting its intracellular pH.

When the zooxanthellae are isolated from their host and incubated in sea water they release some carbohydrate, but the amount increases if homogenized host tissue is added. In the case of zooxanthellae from *Tridacna* (the giant clam) a 16-fold increase in carbohydrate release was obtained in the presence of homogenate of the host. This is not a universal phenomenon, however, because the zoochlorellae from hydra are unaffected by homogenate of their own host animal (Smith *et al.*, 1969).

DIGESTION

Animal food is made up of organic material, which mostly belongs to three major groups, proteins, fats, and carbohydrates. These three kinds of organic compounds completely dominate the make-up of virtually all plants and animals. Most of them have fairly large molecules. Even the smallest (sugars) have molecular weights of a few hundred, fats are larger again, many proteins are over 100 000 to several million in size, and starch and cellulose are polymers, of indeterminate size, of small carbohydrate units. Whether the food is used for fuel or for building and maintenance, the large molecules of the food are first broken down into simpler units, these are then absorbed, and either incorporated into the body or metabolized to provide energy.

The main function of digestion is to break up the large and complex molecules in the food so that they become accessible to absorption and available for use in the body. This breakdown is achieved in the digestive tract with the aid of enzymes.

Intracellular and extracellular digestion

Digestion can take place within cells or outside. In unicellular animals digestion must, of necessity, be inside the cell. A protozoan takes food into the digestive vacuole, and enzymes that aid in the digestion of carbohydrates, fats, and proteins are secreted into the vacuole. Similar intracellular digestion occurs in sponges, and to some extent in coelenterates, ctenophores, and turbellarians. It is found also in a number of more complex animals in conjunction with extracellular digestion. For example, clams often take small food particles into the cells of the digestive gland, where they are digested intracellularly. Some animals that feed on larger pieces of food, such as coelenterates, have partly extracellular and partly intracellular digestion. Digestion begins in their digestive cavity, the *coelenteron;* and fragments of partly digested food are then taken into the cells of the cavity wall where they are digested completely.

Extracellular digestion has one obvious advantage, it permits the ingestion of large pieces of food. Intracellular digestion is limited to particles so small that they can be taken into the individual cells of the organism.

Extracellular digestion is usually associated with a well-developed digestive tract which permits secreted enzymes to act on the food material. The digestive tract may have one opening, as in coelenterates, brittlestars, and flatworms. In these animals, any undigested material is expelled through the same opening that served as a mouth. In more complex animals the digestive tract has two openings, a mouth and an anus. This permits an assembly-line type of digestion. Food is ingested through the mouth and is passed on and acted upon by a series of digestive enzymes, the soluble products of digestion are absorbed, and in the end undigested material is expelled through the anus without interference with food intake. In this way food intake can continue while digestion takes place, and the passage of food through the digestive tract can continue uninterrupted.

All coelenterates are carnivorous, and they have capture devices, tentacles, which help them catch and paralyze their prey. The tentacles have specialized cells, *nematocysts,* which, when suitable prey comes in contact with them, fire a tiny hollow spine which penetrates into the prey, a poison is ejected from the cell through the spine, and this paralyzes the prey. The tentacles draw the prey into the coelenteron for digestion.

Flatworms have a mouth which leads into a gastrovascular cavity which is widely branched and ramifies throughout the body. This cavity, because of its distribution within the body, not only serves in digestion, but at the same time can transport food to all parts of the body. The extensive branching also increases the total surface of the gastrovascular cavity, thus aiding in the absorption of digested food. In the planaria extracellular digestion aids in breaking down the food, but most of the food particles are taken up by the cells that line the cavity and are digested intracellularly.

Enzymatic digestion

Most food compounds are very large (proteins), highly insoluble in water (fats), or large as well as insoluble (starch, cellulose). Before the elements of the food can be absorbed and utilized, they must be brought into soluble form and broken down into smaller component units.

Proteins, starches, and cellulose are polymers formed from simpler building blocks by removal of water, and fats are esters of fatty acids and glycerol, also formed by removal of water. The breakdown of these food compounds into simpler components involves the uptake of water, and is therefore called *hydrolysis.* Hydrolysis is a spontaneous reaction, and energy in the form of heat is released in the process. Spontaneous hydrolysis of food compounds, however, proceeds at an immeasurably slow rate, but the process can be speeded up by catalysts. Catalysts produced by living organisms are called *enzymes,* and enzymes are essential in all digestive processes. Virtually all metabolic processes inside the cells are also catalyzed by enzymes.

All enzymes are proteins, many have been isolated in pure form and their action has been studied in great detail. Many enzymes have been named according to the substrate on which they act, with the attachment of the ending -ase. In the field of digestion the enzymes frequently act on a broad group of substrates, and many of them are

best known under common or trivial names although they are given
more precise names according to internationally accepted nomenclature. For the simple reason that the digestive enzymes from only a
few organisms are known in sufficient detail, we shall discuss most of
the digestive enzymes under their common names, rather than at-
tempting to adhere to the technical nomenclature.

After being described by the substrate it acts on, or by the reaction
it catalyzes, the most important measure of an enzyme is its activity,
that is, the rate with which the catalyzed reaction proceeds. Many
factors influence the activity, and if the protein nature of the enzyme
is changed, it usually loses its activity. High temperature, strong
chemicals, other enzymes which split protein, and heavy metals that
bind to the protein, may all inactivate the enzymatic activity.

The activity of all enzymes is greatly influenced by the pH of the
solution. Most enzymes show their highest activity in a narrow range
of pH, called the optimum pH for the enzyme in question. Two
well-known protein-digesting enzymes are *pepsin*, found in the vertebrate stomach, and *trypsin* which is produced in the vertebrate
pancreas. Pepsin has a pH optimum in highly acidic solutions, about
pH 2, and trypsin has its optimum in slightly alkaline solutions,
about pH 8. The optimum pH for a given enzyme is not necessarily
identical for all substrates on which it acts. Pepsin, for example,
shows an optimum pH of 1.5 for digestion of egg albumin, 1.8 for
casein, and 2.2 for hemoglobin digestion (White, Handler and
Smith, 1964). On both sides of the optimum pH the activity is decreased, and at a very different pH the activity is nil. Pepsin will still
have some activity at pH = 4, and will again resume full activity if the
pH is returned to the optimum. At pH = 8, however, pepsin is destroyed and permanently inactivated.

The rate of enzymatic reactions is greatly influenced by the temperature. An increased temperature speeds up the rate. Since heat
usually makes proteins coagulate, it also inactivates enzymes, and at
temperatures above 50 °C or so, most enzymatic action is completely
destroyed. Therefore, a moderate increase in temperature gives an
increased rate of reaction, but when the temperature increases further, thermal destruction of the enzyme catches up with the accelerated reaction rate. At some temperature we will observe a maximum
rate, but since the thermal destruction increases with time, we will
find that the optimum temperature depends on the duration of the
experiment. In a long-lasting experiment the temperature effect on
the enzyme will have proceeded further, and the observed temperature optimum will therefore be lower, and in a short-lasting experiment with the same enzyme the temperature optimum will be

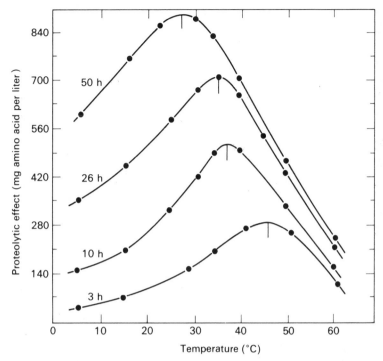

Fig. 5.1. The effect of temperature on a protein-splitting enzyme (protease) from the ascidian *Halocynthia*. The enzyme effect seems to have a temperature 'optimum', but this 'optimum' is lower the longer the duration of the experiment. This is explained in the text (Berrill, 1929).

higher. Thus the temperature optimum is not a specific characteristic of an enzyme, it depends on the duration of an experiment (fig. 5.1).

Most enzymes are rapidly inactivated by temperatures above 45 or 50 °C. This is the temperature range that for most complex organisms forms the upper limit for life, except for some resting stages and certain thermophilic bacteria. A few enzymes, however, are heat resistant, for example, the protein-splitting enzyme papain from the papaya fruit and a similar enzyme from pineapple. These enzymes are used as meat tenderizers because their proteolytic action continues at cooking temperatures as high as 70 or 80 °C, although at boiling temperature they are rapidly inactivated.

Protein digestion

Proteins are polymers consisting of amino acids bound together by peptide bonds. An amino acid is a relatively simple organic acid which on the carbon atom next to the acid group (or carboxyl group, -COOH), carries an amino group (-NH₂). A peptide bond is formed by removal of water between a carboxyl and an amino group. There are about 20 common amino acids which in various proportions occur universally in virtually all living organisms. This means that when proteins are hydrolyzed into amino acids, these amino acids are the building blocks required for synthesis of nearly any specific protein.

The enzymes that digest proteins are divided into two groups, according to where on the protein molecule they act. If they hydrolyze a terminal peptide bond in a long peptide chain, they are called *exopeptidases*, while the *endopeptidases* act in the interior of the peptide chain. The two best known endopeptidases in vertebrate digestion are pepsin and trypsin.

Pepsin is secreted by the stomach as an inactive precursor, *pepsinogen*. The stomach also secretes hydrochloric acid, which gives its contents a low pH. In an acid medium, below pH 6, pepsinogen is activated autocatalytically to form the active enzyme, pepsin. Pepsin specifically hydrolyzes peptide bonds between an amino acid which carries a phenyl group (tyrosin or phenylalanine) and a dicarboxylic acid (glutamic or aspartic acid). Pepsin therefore attacks only a few of the peptide bonds in a large protein and not others, and the result of pepsin digestion is a series of shorter chains or fragments which are not further broken down by pepsin.

Trypsin is secreted by the pancreas in an inactive form, *trypsinogen*. It is activated in the intestine by the enzyme enterokinase, which is secreted by glands in the intestinal wall. Trypsinogen is also activated by active trypsin, i.e. trypsinogen is activated more and more rapidly as more trypsin is formed. This is known as *autocatalytic activation*.

Trypsin acts best at slightly alkaline reaction, between pH 7 and 9. It hydrolyzes peptide bonds that are adjacent to a basic amino acid that carries two amino groups (lysine or arginine) (fig. 5.2).

The protein fragments and peptide chains formed by the action of pepsin and trypsin are further digested with the aid of exopeptidases, which act only on terminal peptide bonds. Carboxypeptidase, secreted by the pancreas, hydrolyzes terminal peptide bonds next to a free carboxyl group, and aminopeptidase, secreted by the intestine, hydrolyzes peptide bonds next to a free amino group. Finally, dipeptidases, also secreted by the intestine, hydrolyze the peptide bond of fragments that consist of only two amino acids.

Protein digestion in invertebrates, whether it is extracellular or intracellular, in principle proceeds as in vertebrates, except that inver-

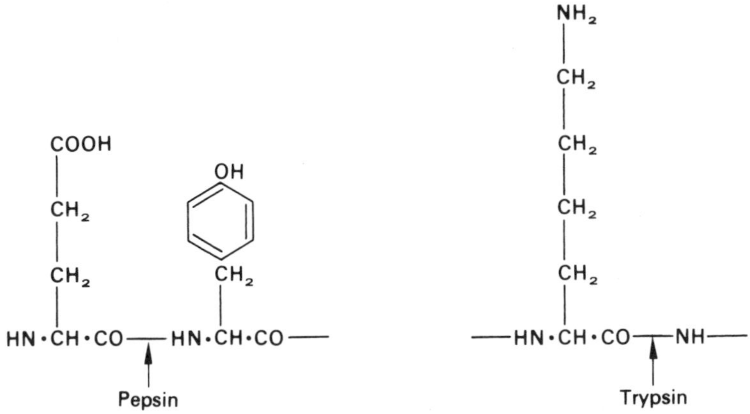

Fig. 5.2. The protein-digesting enzyme pepsin splits peptide bonds between a dicarboxylic and a cyclic amino acid, while trypsin acts on peptide bonds adjacent to an amino acid with two amino groups. Arrows show points of attack (further details in text).

tebrates generally have no pepsin-like enzymes that act in acid solution. Otherwise, although the enzymes are not identical in their structure to the vertebrate enzymes, their action is similar.

Pepsin-like enzymes are not an exclusive characteristic of vertebrates, however. The maggot of the ordinary house-fly (*Musca domestica*) secretes a protease which is active in the pH range of 1.5 to 3.5 with an optimum at 2.4. Whether this enzyme is of importance in the digestion of the maggot depends on whether a suitably acid medium can be produced, and this has not yet been completely clarified (Greenberg and Paretsky, 1955; Lambremont, Fisk and Ashrafi, 1959).

Fat digestion

Digestion of fats is similar in vertebrates and invertebrates. Most fats, whether of plant or animal origin, consist of esters between glycerol (a trivalent alcohol) and long-chain fatty-acids. They are highly insoluble in water and as a result they are not easily hydrolyzed by enzymes. The vertebrate pancreas secretes a fat-hydrolyzing enzyme, *lipase,* but to bring it into contact with the fat, the aid of a detergent is needed. The bile acids, secreted by the liver, serve this function. With the aid of the bile acids the fats are emulsified in the intestine, and as hydrolysis proceeds, the resulting fatty acids (also relatively insoluble) are kept in solution with the aid of the bile salts. Thus they remain in solution and can be absorbed. The glycerol (which is a normal intermediary in virtually all cell metabolism), is water soluble and is easily absorbed and metabolized. Fats can, however, to some extent be absorbed by the intestinal epithelium without preceding hydrolysis; the tiny droplets formed when fats are emulsified with the aid of the bile are taken up directly by the epithelial cells.

Some fatty substances are not hydrolyzed at all by ordinary lipase. This specifically applies to beeswax, which, with one exception, remains undigested by vertebrates and therefore has no nutritive value. Even though no vertebrate produces wax-digesting enzymes, beeswax is a main food item for a small group of birds, the South African honeyguides. These birds, related to the woodpeckers, are known for their habit of guiding the ratel (a badger) or human beings to the nest of wild bees. The bird attracts attention by chattering and noisy behavior, and when it is being followed, it leads the way to a bee's nest. It waits while the nest is plundered, and then eats the wax that has been left by the predator. The honeyguide prefers to eat wax rather than honey, and will also eat pure wax as well as honeycomb. The honeyguide can digest wax because its intestinal tract contains symbiotic bacteria which produce the necessary enzymes for attacking the wax. Apart from honeyguides, wax is indigestible to virtually all other animals, vertebrate and invertebrate alike. Only the larvae of the wax moth (*Galleria*), which is a common parasite in beehives, can live on wax (Friedmann and Kern, 1956*a;* 1956*b*).

The hypothesis that symbiotic bacteria are responsible for the wax digestion of the honeyguide is confirmed by transferring to domestic chicks pure cultures of bacteria obtained from the digestive tract of honeyguides. Normal chicks are completely unable to digest wax, but if wax is fed together with cultures of the isolated micro-organisms from the honeyguide, the chickens can digest and metabolize wax (Friedmann, Kern and Hurst, 1957). The digestion of wax with the aid of symbiotic bacteria is a rare curiosity. In contrast, symbiotic digestion of cellulose, one of the commonest plant materials, is of immense importance to herbivorous animals, as we shall see below.

Carbohydrate digestion

There are no great differences between vertebrates and invertebrates in how carbohydrates are digested. Simple sugars such as glucose and fructose are absorbed without any change and directly utilized in ordinary metabolic pathways. Disaccharides, such as sucrose (of plant origin) or lactose (in milk) are broken down to mono-saccharides before they are absorbed and can be utilized. The enzyme sucrase is secreted in the intestine, but it is absent from the cellular machinery of animals. Therefore, if sucrose is injected into the body of vertebrates, it is completely excreted again in the urine without any change.

A large number of plants store starch as a major energy reserve. Starch is a polymer of single glucose units. It is relatively insoluble, but it is hydrolyzed by the enzyme *amylase* (from Latin *amylum* = starch), secreted in the saliva of man (and some but not all other mammals), and in larger amounts by the pancreas. Although starch is relatively insoluble, it is well digested by many animals. It is more easily attacked by amylase, however, if it first has been heated, and this is the major reason that man prepares starchy foods by cooking (potatoes, bread, etc.). It is quite probable that in many animals which readily digest uncooked starch, bacteria aid in the initial breakdown of the starch, which is then more easily attacked by the amylase.

Symbiotic digestion of cellulose

The most important structural material of plants is cellulose, a carbohydrate polymer which is extremely insoluble and refractory to chemical attack. Cellulose digesting enzymes, *cellulases*, are completely absent from the digestive secretions of vertebrates, and yet, many vertebrates digest cellulose and depend on it as their main energy source. True cellulases have been reported from the intestinal tract of several invertebrates that feed on wood and similar plant products, but in many cases the cellulose digestion is carried out by *symbiotic micro-organisms* which live in the digestive tract of their host.

Invertebrate cellulose digestion

The question of whether a cellulase is produced by an animal or by symbiotic micro-organisms is often difficult to decide with certainty. A few examples will explain the difficulties. One case is the garden snail *Helix pomatia*. It has long been accepted that this animal digests cellulose and actually secretes a cellulase. However, a careful search for cellulase activity in extracts of the digestive gland and of the intestinal wall have failed to reveal any cellulase activity, although cellulase is readily demonstrated in the intestinal contents (Florkin and Lozet, 1949). This seems to indicate a major role of intestinal micro-organisms, but the fact that an extract of digestive organs does not reveal the enzyme does not necessarily exclude the possibility that the animal may produce the enzyme.

Another organism that has been accepted as able to digest cellulose is the 'shipworm' (*Teredo*), actually a wood-boring mollusc. True cellulase has been found in the gut of the shipworm, and gut extracts liberate sugar from cellulose, although no cellulose-digesting bacteria or protozoa could be isolated from the digestive system (Greenfield and Lane, 1953). In this case it could be argued that the apparent absence of cellulose-digesting micro-organisms is no proof that these do not exist in the living animal, and the active extracts from the intestine were made with pieces of gut from which the contents had not been removed.

There are, however, reports of true cellulases obtained from some animals under conditions that seem to exclude any possibility of symbiotic micro-organisms. The silverfish (*Ctenolepisma lineata*) digests cellulose and can survive on a diet of cellulose alone, although this is not satisfactory for prolonged feeding. The gut of silverfish contains many micro-organisms, but no cellulose-digesting species has been isolated. More importantly, bacteria-free silverfish have been obtained by washing the eggs in a solution of mercuric chloride and ethanol and raising the nymphs on sterilized rolled oats and vitamins under aseptic conditions. Such bacteria-free silverfish digest cellulose marked with ^{14}C and exhale ^{14}C-labelled carbon dioxide. Finally a cellulase has been demonstrated in extracts of the mid-gut, and it therefore seems certain that silverfish do produce the cellulase (Lasker and Giese, 1956).

Termites. In the case of termites, which live virtually exclusively on wood, the role of symbiotic organisms is beyond doubt. The intestinal tract of wood-digesting termites is virtually crammed full of flagellates as well as a number of bacteria. Several of the flagellates have been isolated and kept in pure culture, and this has permitted a careful study of their role in cellulose digestion.

The symbiotic flagellates from termites are obligate anaerobic organisms, i.e. they can live only in the absence of free oxygen. This sensitivity to oxygen can be used to remove the flagellates from the host animal, and thus to obtain protozoan-free termites. If termites are exposed to oxygen at 3.5 atm pressure, the flagellates are selectively killed in about half an hour, while the termites survive unharmed (Cleveland, 1925). Such defaunated termites cannot survive if they are fed as usual on wood. They still have bacteria in the intestine, and this indicates that the protozoan symbionts, rather than the bacteria, are responsible for cellulose digestion. When the termites are reinfected with the proper flagellates, they can again digest cellulose (Hungate, 1955).

The hypothesis that the flagellates carry out the cellulose digestion is further supported by studies of these organisms in isolated culture. One such flagellate (*Trichomonas termopsidis*, from the termite *Termopsis*), which has been kept in culture for as long as three years, yields extracts that contain cellulase (Trager, 1932). In fact, this flagellate cannot utilize any carbon source other than cellulose, so there is no doubt about the role of this organism in the cellulose digestion of termites (Trager, 1934).

Although the cellulose from wood can give a substantial amount of energy, wood has a very low nitrogen content. Since termites thrive on wood, it has been suggested that the intestinal microbes could be nitrogen-fixing bacteria, and thus be of aid in the protein supply. This hypothesis seems plausible. Many earlier attempts at demonstrating nitrogen fixation in the overall nitrogen balance of termites were unsuccessful (Hungate, 1941), but recent evidence favors the hypothesis (Benemann, 1973).

Mammalian cellulose digestion

A large number of mammals are herbivores, and most of them live on a diet that makes cellulose digestion essential to them. The ruminants, which include some of our most important domestic meat and milk producing animals, cattle, sheep, and goats, have a specialized digestive tract which is highly adapted to symbiotic cellulose digestion. However, a large number of non-ruminant mammals as well depend on symbiotic micro-organisms for cellulose digestion, although their anatomical adaptations differ from those of the true ruminants.

Ruminants. In ruminants the stomach consists of several compartments (see fig. 5.3); or, to be more precise, the true digestive stomach is preceded by several large compartments, the first and largest called the *rumen*. The rumen serves as a large fermentation vat in which the food, mixed with saliva, undergoes heavy fermentation. Both bacteria and protozoans are found in the rumen in large numbers. These micro-organisms are responsible for the breakdown of cellulose and make it available for further digestion. The fermentation products (mostly acetic, propionic, and butyric acids) are absorbed and utilized, while carbon dioxide and methane (CH_4) formed in the fermentation process are permitted to escape by

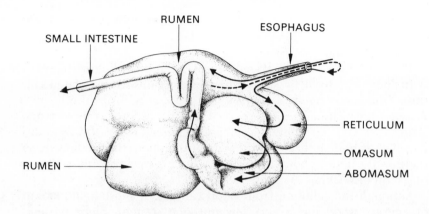

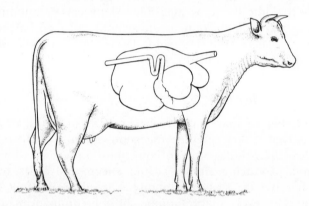

Fig. 5.3. In the ruminant animal the true stomach (abomasum) is preceded by several other compartments. The first and largest of these, the rumen, serves as a giant fermentation vat which aids in cellulose digestion.

belching (eructation). Rumination, or 'chewing the cud,' is regurgitation and re-mastication of undigested fibrous material, which is then swallowed again. This process, rumination, has given the ruminant group its name. As the food re-enters the rumen it undergoes further fermentation. Broken-down food particles gradually pass to the other parts of the stomach where they are subjected to the usual digestive juices in the fourth stomach (which corresponds to the digestive stomach of other mammals).

The fermentation products are mostly short-chain organic acids, and the copious amounts of saliva secreted by ruminants serve to buffer these fermentation products in the rumen. The saliva of ruminants is little more than a dilute solution of sodium bicarbonate, which serves both as a buffer and a proper fermentation medium for the micro-organisms.

The amount of saliva secreted by the ruminant is impressive. The total secretion of saliva per day has been estimated at 6 to 16 liters in sheep and goats, and at 100 to 190 liters in cattle (Blair-West, Coghlan, Denton and Wright, 1967; Bailey, 1961). Since sheep and goats weigh perhaps 40 kg, and cattle some 500 kg, the daily production of saliva may reach roughly one-third of the body weight. Since two-thirds of the body weight is water, about half of the total body water passes through the salivary glands (and the rumen) per day.

The rumen protozoans include ciliates which superficially resemble free-living forms such as paramecium. They occur in numbers of several hundred thousand per milliliter of rumen contents. Many of the rumen organisms have been cultured in the laboratory, and extracts from pure cultures show cellulase activity (Hungate, 1942). The ciliates are obligate anaerobic organisms and they must meet their energy requirements through fermentation processes. This gives a relatively low energy yield for the micro-organisms, but it means that the fermentation products become available to the host animal, which in turn uses them in oxidative metabolism. Since symbiotic cellulose digestion is the only way that cellulose becomes available to mammals, the relatively small loss to the symbiotic micro-organisms is quantitatively unimportant. What is important, however, is that the rumen micro-organisms contribute in several other ways to the nutrition of the host.

There is the important fact that the rumen micro-organisms can synthesize protein from inorganic nitrogen compounds, such as ammonium salts (McDonald, 1952). What is particularly useful is that urea, which is normally an excretory product eliminated in the urine, can be added to the feed of ruminants and increase the protein synthesis. This has been used in the dairy industry, for urea can be synthesized cheaply, and it is therefore less expensive to supplement the diet of milk cows with urea than to use more expensive high-protein feed.

The rumen contents of a cow may weigh as much as 100 kg, with a total weight of protozoans of about 2 kg, and these again contain 150 g protozoan protein. The propagation time for these protozoans makes it possible to estimate that 69% of the protozoan population passes into the omasum each day, and that this contributes a protein supply of over 100 g per day (Hungate, 1942).

The microbial protein synthesis in the rumen is of special importance when the animals are fed on low-grade feed. It has been found that camels fed on a nearly protein-free diet (inferior hay and dates) excrete virtually no urea in the urine. Urea continues to be formed in metabolism, but instead of being excreted this 'waste' product re-enters the rumen, in part through the rumen wall and in part through the saliva. In the rumen the urea is hydrolyzed to carbon dioxide and ammonia, the latter being used for resynthesis of protein. In this way a camel on low-grade feed can recycle much of the small quantity of protein-nitrogen that it has available (Schmidt-Nielsen, Schmidt-Nielsen, Houpt and Jarnum, 1957).

Similar re-utilization of urea nitrogen in animals fed on a low-protein diet has been observed in sheep (Houpt, 1959), and under certain conditions the rabbit (a non-ruminant) can utilize urea to a significant extent in its nitrogen metabolism (Houpt, 1963).

If inorganic sulfate is added to the diet of ruminants, the microbial synthesis of protein is improved, and what is particularly important is that the sulfate is incorporated in the important amino acids cysteine and methionine, which are both essential amino acids (Block, Stekol and Loosli, 1951). Thus, the rumen micro-organisms contribute both to protein synthesis and to the quality of the protein. Because the microbes in the rumen can synthesize all the essential amino acids, ruminants are nutritionally independent of these, and the quality of the protein they receive in their feed is therefore of minor importance.

Another nutritional advantage of ruminant digestion is that some important vitamins are synthesized by the rumen micro-organisms. This applies to several of the vitamin B group, and in particular, the natural supply of vitamin B_{12} for ruminants is obtained entirely from micro-organisms.

Non-ruminant mammals. Cellulose digestion in many non-ruminant herbivorous mammals is also aided by micro-organisms. Cellulose-containing feed is mostly bulky, and fermentation is relatively slow and takes time. Much space is required, and the part of the digestive tract used for fermentation is therefore of considerable size. In some animals the stomach is large and has several compartments, and the digestion has obvious similarities to ruminant digestion. In others the major fermentation of cellulose takes place in a large diverticulum from the intestine, the *caecum*.

Multiple-compartment stomachs are found not only in some non-ruminant ungulates, but in very far removed animals such as the sloth (an edentate) and the langur monkey (Bauchop and Martucci, 1968) (see fig. 5.4). Even among marsupials there are animals that have a rumen-like stomach. One is the rabbit-sized quokka, which weighs 2 to 5 kg. Its large stomach, which harbors micro-organisms

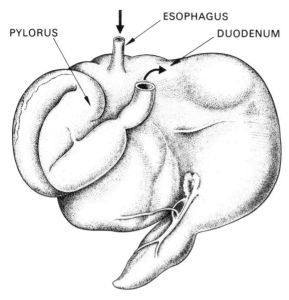

PYLORUS ESOPHAGUS DUODENUM

Fig. 5.4. The stomach of the sloth (*Bradypus tridactylus*) is complex and highly reminiscent of the stomach of ruminants (Grassé, 1955).

that participate in cellulose digestion (Moir, Somers and Waring, 1956), may contain nearly half a kilogram of moist material. For an animal weighing about 3 kg, this equals 15% of the body weight, an amount similar to that which is often found in ruminants.

Microbial fermentation in the caecum has considerable similarity to fermentation in the rumen, but the rumen has two definite advantages above the caecum. One is that rumen fermentation takes place in the anterior portion of the gastrointestinal tract, so that the products of digestion can pass through the long intestine for further digestion and absorption. The other advantage of the ruminant system is that the mechanical breakdown of the food can be carried much further, for coarse and undigested particles can be regurgitated and masticated over and over again. This difference is clearly visible if we compare the fecal material of cattle (ruminants) and horses (non-ruminants); horses have coarse fragments of the food intact in the feces, while cow feces are a well ground-up smooth mass with few large visible fragments.

An improvement in the digestion of plant materials through fermentation is not restricted to mammals. For example, the willow ptarmigan in Alaska subsists during several winter months on a diet of only willow buds and twigs. Most gallinaceous birds have two large caecae suitable for cellulose fermentation. In the willow ptarmigan the fermentation products were found to be ethanol and acetic, propionic, butyric, and lactic acids in various concentrations. The contribution of the caecal fermentation to the basal energy requirement was found to be up to 30%, a very substantial contribution indeed (McBee and West, 1969).

Coprophagy. The disadvantage of locating the cellulose fermentation in the posterior part of the intestinal tract can be circumvented in an interesting way. A number of rodents as well as rabbits and hares form a special kind of feces from the contents of the caecum, and these they re-ingest so that the food passes through the entire digestive tract a second time. There are, in fact, two kinds of feces, the well-known ordinary, firm, dark rodent fecal pellets, and a softer, larger and lighter kind which are not dropped by the animal but are eaten directly from the anus. This latter type of feces is kept separate from the ordinary droppings in the rectum, and their reingestion permits a more complete digestion and utilization of the food.

Coprophagy (from the Greek *copros* = excrement and *phagein* = to eat) is common in rodents, and it is of great nutritional importance. For example, if coprophagy is prevented, rats require supplementary dietary sources of vitamin K and biotin. In addition, deficiencies

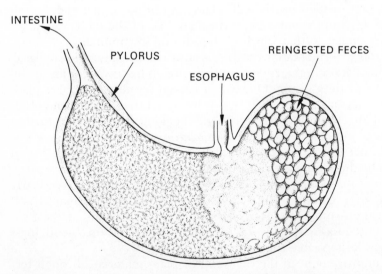

Fig. 5.5. In the stomach of the rabbit ingested food is located in the pyloric part (left) which contains digestive glands. Re-ingested fecal pellets, on the other hand, become located in the large fundus (right) where they remain separate from the food material while fermentation continues (Grassé, 1955; Harder, 1949).

of other vitamins develop more rapidly. What is more interesting is that, even if the rats are provided with a variety of dietary supplements, the growth rate is still depressed by about 15 to 25%. Even if the rats are given access to eating normal rat feces, normal growth is not attained by rats prevented from coprophagy (Barnes, Fiala and Kwong, 1963).

In rabbits prevention of coprophagy leads to a decrease in the digestibility of the food, as well as a decrease in protein utilization and nitrogen retention. When coprophagy is permitted again, there is a corresponding increase in the digestibility of cellulose (Thacker and Brandt, 1955).

The special 'soft' feces which rabbits re-ingest originate in the caecum. When ingested, these feces are not masticated and mixed with other food in the stomach, they tend to lodge separately in the fundus of the stomach (see fig. 5.5). The soft feces are covered by a membrane, and they continue to ferment in the stomach for many hours, one of the fermentation products being lactic acid (Griffith and Davies, 1963). In this way the fundus of the stomach serves as a fermentation chamber, analogous to the rumen of sheep and cattle, and thus provides essential nutritional advantages to the animal.

NUTRITION

The mechanical properties of the food must correspond to the feeding apparatus, and the chemical nature of the food must be suited to the available digestive mechanisms. Once these requirements have been met, the next question is whether the food is adequate in regard to the needs of the organism. Nutrition deals specifically with this question.

The question can conveniently be divided into two parts, the quantity of food and the quality of food. (1) The organism needs energy for external activity and internal maintenance, and (2) it needs a supply of specific substances for building and upkeep. These specific needs include amino acids, vitamins and certain other essential nutrients, and various minerals and other elements, some of them in such minute quantities that it has been difficult to establish whether they are really needed.

The following discussion will first deal with the need for energy and energy-yielding components of the diet. It will then proceed to specific substances that are needed by the organisms. Finally, it will mention some less desirable compounds that may occur in the food.

Energy supply, fuel

The bulk of organic material used for food consists of proteins, carbohydrates, and fats. The main digestion products of these compounds are amino acids (from proteins), various simple sugars (present in the food or derived from starch digestion), short-chain fatty acids (primarily from cellulose fermentation), and long-chain fatty acids (from fat digestion). The oxidation of these digestion products yields virtually all the chemical energy needed by animal organisms. They are all quite simple organic products, consisting mostly of carbon, hydrogen, and oxygen in various proportions, except that the amino acids also contain nitrogen (in the amino group, $-NH_2$). If amino acids are used to supply chemical energy, the amino group is first removed by deamination, and the remaining relatively small fragments (various short-chain organic acids) enter the usual pathways of intermediary metabolism.

These various organic compounds are, within wide limits, interchangeable in the energy metabolism, although there are certain limitations. For example, the human brain needs carbohydrate (glucose), which is the only energy source it can use. Most other organs can use fatty acids, which is the main fuel for mammalian muscle metabolism. In contrast, the flight muscles of the fruit fly, *Drosophila*, require carbohydrate, and when the supply is exhausted, the fruit fly cannot fly, although stored fat is used for other energy-requiring processes (Wigglesworth, 1949). Not all insects need glucose for flight, however, the migratory locust uses primarily fat for its long migratory flights (Weis-Fogh, 1952).

For an animal to remain in a steady state, the total energy expenditure must be covered by an equal intake of food. If the amount of food is insufficient, the remaining energy requirement is covered by the consumption of body substance, primarily stored fat. If the food intake exceeds the energy used, most animals store the surplus as fat, irrespective of the nature of the food. This is the reason that domestic livestock, such as beef and pigs, can be fattened by feeding on grain or corn which contain mostly starch. Some animals, however, store primarily carbohydrate in the form of glycogen. This is important, for glycogen can provide energy in the absence of oxygen (see p. 218–9).

Regulation of food intake. In many animals the intake of food is remarkably well adjusted to the energy expenditure (Mayer and Thomas, 1967). If the energy needs are increased due to physical activity, the food intake is adjusted accordingly. The regulatory mechanisms for the food intake of mammals are located in the hypothalamic area of the brain, but specific peripheral influences play a major role in the momentary level of hunger and food intake. One of the important influences is the blood sugar level. It has been demonstrated that hunger pangs coincide with contractions of the stomach, and that these occur especially when the level of blood glucose decreases. Also, the degree of filling of the stomach is of obvious importance in inducing a feeling of satiation and termination of food intake.

These factors can only be of minor importance in the regulation of long-term food intake and maintenance of body weight. If the bulk or volume of the food is increased by adding non-nutritive material to the food, it has a considerable effect on both gastric filling and motility, but the animal nevertheless rapidly adjusts by increasing the food intake. On a calorically dilute diet rats already after the first day maintain their caloric intake by ingesting larger amounts. On a calorically concentrated diet, the caloric intake remains too high for several days, but then decreases to the basic level (fig. 5.6). The exact mechanism for this control is not fully understood, but it is known that the hypothalamic region of the brain is important. If specific areas in the hypothalamus are destroyed, the food intake is greatly altered, and may be either reduced or increased. Carefully placed hypothalamic lesions may lead to such tremendous increases in food intake that an animal eats continuously and grows so fat that it is unable to move.

Regulation of the food intake is not limited to higher vertebrates. Goldfish, which continue to grow throughout their lifetime (although at a decreasing rate), have been used in similar experiments with caloric dilution of their diet. If permitted to control their food intake, goldfish respond to caloric dilution by increased food intake. Also, a decrease in the water temperature from 25 to 15 °C causes a decrease in food intake by a half to one-third. Goldfish therefore seem to eat for calories, for a 10 °C drop in temperature causes roughly a two- to three-fold decrease in metabolic rate and energy requirement (Rozin and Mayer, 1961).

Specific nutritional needs

A diet that furnishes adequate fuel, measured in terms of calories, can be utterly inadequate in regard to raw materials for growth and building and maintaining the cellular and metabolic machinery. For most animals the energy-requiring processes can be fully satisfied by glucose alone, but glucose contributes nothing towards other needs.

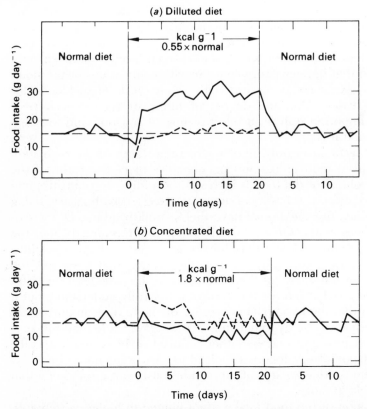

Fig. 5.6. Changes in amount of food eaten per day by adult rats when the energy content of the food is changed. In (*a*) the energy content per unit weight of food was reduced, and in (*b*) it was increased. Solid line, weight of food consumed daily; dashed line, energy content of food eaten (expressed as equivalent weight of initial diet) (Hervey, 1969).

There are three major categories of such needs, proteins and amino acids, vitamins and related compounds, and minerals and trace elements.

Proteins and amino acids

When an animal grows, protein is continually synthesized and added to the organism. In the adult organism, however, the amount of protein remains much the same throughout the lifetime. It might therefore seem that once an organism has reached its adult size, dietary protein would be less important. This is not at all true, an inadequate supply of protein leads to serious malnutrition.

Why is this so? It would appear that, once proteins have been incorporated into an organ (a muscle, for example), they would remain as part of the permanent structure. It was therefore a great surprise when the biochemist Rudolf Schoenheimer found that the body proteins are constantly broken down and resynthesized. Schoenheimer introduced in the diet of adult rats amino acids labelled with heavy nitrogen, and these amino acids were invariably found incorporated in the tissue proteins. Thus, it was shown that cell components that had been regarded as stable structures have a high turnover rate of proteins (Schoenheimer, 1964).

Not all proteins have the same turnover rate. In man, for example, the half-life of serum proteins is approximately 10 days, but connective tissue has a low rate of protein turnover, showing that once these structural proteins are formed, they remain relatively stable compared to proteins of metabolically active tissues such as blood, liver, muscle, and so on.

New protein is synthesized from single amino acids and not from partially degraded fragments of protein molecules. An adult man synthesizes and degrades about 400 g protein per day, and this far exceeds his total protein intake, which on an average diet may be between 100 and 150 g per day. Thus, the synthesis utilizes not only amino acids from the food, but also, in part, those available from protein breakdown in the body.

Each organism contains a large number of different proteins as cell components, enzymes, and so on. Furthermore, every animal or plant species has proteins that are specific for that species, and even individuals within a species may have characteristically different proteins. For example, the blood groups of man depend on the different structures of some of the blood proteins.

Only about 20 different amino acids occur commonly in proteins, but since a protein molecule contains from less than 100 to many thousand amino acids, there is virtually no limit to the number of different proteins that can be built from the 20 basic building blocks.

The amino acids needed by the body are usually supplied by the protein of the food, but they can be replaced by chemically purified amino acids in suitable proportions. However, it is not necessary to supply every one of the 20 different amino acids. Some of them can be formed in the body from other amino acids, but others cannot be synthesized by the organism and must be supplied in the diet. These are called *essential amino acids.*

The terms *essential* and *non-essential* describe the dietary requirements for amino acids, but say nothing about how important these amino acids are. The non-essential amino acids are just as important as structural components of body proteins. In fact, they are probably so important that the dietary supply cannot be trusted, and the organism cannot afford to be without a mechanism for their synthesis.

Of the 20 or so common amino acids, ten are essential to rats (table 5.3) (Rose, 1938). The exclusion of any one of them (except arginine) gives profound nutritional failure. The absence of arginine gives a decreased rate of growth, but it can be synthesized, although not at a rate sufficient for optimal growth. Man and other mammals have the same requirements, except that histidine is not needed by adult humans although it seems to be required by all other higher animals tested (Rose, 1949). It is particularly striking that fish, many insects, and even protozoans have practically overlapping needs for essential amino acids (Tarr, 1958; Taylor and Medici, 1966).

Table 5.3. *Classification of the amino acids with respect to dietary needs of rats*

ESSENTIAL	NON-ESSENTIAL
Lysine	Glycine
Tryptophane	Alanine
Histidine	Serine
Phenylalanine	Norleucine
Leucine	Aspartic acid
Isoleucine	Glutamic acid
Threonine	Proline
Methionine	Citrulline
Valine	Tyrosine
Arginine *	Cystine

* Arginine can be synthesized, but not at a sufficiently rapid rate to meet the demands of normal growth.

Plants, in contrast to animals, do synthesize protein from inorganic nitrogen (ammonia salts or nitrate), and some microbes can even use molecular nitrogen, Such *nitrogen-fixing bacteria* are found in the root nodules of clover and other leguminous plants which are therefore particularly important in increasing the protein yield from agriculture.

Ultimately, all animal proteins are of plant origin. This even holds for the protein synthesized from ammonia or urea in the rumen of ruminants, for without microbes this synthesis does not take place. Microbial protein synthesis in the rumen has an interesting consequence; it makes the ruminant animal independent of essential amino acids, for these are synthesized by the micro-organisms. Ruminants therefore have no dietary need for essential amino acids, although metabolically they are indispensable (Loosli *et al.*, 1949; White *et al.*, 1964).

Accessory foodstuffs, vitamins

Early in this century it was discovered that a diet of carbohydrates, fats, and proteins is insufficient in regard to a number of minor components that have become known as *vitamins*. The quantities needed are small, they are in the range of milligrams or micrograms. For this reason these compounds are also known as *accessory foodstuffs*. The initial discovery of a vitamin was due to the human disease beri-beri, which is caused by the absence of thiamine (vitamin B_1) in the diet. Originally, the vitamins were given letters of the alphabet, but as their chemical structure and exact biochemical roles have been clarified, it has become increasingly common to use their chemical names instead. Detailed lists of the vitamins and their metabolic roles can be found in any adequate textbook of nutrition and will therefore be omitted here.

The vitamin requirements are not the same for all animals. For example ascorbic acid (vitamin C) is needed by man, but not by other mammals (except some primates, a fruitbat, and the guinea pig). Other mammals synthesize ascorbic acid and thus do not need a dietary source.

A striking difference in synthetic ability exists in regard to cholesterol, which vertebrates readily synthesize. For man, cholesterol is in fact considered undesirable because it is a major factor in the development of arteriosclerosis. Insects, however, cannot synthesize cholesterol, and it must therefore be supplied in the food, that is, for insects cholesterol is a vitamin (Horning, 1958). All insects so far studied need a dietary source of sterols, but for protozoans it is required by some but not by others.

Symbiotic organisms are often important sources of vitamins. For example, vitamin K, which is necessary for normal blood coagulation in vertebrates, is synthesized by bacteria in the digestive tract of mammals. Its physiological role was discovered in birds fed on a cholesterol-free diet for the purpose of studying cholesterol synthesis. The birds developed severe bleedings, which were traced to a vitamin deficiency. Only then was it established that this vitamin is needed also by mammals. Its discovery, however, depended on the accidental discovery of its importance in birds, who require it in their food.

Symbiotic synthesis of vitamins is particularly important for ruminants, which appear not to need several of the vitamins in the B-group. Again, it is the microbial synthesis which makes these animals independent of a dietary source (McElroy and Goss, 1940).

Information about the vitamin requirements of lower vertebrates and invertebrates is very inadequate. Studies are difficult, for it requires the use of diets composed of purified components, and many animals will not feed on such diets in the laboratory. Broadly based nutritional studies have therefore mostly been confined to protozoans, insects, and higher vertebrates. Also, to establish nutritional requirements with certainty, it is necessary to know whether symbiotic micro-organisms contribute to meet the vitamin needs. This can be done with the aid of germ-free animals, but the procedures then become very involved as well as expensive.

Minerals and trace elements

How many elements are essential to life? The answer to this question is known for no more than a few laboratory animals. Of the over ninety naturally occurring elements, only 24 are known to be necessary for rats and/or chicks.

The four commonest elements, oxygen, carbon, hydrogen, and nitrogen, make up 96% of the total body weight of a mammal (table 5.4). The seven next most abundant elements make up very close to

Table 5.4. *Approximate elementary composition of the body of a 70 kg man*

	Weight (g)	Per cent body weight
Oxygen	45 500	65.0
Carbon	12 600	18.0
Hydrogen	7 000	10.0
Nitrogen	2 100	3.0
Calcium	1 350	1.93
Phosphorus	785	1.12
Potassium	245	0.35
Sulfur	175	0.25
Sodium	105	0.15
Chlorine	105	0.15
Magnesium	35	0.05
Total	70 000	100.00

the remaining 4%; they are calcium, phosphorus, potassium, sulfur, sodium, chlorine, and magnesium, in that order. Thirteen additional elements are required, but their total combined amount in the mammalian body is less than 0.01% of the body weight.

Water and the organic building blocks of the body, proteins, nucleic acids, fats, and carbohydrates, contain most of the oxygen, hydrogen, carbon, and nitrogen, plus part of the sulfur and phosphorus. The mineral compounds of skeletons together with dissolved ions make up the remaining few per cent. The quantitatively most important ions in animals are sodium, potassium, chloride, and bicarbonate. The maintenance of normal concentrations of sodium and chloride ions are necessary for the osmotic balance of animals, as will be discussed in a later chapter (p. 371). While sodium is usually the major cation in blood and extracellular fluids, potassium is the dominant cation inside the cells, where sodium is correspondingly lower in concentration. Chloride also is usually higher in the extracellular and lower in the intracellular fluid. Bicarbonate ion is intimately connected with respiration and carbon dioxide transport, and is of major importance in the regulation of the pH of the organism, as has been described in previous chapters.

Calcium, magnesium, phosphate, and sulphate are also important ions, they are all necessary for many physiological processes. Any great change in their concentration is deleterious or fatal. Thus calcium ions are necessary for normal cell function, for nerve conduction, for muscle contraction, and for blood coagulation. If the calcium concentration in the blood of mammals falls to about half its normal value (about 5 meq liter^{-1}) the result is severe or fatal tetanic cramps.

Calcium is also an important constituent of skeletons and many other rigid mechanical structures. Vertebrate bones and teeth consist mainly of calcium phosphate (hydroxyapatite). Invertebrate mineralized structures, such as clam shells and coral, are mainly calcium carbonates.

Not all structural substances in animals are mineralized. The exoskeleton of insects consists of the organic material chitin. In crustaceans the exoskeleton may be entirely organic, but in many larger forms (such as crabs and lobster), it is encrusted with calcium carbonate. Invertebrate skeletons frequently contain some magnesium carbonate in addition to the calcium carbonate, and often minor amounts of calcium sulfate as well. A peculiar exception is the skeletal material of the radiolarian *Acantharia* (a protozoan), which consists entirely of strontium sulfate (Odum, 1951; 1957).

The supporting structures of many plants, in addition to cellulose, often contain amorphous silicon oxide (silica). Of animals some sponges have skeletons which consist of spicules of silica (SiO_2).

Trace elements

Let us now consider the 13 elements that make up less than 0.01% of the body of mammals. These elements occur in such small amounts that, although early investigators could demonstrate that they were present in the body, the analytical methods were not sufficiently accurate for a precise analysis. These elements therefore were mentioned as occurring in 'trace' amounts, and they became known as *trace elements*. With better analytical methods, and in particular with the aid of radioisotopes, it has become possible not only to determine these elements more accurately, but also to clarify their physiological and biochemical roles.

The trace elements can be divided into three groups: (1) those known to be essential requirements; (2) those which have metabolic effects but have not been shown to be indispensable; and (3) those which occur widely in living organisms but seem to be only incidental contaminants. Some of the trace elements are needed in such small amounts that it is difficult to devise a diet that contains not even a trace of a particular element, and it is probable that as research progresses, some elements of the second group will be moved to the first group, and even some of those that are now in the third group could possibly prove to be indispensable.

The mere occurrence of a trace element gives no information about its function, and the chemical demonstration of its presence is therefore rather uninformative, for virtually every element in the periodic table can be found in some minute quantity. An understanding of the role of trace elements is obtained mainly in two ways: (1) feeding of synthetic diets composed of highly purified components and designed to lack the element under study, and (2) the study of diseases, both in man and animals, which are suspected to result from a deficiency of a trace element. Such deficiency diseases are often associated with an exceptionally low concentration of certain elements in the soil in specific geographical areas. The first method has the disadvantage that it is difficult to eliminate every trace of a given element from the diet, and a highly purified diet also readily becomes deficient in other respects. For example, should a diet be inadequate with regard to some vitamin as well as the trace element under study, the animal will not show the expected improvement if the trace element alone is added. Therefore, progress in the study of trace elements was difficult before vitamin research was sufficiently advanced to design adequate test diets.

Iron [26].* The presence of iron in living animals has been known for so long that usually it is not considered a trace element. Nevertheless, it is essential. It is grouped in the periodic system among the 'transition elements' (see fig. 5.7), which encompass a total of eight known trace elements. Another five trace elements are located in the periodic table in close proximity to those non-metallic elements which are the most abundant in living organisms.

1a	2a	3a	4a	5a	6a	7a	8a			1b	2b	3b	4b	5b	6b	7b	8b
1 *H*																	2 He
3 Li	4 Be											5 B	6 *C*	7 *N*	8 *O*	9 F	10 Ne
11 *Na*	12 *Mg*			Transition elements								13 Al	14 *Si*	15 *P*	16 S	17 *Cl*	18 Ar
19 *K*	20 *Ca*	21 Sc	22 Ti	23 *V*	24 *Cr*	25 *Mn*	26 *Fe* 27 *Co* 28 Ni			29 *Cu*	30 *Zn*	31 Ga	32 Ge	33 As	34 *Se*	35 Br	36 Kr
37 Rb	38 Sr	39 Y	40 Zr	41 Nb	42 *Mo*	43 Ma	44 Ru 45 Rh 46 Pd			47 Ag	48 Cd	49 In	50 *Sn*	51 Sb	52 Te	53 *I*	54 X
55 Cs	56 Ba	57-71 ΣLa	72 Hf	73 Ta	74 W	75 Re	76 77 78 Os Ir Pt			79 Au	80 Hg	81 Tl	82 Pb	83 Bi	84 Po	85 At	86 Rn
87 Fr	88 Ra	89 Ac	90 Th	91 Pa	92 U												

Fig. 5.7. Periodic system of the naturally occurring elements. Those elements that are marked in bold are known to be essential for animals.

Iron is a constituent of hemoglobin and of several intracellular enzyme systems, notably the cytochromes. The total amount of iron in the organism is not very great, in an adult human it is about 4 g. Of this amount 70% is found in hemoglobin, 3.2% in myoglobin, 0.1% in cytochromes, 0.1% in catalase, the remainder being in storage compounds mostly in the liver (O'Dell and Campbell, 1971). The dietary requirement for an adult mammal is very small, for iron from the breakdown of hemoglobin is stored in the liver and used again for hemoglobin synthesis. Growing animals need more iron, and adult females need to replace that which is lost in reproductive processes (growth of fetus, menstruation).

* The number given in brackets after the name of the element is the atomic number of that element in the periodic table.

Cobalt [27]. Cobalt is unique among the trace elements because the cobalt ion as such is not known to be needed by any animal organism. All higher animals, however, need cobalt in the specific form of vitamin B_{12}, which is required for blood formation. Deficiency leads to severe anemia. In ruminants vitamin B_{12} is formed in the rumen by the microbial flora, provided that a sufficient amount of cobalt is present in the diet.

The first conclusive evidence that cobalt is an essential trace element came from Australia, where a serious disease of cattle and sheep caused great economic losses. Addition of small amounts of cobalt prevents this disease. The use of cobalt in fertilizers is expensive because fairly large amounts are needed. Instead it has been found practical to make each sheep swallow a small cobalt-containing ceramic ball which remains in the rumen and slowly releases cobalt over a period of several years.

Adjacent to cobalt in the periodic table is *nickel* [28], which is suspected of being essential. Although it is known that nickel influences the growth of feathers in chickens, it is not certain that it is essential for life. The problem is that small amounts of nickel are nearly always associated with iron, and therefore are extremely difficult to eliminate completely from a diet.

Copper [29]. Copper occurs widely in soil, but in certain areas the concentration is so low that both plants and animals suffer a deficiency. One of the most conspicuous symptoms is anemia. Copper is not part of the hemoglobin molecule, but somehow copper deficiency prevents the formation of hemoglobin. The exact mechanism of this effect is not known.

Copper is also a constituent of a dozen or more enzymes, one of the most important being cytochrome oxidase, which contains both iron and copper.

Zinc [30]. Zinc is a constituent of many important enzymes, among these carbonic anhydrase and several peptidases. Deficiency symptoms can be produced by feeding a purified diet.

Vanadium [23]. It has recently been found that vanadium exerts a pronounced growth-promoting effect on rats which are kept in an all-plastic isolated environment on a highly purified amino acid diet. The amounts needed, around 1 μg per day for a rat, are within the range of concentrations found naturally in food and in tissues of higher animals (Schwarz and Milne, 1971).

It is peculiar that some species of tunicates accumulate very high concentrations of vanadium in their blood, although other tunicates have no unusual concentrations of this element. The function of this vanadium enrichment, if any, remains uncertain.

Chromium [24]. Chromium also is a recent addition to the list of essential trace elements. It is needed for insulin to exert its maximal effect in promoting glucose uptake by the tissues (Mertz, 1969). So far there is no evidence that chromium is a component of any enzyme system or serves as an enzyme activator.

Molybdenum [42]. Molybdenum is required by all nitrogen-fixing micro-organisms, by higher plants, and by animals. Molybdenum is a constituent of the enzyme xanthine oxidase, and thus has a role in the oxidation of purine to uric acid.

Manganese [25]. Manganese is very widespread in soil and is found in all living plants and animals. It was established rather late that manganese is an essential dietary component. Deficiency symptoms include abnormal development of the bones, and the element is also needed for the function of ovaries and testes. Its biochemical effect has become known only recently, it functions as an activator of certain enzyme systems, but the connection with the deficiency symptoms is not entirely clear.

Silicon [14]. It has long been known that silicon, which next to oxygen is the most abundant element in the earth's crust, is essential for plants. Certain knowledge that it is essential for growth and normal development in animals is, however, of quite recent date (Carlisle, 1972). Its precise role or mechanism of action, on the other hand, is not known.

Tin [50]. Located in the same group of the periodic table as carbon and silicon, tin is likewise among the most recent additions to the list of essential trace elements. Inorganic tin salts are necessary for normal growth of rats kept on highly purified diets with other known trace elements added (Schwarz, Milne and Vinyard, 1970).

Selenium [34]. Selenium has, because of its toxicity, been of greater practical importance in animal nutrition than as an essential element. In some areas selenium is found in the soil in such high concentrations that plants become toxic to livestock, and grazing animals then may suffer from acute selenium poisoning. Selenium, however, is also an essential trace element, with deficiency symptoms that include liver necrosis and muscular dystrophy. The biochemical role of selenium seems to be related to the function of vitamin E, which is necessary for growth and normal reproduction in rats.

Fluorine [9]. Until recently fluorine was not classed as essential, although some of its physiological effects were well known. This has changed, however, when it was found to be essential for the normal growth of rats (Schwarz and Milne, 1972).

It has long been known that fluorine is present in small amounts in teeth and bone, where the fluoride ion is incorporated into the crystal lattice of calcium phosphate (apatite), which is the main mineral component. Whether it is essential in this role is doubtful, but it makes the teeth more resistant to dental caries. It has therefore been recommended as a public health measure that fluorine should be added to drinking water to reduce the incidence of tooth decay. The effectiveness is beyond doubt, and the amounts added are so small that there are no known toxic effects. In some geographical areas, however, fluorine is present in the natural water supply in sufficiently high concentrations to cause abnormalities of the tooth enamel, known as 'mottled enamel.' Cosmetically, mottled enamel is undesirable, but mottled teeth are particularly resistant to decay. Even higher intakes of fluorine leads to deformation of bones and serious poisoning, and this has occurred near industrial plants and smelteries where fluorine is discharged to the atmosphere.

Iodine [53]. The biochemical significance of iodine relates to its role as a constituent of the thyroid hormones. These accelerate all cellular metabolism, and are necessary for the control of metabolic rate and for normal growth of vertebrates. It has long been known that the incidence of human goiter (an enlargement of the thyroid gland) is concentrated in geographical regions where the iodine content of the soil is particularly low. This enlargement of the thyroid appears to be a response to iodine deficiency, and the addition of small amounts of iodine to the food has virtually eliminated this rather serious public health problem (usually achieved by adding a minute quantity of sodium iodide to ordinary table salt). Iodine-deficient regions are mostly inland and mountainous areas far from the ocean. Sea water contains a small amount of iodine, and spray that enters the atmosphere from wave action is precipitated with rain and meets the needs of both animals and man, even several hundred kilometers from the coast.

Many additional elements, lithium, rubidium, beryllium, strontium, aluminum, boron, germanium, arsenic, bromine, and others, are widely present in living material and may have functional significance, although definite evidence is lacking. Some of them are highly toxic, but this does not exclude that they may yet be needed in trace amounts. Furthermore, at present there is no reason to believe that all animals have exactly the same requirements for trace elements.

Noxious compounds

Among the myriad of plant species which animals feed on, many are highly toxic to man and domestic animals, yet they are harmless to many other animals, especially insects. The many toxic plants will not be discussed here, but among the fairly small number of plants that constitute the foundation for human consumption and food production, there are some which contain noxious compounds with profound physiological effects.

Soybeans and trypsin inhibitors. A number of plants, especially of the pea family (Leguminosae) contain substances which inhibit the action of trypsin, the important proteolytic enzyme of the pancreas. Such trypsin inhibitors are found in soybeans, lima beans, peanuts, and also in a number of plants from other families. Most of the trypsin inhibitors are heat sensitive, and because man mostly eats cooked food, they are of minor importance in human nutrition. However, they reduce the protein utilization of animals that eat them, and are therefore of importance in animal husbandry. Usually heat treatment reduces or eliminates the deleterious effect (Rackis, 1965).

Cabbage and antithyroid effects. Goiter may be induced in laboratory animals by diets containing large amounts of cabbage leaves or turnips. These plants belong to the Cruciferae, which also include mustards and radishes. Several substances with an anti-thyroid effect have been isolated from cruciferous plants, and they are related to drugs that already are known to have anti-thyroid effects, such as thiourea.

It is well known that some persons find the compound phenyl thiourea extremely bitter, while others find it completely tasteless. The non-tasting condition is inherited as a recessive gene. One of the anti-thyroid substances from turnips and cabbage is similarly tasteless to some and bitter to others, with tasters and non-tasters matching the tasters of phenyl thiourea (Boyd, 1950). This suggests a possible physiological role for the gene that is responsible for this tasting ability.

Cyanide. It is widely known that bitter almonds contain substantial amounts of cyanide, and also that the kernels of apricots, prunes, and cherries contain smaller amounts of this highly toxic compound. It is perhaps less well known that a major staple food in tropical countries, cassava (manioc) contains so much cyanide that even the preparation of the tubers for eating involves a risk of fatal poisoning. The tubers are harvested and shredded with large quantities of water, and after the starch granules have settled, the poisonous liquid is poured off. In certain central American tribes the process of preparing cassava is considered so dangerous that only older people are permitted to carry it out, presumably because they are more expendable than young persons.

It is interesting that to some persons cyanide smells strongly of bitter almonds, and to others is completely odorless. This difference is not due to an inability to smell 'bitter almond,' for benzaldehyde, which to cyanide 'smellers' smells very similar to cyanide, to non-cyanide smellers also has a characteristic 'bitter almond' smell. Cyanide non-smellers who are quite common, should be careful if they use this dangerous substance.

Spinach and rhubarb. These two common plants belong to the same family, the Chenopodiaceae, and they may contain as much as 10% oxalic acid in the dry matter. Oxalic acid is a highly poisonous substance, for it precipitates calcium in the form of insoluble calcium oxalate. Spinach probably cannot be consumed in sufficient quantities to cause acute poisoning, but rhubarb leaves used as a vegetable have caused several recorded deaths from oxalate poisoning. However, if large amounts of spinach are eaten daily, the precipitation of calcium oxalate in the intestine will reduce the calcium available for absorption, and may over a period of time cause calcium deficiency.

REFERENCES

BAILEY, C. B. (1961). Saliva secretion and its relation to feeding in cattle. 3. The rate of secretion of mixed saliva in the cow during eating, with an estimate of the magnitude of the total daily secretion of mixed saliva. *Brit. J. Nutr.,* **15,** 443–51.

BARNES, R. H., FIALA, G. and KWONG, E. (1963). Decreased growth rate resulting from prevention of coprophagy. *Fed. Proc.,* **22,** 125–33.

BAUCHOP, T. and MARTUCCI, R. W. (1968). Ruminant-like digestion of the langur monkey. *Science,* **161,** 698–700.

BENEMANN, J. R. (1973). Nitrogen fixation in termites. *Science,* **181,** 164–5.

BERRILL, N. J. (1929). Digestion in ascidians and the influence of temperature. *J. Exp. Biol.*, **6**, 275–92.

BLAIR-WEST, J. R., COGHLAN, J. P., DENTON, D. A. and WRIGHT, R. D. (1967). Effect of endocrines on salivary glands. *Handbook of Physiology*, sect. 6: *Alimentary Canal*, vol. II, *Secretion* (c.f. code, ed.), pp. 633–64, Washington, D.C.: American Physiological Society.

BLOCK, R. J., STEKOL, J. A. and LOOSLI, J. K. (1951). Synthesis of sulfur amino acids from inorganic sulfate by ruminants. II. Synthesis of cystine and methionine from sodium sulfate by the goat and by the microorganisms of the rumen of the ewe. *Arch. Biochem. Biphys.*, **33**, 353–63.

BOYD, W. C. (1950). Taste reactions to antithyroid substances. *Science*, **112**, 153.

CARLISLE, E. M. (1972). Silicon: an essential element for the chick. *Science*, **178**, 619–21.

CLEVELAND, L. R. (1925). Toxicity of oxygen for protozoa in vivo and in vitro: animals defaunated without injury. *Biol. Bull.*, **48**, 455–68.

FLORKIN, M. and LOZET, F. (1949). Origine bacterienne de la cellulase du contenu intestinal de l'escargot. *Arch. Int. de Physiol.*, **57**, 201–7.

FRIEDMANN, H. and KERN, J. (1956a). The problem of cerophagy or wax-eating in the honey-guides. *Quart. Rev. Biol.*, **31**, 19–30.

FRIEDMANN, H. and KERN, J. (1956b). *Micrococcus cerolyticus*, nov. sp., an aerobic lipolytic organism isolated from the African honey-guide. *Can. J. Microbiol.*, **2**, 515–17.

FRIEDMANN, H., KERN, J. and HURST, J. H. (1957). The domestic chick: a substitute for the honey-guide as a symbiont with cerolytic microorganisms. *Am. Nat.*, **91**, 321–5.

GRASSÉ, P.-P. (1955). *Traité de Zoologie. Anatomie, Systématique, Biologie.* **17,** *fasc. 2: Mammifères*, pp. 1173–2300, Paris: Masson et Cie.

GREENBERG, B. and PARETSKY, D. (1955). Protcolytic enzymes in the house fly, *Musca domestica* (L.). *Ann. Ent. Soc. Am.*, **48**, 46–50.

GREENFIELD, L. J. and LANE, C. E. (1953). Cellulose digestion in *Teredo. J. Biol. Chem.*, **204**, 669–72.

GRIFFITHS, M. and DAVIES, D. (1963). The role of the soft pellets in the production of lactic acid in the rabbit stomach. *J. Nutr.*, **80**, 171–80.

HARDER, W. (1949). Zur Morphologie und Physiologie des Blinddarmes der Nagetiere. *Verhandlungen der Deutschen Zoologen*, **2**, 95–109.

HERVEY, G. R. (1969). Regulation of energy balance. *Nature, Lond.*, **222**, 629–31.

HORNING, M. G. (1958). The sterol requirements of insects and of protozoa. In *Cholesterol. Chemistry, Biochemistry, and Pathology* (R. P. Cook, ed.), pp. 445–55, New York: Academic Press.

HOUPT, T. R. (1959). Utilization of blood urea in ruminants. *Am. J. Physiol.*, **197**, 115–20.

HOUPT, T. R. (1963). Urea utilization by rabbits fed a low-protein ration. *Am. J. Physiol.*, **205**, 1144–50.

HUNGATE, R. E. (1941). Experiments on the nitrogen economy of termites. *Ann. Entomol. Soc. Am.*, **34**, 467–89.

HUNGATE, R. E. (1942). The culture of *Eudiplodinium neglectum,* with experiments on the digestion of cellulose. *Biol. Bull.,* **83,** 303–19.

HUNGATE, R. E. (1955). Mutualistic intestinal protozoa. In *Biochemistry and Physiology of Protozoa,* vol. II (S. H. Hutner and A. Lwoff, eds), pp. 159–99, New York: Academic Press.

JOHANNES, R. E., COWARD, S. J. and WEBB, K. L. (1969). Are dissolved amino acids an energy source for marine invertebrates? *Comp. Biochem. Physiol.,* **29,** 283–8.

LAMBREMONT, E. N., FISK, F. W. and ASHRAFI, S. (1959). Pepsin-like enzyme in larvae of stable flies. *Science,* **129,** 1484–5.

LASKER, R. and GIESE, A. C. (1956). Cellulose digestion by the silverfish *Ctenolepisma lineata. J. Exp. Biol.,* **33,** 542–53.

LOOSLI, J. K., WILLIAMS, H. H., THOMAS, W. E., FERRIS, F. H. and MAYNARD, L. A. (1949). Synthesis of amino acids in the rumen. *Science,* **110,** 144–5.

MCBEE, R. H. and WEST, G. C. (1969). Cecal fermentation in the willow ptarmigan. *Condor,* **71,** 54–8.

MCDONALD, I. W. (1952). The role of ammonia in ruminal digestion of protein. *Biochem. J.,* **51,** 86–90.

MCELROY, L. W. and GOSS, H. (1940). A quantitative study of vitamins in the rumen contents of sheep and cows fed vitamin-low diets. I. Riboflavin and vitamin K. *J. Nutr.,* **20,** 527–40.

MATTHEWS, L. H. and PARKER, H. W. (1950). Notes on the anatomy and biology of the basking shark (*Cetorhinus maximus* (Gunner)). *Proc. Zool. Soc. Lond.,* **120,** 535–76.

MAYER, J. and THOMAS, D. W. (1967). Regulation of food intake and obesity. *Science,* **156,** 328–37.

MERTZ, W. (1969). Chromium occurrence and function in biological systems. *Physiol. Rev.,* **49,** 163–239.

MOIR, R. J., SOMERS, M. and WARING, H. (1956). Studies on marsupial nutrition. I. Ruminant-like digestion in a herbivorous marsupial (*Setonix brachyurus* Quoy and Gaimard). *Aust. J. Biol. Sci.,* **9,** 293–304.

O'DELL, B. L. and CAMPBELL, B. J. (1971). Trace elements: metabolism and metabolic function. *Comprehensive Biochemistry,* **21,** 179–266.

ODUM, H. T. (1951). Notes on the strontium content of sea water, celestite radiolaria, and strontianite snail shells. *Science,* **114,** 211–13.

ODUM, H. T. (1957). Biogeochemical deposition of strontium. *Univ. Texas, Publ. Inst. Mar. Sci.,* **4,** 38–114.

RACKIS, J. J. (1965). Physiological properties of soybean trypsin inhibitors and their relationship to pancreatic hypertrophy and growth inhibition of rats. *Fed. Proc.,* **24,** 1488–1500.

ROSE, W. C. (1938). The nutritive significance of the amino acids. *Physiol. Rev.,* **18,** 109–36.

ROSE, W. C. (1949). Amino acid requirements of man. *Fed. Proc.,* **8,** 546–52.

ROZIN, P. and MAYER, J. (1961). Regulation of food intake in the goldfish. *Am. J. Physiol.,* **201,** 968–74.

SCHMIDT-NIELSEN, B., SCHMIDT-NIELSEN, K., HOUPT, T. R. and JARNUM, S. A. (1957). Urea excretion in the camel. *Am. J. Physiol.,* **188,** 477–84.

SCHOENHEIMER, R. (1964). *The Dynamic State of Body Constituents*, New York: Hafner Publ. Co., 78 pp.

SCHWARZ, K. and MILNE, D. B. (1971). Growth effects of vanadium in the rat. *Science*, **174**, 426–8.

SCHWARZ, K. and MILNE, D. B. (1972). Fluorine requirement for growth in the rat. *Bioinorg. Chem.*, **1**, 331–8.

SCHWARZ, K., MILNE, D. B. and VINYARD, E. (1970). Growth effects of tin compounds in rats maintained in a trace element-controlled environment. *Biochem. Biophys. Res. Comm.*, **40**, 22–9.

SMITH, D., MUSCATINE, L. and LEWIS, D. (1969). Carbohydrate movement from autotrophs to heterotrophs in parasitic and mutualistic symbiosis. *Biol. Rev.*, **44**, 17–90.

STEPHENS, G. C. (1962). Uptake of organic material by aquatic invertebrates. II. Uptake of glucose by the solitary coral, *Fungia scutaria*. *Biol. Bull.*, **123**, 648–59.

STEPHENS, G. C. (1964). Uptake of organic material by aquatic invertebrates. III. Uptake of glycine by brackish-water annelids. *Biol. Bull.*, **126**, 150–62.

TARR, H. L. A. (1958). Biochemistry of fishes. *Ann. Rev. Biochem.*, **27**, 223–44.

TAYLOR, M. W. and MEDICI, J. C. (1966). Amino acid requirements of grain beetles. *J. Nutr.*, **88**, 176–80.

THACKER, E. J. and BRANDT, C. S. (1955). Coprophagy in the rabbit. *J. Nutr.*, **55**, 375–85.

TRAGER, W. (1932). A cellulase from the symbiotic intestinal flagellates of termites and of the roach, *Cryptocercus punctulatus*. *Biochem. J.*, **26**, 1762–71.

TRAGER, W. (1934). The cultivation of a cellulose-digesting flagellate, *Trichomonas termopsidis*, and of certain other termite protozoa. *Biol. Bull.*, **66**, 182–90.

WEIS-FOGH, T. (1952). Fat combustion and metabolic rate of flying locusts (*Schistocerca gregaria* Forskål). *Phil. Trans. Roy. Soc. Lond. B*, **237**, 1–36.

WHITE, A., HANDLER, P. and SMITH, E. L. (1964). *Principles of Biochemistry*, 3rd edn, New York: McGraw-Hill Book Co., 1106 pp.

WIGGLESWORTH, V. B. (1949). The utilization of reserve substances in *Drosophila* during flight. *J. Exp. Biol.*, **26**, 150–63.

WRIGHT, R. T. and HOBBIE, J. E. (1966). Use of glucose and acetate by bacteria and algae in aquatic ecosystems. *Ecology*, **47**, 447–64.

6
Energy metabolism

In the preceding chapter we dealt with food and feeding, and in this chapter we will be concerned with the use of food to provide energy. Animals need a supply of chemical energy to carry out their various functions, and their over-all use of chemical energy is often referred to as their *energy metabolism.**

Animals obtain energy mostly through the oxidation of foodstuffs, and their consumption of oxygen can therefore be used as a measure of their energy metabolism. Much of this chapter will be concerned with the rate of oxygen consumption, which we often take to mean the rate of energy metabolism. However, chemical energy can also be obtained in other ways, without the use of oxygen, and oxygen consumption is therefore not always a measure of energy metabolism. For example, some animals can live in the complete absence of free oxygen; they still utilize chemical energy for their energy needs, although the metabolic pathways are different. We say that such animals obtain their energy needs through *anaerobic metabolism.* This situation is normal for quite a few animals that live in oxygen-poor environments and/or tolerate prolonged exposure to lack of oxygen. They still need energy, which they obtain without utilizing oxidation processes (see p. 218).

The energy-requiring processes and reactions in the living organism use a common source of energy, *adenosine triphosphate (ATP).* This ubiquitous compound, through the hydrolysis of an 'energy-rich' phosphate bond, seems to be the immediate source of chemical energy for energy-requiring processes such as muscle contraction,

---

* The term metabolism is used in several different ways. For example, in referring to iron metabolism, the term applies to the total physiological role of iron, its intake, absorption, storage, synthesis of hemoglobin and cytochromes, excretion, etc. The term intermediary metabolism has come to mean a description of all the biochemical reactions and interconversions that take place in the organism. Thus the term metabolism is poorly defined. In this chapter it will mean energy metabolism.

ciliary movement, firefly luminescence, discharge of electric fish, cellular transport processes, all sorts of synthetic reactions, and so on. ATP is formed at the various energy-yielding steps in the oxidation of foodstuffs, and also in anaerobic energy-yielding processes, but in smaller amounts. Thus ATP is a universal intermediate in the flow of the chemical energy of the food to the energy-requiring processes of metabolism in both aerobic and anaerobic organisms.

Metabolic rate

Metabolic rate refers to the energy metabolism per unit time. It can, in principle, be determined in three different ways.

(1) By determination of the difference between the energy value of all food taken in and the energy value of all excreta (primarily feces and urine). This method assumes that there is no change in the composition of the organism, and this excludes its use for growing organisms and organisms that have an increase or a decrease in storage of fat, etc. The method is technically cumbersome and is accurate only if carried out over a sufficiently long period of observation to assure that the organism has not undergone significant changes in size and composition.

(2) From the total heat production of the organism. This method should give information about all fuel used, and in principle it is the most accurate method. In practice, determinations are made with the organism inside a calorimeter. This can yield very accurate results, but technically it is a complex procedure. Such items as heating of ingested food, vaporization of water, etc. must be entered into the total heat account. Also, if the organism performs any external work that does not appear as heat, this work must be added to the account. Assume that a man is located inside a large calorimeter and lifts a ton of bricks from the floor on to a table. The increase in potential energy of the bricks is derived from the energy metabolism of the man, and could be released as heat if the bricks were permitted to slide down again.

(3) The amount of oxygen used in oxidation processes can be used to determine the metabolic rate, provided information is available about which substances have been oxidized (and there is no anaerobic metabolism.

The determination of oxygen consumption is technically easy, and is so commonly used for estimation of metabolic rates that the two terms are often used interchangeably. This is obviously not correct; for example, a fully anaerobic organism has zero oxygen consumption but certainly does not have a zero metabolic rate!

The reason that oxygen can be used as a practical measure of metabolic rate is that the amount of heat produced for each liter of oxygen used in metabolism remains nearly constant, irrespective of whether fat, carbohydrate, or protein is oxidized (table 6.1, column *c*). The highest figure (5.0 kcal per liter O_2 for carbohydrate metabolism) and the lowest (4.5 kcal per liter O_2 for protein) differ by only 10%, and it has become customary to use an average value of 4.8 kcal per liter O_2 as a measure of the metabolic rate.* The largest error resulting from the use of this mean figure would be 6%, but if the metabolic fuel is a mixture of the common foodstuffs, carbohydrate, fat, and protein, the error is usually insignificant. Most determinations of metabolic rates are not very precise anyway, not because the available apparatus is inherently inaccurate, but because the amounts of animal activity and other physiological functions tend to vary a great deal.

Although one liter of oxygen gives similar amounts of energy for the three major foodstuffs, each gram of foodstuff gives a very different amount. The energy derived from the oxidation of one gram of fat is more than twice as high as for the oxidation of one gram of carbohydrate or protein (table 6.1, column *a*). The high energy value of fat is widely known; the biologically most important consequence is that fat stores energy with a smaller addition of weight and bulk than otherwise possible.†

The energy value given for protein, 4.3 kcal g^{-1}, differs from the combustion value determined for proteins in the chemical laboratory, which is usually given as 5.3 kcal g^{-1}. The reason is that protein is not completely oxidized in the body. Mammals excrete the nitrogen as urea, $CO(NH_2)_2$, an organic compound whose combustion value accounts for the difference. This difference must be considered if the energy balance of an animal is calculated from combustion values determined by employing a bomb calorimeter.

The amount of oxygen needed to oxidize one gram of fat is more than twice the amount needed for oxidation of carbohydrate or pro-

---

* One calorie (cal) is defined as the amount of heat necessary to heat one gram of water from 14.5 °C to 15.5 °C. One kcal (kilocalorie) equals 1000 cal. Often, especially in elementary books and popular writings, the calorie and kilocalorie are confused. At times, a kilocalorie has been called a 'large calorie' and has been written as 'Cal'. Since the calorie is a precisely defined unit, it is meaningless to refer to 'large' and 'small' calories. The metric system of units recognizes only one calorie (cal) which can be used with the usual prefixes of the metric system, kilo-, milli-, etc.

† It may be of interest that the energy value of alcohol is nearly 7 kcal g^{-1}. This is one reason that persons who wish to lose weight by 'dieting' usually avoid alcohol, another reason is that alcohol readily weakens the determination to restrict food intake and thus contributes further to the lack of success in controlling energy intake.

Table 6.1. *Heat produced and oxygen consumed in the metabolism of the common foodstuffs. The values for protein depend on whether the metabolic end product is urea or uric acid. The ratio between carbon dioxide formed and oxygen used is known as the respiratory quotient (RQ).* (Data based on Lusk, 1931, and King, 1957)

	(a) kcal g^{-1}	(b) liter O$_2$ g^{-1}	(c) kcal per liter O$_2$	(d) RQ = $\dfrac{CO_2\ \text{formed}}{O_2\ \text{used}}$
Carbohydrate	4.2	0.84	5.0	1.00
Fat	9.4	2.0	4.7	0.71
Protein (urea)	4.3	0.96	4.5	0.81
Protein (uric acid)	4.25	0.97	4.4	0.74

tein (table 6.1, column *b*). In combination with the high energy yield from fat, the result is that the energy per liter oxygen is approximately the same for fat as for carbohydrate and protein (see above).

The last column in table 6.1 shows the ratio between the carbon dioxide formed in metabolism and the oxygen used. This ratio, known as the *respiratory quotient* or *RQ*, is an important concept in metabolic physiology. The RQ gives information about the fuel used in metabolism. Usually the RQ is between 0.7 and 1.0. If the RQ is near 0.7, it suggests primarily fat metabolism, and an RQ near 1.0 suggests primarily carbohydrate metabolism. For an intermediate RQ, it is more difficult to say what foodstuffs have been metabolized, it could be protein, or a mixture of fat and carbohydrate, or a mixture of all three. It is possible, however, to determine the amount of protein metabolized from the nitrogen excretion (urea in mammals, uric acid in birds and reptiles), and from this knowledge we can calculate that fraction of the oxygen consumption and carbon dioxide production which is due to protein metabolism. The remainder of the gas exchange is now due to fat and carbohydrate metabolism, and the amount of each can then be calculated. From determinations of oxygen consumption, carbon dioxide elimination, and nitrogen excretion, it is thus possible to calculate the separate amounts of protein, fat, and carbohydrate which have been metabolized in a given period of time.

Could the carbon dioxide production be used for determinations of metabolic rate equally as well as oxygen consumption? In theory the answer is yes, but there are difficulties that make it less practical, and in fact far less accurate. There are two main reasons, one is that the body contains a large pool of carbon dioxide, the other has to do with the caloric equivalent of one liter of carbon dioxide.

The large pool of carbon dioxide that is always present in the body changes easily. For example, hyperventilation of the lungs causes large amounts of carbon dioxide to leave in the exhaled air. This carbon dioxide, of course, is no measure of the metabolic rate. Hyperventilation will be followed by a period of reduced ventilation during which the carbon dioxide will again build up to its normal level, and during this period the reduced amount of carbon dioxide given off again is no measure of metabolic rate. Other circumstances also influence the pool of carbon dioxide. In heavy exercise, for example, lactic acid is formed in the muscles as a normal metabolic product. As the lactic acid enters the blood, it has the same effect as pouring acid on a bicarbonate solution, carbon dioxide is driven out of the blood and released in the lung in excessive amounts. Again, the amount of carbon dioxide is not a measure of the metabolic rate.

The other reason that carbon dioxide is less suitable than oxygen for measuring metabolic rates is that the different foodstuffs give rather different amounts of energy for each liter of carbon dioxide produced. Thus, carbohydrate gives 5.0 kcal per liter CO_2 formed, and fat gives 6.7 kcal per liter CO_2. The value for fat is therefore one third higher, and unless we know whether fat or carbohydrate is metabolized, the error can be substantial. For these reasons the carbon dioxide production is not a suitable measure of metabolic rate.

Energy storage, fat

For most adult animals the food intake and the energy expenditure remain approximately equal. If the energy expenditure exceeds food intake, the excess energy is covered by the consumption of body substance, primarily *fat*. If the food intake exceeds the energy used, the surplus is stored as fat, irrespective of the composition of the food. For example, pigs are fattened on feed consisting mostly of carbohydrate (grains). During fattening, if large amounts of carbohydrate are changed into fat, the RQ will exceed 1.0. This is because fats contain relatively less oxygen than do carbohydrates, i.e. some of the 'excess' oxygen from the carbohydrate can be used in metabolism, thus reducing the respiratory oxygen uptake and increasing the respiratory carbon dioxide : oxygen ratio.

Since fat yields more than twice as much energy as carbohydrates, it is better suited for energy storage. However, there are exceptions to this rule. For animals that do not move about, such as oysters and clams, the weight is of minor consequence, and *glycogen* is used for storage. Also, many intestinal parasites, such as the roundworm *Ascaris,* store glycogen. In these animals glycogen is probably a more suitable storage substance than fat, for they are frequently exposed to conditions of low or zero oxygen, and in the absence of oxygen, glycogen can yield energy by breakdown to lactic acid. Therefore, since bivalves frequently close their shells for long periods and intestinal parasites live in an environment virtually devoid of oxygen, it is advantageous for these animals to store glycogen rather than fat. Sessile animals and parasites have little need for weight economy, and from this viewpoint the use of glycogen instead of fat is no disadvantage.

For animals that move about, however, weight economy is of great importance. It is particularly important for migrating birds, which may fly non-stop for more than 1000 km. To have enough fuel, migrating birds may carry as much as 40 to 50% of their body weight as fat; if the fuel were heavier, the weight would be excessive, and long-distance migration would be impossible because much of the available energy would merely go into carrying the heavier body.

Storage of glycogen involves much more weight, not only because of the lower energy content of carbohydrates as compared to fat, but also because glycogen is deposited in the cells with a considerable amount of water. It has been estimated that the deposition of glycogen in the cells of liver and muscle is accompanied by about 3 g water for each gram of glycogen stored. Some investigations even indicate that the amount of water may be as high as 4 or 5 g water per gram glycogen (Olsson and Saltin, 1968). With the extra weight of water, the carrying of energy in glycogen turns out to be some ten times as heavy as carrying the same energy in fat.

In spite of this weight difference, glycogen is an important storage form for energy. Its advantage is two-fold, it can provide fuel for carbohydrate metabolism very quickly while the mobilization of fat is slow, and perhaps more importantly, glycogen can provide energy under anoxic conditions. This is commonly the case during heavy muscular exercise when the blood does not deliver sufficient oxygen to meet the demands. For long-term storage of large amounts of energy, however, glycogen is unsuited, and fat is the preferred substance.

The fact that water is stored with the glycogen is probably the explanation for a puzzling observation made by many persons when they begin on a reducing diet. During the first few days they lose weight rapidly, often far in excess of what can be due to the loss of fat alone. In this initial period the glycogen stores become depleted, and the excess water is excreted. Much of the weight loss is therefore only loss of water. After a few days, even if the food intake is less than what is metabolized, the glycogen stores are replenished. When metabolism now turns to the use of stored fat there is no longer a rapid decrease in body weight, on the contrary, the body weight may increase again, although the person remains in negative calorie balance and continues to lose fat. These unexpected changes in weight, which are rarely adequately explained, cause many persons on a reducing diet to be over-optimistic during the first days, and then rapidly become discouraged again when they lose weight less quickly or even increase in weight in spite of their continued sacrifice of food intake.

Effect of oxygen concentration

It is often assumed that oxygen consumption (or metabolic rate) within wide limits is independent of the oxygen concentration. For example, if we replace the ordinary air with pure oxygen, a mammal will continue to consume oxygen at the same rate, although oxygen is present in nearly five times the usual concentration. The reverse is also true, if we reduce the oxygen concentration to half, which can be done by reducing the total atmospheric pressure to half (equivalent to about 6000 meters altitude), the rate of oxygen consumption likewise remains virtually unaffected.

This independence of oxygen concentration is not universal, the situation may be just the opposite. Fig. 6.1 shows how the oxygen uptake of some fish depends on the oxygen concentration in the water (Hall, 1929). For two of the species (scup and puffer) the oxygen uptake decreases by about one-third as oxygen in the water is reduced, but at very low oxygen concentration (less than about 15 mm Hg) both species are unable to survive. The toadfish is different, its oxygen consumption decreases linearly with a decrease in available oxygen, giving a straight line relationship. The toadfish can survive long periods in oxygen-free water, and in this situation its metabolic processes must be completely anaerobic with zero oxygen uptake.

Some invertebrates, the common lobster, for example, respond in a similar way to changing oxygen (see fig. 6.2). At high temperatures the oxygen consumption is higher than at low temperature, but at any one temperature the oxygen uptake is linearly related to the available oxygen (Thomas, 1954).

If the oxygen uptake is studied over a wider temperature range in an animal such as the goldfish, we obtain the relationship shown in

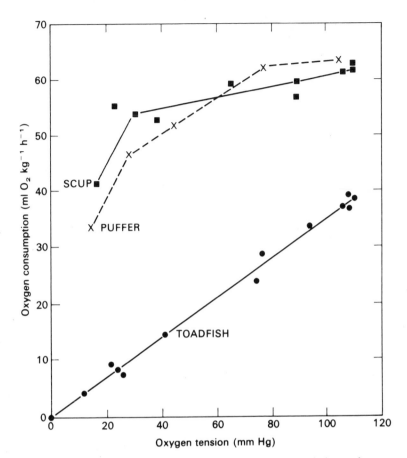

Fig. 6.1. The oxygen consumption of three species of fish in relation to the oxygen tension in the water. For two of the fish, scup and puffer, the oxygen consumption decreases only moderately until the oxygen concentration is quite low; for the third species, the toadfish, the oxygen consumption changes in proportion to the change in oxygen throughout the range (Hall, 1929).

fig. 6.3. At relatively high oxygen tension the oxygen uptake of gold-fish is independent of the available oxygen, but at lower oxygen tensions there is a linear relationship. The inflection point which indicates where the oxygen uptake changes from concentration dependence to independence is lower at low temperature than at high temperature. Thus, in fig. 6.3 the inflection point changes from less than 20 mm O_2 at 5 °C to about 40 mm O_2 at 35 °C (Fry and Hart, 1948).

If we return briefly to figs. 6.1 and 6.2 we can see that the straight line relationships observed for toadfish and for lobster may indicate

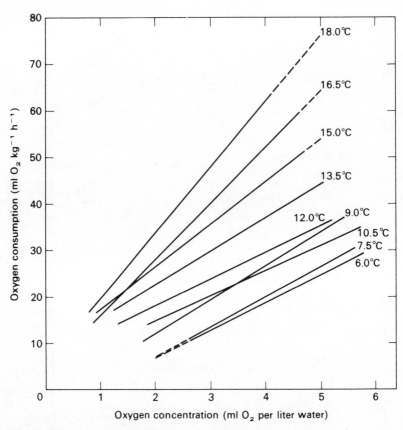

Fig. 6.2. The oxygen uptake of lobsters is higher at high temperatures, but at any given temperature it varies linearly with the oxygen concentration in the water (Thomas, 1954).

that we are looking at only the lower part of a curve of the same general shape as the goldfish curves. If this is so, we can expect that these animals at some unknown higher oxygen concentration would level off and become independent of further increases in oxygen.

Even mammals, whose metabolic rates show little dependence on available oxygen, may be more dependent on oxygen concentration than we usually assume. The oxygen consumption of isolated muscles of rats and rabbits is reversibly depressed at low oxygen, without alteration of other cell functions (Whalen, 1966). This is of importance to our evaluation of the observation that diving animals, after a dive do not increase their oxygen consumption enough to equal the oxygen debt they are expected to have acquired during the time of the dive (see pp. 232–4). Since the muscle mass is the major

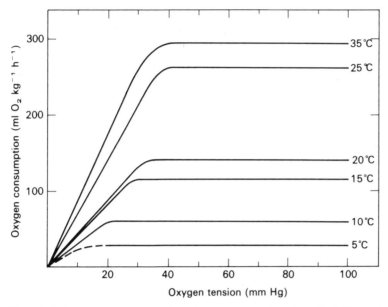

Fig. 6.3. At low oxygen concentrations the oxygen consumption of goldfish varies linearly with the oxygen content of the water, but at higher oxygen it is independent of the oxygen content (Fry and Hart, 1948).

oxygen consuming tissue, and oxygen-dependent metabolic rate in this tissue would yield exactly the result observed in many diving animals, the conclusion must be that oxygen-independent metabolism as described above is a special case, and that oxygen dependence probably is the general rule.

Acclimation to low oxygen

Many fast-swimming and active fish are quite sensitive to low oxygen in the water. This is true of most salmonid fish, and as an example we can examine the speckled trout (*Salvelinus fontinalis*). For this fish the lethal level for oxygen in the water depends on previous exposure to low but tolerable oxygen levels. If the fish has been kept in relatively oxygen-poor water, it becomes more tolerant to low oxygen (see fig. 6.4). This improved tolerance is probably due to an enhancement of the ability to extract oxygen from the water. If there was a true increase in the ability of the tissues to tolerate anoxic conditions, it could be expected that a fish acclimated to low oxygen should survive longer than a non-acclimated fish when placed in completely anaerobic conditions. However, there is no evidence of such a difference (Shepard, 1955). The improved oxygen

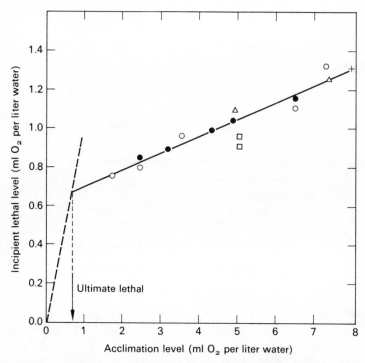

Fig. 6.4. If speckled trout have already been exposed to a reduced oxygen concentration in the water (abscissa), they become more tolerant to low oxygen and have a lowered lethal oxygen limit (ordinate) (after Shepherd, 1955).

extraction could be due to an increased volume of water pumped over the gills, or an increased ability to remove oxygen from the water flowing over the gills, or an improved oxygen transport capacity in the blood, or an improved function of the heart, or any combination of all these.

Anaerobic metabolism

Animals that tolerate prolonged exposure to zero oxygen must of necessity obtain their metabolic energy from non-oxidative reactions. This holds true for many intestinal parasites, for animals that live in the oxygen-free mud of lakes and ponds, for bivalves that remain closed during long periods of time, etc.

One common reaction available for anaerobic metabolism is the breakdown of carbohydrate into lactic acid. One mole of glucose can anaerobically be broken down to two molecules lactic acid ($C_6H_{12}O_6 \rightarrow 2\ C_3H_6O_3$). This process, known as glycolysis, occurs commonly in vertebrate muscle when the energy demands are high and exceed the available oxygen. The process, which has a free energy yield (ΔF) of 50 kcal per mole glucose, makes available only about 7% of the energy yield from complete oxidation of glucose (691 kcal per mol).

Most anaerobic glycolysis depends on glycogen, rather than glucose, as the substrate. For glycogen the molar energy is 7 kcal higher than for glucose, and lactate formation now yields 57 kcal. Thus, the energy available from glycogen in anaerobic glycolysis is slightly higher than for glucose. Since ATP is the common energy source for energy-requiring metabolic processes, we can consider, instead of the free energy yields, the amounts of ATP formed in aerobic and anaerobic metabolism, respectively. Glycolysis (formation of 2 moles lactic acid from 1 mole glucose) yields 2 moles ATP. (If glycogen is the substrate, the yield is 3 ATP per 2 moles lactic acid.) In contrast, the complete oxidation of 1 mole glucose to carbon dioxide and water yields 36 moles ATP (6 in the formation of 2 moles pyruvic acid, and 30 in the complete oxidation of the pyruvic acid via the Krebs cycle).

Thus, if an animal is totally anaerobic, the yield of ATP from glycolysis is small, but an intestinal parasite probably doesn't care much about economy. However, animals that use glycolysis only intermittently can be expected to use lactic acid as a substrate for further oxidation (in the Krebs cycle) and thus in the end gain the full energy value of the original carbohydrate substrate.

The tolerance to anoxic conditions sometimes is amazing. It has been reported that the crucian carp (*Carassius carassius*) can live for 5½ months under the ice of frozen lakes when the water under the ice is oxygen-free because of fermentation of dead plant material, and hydrogen sulfide is present (Blazka, 1958). (Hydrogen sulfide is poisonous because it binds to and inactivates cytochrome oxidase, but since no oxygen is present to permit oxidative metabolism, this does not poison the fish). These carp produce practically no lactic acid in the tissues, and other metabolic processes must therefore take place. It was suggested that in these carp the formation of fatty acids is the end result of metabolism, represented as an increase in the total amount of fat in the fish during the long period of anaerobic conditions. However, several other reactions are available for anaerobic metabolism, and a number of pathways are well known and established for a variety of organisms that can tolerate long periods or permanent absence of free oxygen (Hochachka and Mustafa, 1972).

We shall now move on to animals that, because of their diving habits, are intermittently subjected to complete exclusion from atmospheric oxygen.

Diving mammals and birds

All aquatic mammals and birds have retained lungs and breathe air, but a successful aquatic life requires a number of modifications in the usual terrestrial pattern. Perhaps the most obvious one is that lengthy dives require effective utilization of a limited oxygen supply, but other functional modifications also contribute importantly to a successful aquatic life.

Adaptation to an aquatic life is found in all major orders of mammals, except the Chiroptera (bats), Lagomorpha (rabbits and hares), primates, and marsupials (table 6.2). A similar list of aquatic and non-aquatic birds shows that the successful ones are phylogenetically quite diverse (table 6.3). The physiological adaptations which these animals employ to cope with the special demands of their environment, however, are similar. The fact that all divers have employed common mechanisms, and that to a lesser degree such mechanisms are present in non-diving vertebrates as well, suggest that the successful divers have taken advantage of, and developed, already existing mechanisms.

Some of the problems that must be solved by diving animals can readily be explained by reference to the problems encountered by a

Table 6.2. *Most of the major orders of mammals have a few aquatic representatives, and some are exclusively aquatic*

MAMMALIAN ORDER	AQUATIC REPRESENTATIVES
Monotremata	Platypus (*Ornithorhynchus*)
Marsupialia	None
Insectivora	Water shrew
Chiroptera	None
Primates	None
Rodentia	Beaver, Muskrat
Lagomorpha	None
Cetacea	Whales (all)
Carnivora	
Fissipedia	Sea otter
Pinnipedia	Seals (all)
Sirenia	Sea cows, manatee, dugong (all)
Ungulata	
Perissodactyla	Tapir
Artiodactyla	Hippopotamus

Table 6.3. *Many bird families are almost completely aquatic (except for their reproduction activities), others are primarily or exclusively non-aquatic*

PRIMARILY AQUATIC	PRIMARILY NON-AQUATIC
Penguins	Ratites (ostrich, rhea, etc.)
Loons	Hawks
Grebes	Fowls and pheasants
Petrels and albatrosses	Pigeons
Pelicans	Parrots
Herons	Owls
Ducks	Nightjars
Waders	Hummingbirds
Gulls	Woodpeckers
Auks	Passeriformes (finches, thrushes, swallows, etc.)

man who dives, either with or without equipment. Some of man's difficulties have no direct bearing on diving animals; others are directly applicable to them. All air-breathing animals that dive must cope with the problem of oxygen supply, whether they are shallow or deep divers, but the latter encounter additional problems which fall into four main groups, (1) bends or diver's disease, (2) oxygen toxicity, (3) narcotic effect of gases, and (4) direct effects of high pressures.

The bends. This dangerous syndrome is known under several names, diver's disease, caisson disease, and aeroembolism. It occurs when a human diver returns to the surface after spending a prolonged time at considerable depth, below 20 meters or so, and is increasingly severe with increasing depth and diving time. It is caused by gas bubbles in the tissues and the blood stream, formed in the same way that bubbles form in soda-water when the cap is removed from the bottle. In both cases bubbles appear when the pressure is reduced over a liquid that is saturated with gas at a higher pressure. The bends may also occur during rapid ascent in aircraft (balloons or airplanes) with non-pressurized cabins; in these the danger of bends occurs when the pressure is reduced to about 0.5 atm (above 6000 meters).

In soda-water the dissolved gas is carbon dioxide. The gas causing the bends is always nitrogen (unless the diver has been breathing artificial gas mixtures that contain other inert gases such as helium). Even when the diver is at great depth, the carbon dioxide tension does not increase much above the physiologically normal of about 40 mm Hg, and to cause the bends, a supersaturation of some 2 atm would be necessary. Oxygen will not cause the bends because it is rapidly used up by the tissues, and, as we shall see later, oxygen at a pressure of more than 1 atm is so toxic that higher concentrations must be avoided.

The pressure in water increases by about 1 atm for each 10 meters increase in depth. To keep an ordinary diving suit inflated with air at 10 meters depth therefore requires 1 atm pressure above normal atmospheric pressure, or a total air pressure of 2 atm. If a diver at 10 meters depth is supplied with ordinary air, he will be breathing air which has 0.4 atm oxygen pressure and 1.6 atm nitrogen pressure. The blood dissolves nitrogen at the higher pressure, and gradually all the tissue water will become equilibrated with nitrogen at this pressure. Furthermore, the body fat, which in a normal lean male constitutes about 15% of his body weight, dissolves about five times as much nitrogen per unit weight as does water. Our standard 70 kg man contains nearly 50 liters water and 10 kg fat, the total amount of nitrogen dissolved in the fat therefore about equals the amount dissolved in the body water. It takes considerable time before a man at a given depth reaches full saturation with nitrogen. During short dives this is an advantage, but after a long dive, when the tissues are nearly saturated with nitrogen at high pressure, the danger increases. When the diver returns to the surface, it takes up to several hours to eliminate the nitrogen, for it is only slowly carried from the tissues, dissolved in the blood, and released in the lungs.

The danger of bubble formation is greater when gas is dissolved at a higher pressure. Increased movements, muscular effort, and increased circulation tend to increase bubble formation, analogous to the effect of shaking a beer can or a soda-water bottle before opening. Bubbles often form in the joints, which causes considerable pain. If bubbles occur in the blood stream, they block the finer blood vessels, and when this happens in the central nervous system it is particularly dangerous and may cause rapid death. The only possible treatment for the bends is quickly to increase the pressure again, so that the bubbles re-dissolve. This can be done either by returning the diver to the depth from which he ascended, or by transferring him to a pressure chamber where the air pressure can be increased to the desired level. The way to prevent the bends is to return a diver slowly to the surface by stages. It is reasonably safe to let a diver ascend to a point where the pressure is half that at which he was working, and keep him there until a considerable amount of the nitrogen has been eliminated, say 20 to 30 min. It is then feasible to let him ascend part of the way again, and thus continue the process of staging. Staging tables which give precise schedules for the fastest possible safe ascent from a given depth are widely used.

If a diver uses a tank of compressed air, he must breathe air at the pressure of the surrounding water, or his chest would be collapsed by the water pressure. The scuba diver is therefore in the same danger of developing the bends as the diver using the ordinary diving suit.

To some extent the danger of bends can be reduced by using gases other than nitrogen. One widely used gas is helium, which has the advantage that it is less soluble than nitrogen in water and fat, and that it diffuses several times faster. The faster diffusion reduces the time necessary for staging. On the other hand, it also speeds up the saturation of the tissues with helium, and for short dives this counteracts some of its advantages.

Our discussion so far has been concerned only with divers supplied with air under pressure. However, a skin-diver without equipment who repeatedly descends to some considerable depth is also in danger of developing bends. Each time he fills his lungs and descends, the water pressure on his chest compresses the rib cage so that the air pressure in the lungs increases. If he descends to, say, 20 meters, the air pressure in the lung will be 3 atm. Although a single or a few dives to 20 meters depth will not be of any significance, many repeated dives are risky. A well-documented case of a skin diver developing the bends occurred when a Danish physician practised underwater escape techniques in a 20 meter deep training tank. He used the so-called 'bottom drop' technique, in which the diver after a deep breath pushes himself downward from the edge of the tank. After descending two or three meters the chest is compressed so much that he continues to drop to the bottom with increasing speed. After five hours of spending about two minutes at the bottom during each of 60 dives, he developed pain in the joints, breathing difficulties, blurred vision, and abdominal pains. He was placed in a recompression chamber at 6 atm pressure, whereupon the symptoms rapidly disappeared. He was then 'staged' to 1 atm pressure, according to the US Navy staging tables, which required 19 h 57 min. This treatment was completely successful, but in many other cases this has not been so (Paulev, 1965).

If a man develops the bends by repeated breath-hold dives, how can seals and whales which dive over and over again avoid this danger? Often they come to the surface for seconds only, and they may dive to considerable depths.

The best diver among the seals is the antarctic Weddell seal, which voluntarily will dive to below 600 meters and stay under water for as long as 43 min (Kooyman, 1966). The deepest diving whale is the sperm whale (*Physeter catadon*), for which the known record is 1134 meters. This curious record was obtained because a trans-Atlantic cable was broken at this depth, and when the cable was taken up for repair, a dead sperm whale had its lower jaw entangled in the cable. The way the jaw was caught showed that the whale had been swimming along the bottom when hitting the cable and had tried to free itself.

Why can seals and whales tolerate deep dives and rapid ascent without developing the bends? There are three possibilities, (1) they are tolerant to bubbles, (2) they have mechanisms to avoid bubble formation in spite of supersaturation, or (3) there is no supersaturation. The information we have, although sparse, points to the third possibility.

The most important difference between seals and whales on the one hand and a man diving with equipment on the other is that the animals do not receive a continuous supply of air. However, like man, many repeated dives with air-filled lungs should eventually give bends. In this regard it is significant that seals exhale at the beginning of each dive, rather than diving with filled lungs as a man usually does. Whether whales also exhale before diving is uncertain, but the best divers among the whales have relatively small lung volumes. In man the filled lungs occupy about 7% of the body volume, in a fin whale less than 3%, and in the bottlenose only about 1% (Scholander, 1940). As a whale dives, the increasing water pressure compresses the lungs and forces air into the trachea, which is large and rigidly reinforced with circular bone rings. If the volume of the non-compressible trachea is one-tenth of the lung volume, the lung will be completely collapsed and all the air forced into the trachea at 100 meters depth. With no air in the lung, nitrogen cannot enter the blood, as it does in man. Another reason that nitrogen does not enter the blood is that the blood flow to the lung is probably arrested during the dive (see p. 228ff); thus, even if the lung contains some air before the animal reaches the depth where the lung is collapsed, no nitrogen will be carried from the lung.

Oxygen toxicity. Pure oxygen at 1 atm pressure is harmful to most warm-blooded animals. A man can safely breathe 100% O_2 for 12 hours, but after 24 hours there is substernal distress and increasing irritation of the lungs. Rats that are kept in pure oxygen will die with symptoms of severe lung irritation after a few days. If the oxygen pressure is higher than 1 atm, nervous symptoms develop before the general irritation of the respiratory organs sets in. Continued breathing of oxygen at 2 atm pressure causes convulsions, and exercise reduces the tolerance. A pressure of 3 atm oxygen can be tolerated by man for a few hours, but longer exposure must be avoided.

The importance of oxygen toxicity to diving is obvious. If a diver descends to 40 meters, being supplied with compressed air, he breathes at a total pressure of 5 atm. Since one-fifth of the air is oxygen, the partial pressure of oxygen is 1 atm; i.e. near the toxic limit. Dives with compressed air at even greater depths will of course be increasingly harmful as the oxygen partial pressure exceeds one atmosphere.

The only practical way to avoid oxygen toxicity for divers working at 40 meters or more is to decrease the oxygen content of the air they breathe. A diver working at, say, 40 meters (5 atm) could be supplied with a gas containing 10% O_2 and 90% N_2. He will then breathe oxygen at a partial pressure of 0.5 atm, which is completely safe. At greater depth the oxygen content of the gas mixture should be further reduced, and to reduce the danger of bends during the decompression, nitrogen may be partly or completely replaced by other gases such as helium.

Even so, there are limits to the depth to which a man can dive safely, for the inert gases have other effects which appear at high pressure. Nitrogen at several atmospheres pressure has a *narcotic effect*, similar to that of nitrous oxide ('laughing gas'), but nitrogen requires higher pressures. The narcotic effect requires some time to develop, but at a depth of about 100 meters it is so severe that nitrogen–oxygen mixtures fail to be acceptable to divers. It is therefore necessary to use other gases, helium again being favored. Hydrogen is another possibility, although much more hazardous because of the danger of explosions. Eventually, helium under high pressure also has serious physiological side effects, thus limiting the practical depth to which it is possible to dive, even for the well-equipped man, unless he is within a completely rigid structure, such as a bathyscaphe, in which a low pressure can be maintained.

Diving animals are not subject to the dangers of oxygen toxicity or narcotic effects of inert gases for the simple reason that they do not breathe a continuous supply of air during the dive.

Oxygen supply during diving

For a man who swims under water without equipment, the most immediate need is to obtain oxygen. He usually fills his lungs to increase the amount of air he carries with him, thus postponing the moment he must surface to breathe again. There are, however, many possible solutions for extending diving time (see table 6.4). Some of these are not physiologically feasible or are not known to be used by animals. A combination of several possible avenues has been adopted by most diving animals, often resulting in striking diving performances when measured against human standards.

Table 6.4. *Possible solutions for extending under-water time for diving animals. Solutions marked by an asterisk are not widely used, or are physiologically improbable or impossible*

1. Increased O₂ storage
    (*a*) Increased lung size *
    (*b*) Blood, higher blood volume and higher hemoglobin content
    (*c*) Tissues, higher muscle hemoglobin
    (*d*) Other 'unknown' O₂ storage *
2. Use of anaerobic processes
    (*a*) Lactic acid
    (*b*) 'Unknown' hydrogen acceptors *
3. Decreased O₂ consumption
    i.e. decreased metabolic rate
4. Aquatic respiration
    (*a*) Cutaneous respiration (frog)
    (*b*) Esophageal or rectal respiration (some turtles)
    (*c*) Breathing water (only tried experimentally) *

Oxygen storage. Increasing the amount of oxygen in the lung seems a simple solution to the problem of carrying more oxygen during the dive. However, diving animals have no greater lung volumes than other mammals, and many of the best divers have conspicuously small lung volumes. A moment's thought helps to explain why this is so. First of all, a large amount of air in the lung may permit more nitrogen to invade the body fluids and thus increase the hazards of the bends. Furthermore, if the lungs are particularly large, an animal will have difficulty in initially becoming submerged. This may be a minor matter, however, for once an animal with a large lung were successfully submerged, compression of the lung would decrease the buoyancy and there would be no difficulty in remaining at depth.

Let us now examine what other oxygen supplies may be available in a seal at the beginning of a dive, and see how long this oxygen may last. Such an account has been given by the Norwegian physiologist, P. F. Scholander, who compared the oxygen stores in a small seal and a man (table 6.5).

The most striking difference is the small amount of oxygen the

Table 6.5. *Oxygen stores in seal and man. The figures are approximate; for example, the blood oxygen content will be lower than listed because a large part of the blood is located in the venous system and not fully saturated with oxygen.* (Data for seal from Scholander, 1940)

		(ml O₂)
SEAL	Air in lungs, 350 ml, 16% O₂	50
(30 kg)	Blood, 4.5 liter, 25 ml O₂ per 100 ml	1000
	Muscle, 6 kg, 4.5 ml O₂ per 100 g	270
	Tissue water, 20 liter, 5 ml per liter	100
	Total:	1520
	ml O₂ per kg body weight:	50
MAN	Air in lungs, 4.5 liter, 16% O₂	800
(70 kg)	Blood, 5 liter, 20 ml O₂ per 100 ml	1000
	Muscle, 16 kg, 1.5 ml O₂ per 100 g	240
	Tissue water, 40 liter, 5 ml per liter	200
	Total:	2240
	ml O₂ per kg body weight:	32

seal carries in its lungs. Rather than filling the lungs at the beginning of a dive, all seals seem to exhale and begin the dive with a minimal volume of air in the lungs. This is the reason for the very small amount of oxygen in the lung-air of the seal, as compared to man, who fills his lung with air when he dives.

The amount of oxygen in the blood, however, is a different matter. Although the seal weighs less than half as much as the man, its total blood volume is almost as great as in a man. Furthermore, seal blood has a higher oxygen capacity than human blood, and as a result the total amount of oxygen carried in the blood of the seal exceeds that of the man. The high oxygen capacity of seal blood is characteristic of good divers. Although the oxygen capacity of human blood, 20 ml O₂ per 100 ml blood, is in the higher range of terrestrial mammals, seals and whales have oxygen capacities ranging up to between 30 and 40 ml per 100 ml blood. There is a limit to how much the oxygen capacity can be increased, however. The red blood cells are already about as densely packed with hemoglobin as possible, and if the red cells make up more than about 60% of the total blood volume the blood will not flow freely, it becomes so viscous that the heart must work excessively to pump the blood through the vascular system. Increasing the blood volume is one way of increasing the total amount of hemoglobin in the body without increasing blood viscosity. This solution is common among successful divers, but again, there are limits to how much blood it is feasible to have in the body.

In the seal the second largest amount of oxygen is stored in the muscle mass. The red color of mammalian muscle is due to hemoglobin (myoglobin). It is located within the muscle cell, and therefore remains even after all blood has been drained out. The hemoglobin content of muscle is much lower than in blood, but due to the large muscle mass of the body, about 20% of the body weight,* the muscle hemoglobin carries a substantial amount of oxygen. Seal and whale muscle have much higher hemoglobin content than muscle of terrestrial mammals (their meat is visibly darker than other mammalian muscle), and as a result the 30 kg seal has in its muscles about the same amount of stored oxygen as in the 70 kg man.

All body water will contain some dissolved oxygen. The amount depends on the water content of the body, which is similar in man and seal, but the total amount is small. Some oxygen is dissolved in the body fat, but it is unimportant as an oxygen store for the simple reason that fat is poorly vascularized and probably excluded from circulation during diving.

If we add up all the oxygen we can account for in table 6.5, it shows that a seal may have, per kilogram body weight, substantially more oxygen than a man, although not quite twice as much.

The stored oxygen, if used up at the normal metabolic rate of the seal at rest, would last at most five minutes. However, the seal is able to dive for at least three times as long. The total oxygen store in man would be sufficient to supply his metabolic rate at rest for perhaps four minutes, but this exceeds the maximum diving time for man, which is about two minutes (excepting unusual record performances). The reason is simple, a man cannot tolerate full depletion of all free oxygen in the body, and the maximum diving time is therefore shorter than that calculated from the total oxygen stores. Why, then, can a seal dive for much longer?

We cannot *a priori* exclude some 'unknown' oxygen store in the seal, but on the basis of present knowledge, it seems most unlikely. What we know about diving mammals indicates that the performance of even the best divers can be fully accounted for without invoking new and so far unknown mechanisms.

Circulation during diving. One of the most characteristic physiological responses observed in a diving seal is that the heart rate at the

---

* This figure is low because of the large amount of fat in the body. The muscles make up a much larger fraction of the fat-free body mass.

beginning of submersion decreases precipitously to rates as low as one-tenth of the normal value or even less (see fig. 6.5). This sudden drop in heart rate must be due to a nerve reflex, for if it were a response to a gradual depletion of oxygen it would develop more slowly.

In spite of the decrease in heart rate, the arterial pressure is maintained. This can only happen if the peripheral resistance in the vascular system is increased through vasoconstriction. With the reduction in cardiac output, only the most vital organs receive any blood, these are the central nervous system and the heart itself. There is a complete redistribution of blood flow, and muscles as well as abdominal organs receive little or no blood during the dive. In a way this makes the seal into a much smaller animal, the oxygen is reserved for a small part of the body, i.e. the organs for which oxygen is most essential. The oxygen present in the circulating blood at the beginning of the dive is therefore used up only slowly, and the duration of a dive approximately coincides with the time it takes to deplete fully the oxygen in the blood.

More accurate information about the distribution of blood flow during a dive has been obtained by injection of a radioactive isotope of rubidium, which rapidly exchanges with potassium inside the cells and thus gives an estimate of the amount of blood reaching an

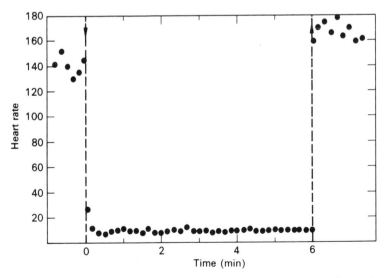

Fig. 6.5. When a harbor seal is submerged, its heart rate drops precipitously from about 140 to less than 10 beats per minute. Beginning and end of dive indicated by arrows (Elsner, 1965).

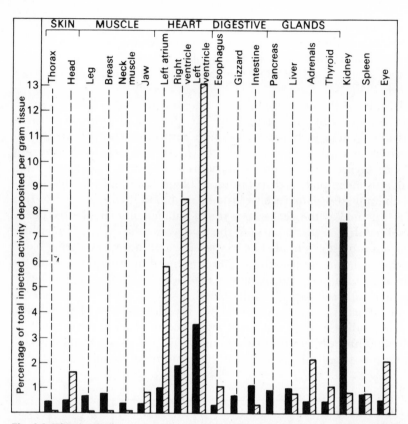

Fig. 6.6. The circulation to various organs of a diving duck. Black bars indicate values during normal breathing of air, cross-hatched bars values obtained during submersion. The circulation was determined with the aid of radioactive rubidium, and the height of a bar indicates the percentage of the total injected amount deposited per gram tissue (after Johansen, 1964).

organ. Fig. 6.6 shows the result of an experiment on a duck. During the dive the circulation is increased especially to head, heart, and eye, and the greatest decrease is for the kidney, which in the non-diving state receives a large fraction of the total blood supply (Johansen, 1964).

Most muscles, whether used vigorously or not, receive virtually no blood during the dive. The work of the muscles is based entirely on anaerobic processes, which lead to the formation of lactic acid in the muscles. Since there is no blood flow to the muscles, this lactic acid accumulates in the muscle during the dive, and lactate concentration in the blood remains low. Immediately after the dive, however, when circulation to the muscles is resumed, the blood lactic acid concentration shoots up. (The ability of seal muscle to work anaerobically is not unique; lactic acid is normally formed in the muscles of man during strenuous exercise or athletic performances.)

The muscle hemoglobin in a diving seal provides some oxygen, but this oxygen is depleted within a period much shorter than the normal duration of a dive. The amount of oxygen stored in the muscle hemoglobin is much too small to explain the long duration of a dive, it is the circulatory adjustments and the ability of muscle to work anaerobically that are the major factors.

Slowing of the heart beat during diving has been recorded in a large number of mammals, seals and whales, beaver, muskrat, hippopotamus, ducks, and so on. In some of these animals the heart slows down more gradually than in fig. 6.5, but a substantial drop seems to be universal in all divers. It is interesting that a fish taken out of water shows a similar response, its heart rate immediately decreases. This has been observed in both teleosts and elasmobranchs. An air-breathing fish, such as the mud-skipper (*Periophthalmus*) (see p. 48), differs from fully aquatic fish in that its heart rate decreases when it escapes into its water-filled burrow. In this respect the mud-skipper reacts as a terrestrial air breather which performs a dive (Gordon *et al.*, 1969).

A record of respiration and blood gases in a diving seal are shown in fig. 6.7. The top graph (fig. 6.7a) shows that the blood oxygen gradually is depleted during the dive while the carbon dioxide shows a corresponding increase. The blood lactic acid changes only a little during the dive, but after the dive there is a wave of lactic acid appearing in the blood as soon as air breathing is resumed. This lactic acid, which signifies an oxygen debt, is afterwards gradually removed from the blood. Part of the lactic acid is re-synthesized into glycogen by the liver and the muscles, and a smaller fraction is completely oxidized to carbon dioxide and water. The center graph (fig. 6.7b) shows the ventilation volume, which during the dive is zero, followed by an increased ventilation of the lung immediately after the dive. The lower graph (fig. 6.7c) shows the corresponding respiratory oxygen uptake, which, immediately after the dive, shows a sharp increase. Most of the time the carbon dioxide elimination is of the same magnitude as the oxygen uptake, although a little lower, except immediately after the dive, when the carbon dioxide output is greatly increased. This is because the surge of lactic acid in the blood drives out carbon dioxide from the bicarbonate in the blood, as vinegar poured on baking soda drives off the carbon dioxide.

Decreased metabolic rate. Fig. 6.7c shows that the 'oxygen debt' acquired during the dive is followed by an increased oxygen uptake.

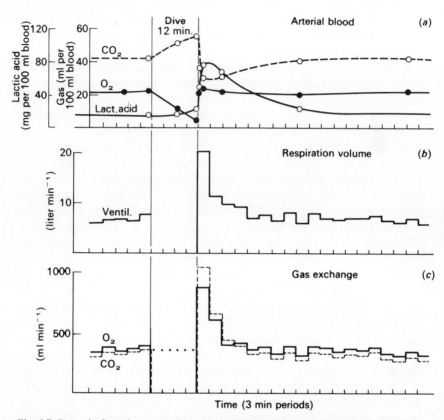

Fig. 6.7. Record of respiratory exchange in a 29 kg seal during a 12 minute experimental dive. (*a*) Blood concentrations of oxygen, carbon dioxide, and lactic acid. (*b*) Ventilation volume. (*c*) Respiratory oxygen uptake and carbon dioxide release (after Scholander, 1940).

The expected oxygen debt is represented by the area under the dotted line (which indicates the normal metabolic rate of the seal). The increased oxygen uptake after the dive indicates that the oxygen debt acquired during the dive is being paid off, and the after-dive increase should equal the debt. If the area under the after-dive oxygen uptake curve is compared with the oxygen debt, however, the amount repaid seems to be smaller. This could be due to a long-lasting extension in time of the processes used to repay the oxygen debt which cannot easily be distinguished from the normal resting metabolic rate. It could, however, be that there is a true decrease in metabolic rate during the dive, and that the true oxygen debt is therefore of a smaller magnitude than indicated in the graph.

Further comparisons of the after-dive oxygen uptake and the ex-

pected oxygen debt show that a reduced metabolic rate is the probable explanation. In the duck the observed after-dive increase in
oxygen consumption is about one-quarter or one-third of the expected oxygen debt; this can be observed in consecutively repeated
dives, although the duck shows no signs of an accumulating oxygen
debt. A decrease in metabolism to values below the normal resting
rates is unexpected, for we are accustomed to thinking of metabolic
rate as being very constant. This is not necessarily so. It should be
remembered that during the dive many organs are very poorly circulated, or not circulated at all. We saw that this is the case with the
kidney, which normally has a high blood flow and a high oxygen
consumption. During a dive circulation to the kidney is practically
nil, and there is an immediate cessation of glomerular filtration and
of urine production (Murdaugh *et al.*, 1961). Other organs are also
virtually excluded from circulation during diving, and we can expect
that their activity practically ceases, thus decreasing the amount of
oxygen needed to repay an oxygen debt after the dive.

It is not easy to measure accurately the *true* metabolic rate during
the dive of a mammal or a bird. However, aquatic turtles are highly
tolerant to prolonged submersion and decreased oxygen, and can be

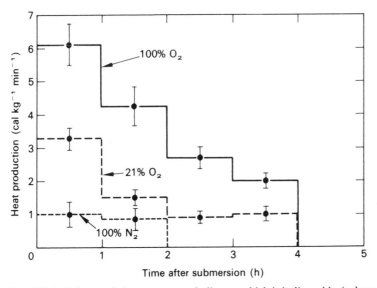

Fig. 6.8. A diving turtle has a true metabolic rate which is indicated by its heat production. During the dive, the metabolic rate of a turtle that has been breathing air
gradually decreases to a level of about 1 cal per kg body weight per minute, the same
level as when the turtle has been breathing pure nitrogen before the dive (Jackson
and Schmidt-Nielsen, 1966).

used for studies of the metabolic rate during diving. Oxygen consumption is obviously not a measure of metabolic rate under anaerobic conditions, but heat production accurately reflects the metabolic rate (see p. 209). In a diving turtle the metabolic rate, measured as heat production, gradually decreases, as shown by the dashed curve in fig. 6.8. After two hours' submersion, metabolic heat production has reached a minimum value which is similar to the level obtained if the turtle is submerged after breathing nitrogen. This heat production indicates the level of totally anaerobic processes which is sustained in the complete absence of oxygen. If, on the other hand, the turtle is given pure oxygen to breathe before a dive, its heat production during submersion is initially very high, and gradually decreases as the oxygen is depleted. In the turtle, then, metabolic rate is clearly dependent on the available oxygen, and gradually decreases as the oxygen supply is depleted. These results are fully in accord with the fact that the oxygen consumption of resting muscle is not constant but is dependent on the blood flow (i.e. the oxygen supply) (Whalen, Buerk and Thuning, 1973).

Metabolic rate is reduced thus O_2 debt not as high as expected

Addendum: 'liquid breathing'

What would happen if a terrestrial animal would attempt to breathe water? We know that for man this is impossible, he drowns. We also know that one liter of water contains only a small fraction of the amount of oxygen contained in one liter of air. It is therefore reasonable to ask what would happen if we were to increase the amount of oxygen dissolved in water, which can be achieved by equilibrating the water with oxygen under high pressure.

This has been tried. If water is equilibrated with pure oxygen at 8 atm pressure, it will dissolve about 200 ml O_2 per liter (at 37 °C); in other words, it will contain about the same amount of oxygen as one liter of ordinary air. Another measure must also be taken, instead of using water to breathe, we must use a balanced salt solution similar to the blood plasma. If just water were used, it would result in a rapid osmotic uptake of water into the blood in the lung, and cause hemolysis of the red corpuscles and serious loss of salts. In fact, the actual cause of death in fresh-water drowning is usually due to this effect of inhaled water. (For this reason artificial respiration is often ineffective in saving victims of fresh-water drowning and more successful in salt-water drowning.)

Complete submersion in oxygen supersaturated water has been tried with success with both mice and dogs, and the animals have survived for several hours. Mice introduced into a pressure chamber and submerged in water supersaturated with oxygen first try to reach the surface, but if they are kept under, they will inhale the oxygenated liquid, and instead of drowning they continue to breathe the oxygenated saline solution.

A major drawback in liquid breathing is that water is approximately 50 times more viscous than air, and the work of breathing is therefore correspondingly increased. Another problem is that the normal surfactants that line the lung are washed away during liquid breathing; this causes no difficulty during the experiment, but after return to air breathing there is a tendency for the lungs to collapse.

A more important limitation on liquid breathing is that the solubility of carbon dioxide in a saline solution at 40 mm Hg P_{CO_2} (the normal P_{CO_2} of blood), requires a larger volume of water than of air. Animals used in experiments with liquid breathing show this quite clearly. Although the arterial blood is fully saturated with oxygen, the carbon dioxide concentration increases greatly. This problem will be very difficult to overcome (Kylstra, 1968).

Another approach to liquid breathing has been tried. Instead of using water, certain synthetic liquids, such as fluorocarbons, have been used. Oxygen is extremely soluble in some of these organic liquids and this alleviates the problem of saturation at high oxygen pressure. The solubility of carbon dioxide, however, is not as great, and this will cause difficulties in the practical application of fluorocarbons. Such liquids have been tried with animals, which have survived several hours' liquid breathing without complications. For man, however, this does not seem to be a promising avenue for practical applications.

Liquid breathing has one potential advantage that under certain very limited circumstances could be of value. Since only liquid and no gas is inhaled, there is no need for a carrier gas such as nitrogen or helium. This gives a theoretical possibility for using liquid breathing at very high pressures without supersaturating the body fluids with an inert gas, and the danger of bends could therefore in theory be avoided. The practical use of this, if any, will be very limited.

Metabolic rate and body size

Among the diving mammals we have just discussed, the large whales can remain submerged for perhaps a couple of hours or so, but a very small diver, such as the water shrew, makes dives that rarely exceed half a minute. On the whole, the larger a diving mammal, the longer lasting are the dives it can perform. Why this difference? The answer is simple, the rate of oxygen consumption per gram body mass of the small mammal is much higher than that of the larger mammal. Let us now compare small and large animals in this regard.

The rates of oxygen consumption of a variety of mammals are listed in table 6.6. The largest animal, the elephant, is a million times larger than the smallest, the shrew, and its total oxygen consumption obviously must be much higher. However, we do not get a fair comparison of the two animals by comparing their total oxygen consumption. If instead we calculate the rate of oxygen consumption per unit body mass (last column), we get a striking picture of the relationship between body size and oxygen consumption.

Table 6.6. *Observed rates of oxygen consumption in mammals of various body sizes*

ANIMAL	Body mass, M_b (g)	Total O_2 consumption, $\dot{V}_{O_2}$ (ml O_2 h^{-1})	O_2 consumption per gram, $\dot{V}_{O_2}/M$ (ml O_2 g^{-1} h^{-1})
Shrew (a)	4.8	35.5	7.40
Harvest mouse (b)	9.0	22.5	2.50
Kangaroo mouse (c)	15.2	27.3	1.80
Mouse (d)	25	41	1.65
Ground squirrel (e)	96	98.8	1.03
Rat (d)	290	250	0.87
Cat (d)	2 500	1 700	0.68
Dog (d)	11 700	3 870	0.33
Sheep (d)	42 700	9 590	0.22
Man (d)	70 000	14 760	0.21
Horse (d)	650 000	71 100	0.11
Elephant (d)	3 833 000	268 000	0.07

(a) Hawkins, Jewell and Tomlinson (1960).   (d) Brody (1945).
(b) Pearson (1960).   (e) Hudson (1962).
(c) Bartholomew and MacMillen (1961).

We see that the rate of oxygen consumption per gram decreases consistently with increasing body size. This is even more apparent if the data are plotted graphically, as in fig. 6.9. In this graph the abscissa has a logarithmic scale; otherwise all values for the medium-to-small animals would be crowded together at the left part of the graph. The ordinate, on the other hand, is on a linear scale, and we can now see that one gram of shrew tissue consumes oxygen at a rate some 100-fold as great as one gram elephant tissue. This tremendous increase in oxygen consumption of the small animal necessi-

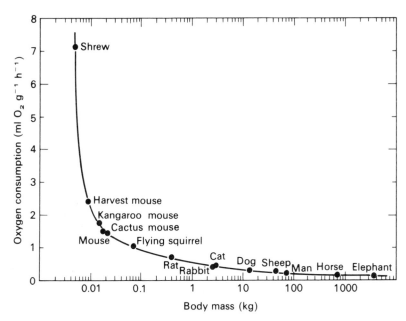

Fig. 6.9. Observed rates of oxygen consumption of various mammals. The oxygen consumption per unit body mass increases rapidly with decreasing body size. Note that the abscissa is logarithmic while the ordinate has an arithmetic scale.

tates that the oxygen supply, and hence the blood flow, to one gram shrew tissue must be about one hundred times greater than in the elephant. Other physiological variables, heart function, respiration, food intake, and so on must of course be similarly affected.

If we plot the data from table 6.6 on a graph with logarithmic scales on both coordinates, the points fall more or less along a single straight line (fig. 6.10). This regression line represents the generalization that the oxygen consumption of mammals per unit body weight decreases regularly with increasing body size, and it also gives a quantitative expression of the magnitude of the decrease.

We could use this graph to read off the 'expected' oxygen consumption for a mammal of any given size. For example, a typical mammal of 1 kg size can be expected to have an oxygen consumption of 0.6 ml O_2 g^{-1} h^{-1}. If we find that a mink which weighs 1 kg has a rate of oxygen consumption twice as high as indicated for a 1 kg mammal on the graph, we say that the mink has a higher oxygen consumption than expected for a typical mammal of its size.

Instead of using the graph, the regression line can be represented by the equation:

$$\dot{V}_{O_2}/M_b = 3.8 \times M_b^{-0.25}$$

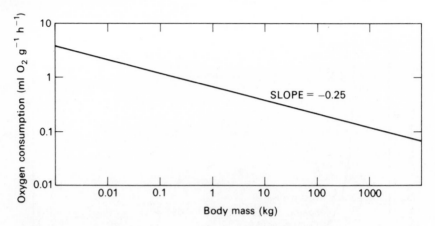

Fig. 6.10. The rates of oxygen consumption of various mammals, when calculated per unit body mass and plotted against body mass on logarithmic coordinates, tend to fall along a straight regression line with a slope of -0.25. Data from table 6.6.

in which $\dot{V}_{O_2}/M_b$ is oxygen consumption in ml O_2 per gram body mass per hour (ml O_2 g^{-1} h^{-1}) and M_b is body mass in grams. This general equation is based on numerous observations on a much greater variety of mammals than listed in table 6.6, and ranging in size from a few grams to several tons. Observations on a single species may deviate more or less from the line, which only represents the best fitting straight line calculated from all available data. Various authors have arrived at slightly different numerical constants, for example, an exponent or slope of -0.27, but these differences are too small to be statistically valid (Kleiber, 1961).

It is easier to use the above equation for calculations if it is written in the logarithmic form:

$$\log \dot{V}_{O_2}/M_b = \log 3.8 - 0.25 \log M_b.$$

This equation is of the general form $y = b + ax$, which represents a straight line with the slope a. In the metabolic equation above the slope is negative ($a = -0.25$) (see also appendix 3).

If we return to table 6.6 and consider the oxygen consumption of the entire animal (column 2) we find, of course, that a whole elephant consumes more oxygen than a mouse. If the data are plotted on logarithmic coordinates, we obtain a straight line with a positive slope of 0.75. The equation for the regression line is

$$\dot{V}_{O_2} = 3.8 \, M_b^{0.75}.$$

This equation can be derived from the preceding equation by multiplying both sides by body mass ($M_b^{1.0}$). For arithmetic calculations,

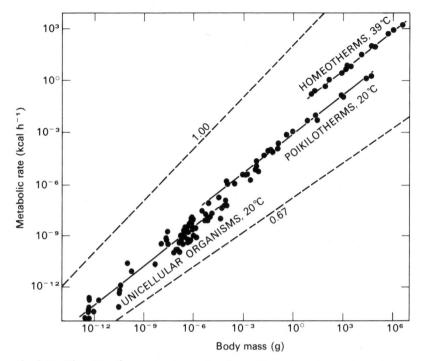

Fig. 6.11. The rates of oxygen consumption for a wide variety of organisms when plotted against body mass (log coordinates) tend to fall along regression lines with a slope of 0.75. Note that each division on the coordinates signifies a thousand-fold change (Hemmingsen, 1960).

it is again easier to use the equation in its logarithmic form, $\log \dot{V}_{O_2} = \log 3.8 + 0.75 \log M_b$, as described above.

Data which relate oxygen consumption to body size have been compiled for all sorts of animals, both vertebrate and invertebrate. Oxygen consumption is mostly lower for the cold-blooded than for the warm-blooded vertebrates, but the data again tend to fall on straight lines with slopes of about 0.75 (fig. 6.11). Many invertebrate animals also have rates of oxygen consumption which fall on the same or similar regression lines, although there are some exceptions to this general rule. For example, some insects, pulmonate snails, and a few other groups have oxygen consumption rates that fall on regression lines with a slope closer to 1.0. A slope of 1.0 means that the rate of oxygen consumption is directly proportional to the body weight, i.e. an animal twice as big has twice as high oxygen consumption, and so on.

The oxygen consumption rates of micro-organisms also fall on lines with a similar slope, and even some trees show the same relationship to size (Hemmingsen, 1960). The fact that the oxygen consumption of so many different organisms bear the same relationship to their size is such a common phenomenon that it seems to represent a general biological rule.

Several physiologists who have worked with these problems have attempted to reach a rational explanation for the regular relationship between oxygen consumption, or metabolic rate, and body size.

Nearly 100 years ago the German physiologist Max Rubner examined the metabolic rates of dogs of various sizes. He found that smaller dogs had a higher metabolic rate per unit body weight than larger dogs, a finding that is entirely in accord with the above discussion. Rubner had the clever idea that this relationship might be due to the fact that a smaller animal has a larger body surface relative to the body weight than a large animal. Since small and large dogs have the same body temperature, in order to keep warm they must produce metabolic heat in relation to their heat loss. Small dogs, due to their larger relative surface must therefore produce a greater amount of heat per unit body weight. Rubner then calculated the heat production per square meter body surface, and found that it

Plate 4. *Body size.* This high-speed photograph shows a shrew which instantly attacks a cockroach thrown to it for food. It gives a good impression of the relative body size of one of the smallest mammals and a moderately large insect (Peter Morrison, University of Alaska; from *Physiological Zoology*, 1959, **32,** 263, © University of Chicago Press).

was approximately 1000 kcal per m² body surface per day, both in large and small dogs. He therefore thought that his theory had been confirmed, and that metabolic rate was determined by surface area and the need to keep warm, a conclusion which became known as Rubner's surface rule.

Unfortunately, the surface rule cannot be explained as being due to problems of heat loss, for temperature regulation is not a problem for fish or crabs or beech trees, which follow the same metabolic relationship to body size as mammals do. Furthermore, if the metabolic rates were really proportional to surface, the slopes of the regression lines in fig. 6.11 should be 0.67.* Actually the slopes are closer to 0.75. The fact that cold-blooded vertebrates as well as many invertebrates (and at least some plants) have regression lines with the same slope, excludes the possibility that temperature regulation is a primary cause of the regularity of the regression lines.

The regular relationship between metabolic rate and body size, and the value of 0.75 for the slope of the regression lines, is not easy to explain. We can say, however, that it would be virtually impossible to design mammals of widely different body sizes that would follow a metabolic regression line with a slope of 1.0, that is, with metabolic rates directly proportional to body size. It has been calculated by Kleiber (1961) that if a steer were designed with the weight-related metabolic rate of a mouse, to dissipate heat at the rate it is produced, its surface temperature would have to be well above the boiling point. Conversely, if a mouse were designed with the weight-related metabolic rate of a steer, to keep warm it would need as insulation a fur at least 20 cm thick.

It is important to realize that many physiological processes, not only heat loss, are surface related. In fact, to design a workable organism it would be necessary to include a careful consideration of surface areas. A large number of physiological processes are surface related, the uptake of oxygen in the lungs or in the gills depends on the area of these organs, the diffusion of oxygen from the blood to the tissues takes place across the capillary walls, again a surface function, food uptake in the intestine depends on the surface area of the intestine, and so on. In fact, all cells have surfaces through which nutrients and oxygen must enter the cell and metabolic products

---

* If two bodies of different size are geometrically similar, their surface areas will be related as the square of a corresponding linear dimension, and their volumes as the cube of the linear dimension. Their areas will therefore be related as their volumes raised to the power 2/3, or 0.67. A few minutes spent in relating edges, surfaces, and volumes of cubes of different sizes will clarify this rule, which applies to any geometrically similar bodies.

leave, the cells maintain a different ionic composition from the extracellular fluid, and the difference must again be maintained by surface or area-related processes. It is therefore easy to understand that over-all metabolism cannot be independent of surface considerations; it is more difficult to explain why it deviates in such a regular fashion that the slope of the regression lines we have discussed often is 0.75 or very close to this value.

Metabolic rate of birds. We have discussed the metabolic rates of mammals, and it would be interesting to know whether the other major group of warm-blooded vertebrates, the birds, have similar metabolic rates. Passerine birds (sparrows, finches, crows, etc.) mostly have somewhat higher metabolic rates than non-passerine birds, and we will therefore divide the birds into two separate groups. Metabolic rates have been compiled for 58 species of non-passerine birds, ranging in size from the 3 g hummingbird to the 100 kg ostrich (fig. 6.12). The equation for the regression line corresponds to $\dot{V}_{O_2} = 4.6 \times M_b^{0.723}$, which is similar to the equation for mammals. This similarity between mammals and birds means that a mammal

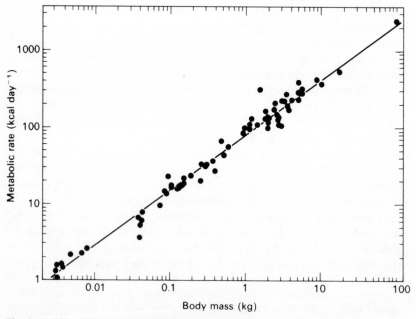

Fig. 6.12. The resting oxygen consumption (metabolic rate) of birds increases with body size. For a non-passerine bird of any given size, it can be expected that the oxygen consumption is similar to that of a mammal of the same size. Passerine birds as a group have higher metabolic rates. (Lasiewski and Dawson, 1967).

and a non-passerine bird of the same body size are likely to have nearly the same metabolic rates.

The available data for 36 species of passerine birds, ranging in size from 6.1 g to 866 g, give the equation $V_{O_2} = 7.54 \times M_b^{0.724}$. The slope of the regression line for passerine and non-passerine birds thus is the same, but the metabolic rate for a passerine bird can be expected to be higher than that of a non-passerine bird of the same size by an amount of about 65% (the difference between 7.54 and 4.6). Of course, any individual species may deviate more or less from the regression lines, which represent the best fit for all available data (Lasiewski and Dawson, 1967).

Metabolic rate of marsupials and monotremes. The body temperature of most marsupials (about 35 °C) is somewhat lower than for eutherian mammals (about 38 °C). This fact has been widely quoted, and is often taken to mean that marsupials are physiologically intermediate between the 'lower' monotremes (which have even lower body temperatures, about 30 °C) and the 'higher' eutherian mammals. There is no *a priori* reason to regard a lower body temperature as physiologically inferior, for marsupials regulate their body temperature effectively over a wide range of external conditions. If a low body temperature were to indicate 'lower' in the sense of 'inferior', birds would be 'superior' to mammals because they have higher body temperatures (usually about 40–42 °C).

These differences in body temperature, however, make it interesting to compare the energy metabolism of the various groups and see whether the marsupials as a whole adhere to a pattern similar to the other warm-blooded vertebrates. A study of Australian marsupials, ranging in size from 9 to 53 970 g, showed that their metabolic rates varied with body size in a way similar to eutherian mammals, but at an over-all somewhat lower level (Dawson and Hulbert, 1970).

Expressed with the same units as used for eutherian mammals, the equation for marsupials would be $\dot{V}_{O_2} = 2.3 \times M_b^{0.75}$. The data fall on a line with the same slope as the general equation for eutherian mammals ($\dot{V}_{O_2} = 3.8 \times M_b^{0.75}$), as shown in fig. 6.13. The consistently lower metabolism in the marsupials is evident. It is interesting that this lower metabolism of marsupials also is associated with a 3 °C lower body temperature. However, it is not possible to say whether the lower body temperature is a result of the lower metabolism, or vice versa.

If, however, we were to estimate the metabolism of marsupials in the event that their body temperature were raised to 38 °C, we would find that their 'corrected' metabolic rate would be similar to

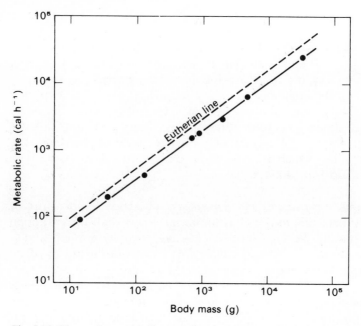

Fig. 6.13. The resting metabolic rate of marsupials is consistently somewhat lower than for eutherian mammals. The increase with increasing body size, however, follows regression lines of identical slope (Dawson and Hulbert, 1970).

that of eutherian mammals. This calculation,* carried out for the three major groups of mammals and for birds shows that monotremes, marsupials, eutherian mammals, and non-passerine birds all have similar 'temperature-corrected' metabolic rates. Passerine birds, however, fall distinctly outside the common range; their metabolic rates are, on the whole, more than 50% higher, even if recalculated to a lower body temperature than they normally have (Dawson and Hulbert, 1970).

Metabolic cost of locomotion

Swimming, running, and flying requires more energy than sitting still, but how do the different kinds of locomotion compare? Walking and running are more familiar to us than swimming and flying, and of more immediate interest, and we therefore know more about these ways of moving. Also, man is our best experimental animal, for he is highly cooperative and at times even highly motivated. This is of great help in studies of maximum performance, for when an

* For information on how to calculate the effect of temperature change on metabolic rate, see page 263ff.

animal refuses to run faster or longer, is it because he will not or can not? Dogs have also been extensively studied, for they are often co-operative and easy to work with – at times they even seem to be highly motivated. Flying is a different matter, for man has minimal experience in flying under his own power, and he is a clumsy and ineffective swimmer compared to fish and whales.

When we compare the different kinds of locomotion, we should realize that the greatest differences between moving in water, on land, and in the air, are due to the different physical qualities of these media. The two most important differences are in the support of the animal body that the medium provides, and in the different resistance to movement through water and air.

Most swimming animals are in near-neutral buoyancy, their weight is fully supported by the surrounding water, and no effort goes into supporting the body. Running and flying animals are in a very different situation, they must support the full weight of their bodies. The running animal has solid support under its feet, but the flying animal must continuously support its weight against a fluid medium of low density and low viscosity.

The resistance of the medium is of great importance to the swimming animal, for water has a high viscosity and density. In contrast, running and flying animals have the advantage of moving through a medium of low viscosity and low density. The physical characteristics of the medium have profound effects on the structural adaptations that animals show in their mode of locomotion. Aquatic animals have a streamlined body and propel themselves with fins and tail, or modifications of these structures. Birds have a streamlined body and wings, which basically function on the same principles of fluid dynamics as the fish tail. The smaller the size of the animal, however, the less effective is streamlining in reducing drag, and small insects are not obviously streamlined.

Running animals use their extremities as levers for moving over the solid substratum. Air resistance is of little importance, and running animals are not particularly streamlined. Streamlining is much more important for birds, for they move much faster than running mammals, and air resistance increases approximately with the square of the speed. Most mammals are quadrupedal, but an assortment of oddities such as man and kangaroos are primarily bipedal.

Running

Let us first consider the oxygen consumption of an animal, for example, a rat, as it runs at various speeds (fig. 6.14). The oxygen

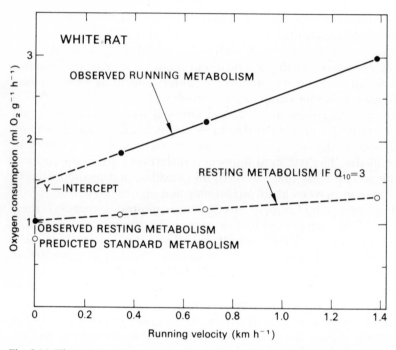

Fig. 6.14. The oxygen consumption of running rats increases linearly with increasing speed (Taylor, Schmidt-Nielsen and Raab, 1970).

consumption increases with the running speed, and within a wide range of speeds the increase is linear. (At even higher speeds the line may curve upwards and the cost of running increase out of proportion to the increase in speed, an experience familiar to a man who runs a high-speed dash.) If we extend the straight line back to the ordinate, corresponding to zero running speed, the intersect is higher than the resting oxygen consumption. Therefore, if we were to draw a complete curve, covering all running speeds from zero to the maximum, we would not obtain a straight line. However, in the intermediate range of speeds, the curves for a wide variety of running mammals, including man, follow straight lines.

For a man, walking at moderate speed is less expensive than running, in terms of energy cost (fig. 6.15). With increasing speed, walking becomes increasingly more expensive, and where the curve for walking intersects the curve for running it is cheaper to change to running. Fig. 6.15 also shows the well-known fact that walking or running up-hill is energetically more expensive than moving on the level, while down-hill locomotion is cheaper.

Let us now return to the running animal in fig. 6.14 and the

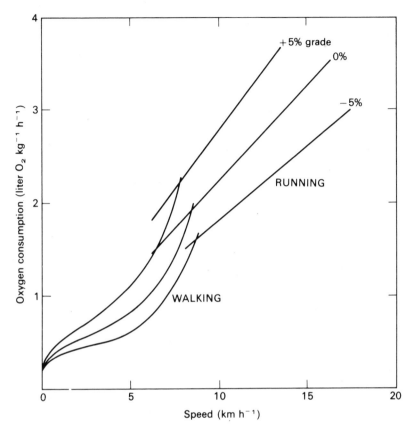

Fig. 6.15. The energy expenditure of a man during walking and running, at three different grades, as indicated on the graph (Margaria, Cerretelli, Aghemo and Sassi, 1963).

straight line relationship between cost and speed. The fact that the regression line does not extrapolate to the resting metabolic rate seems to be due to the metabolic cost of maintaining the posture of running. For example, the metabolic rate of a man who stands quietly instead of running, falls close to the intercept on the ordinate. The simplest interpretation of this is to assume that the difference between the intercept on the ordinate and resting metabolism is due to the cost of maintaining the posture of locomotion, and could therefore be called the 'postural effect'. The linear increase in oxygen consumption above the intercept on the ordinate can therefore be used as a measure of the cost of increasing the speed, for this increase represents the excess that goes into moving at any speed. Thus, the metabolic cost of moving is given by the slope of the line,

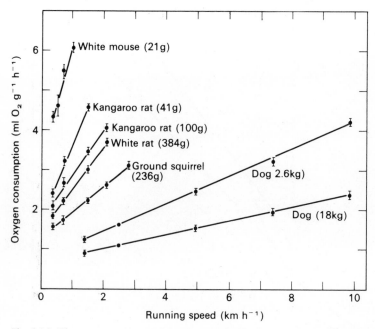

Fig. 6.16. The oxygen consumption of a variety of running mammals increases linearly with speed. The increase, per unit body weight, is smaller the larger the body size of the animal (Taylor *et al.*, 1970).

and since the line is straight the cost (slope) remains constant and does not change with the speed of running.

If we determine the oxygen consumption in relation to running speed for a number of mammals (see fig. 6.16) we find that the smaller mammals have the steeper regression lines, i.e. the cost of running increases more rapidly with increasing running speed.

A simple way of comparing the cost of running for different animals is to calculate how much energy it takes to move one unit of body mass over one unit distance, say, to move one gram body mass over one kilometer (given as ml O_2 g^{-1} km^{-1}). If we now return to figs. 6.14 and 6.16, we see that the slopes of the regression lines give exactly this information. The slope (defined as an increment in the ordinate-value over an increment in the corresponding abscissa-value) will be the increment in oxygen consumption (ml O_2 g^{-1} h^{-1}) over the increment in speed (km h^{-1}), and this cancels out to give the slope in ml O_2 g^{-1} km^{-1}, precisely the units used to define the cost of running at the beginning of this paragraph.

We can now compare directly the cost of running one kilometer per unit body mass for the animals in fig. 6.16, and the result is plotted in fig. 6.17. It is immediately apparent that the cost of running

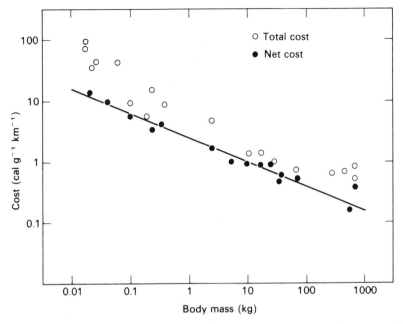

Fig. 6.17. The cost of running for mammals of various body sizes. The net cost (●) designates the cost in moving one gram body mass over a distance of one kilometer, calculated from the increase in metabolism caused by running (and obtained from slopes of regression lines). The total cost (o) includes the total metabolism while running and is therefore somewhat higher (Schmidt-Nielsen, 1972*a*).

decreases quite regularly as the body size increases. The significance of this relationship is that, for a running mammal, it is metabolically less expensive, and thus an advantage, to be of large body size.

There is also another way the metabolic cost of locomotion can be considered. The preceding discussion considered what it costs to run, above the cost of not running. However, an animal while it moves does in fact have to supply energy for its total metabolism. Thus, a sheep that grazes must eat enough food to cover its total metabolic rate, not merely the excess that is due to the act of locomotion *per se*. If we then consider the total metabolic cost while moving, we obtain higher figures, for the non-moving metabolic rate is added in.

Swimming and flying

We can examine the cost of swimming, or of flying, in terms similar to those which we have used for running, and we again find that the cost of moving is related to the body size.

While the oxygen consumption during running (fig. 6.14) increases regularly and linearly with increasing speed, the situation is

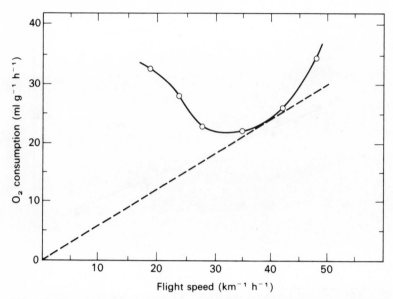

Fig. 6.18. Oxygen consumption of a flying budgerigar (parakeet) in relation to flying speed. The lowest oxygen consumption occurs at about 35 km h^{-1}, but the lowest cost of transport is at about 40 km h^{-1} (see text) (Schmidt-Nielsen, 1972b). Data from V. A. Tucker, J. exp. Biol. 1968, **48**, 67–87.

different for flying birds. Determinations on budgerigars (parakeets) have shown that there is an optimum speed of flight at which their oxygen consumption is at a minimum (fig. 6.18). Flying slower or faster increases the oxygen consumption, so that the metabolic cost of flight relative to speed yields a U-shaped curve.

Although flying horizontally at 35 km h^{-1} gives the lowest metabolic rate for the flying budgerigar, the speed that enables the bird to fly a certain distance most economically will be higher, in the case of the budgerigar about 40 km per h^{-1}. This is because the bird that flies faster covers the distance in less time. The most economical flight speed can be calculated, but it can also be found directly from the graph by taking the tangent to the curve from the origin. The point of contact gives the lowest possible slope for the oxygen consumption relative to flight speed, and this is the cost of transport as previously defined, expressed in ml O$_2$ g^{-1} km^{-1}. (The cost of transport derived by using the tangent to the curve in fig. 6.18 gives the total cost of transport, rather than the net transport as we calculated for the mammals. For birds this is relatively unimportant, since their metabolic rate during flying is about ten times as high as the resting metabolic rate, and subtracting the latter makes only a minor dif-

Plate 5. *Budgerigar*. The budgerigar, or parakeet (*Melopsittacus undulatus*), photographed during flight in a wind tunnel. To determine the rate of oxygen consumption in flight, this trained bird is equipped with a plastic face mask which, through the attached tube, permits the collection of all exhaled air (V. A. Tucker, Duke University; from *Journal of Experimental Biology*, 1968, **48**, 67–87).

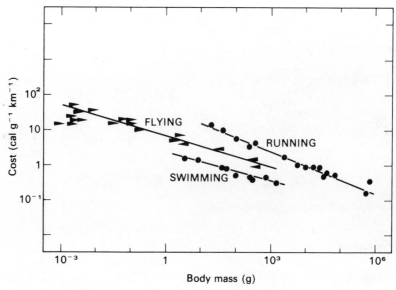

Fig. 6.19. Comparison of the energy cost of moving one unit body weight over 1 km for running, flying, and swimming (Schmidt-Nielsen, 1972*a*), copyright 1972 by the American Association for the Advancement of Science.

ference in the result. For mammals the difference in net and total cost of locomotion is greater.)

Information about the cost of locomotion for swimming, flying, and running has been put together in fig. 6.19. It is interesting that for a given body size, flying is a far less expensive way to move to a distant point than running. At first glance this is surprising, for we have the intuitive feeling that flying must require a great deal of effort, just to stay up in the air. We know, however, that migrating birds fly non-stop for more than 1000 km, and it would be difficult to imagine a small mammal such as a mouse which could run that far without stopping to eat and drink.

Swimming fish use even less energy than birds. This may seem surprising, for if we consider the high viscosity of water, we would expect a considerable resistance to moving a body through water. There are two factors to take into account, however, fish on the whole do not move very fast, and their streamlined bodies are highly adapted to moving in a medium of relatively high density and viscosity. Another factor is that fish are in near-neutral buoyancy, the body is fully supported by the medium, and no effort goes into keeping them from sinking.

For animals less well adapted to swimming the situation is different. If we calculate the cost of transportation for a swimming duck, it turns out to be similar to the cost for running on land for a mammal of the same size. For a man, swimming is even more expensive. Compared to a fish he moves through water with difficulty, his body shape and his appendages make him an ineffective swimmer, and the cost of moving through the water is five or ten times as high as moving the same distance on land. He is, in fact, so unsuited to swimming, that the cost of transportation puts a swimming man well above the range represented by the regression line for running in fig. 6.19.

Body size and problems of scaling

In the preceding discussions we have often related metabolic rates to body size, and we have been able to derive general equations which represent this relationship. Obviously, many other variables can also be related to body size. For example, although the heart of a horse weighs far more than that of a mouse, in both species the heart is about 0.6% of the total body mass (M_b). If we express this in a general equation, it would be:

$$\text{Heart weight} = 0.006 \, M_b^{1.0}.$$

This equation, which is based on the measurement of hundreds of heart weights of different mammals, shows that the weight of the mammalian heart is proportional to the body weight (exponent of 1.0). There is, of course, still room for a certain margin of variation, but the equation gives the general trend.

Similarly, the lung volume of mammals is scaled in simple proportion to body size, as was shown in fig. 2.6 on page 36. The regression line in this figure has a slope of 1.02, which goes to say that mammals have lungs which are very nearly the same proportion of the body. As a general rule, then, all mammalian lungs are scaled similarly and have a volume of 63 ml per kg body weight (again with a certain margin of variation).

If we now examine the surface area of the mammalian lung, rather than the volume, we find that this variable is directly related to the oxygen consumption of the animal, rather than to body size (fig. 6.20). This does indeed make sense, for oxygen consumption is, as we have seen, related to body weight to the 0.75 power, and scaling of the surface area of the lung, or diffusion area for oxygen uptake, should therefore be scaled the same way, also related to the body weight to the 0.75 power.

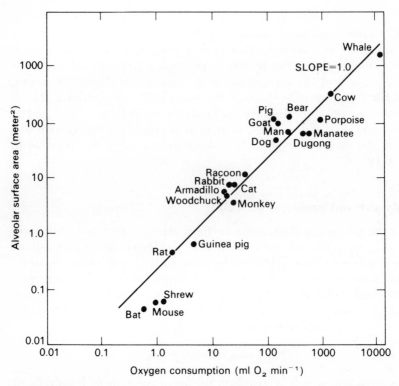

Fig. 6.20. The total surface area of the mammalian lung is proportional to the rate of oxygen consumption as evident from the slope of 1.0 of the regression line that relates these two variables (Tenney and Remmers, 1963).

Table 6.7 shows a number of physiological variables scaled to body size by similar equations. As we saw above, the lung volume is directly related to body size. The tidal volume at rest is also directly related to body size (table 6.7, line 5). If we now divide the equation for tidal volume by the equation for lung volume we obtain an expression for the tidal volume relative to lung volume. The body size exponents are so close that the residual exponent of 0.01 obtained in this division is insignificant, compared to the inherent inaccuracy of the measurements used to establish the original equations. We thus obtain the value 0.0062 : 0.063, or very nearly 0.1. This dimensionless number predicts that mammals in general will have a tidal volume of one-tenth of their lung volume, irrespective of their body size. Although there may be deviations from this general pattern, the statement has a predictive value which is useful. We can consider a rat or a dog and predict their expected physiological parameters, and we can then examine our measurements to see how they fit the general pattern or deviate from it.

Table 6.7. *Relationship between several physiological variables of mammals and their body mass* (M_b *in kg*). (Data for these equations selected from Adolph (1949), Drorbaugh (1960) and Stahl (1967).) *

O₂ consumption (liter min⁻¹)	=	$0.0116 \times M_b^{0.76}$
O₂ consumption per kg (liter min⁻¹ kg⁻¹) =		$0.0116 \times M_b^{-0.24}$
Lung ventilation rate (liter min⁻¹)	=	$0.334 \times M_b^{0.76}$
Lung volume (liter)	=	$0.063 \times M_b^{1.02}$
Tidal volume (liter)	=	$0.0062 \times M_b^{1.01}$
Blood volume (liter)	=	$0.055 \times M_b^{0.99}$
Heart weight (kg)	=	$0.0058 \times M_b^{0.99}$
Respiration frequency (min⁻¹)	=	$53.5 \times M_b^{-0.26}$
Heart rate (min⁻¹)	=	$241 \times M_b^{-0.25}$

* Numerical differences from previous equations may depend on different choice of units, and, to a minor extent, on different source material.

When we continue to examine the equations in table 6.7 we can see other interesting relationships of predictive value. The blood volume of mammals, for example, is universally 5.5% of the body mass, irrespective of body size (exponent 0.99, which can be considered as insignificantly different from 1.0). As we saw before, the heart weight (which is the size of the circulatory pump) is also proportional to body weight. Since the blood supplies oxygen to the tissues at the rate required by metabolism, which is scaled to the 0.75 power of body size, the rate of the pump's beat (heart rate) must be adjusted accordingly, and we do indeed find this to be the case. The heart rate of the small animal is very high, and decreases with body size as indicated by the negative slope of -0.25 of the regression line (table 6.7, last line). This, then, shows that heart rate, and thus cardiac output, is indeed scaled to pump blood at the rate required to supply oxygen for the metabolic rate.

The equations listed in table 6.7 relate to variables that are either directly proportional to body size (exponent = 1.0), or are scaled to the metabolic rate (exponent = 0.75). We cannot assume that all anatomical and physiological variables are scaled in one of these two ways, and indeed, some differ substantially. For example, for the size of kidney and liver the equations are: Kidney weight = $0.021 \times M_b^{0.85}$ and liver weight = $0.082 \times M_b^{0.87}$. Both these organs are metabolically very important, but they are not scaled in proportion to body weight, they are relatively smaller in the larger animal. On the other hand, they are not scaled as metabolic rate either, for their size increases with an exponent that exceeds the 0.75 power for metabolic rate. We can therefore conclude that the scaling of animals, their organs, and their physiological functions, is not always a simple function of body size, and that more complex considerations enter into what is necesary to form an integrated and well-functioning organism.

REFERENCES

ADOLPH, E. F. (1949). Quantitative relations in the physiological constitutions of mammals. *Science,* **109,** 579–85.

BARTHOLOMEW, G. A. and MACMILLEN, R. E. (1961). Oxygen consumption, estivation, and hibernation in the kangaroo mouse, *Microdipodops pallidus. Physiol. Zool.,* **34,** 177–83.

BLAZKA, P. (1958). The anaerobic metabolism of fish. *Physiol. Zool.,* **21,** 117–28.

BRODY, S. (1945). *Bioenergetics and Growth. With Special Reference to the Efficiency Complex in Domestic Animals,* New York: Reinhold Publ. Corp., 1023 pp. Reprinted (1964): Darien, Conn.: Hafner Publ. Co.

DAWSON, T. J. and HULBERT, A. J. (1970). Standard metabolism, body temperature, and surface areas of Australian marsupials. *Am. J. Physiol.,* **218,** 1233–8.

DRORBAUGH, J. E. (1960). Pulmonary function in different animals. *J. Appl. Physiol.,* **15,** 1069–72.

ELSNER, R. (1965). Heart rate response in forced versus trained experimental dives in pinnipeds. *Hvalrådets Skrifter,* nr. **48,** 24–9.

FRY, F. E. J. and HART, J. S. (1948). The relation of temperature to oxygen consumption in the goldfish. *Biol. Bull.,* **94,** 66–77.

GORDON, M. S., BOËTIUS, I., EVANS, D. H., MCCARTHY, R. and OGLESBY, L. C. (1969). Aspects of the physiology of terrestrial life in amphibious fishes. I. The mudskipper, *Periophthalmus sobrinus. J. Exp. Biol.,* **50,** 141–49.

HALL, F. G. (1929). The influence of varying oxygen tensions upon the rate of oxygen consumption in marine fishes. *Am. J. Physiol.,* **88,** 212–18.

HAWKINS, A. E., JEWELL, P. A. and TOMLINSON, G. (1960). The metabolism of some British shrews. *Proc. Zool. Soc. Lond.,* **135,** 99–103.

HEMMINGSEN, A. M. (1960). Energy metabolism as related to body size and respiratory surfaces, and its evolution. *Reports of the Steno Memorial Hospital,* **9,** (2), 1–110.

HOCHACHKA, P. W. and MUSTAFA, T. (1972). Invertebrate facultative anaerobiosis. *Science,* **178,** 1056–60.

HUDSON, J. W. (1962). The role of water in the biology of the antelope ground squirrel *Citellus leucurus. Univ. Calif. Publ. Zool.,* **64,** 1–56.

JACKSON, D. C. and SCHMIDT-NIELSEN, K. (1966). Heat production during diving in the fresh water turtle, *Pseudemys scripta. J. Cell. Physiol.,* **67,** 225–32.

JOHANSEN, K. (1964). Regional distribution of circulating blood during submersion asphyxia in the duck. *Acta Physiol. Scand.,* **62,** 1–9.

KING, J. R. (1957). Comments on the theory of indirect calorimetry as applied to birds. *Northwest Science*, **31**, 155–69.

KLEIBER, M. (1961). *The Fire of Life. An Introduction to Animal Energetics*, New York: John Wiley & Sons, Inc., 454 pp.

KOOYMAN, G. L. (1966). Maximum diving capacities of the Weddell seal, *Leptonychotes weddelli. Science*, **151**, 1553–4.

KYLSTRA, J. A. (1968). Experiments in water-breathing. *Sci. Am.*, **219**, 66–74.

LASIEWSKI, R. C. and DAWSON, W. R. (1967). A re-examination of the relation between standard metabolic rate and body weight in birds. *Condor*, **69**, 13–23.

LUSK. G. (1931). *The Elements of the Science of Nutrition*, 4th edn, Philadelphia: W. B. Saunders, 844 pp.

MARGARIA, R., CERRETELLI, P., AGHEMO, P. and SASSI, G. (1963). Energy cost of running. *J. Appl. Physiol.*, **18**, 367–70.

MURDAUGH, H. V., Jr., SCHMIDT-NIELSEN, B., WOOD, J. W. and MITCHELL, W. L. (1961). Cessation of renal function during diving in the trained seal (*Phoca vitulina*). *J. Cell. Comp. Physiol.*, **58**, 261–6.

OLSSON, K. E. and SALTIN, B. (1968). Variation in total body water with muscle glycogen changes in man. In *Biochemistry of Exercise*, Medicine and Sport Series, vol. III, pp. 159–62, Basel: S. Karger.

PAULEV, P. (1965). Decompression sickness following repeated breath-hold dives. *J. Appl. Physiol.*, **20**, 1028–31.

PEARSON, O. P. (1960). The oxygen consumption and bioenergetics of harvest mice. *Physiol. Zool.*, **33**, 152–60.

SCHMIDT-NIELSEN, K. (1972a). Locomotion: energy cost of swimming, flying, and running. *Science*, **177**, 222–8.

SCHMIDT-NIELSEN, K. (1972b). *How Animals Work*, Cambridge, England: Cambridge University Press, 114 pp.

SCHOLANDER, P. F. (1940). Experimental investigations on the respiratory function in diving mammals and birds. *Hvalrådets Skrifter*, nr. **22**, 1–131.

SHEPARD, M. P. (1955). Resistance and tolerance of young speckled trout (*Salvelinus fontinalis*) to oxygen lack, with special reference to low oxygen acclimation. *J. Fish. Res. Bd. Can.*, **12**, 387–446.

STAHL, W. R. (1967). Scaling of respiratory variables in mammals. *J. Appl. Physiol.*, **22**, 453–60.

TAYLOR, C. R., SCHMIDT-NIELSEN, K. and RAAB, J. L. (1970). Scaling of the energetic cost of running to body size in mammals. *Am. J. Physiol.*, **219**, 1104–7.

TENNEY, S. M. and REMMERS, J. E. (1963). Comparative quantitative morphology of the mammalian lung: diffusing area. *Nature, Lond.*, **197**, 54–6.

THOMAS, H. J. (1954). The oxygen uptake of the lobster (*Homarus vulgaris* Edw.). *J. Exp. Biol.*, **31**, 228–51.

WHALEN, W. J. (1966). Intracellular P_{O_2}: a limiting factor in cell respiration. *Am. J. Physiol.*, **211**, 862–8.

WHALEN, W. J., BUERK, D. and THUNING, C. A. (1973). Blood flow-limited oxygen consumption in resting cat skeletal muscle. *Am. J. Physiol.*, **224**, 763–8.

USEFUL REFERENCE MATERIAL FOR PART II

ANDERSEN, H. T. (ed.) (1969). *The Biology of Marine Mammals,* New York: Academic Press, 511 pp.

BRODY, S. (1964). *Bioenergetics and Growth. With Special Reference to the Efficiency Complex in Domestic Animals,* reprint edn, Darien, Conn.: Hafner Publ. Co, 1023 pp.

CODE, C. F. and HEIDEL, W. (eds.) *Handbook of Physiology.* Sect. 6, *Alimentary Canal,* vols I–V, pp. 1–2896 (1967–1968), Washington, D.C.: American Physiological Society.

FLORKIN, M., and STOTZ, E. H. (eds) (1971). Metabolism of Vitamins and Trace Elements. *Comprehensive Biochemistry,* **21,** Amsterdam: Elsevier Publ. Co. 297 pp.

HALVER, J. E. (ed.) (1972). *Fish Nutrition,* New York: Academic Press, 714 pp.

HEMMINGSEN, A. M. (1960). Energy metabolism as related to body size and respiratory surfaces, and its evolution. *Reports of the Steno Memorial Hospital and the Nordisk Insulinlaboratorium,* **9,** (2), 1–110.

KLEIBER, M. (1961). *The Fire of Life. An Introduction to Animal Energetics,* New York: John Wiley & Sons, Inc., 454 pp.

LIENER, I. E. (1969). *Toxic Constituents of Plant Foodstuffs,* New York: Academic Press, 500 pp.

MCDONALD, P., EDWARDS, R. A. and GREENHALGH, J. F. D. (1973). *Animal Nutrition,* 2nd edn, Darien, Conn.: Hafner Publ. Co., 487 pp.

III

TEMPERATURE

7

Temperature effects

In the preceding chapter we were concerned with the energy metabolism of living animals and the influence of such variables as oxygen concentration, body size, activity, etc. In this chapter we shall discuss the profound influence that temperature has on living organisms and their metabolic processes.

Active animal life is limited to a narrow range of temperatures, from approximately -2 °C to approximately $+50$ °C.* Compared to cosmic temperatures, this is a very narrow range indeed. However, temperatures suitable for life can be found throughout the oceans and over most of the surface of the earth, for at least part of the year. At the lower extreme, in arctic waters there are fish and numerous invertebrates living near -1.8 °C throughout the year. At the other extreme, in hot springs, there are a few animals living at temperatures of about 50 °C. Some primitive plants live at even higher temperatures, and thermophilic bacteria thrive near the boiling point of water.

Outside the temperature range that permits active life, many animals can survive in an inactive or torpid state. In fact, some animals can survive extremely low temperatures, such as that of liquid air (about -190 °C), or even liquid helium (-269 °C). The tolerance to high temperatures is more limited, although some animals when in a 'resting stage' can be quite resistant, and bacterial spores may tolerate temperatures in excess of 120 °C.

Most animals, among them all aquatic invertebrates, have nearly the same temperature as their surroundings. Birds and mammals, in contrast, generally maintain their body temperature nearly constant and independent of large variations in the temperature of their environment. However, quite a few other animals, both invertebrates

---

* We are concerned with the temperature of the organism itself. When we go out in sub-freezing temperatures, our body temperature remains at 37 °C.

and vertebrates, can at times maintain a substantial difference between their own temperature and that of the surroundings.

There is no simple and easy way to classify these differences. Traditionally, birds and mammals have been called 'warm-blooded' and the other animals have been called 'cold-blooded.' These terms are so well established that it is convenient to continue to use them. They are, however, rather inaccurate and may be quite misleading. A 'cold-blooded' animal is not necessarily cold – a tropical fish or a desert lizard or an insect sitting in the sun may have a higher body temperature than a mammal. Furthermore, quite a few mammals and some birds undergo periods of torpor or hibernation during which their temperature may decrease to near the freezing point of water, with no harm to the animal. In that state it seems odd to call them 'warm-blooded.'

The corresponding scientific terms, *poikilothermic* for cold-blooded and *homeothermic* for warm-blooded, are also imprecise. The word 'poikilothermic' (Greek *poikilos* = changeable) refers to the fact that the temperature of cold-blooded animals fluctuates with that of their surroundings. Thus, a fish has the temperature of the water it swims in and an earthworm that of the soil it crawls in. However, a deep sea fish that spends its entire life in water that has barely measureable temperature fluctuations, is truly an animal with a virtually constant body temperature. It might be logical to call such a fish 'homeothermic', but the term *homeothermic* * is used specifically to refer to birds and mammals, animals that in fact have body temperatures that normally fluctuate by several degrees and, in hibernation, may even drop to near 0 °C. Animals that at times have a high and well-regulated body temperature, but at other times are more like cold-blooded animals, are often called *heterothermic* (Greek *heteros* = different).

Homeothermic animals (birds and mammals) usually maintain a high body temperature and remain active in cold as well as in warm surroundings. Most poikilothermic animals become more and more inactive as the temperature decreases. There are some remarkable exceptions, for example, by exposing itself to the sun, a lizard can maintain its body temperature far above that of the surrounding air. To distinguish the lizard from birds and mammals, which maintain their high body temperature by metabolic heat production, we have the terms *ectothermic* and *endothermic*. Endothermic animals are those which are able to maintain a high body temperature by internal heat

---

* The terms *homothermic, homoiothermic,* and *homeothermic* are used interchangeably. They are derived from the Greek, *homos* = like, and *homoios* = similar. Homeothermic is an anglicized spelling form which is most commonly used.

production; ectothermic animals are those which depend on external heat sources, primarily solar radiation. These definitions also have their limitations; as we shall see later, a tuna fish may keep the temperature of its muscles 10 or 15 °C above that of the water. The heat is derived from muscular metabolism, but we do not consider the tuna an endothermic animal in the same sense as birds and mammals. Insects make it even more difficult to apply this terminology, they may sit in the sun and warm up and at the same time produce additional heat by intense muscular contractions. Obviously, this case does not clearly fit one of the terms to the exclusion of the other. The same is true of some mammals and birds that, under appropriate circumstances, take advantage of solar heat to reduce internal heat production.

The choice of terminology is primarily a matter of convenience, and terms that are suitable in every situation may at times be difficult to find. It is not a question of whether a certain terminology is 'right' or 'wrong', but of how useful it is for a given purpose. Nevertheless, the terms we use should always be accurately and precisely defined.

PHYSIOLOGICAL EFFECTS OF TEMPERATURE CHANGE

Temperature change has striking effects on many physiological processes. Within limits, a temperature increase accelerates most processes. For example, let us consider its effect on the rate of oxygen consumption, which is a convenient expression for the over-all metabolic activity of an animal.

Within the temperature range that an animal can tolerate, the rate of oxygen consumption is often found to increase in a fairly regular manner with increasing temperature. In general, a rise of 10 °C in temperatures causes the rate of oxygen consumption to increase about two-fold or three-fold.

The increase in a rate caused by a 10 deg C increase in temperature is called the Q_{10}. If the rate doubles, Q_{10} is 2, if the rate triples, Q_{10} is 3, and so on. This term is used not only for oxygen consumption but for all rate processes that are affected by temperature (see appendix 4).

If an animal has a wide range of temperature tolerance, its rate of oxygen consumption may accelerate vastly as temperature increases. Thus, with a Q_{10} of 2, and a beginning temperature of 0 °C, the rate would double with a temperature increase of 10 °C; it would quad-

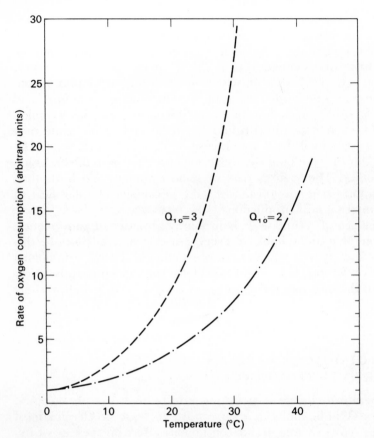

Fig. 7.1. Many temperature-dependent rate processes proceed more and more rapidly as temperature increases. See text for further information.

ruple with an increase to 20 °C, and it would increase eightfold with a temperature increase to 30 °C. With a Q_{10} of 3, the increases in oxygen consumption would be threefold, ninefold, and 27-fold, respectively, for the same temperature intervals.

If we plot these figures on graph paper (fig. 7.1), we obtain rapidly rising curves, known as exponential curves because mathematically they are described by exponential functions * of the general form

$$y = b \cdot a^x.$$

If R_2 and R_1 are the rates at two temperatures T_2 and T_1, and we use the customary symbol Q_{10}, the equation will be

* See also appendix 3.

$$R_2 = R_1 \cdot Q_{10}^{\frac{T_2 - T_1}{10}}$$

Many rate processes, such as acceleration, radioactive decay, growth curves, etc. are described by exponential equations. Calculations which involve exponential equations can conveniently be carried out if we use the logarithmic form

$$\log y = \log b + x \cdot \log a$$

or

$$\log R_2 = \log R_1 + \log Q_{10} \cdot \frac{T_2 - T_1}{10}$$

We can now see that $\log R_2$ increases linearly with the temperature change $(T_2 - T_1)$. This means that if we plot log rate against temperature, we get a straight line. This is commonly done by using graph paper with a logarithmic ordinate and a linear abscissa (so-called 'semi-log' graph paper). A graph representing this form of the equation is given in fig. 7.2.

Obviously it is not necessary to determine two rates exactly 10 °C apart in order to calculate the Q_{10}. Any two temperatures can be used, provided they are sufficiently far apart to give reliable infor-

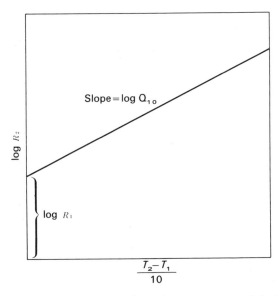

Fig. 7.2. A temperature-dependent rate process of the kind shown in fig. 7.1 will, when plotted on a logarithmic scale, give a straight line. (Note that only the ordinate is logarithmic). For further explanation see text.

mation about the temperature effect. To solve for Q_{10} when the rates at two different temperatures have been observed, we can simply change the equation and use one of the two following forms

$$\log Q_{10} = (\log R_2 - \log R_1) \cdot \frac{10}{T_2 - T_1}$$

or

$$Q_{10} = \left(\frac{R_2}{R_1}\right)^{\frac{10}{T_2 - T_1}}$$

We can thus calculate the Q_{10} for the interval between T_1 and T_2. However, most often Q_{10} does not remain the same throughout the range of temperatures an animal can tolerate, and it is necessary to specify accurately the conditions under which the observations were made.

Let us now turn to a real example. The rates of oxygen consumption of the Colorado potato beetle between 7 °C and 30 °C are given in table 7.1. Throughout this range the oxygen consumption increases with temperature, the Q_{10} for the entire range being 2.17. However, if we calculate the Q_{10} for each temperature interval for which observations were made, we find that up to 20 °C the Q_{10} remains rather constant at nearly 2.5, but at higher temperatures the Q_{10} falls off (table 7.1, last column). If the same data are plotted in graph form, we can more clearly see that the observations, when the temperature exceeds about 20 °C, deviate from a regular exponential function (fig. 7.3).

Table 7.1. *Rates of oxygen consumption of the Colorado potato beetle* (Leptinotarsa decemlineata) *at various temperatures between 7 °C and 30 °C. Before the experiment the animals had been maintained at 8 °C.* (Data from Marzusch, 1952)

Temp. (°C)	O$_2$ consumed (μl g^{-1} h^{-1})	Temp. interval (°C)	Q_{10}
7	61		
10	81	7–10	2.57
15	126	10–15	2.41
20	200	15–20	2.52
25	290	20–25	2.10
30	362	25–30	1.56

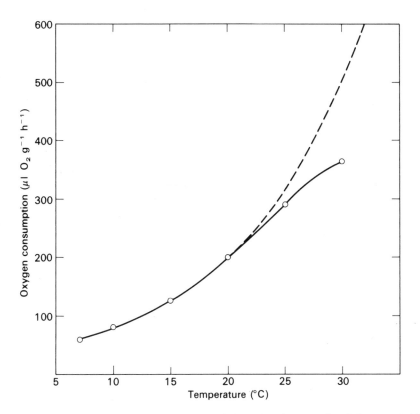

Fig. 7.3. The rate of oxygen consumption of the Colorado potato beetle increases with temperature (circles and solid line). The broken line shows the expected curve, had the Q_{10} remained constant at 2.5. Same data as in table 7.1.

However, it is a bit cumbersome to calculate the Q_{10} for each set of observations, and it is also inconvenient to construct an exponential curve on arithmetic coordinates for comparison with the observed data. If instead we plot the figures on a logarithmic ordinate, we obtain a straight line for that interval in which the temperature effect remains constant and the Q_{10} does not change (fig. 7.4). The straight line in this figure corresponds to a Q_{10} of 2.50, and we can see that the observations up to 20 °C follow this line completely. Above 20 °C, however, the rate of oxygen consumption deviates increasingly from the straight line, i.e., although there is still an increase in rate with temperature, the extent of the effect of temperature is diminished. At even higher temperatures the drop in oxygen consumption becomes more pronounced, and eventually the lethal temperature limit is reached.

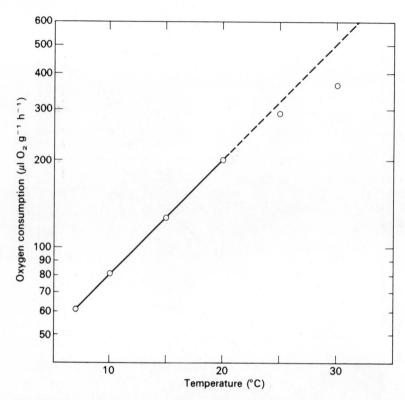

Fig. 7.4. When plotted on a logarithmic ordinate, the rate of oxygen consumption of the Colorado potato beetle shows a linear dependence on temperature up to 20 °C. Above this temperature the rate does not increase as rapidly as expected from a constant Q_{10} of 2.5 (broken line). Same data as in table 7.1 and fig. 7.3.

EXTREME TEMPERATURES: LIMITS TO LIFE

Various animals differ in the range of temperatures that they can tolerate. Some animals have a very narrow tolerance range, others have a wide tolerance range. Furthermore, temperature tolerance may change with time, and a certain degree of adaptation is possible (so that continued exposure to a temperature close to the limit of tolerance often changes the limit). Some organisms may be especially sensitive to extreme temperatures during certain periods of their lives, particularly during the early stages of development.

When discussing tolerance to extreme temperatures, we must distinguish between temperatures that an organism can survive and those at which it can carry out its entire life cycle. It is also important to realize that we cannot determine an exact temperature that is lethal for a given organism, for the duration of exposure is of great importance. A certain high temperature that may be tolerated for several minutes may be lethal if continued for several hours.

One further matter is important, we are concerned with the temperature of the organism itself, not that of its environment. For virtually all aquatic organisms the two are nearly equal, but for terrestrial organisms, as we have indicated, they may differ a great deal. The lizard sitting in the sun may, because of solar radiation, acquire a body temperature of 10 or 20 degrees above the surrounding air temperature. Among the so-called warm-blooded animals, mammals and birds, most can tolerate only a rather narrow range of body temperatures, and this must not be confused with the wide range of temperatures of the environments in which they live, ranging from the arctic to the hottest deserts.

Tolerance to high temperature

No animal is known to carry out its complete life cycle at a temperature over 50 °C (Brock, 1970). Some plants are more heat tolerant than animals, for example, the unicellular blue-green algae *Synechococcus* is found in hot springs at temperatures as high as 73 to 75 °C. This seems to be the upper limit for photosynthetic life. Thermophilic bacteria are even more heat tolerant and have been found to live and grow in the hot springs of Yellowstone National Park, where the temperature is about 92 °C (which is the boiling point of water at the altitude of Yellowstone) (Bott and Brock, 1969).

As was indicated at the beginning of the chapter, animals in a resting stage may be extremely tolerant to high temperatures. For example, a fly larvae (*Polypedilum*, from Nigeria and Uganda), can tolerate dehydration, and in the dehydrated state it can survive a temperature of 102 °C for one minute and afterwards grow and metamorphose successfully (Hinton, 1960). Another example of extreme tolerance is provided by the eggs of a fresh-water crustacean (*Triops*, from Sudan); these eggs survive through winter and early summer in dry mud, where they may be exposed to temperatures up to 80 °C. In the laboratory they withstand temperatures within 1 °C of boiling. With the boiling point of water increased by subjecting the vessel to pressure, they have withstood 103 °C ± 1 °C for sixteen hours, although they are killed within 15 minutes at 106 °C (Carlisle, 1968).

From these examples it should be clear that the upper temperature limit for life cannot be accurately defined.

Determinations of lethal temperatures

When a group of animals is exposed to a temperature close to the limit of their tolerance, some of them may die and others survive. What is the exact *lethal temperature* for this species?

The lethal temperature is commonly defined as that temperature at which 50% of the animals die and 50% survive, often written as T_{L50}. To find this exact temperature by trial and error would require a great deal of experimentation before one experiment comes out with exactly half of the animals dead. Instead, after the approximate lethal temperature is established, the procedure indicated in fig. 7.5 is followed. Several groups (say, four) are exposed for the same time period (say, two hours) to a series of temperatures, and the per cent survival at each temperature is plotted. The T_{L50} can now be readily found from the graph. This procedure gives the lethal temperature only for the exposure time used in the experiment; shorter exposure gives higher survival and longer exposures usually lower survival. The effect of exposure time can be determined by carrying out tests, similar to that in fig. 7.5, for different periods of exposure, say one, two, four, and eight hours. If the various T_{L50} obtained are plotted against exposure time the points will often fall on a straight line on a log scale abscissa (fig. 7.6). This is the reason that the time intervals in this example were chosen so as to correspond directly to a logarithmic scale. We now have a single graph that gives a quite complete picture of the temperature tolerance of the organism in question.

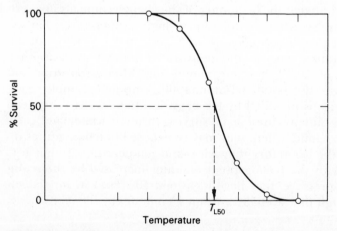

Fig. 7.5. To determine the temperature that is lethal to 50% of a group of organisms (the T_{L50}), experiments are carried out at a range of temperatures, and after plotting the data the temperature for 50% survival is read from the graph.

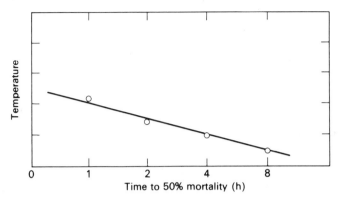

Fig. 7.6. The effect of exposure time on temperature tolerance becomes evident if the T_{L50} is plotted against the duration of each experiment in which the T_{L50} was determined.

Lethal temperatures and cause of heat death

Although active animal life has an upper limit at about 50 °C, many animals die when they are exposed to much lower temperatures. This is particularly true of aquatic animals, and especially marine animals, which usually are not exposed to very high temperatures. Even in tropical seas the temperature is rarely as high as 30 °C, although it may be higher in small enclosed bays and pools. For intertidal animals, however, the situation is different. At low tide, intertidal animals are exposed to warm air and to solar radiation, and their temperature may increase appreciably. To some extent they remain cool by evaporation of water, but while in air their restricted water reserves limit this way of cooling.

Intertidal snails show a clear correlation between heat resistance and their location in the tidal zone (table 7.2). Those snails that are located high in the tidal zone, and are thus out of water and exposed for the longest time, have substantially higher heat resistance than those located in the lower part of the tidal zone, close to the low-water mark.

The most heat-tolerant fish of any are probably the desert pupfish that live in warm springs in California and Nevada. The species *Cyprinodon diabolis* lives in a spring known as Devil's Hole, in which the temperature (33.9 °C) probably has remained almost unchanged for at least 30 000 years (Brown and Feldmeth, 1971). The upper lethal limit for this tiny fish, which as an adult weighs less than 200 mg, is about 43 °C, as high as is known for any species of fish.

In contrast, some arctic and antarctic animals have a surprisingly low heat tolerance. Antarctic fishes of the genus *Trematomus* are par-

Table 7.2. *The heat resistance of snails found in the upper part of the tidal zone is almost 10 deg C higher than for those found near the low-water mark. The temperature given is the highest from which the snails can recover after 1 h exposure.* (Fraenkel, 1968)

Name	Temperature (°C)	Occurrence
Tectarius vilis	48.5	
Planaxis sulcatus	48	Spray zone
Nodilittorina granularis	47	Upper part of intertidal zone
Littorina brevicola	47	
Paesiella raepstorttiana	47	Shallow splash pools
Nerita japonica	46	Middle part of intertidal zone
Nerita albicilla	44	
Lunella coronata	43	Lower part of intertidal zone in
Drupa granulatus	42	sheltered or shadowy places or
Purpura clavigerus	42	under stones
Monodonta labis	42	
Tegula lischkei	39	Lower part of intertidal zone at the water's edge

ticularly heat-sensitive and have an upper lethal temperature of about 6 °C. In nature these fish live in water with an average temperature of -1.9 °C which through the year varies by only about 0.1 °C (Somero and DeVries, 1967).

The extremely low thermal tolerance of these animals is important to the understanding of the mechanism of heat death. Some possible factors that have been suggested as contributing to heat death are:

(1) Denaturation of proteins, thermal coagulation
(2) Thermal inactivation of enzymes at rates exceeding the rates of formation
(3) Inadequate oxygen supply
(4) Temperature effect on lipid membranes
(5) Different temperature effects (Q_{10}) on interdependent metabolic reactions

Let us consider heat denaturation of proteins. It is true that many animals die at temperatures where thermal denaturation of protein takes place, for many proteins are denatured at temperatures above 45 to 55 °C. It is difficult to imagine, however, that thermal denaturation of any protein could take place at +6 °C, the lethal temperature of *Trematomus*.

Another suggestion for the cause of thermal death is thermal inactivation of certain particularly temperature-sensitive enzyme systems. Again, it is difficult to imagine enzymes so thermolabile that they become inactivated at +6 °C. Some enzyme systems from *Trematomus* that have been tested in this regard actually show increased activity up to about 30 °C (Somero and DeVries, 1967).

The third possibility, thermal death being caused by inadequate oxygen supply as the increased temperature accelerates the demand for oxygen, is a possibility that can be excluded in many cases. For example, supplying insects with pure oxygen instead of air does not enable them to survive higher temperatures. Likewise, the trout, a cold-water fish, dies at the same temperature in warm water even if the oxygen content of the water is increased several-fold by aeration with pure oxygen.

The fourth possibility is an old theory, based on speculations about the phase change in lipids that takes place with increasing temperature, but exact information to clarify this point is inadequate.

The last possibility, however, is consistent with much of the information at hand. If various processes in the intermediary metabolism are influenced differently by temperature (have different Q_{10}), it may lead to the depletion or accumulation of certain intermediary metabolic products. Let us consider the following scheme.

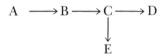

If in this scheme the process C → D is accelerated by increased temperature more than the process B → C, the intermediary C will be depleted at high temperature. If intermediary C is also needed for some other metabolic purpose, E, it will no longer be available in sufficient quantity. A different temperature sensitivity of the several hundred metabolic enzymes that participate in intermediary metabolism may therefore easily lead to a complete derangement of the normal biochemical balance of the organism.

Although the possibility of different thermal sensitivities of enzymatic systems gives a plausible explanation of heat death under some circumstances (for example, the antarctic fish that die at +6 °C), it is not necessarily the only factor in heat death. Any of the other possibilities mentioned on page 272 may be contributing, and for organisms that have a very high lethal temperature, it is more likely that several of the listed factors play a role.

Tolerance to low temperature

The effects of low temperature are at least as perplexing as those of high temperature. Some organisms can tolerate extensive freezing, but most animals cannot. Some can survive submersion in liquid nitrogen at −196 °C or even liquid helium at −269 °C (about 4 K), but on the other hand, some animals are so sensitive to cold that they die far above freezing temperatures. People who keep tropical fish in home aquaria know this; if the heater is off, the fish may be found dead on a cold morning. Guppies (*Lebistes reticulatus*), for example, that have been kept at 23 °C room temperature or above, will die if they are cooled to about 10 °C (Pitkow, 1960). For this fish the evidence indicates that death is due to cold depression of the respiratory center, followed by damage due to anoxia, for an increase in the oxygen content of the water improves survival, and a decrease in the oxygen content increases the susceptibility to cold.

Cold tolerance and freezing tolerance

Animals which live in temperate and cold regions withstand long periods of winter temperatures that are far below the freezing point. They can escape cold injury by two means, *supercooling* and *freezing tolerance*. The former is the lowering of the temperature of a fluid to below its 'freezing point' without formation of ice; the latter is a tolerance to freezing and ice formation in the body. Those animals that cannot tolerate ice formation in the body are called 'freezing-susceptible', and those that survive freezing are 'freezing-tolerant' species.

How does supercooling occur? If water or aqueous solutions are cooled to below the freezing point, freezing does not necessarily take place.* Pure water can be cooled far below 0 °C without any ice formation. The probability that such supercooled water will freeze depends on three important variables, the temperature, the presence of nuclei for ice formation, and time. In the absence of foreign nucleating materials, pure water is readily supercooled to −20 °C before it freezes and if extraordinary precautions are taken, water may be supercooled to near −40 °C. The moment an initial ice nucleus is formed, freezing progresses rapidly throughout the sample.

---

* The term 'freezing point' for water or aqueous solutions is, because of the tendency to supercool, a somewhat misleading term. By freezing point we do not mean the temperature where freezing begins, but rather the temperature at which a minute amount of ice is present in a liquid sample that is in thermodynamic equilibrium. If heat is added to such a sample, ice melts, if heat is removed, more water turns to ice. In the latter case, the solute concentration in the remaining liquid increases, thus lowering its freezing point. The point where the entire sample is frozen throughout is difficult to determine and is only rarely considered. The determination of the freezing point as defined here is frequently carried out by freezing the entire sample, observing it during slow heating, and reading the temperature when the last crystal of ice is about to disappear. This point could therefore more correctly be designated as the 'melting point,' but it is usually referred to as the *freezing point*.

The presence of certain solutes not only greatly lowers the freezing point, but also influences the extent of supercooling that can take place before freezing occurs.

We should now consider the possibility that, although an animal is exposed to temperatures considerably below the freezing point of its body fluids, it may remain supercooled. Such supercooling is in fact of great importance to survival. During an occasional cold night, for example, it may be essential for freezing-susceptible animals that are unable to escape. Experiments with reptiles and various other vertebrates, whose body fluids have a freezing point of -0.6 °C, have shown that they can be supercooled to as much as -8 °C without freezing (Lowe, Lardner and Halpern, 1971).

Other animals survive in spite of extensive ice formation in their bodies. Midge larvae (*Chironomus*) can be frozen to -25 °C and thawed repeatedly without injury. It may seem difficult to decide whether these tiny animals are supercooled or frozen to ice. However, the actual amount of ice can be accurately determined without injury to the animal by an ingenious method which takes advantage of the change in volume as water freezes to ice. By determining the specific gravity of the frozen larvae, it was found that, at -5 °C, 70% of their body water is frozen, and at -15 °C, 90% is frozen. Since no harm is done to the larvae by these determinations, the same specimens that have been frozen can be thawed and tested for survival (Scholander, Flagg, Hock and Irving, 1953).

Although many invertebrate animals at high latitudes are normally exposed to very low temperatures during winter, few are as exposed to as rapidly changing conditions as tidal organisms in the sub-polar regions. In winter, the temperature of the tidal zone regularly alternates between freezing and thawing twice a day. As the water recedes, animals in the tidal zone will have most of their body water frozen to ice, and as the water rises again, they will thaw out. Such tidal animals may be exposed to air temperatures as low as -30 °C for six hours or more, and their internal temperature then approaches that of the air. At -30 °C more than 90% of the body water is frozen and the remaining fluid therefore contains solutes in extremely high concentrations. This means that the cells in addition to losing water to the ice crystals, must be able to tolerate an exceptional increase in osmotic concentration.

Examination of tidal animals show that they indeed freeze rather than become supercooled, and that there is a remarkable distortion of muscles and internal organs by the ice. The ice crystals, however, are generally located outside the cells, which are shrunken and possibly have no ice formed within them. Within a few seconds of thaw-

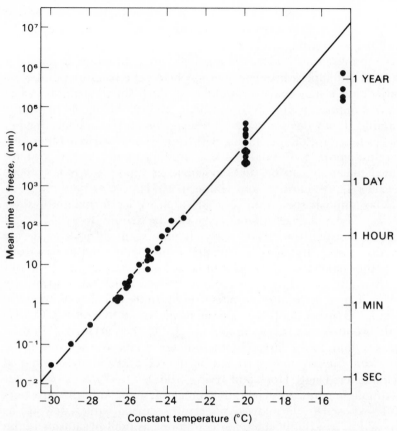

Fig. 7.7. The relation between mean time to freeze for larvae of the wheat stem sawfly. Each point shows the time required for 50% of a group of larvae to freeze [Salt, 1966, reproduced by permission of the National Research Council of Canada from the *Canadian Journal of Zoology*, **37**, 59–69 (1959)].

ing the tissues again assume a normal appearance (Kanwisher, 1959).

Many plants are also highly tolerant to freezing, and on cold winter days the water in the plants freezes without causing permanent damage. The tolerance changes with the seasons, so that a plant, that in summer is killed at temperatures a few degrees below freezing, may survive cooling to −196 °C in winter (Weiser, 1970). We said that, in addition to the degree of cooling, the time of exposure to a given low temperature is important for whether or not freezing takes place. This time-dependence is shown in fig. 7.7 for hibernating larvae of the wheat stem sawfly, a plant feeding wasp, *Cephus cinctus*. When these larvae are kept in constant sub-freezing temperature, no freezing takes place at temperatures above $-15\,°C$. As the temperature is lowered further, freezing takes place more and more rapidly, so that at $-30\,°C$, freezing occurs in about 1.2 s, while at $-17\,°C$ it would take more than a year for 50% of the larvae in a large sample to freeze.

It has long been known that *glycerol* protects red blood cells and mammalian spermatozoa from injury caused by freezing. Glycerol is widely used for this purpose, and samples of human or bull sperm can be kept frozen and remain viable for several years if glycerol is added in a suitable concentration before freezing. (Without such treatment, freezing is lethal to spermatozoans.) Because of its well-known protective action, it has been suggested that the natural occurrence of high concentrations of glycerol in some insects explains their ability to survive low temperatures. Glycerol could influence the cold-resistance of insects in two ways: (1) by its protective action against freezing damage, glycerol could help insects that do freeze to survive, and (2) by lowering the freezing point and increasing the degree of supercooling, glycerol could increase the probability that an insect will completely avoid ice formation.

In many insects the glycerol concentration increases before winter. In the parasitic wasp *Bracon cephi*, the concentration becomes as high as 5 molal (roughly 30%), and decreases again in the spring. This large amount of solute depresses the freezing point of the blood to $-17.5\,°C$. The supercooling points are lowered even more than the freezing points, so that larvae of *Bracon* can be super cooled without ice formation to as low as $-47.2\,°C$. The relationship between supercooling points and freezing points (actually melting points) in the *Bracon* larvae is shown in fig. 7.8. In this figure the solid line represents the calculated regression line for all observations. The broken line indicates a constant degree of supercooling of $29\,°C$ below the actual freezing point. Nearly all the observations are located below this line, so that in these animals virtually every individual tested showed a degree of supercooling in excess of $30\,°C$, and the degree of observed supercooling actually increased with decreasing melting points.

Glycerol, however, is not unique in protecting insects against cold and freezing injury. Of eleven insect species collected in Alberta, Canada, three species were freeze-tolerant, although one of these contained less than 3% glycerol and another no glycerol at all. Eight other species, most of which contained more than 15% glycerol, were killed by freezing. It is therefore evident that glycerol alone cannot protect insects against freezing injuries. There is also a considerable degree of adaptation taking place, for when insects are ex-

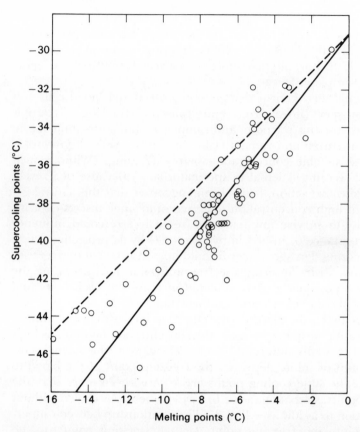

Fig. 7.8. Relation between observed supercooling and the melting point for larvae of a parasitic wasp, *Bracon cephi*. Solid line is the calculated regression line for all observations, broken line indicates a constant degree of supercooling of 29 °C [Salt, 1959, reproduced by permission of the National Research Council of Canada from the *Canadian Journal of Zoology*, **37**, 59–69 (1959)].

posed to moderate cold, they afterwards show an increased tolerance to freezing temperatures and an increase in the degree of super-cooling. This happens both in glycerol-forming species and in non-glycerol-forming species. Thus, the effect of glycerol does not fully explain cold-tolerance, and other solutes in the hemolymph must contribute (Sømme, 1964; 1967).

Anti-freeze in fish

Teleost fish have an osmotic concentration in their body fluids of about 300–400 mosm, and this corresponds to a freezing point of about − 0.6 to − 0.8 °C. Sea water in the polar regions often has a temperature of about − 1.8 °C, yet a variety of fish live in these waters.* The question is, when fish swim in water at − 1.8 °C, why don't they freeze? Do they have a lower freezing point than ordinary fish, or do they remain supercooled throughout life? The answer is that both possibilities seem to have been realized.

These problems have been studied in the Hebron Fjord in Northern Labrador, where the temperature of the surface water in summer is several degrees above the freezing point, while the water at the bottom remains at − 1.73 °C throughout the year. Several species of fish live in this fjord throughout the year, both near the surface and at the bottom, and they provide excellent material for studies of the freezing problem.

In summer all the fish in the Hebron Fjord have approximately the same freezing point in their blood as teleost fish in general, about − 0.8 °C (see fig. 7.9). For the fish at the surface this causes no problem, but those from the deeper water evidently remain supercooled. If they are caught and removed to the laboratory, they can be supercooled to the temperature of the water in which they were caught (− 1.73 °C), and will not freeze. If, however, they are touched with a piece of ice, ice formation rapidly spreads in their bodies and they die almost immediately. It can therefore be assumed that the bottom fish survive throughout their lifetime in the supercooled state. Invertebrate animals, of course, have no special problem. Marine invertebrates in general are in osmotic equilibrium with the sea water in which they live, and therefore avoid freezing by just remaining in the water.

In winter the bottom fish have the same problem as in summer, their freezing point is still about − 1.0 °C. Since the water temperature remains unchanged at − 1.73 °C, they must remain supercooled, for ice formation would kill them immediately.

The fish near the surface, however, will in winter come into contact with ice crystals, and if they were only supercooled, they would have no means of avoiding freezing. The explanation is that in winter the freezing point of the surface fish is lowered to about that

* The salt content of sea water corresponds to a concentration of nearly 1.0 osmolar. The molar freezing point depression for water is 1.86 °C, and sea water therefore does not freeze until its temperature is this low. When ice begins to form, it floats at the surface, both in fresh water and sea water. Fresh water has a point of highest density at +4 °C, and when ice forms at the surface the bottom water in a lake may still be at +4 °C. Sea water, however, does not exhibit this peculiarity, and the temperature of the polar seas, when there is ice at the surface, will be at about − 1.8 °C throughout.

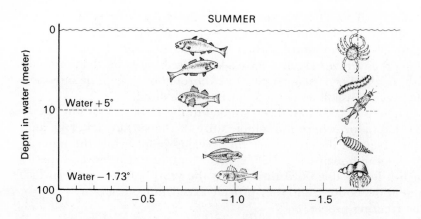

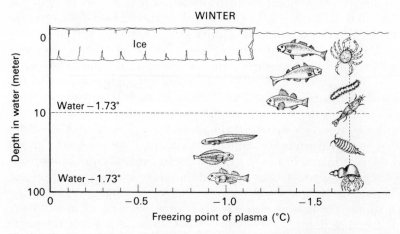

Fig. 7.9. In summer, the surface fish in the Hebron Fjord, Laborador, have no freezing problem. The fish that live deeper, however, where the water is at − 1.73 °C, have a freezing point in their body fluids of − 1.0 °C and must remain supercooled (Scholander *et al.*, 1957).

of the water in which they swim. These fish show only small increases in the major dissolved ions, sodium and chloride, the most important contribution to lowering the freezing point being made by an unidentified anti-freeze substance (Gordon, Amdur and Scholander, 1962).

The anti-freeze in the blood of an antarctic fish, *Trematomus*, however, has been isolated and its chemical nature studied in detail. Chemically, this anti-freeze is a glycoprotein which occurs in the blood with three distinct molecular weights, 10 500, 17 000, and 21 500. At low concentrations (6 g liter^{-1}) this glycoprotein is more

Plate 6. *Ice fish.* The blood of the antarctic fish *Trematomus borchgrevinki* contains a glycoprotein which acts as an antifreeze substance. This permits this fish to swim in sea water at a temperature of $-1.8\,°C$, although the osmotic pressure of its blood, in the absence of the antifreeze, is insufficient to prevent ice formation at this temperature (A. L. DeVries, University of California, San Diego).

effective than sodium chloride in preventing formation of ice in water (Komatsu, DeVries and Feeney, 1970). This is amazing, for the glycoprotein molecule is several hundred times larger than sodium chloride. This means that, if we dissolve equal weights of the glycoprotein and of NaCl in water, the molar concentration of the glycoprotein is several hundred-fold less, and the freezing point depression is normally determined by the number of dissolved particles. On a molar basis the glycoprotein is therefore abnormally effective in preventing the ice formation. Fig. 7.10 shows the freezing point depression caused by various concentrations of glucose and of sodium chloride. For both sodium chloride and glucose the freezing point depression increases linearly with the concentration, but sodium chloride, because of its dissociation into sodium and chloride ions, has twice as high freezing point depression as glucose. In contrast, even a minute concentration of the anti-freeze substance prevents the formation of ice in water about 200 to 500 times as effectively as sodium chloride.

Chemically the glycoproteins from *Trematomus* consist entirely of repeating units of the two amino acids alanin (23%) and threonin (16%), with a disaccharide derived from galactose attached to the threonin unit. This seems to be the first case where the entire struc-

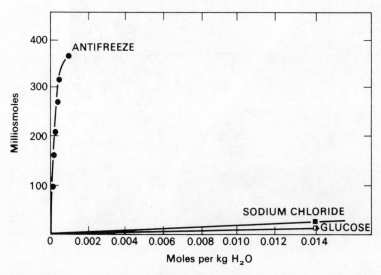

Fig. 7.10. The antifreeze substance from the antarctic fish *Trematomus* has an abnormally high effect in preventing formation of ice in an aqueous solution. On a molar basis it is several hundred times as effective in this regard as expected. The abscissa gives the molal concentration of the substance and the ordinate, the apparent osmolal concentration as determined from freezing point depression (DeVries, 1970).

ture of a naturally occuring glycoprotein has been completely clarified (DeVries, Komatsu and Feeney, 1970; Shier, Lin and DeVries, 1972).

There are several ways in which the glycoproteins could act to depress the freezing point of aqueous solutions. One is that the water is structured by hydrogen bonding to the glycoprotein, and another is that the glycoprotein inhibits the growth of ice crystals by being absorbed to their surfaces, thus hindering the addition of further water molecules to the crystal lattice. The latter possibility is most strongly supported by the finding that once a solution of the glycoprotein has been frozen, its melting point differs from the freezing point by being much higher. The melting point thus is not depressed, which could be expected if structuring of the water were responsible; it is better explained by the model that the glycoproteins bind to the crystal surface and impede its growth, thus lowering the freezing point of the solution (Lin, Duman and DeVries, 1972).

The notion of inhibition of the growth of ice crystals is supported by information on a larger number of arctic and antarctic fishes. Those fish that live in deep water and near the bottom where they can never contact the ice, are supercooled by about 1.0 °C. Shallow water fishes, which live in contact with sea ice at −1.8 °C, also have freezing points about −0.8 °C, but their blood plasma is much more resistant to freezing because ice propagation is inhibited. They have higher plasma protein concentrations than most other vertebrates, and it appears that the proteins impede the growth of ice crystals so that the fish are not readily subjected to extensive ice formation (Hargens, 1972).

PHYSIOLOGICAL TEMPERATURE ADAPTATION

The limits of temperature tolerance for a given animal are not fixed. Exposure to a near-lethal temperature often leads to a certain degree of adaptation so that a previously lethal temperature becomes tolerable. Very frequently the range of tolerance is different for the same species in summer and in winter. A winter animal often tolerates and even is active at a temperature so low that it is lethal to a summer animal, and conversely, the winter animal is less tolerant to high temperature than a summer animal. Such changes in the temperature tolerance with climatic changes are called *acclimatization*. Similar effects can be simulated in laboratory experiments by keeping animals for some time at given temperatures. To distinguish the adaptations or adjustments that take place in laboratory experiments from natural acclimatization, the response to experimental conditions is often described by the term *acclimation* (Hart, 1952; Prosser, 1973).

Acclimatization and acclimation can take place in response to many environmental factors, such as temperature, oxygen tension, nature of the food, moisture, etc. Here we shall be concerned primarily with responses to temperature, a subject which has been better studied and about which more information is available than the responses to other environmental factors.

Geographical differences and seasonal adjustments

It is self evident that many arctic animals live, breathe, and are active at temperatures close to freezing, and in a range that would completely inactivate or kill similar animals from the tropics. Arctic fish and crustaceans are active in ice-cold water, and are killed when the temperature reaches 10 to 20 °C. Comparable tropical forms die below 15 to 20 °C. Thus, there are obvious differences in the temperature range that can be tolerated by animals from different geographical areas (Scholander, Flagg, Walters and Irving, 1953).

Closely related animals, such as various species of frogs of the genus *Rana,* have a geographical distribution which is closely related to their temperature tolerance. Fig. 7.11 shows that of four species of North American frogs, the species that has the northernmost geographical limit is also the most cold tolerant.

The correlation between distribution and temperature conditions extends to the temperature of the water at the time of breeding, and

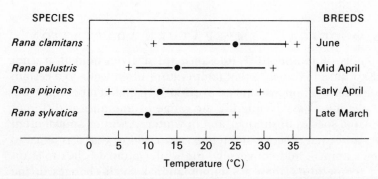

Fig. 7.11. The relation between temperature tolerance and breeding in four species of North American frogs. The horizontal lines show the temperature range for normal development, with the average water temperature at the time of egg-laying indicated by a black dot. The crosses indicate lethal temperatures (after Moore, 1939).

Table 7.3. *Comparison of breeding and development of various species of the frog* Rana. (Moore, 1939)

	R. sylvatica	R. pipiens	R. palustris	R. clamitans
Order of breeding	1	2	3	4
Water temperature at time of breeding (°C)	10	12	15	25
Most northern record	67°N	60°N	55°N	50°N
Lower limiting embryonic temperature (°C)	2.5	6	7	11
Upper limiting embryonic temperature (°C)	24	28	30	35
Time between stages 3 and 20 at 18.5 °C (h)	87	116	126	138

to the lower as well as the upper limiting temperature for embryonic development. Thus, frogs which breed at relatively low environmental temperatures have low minimal and maximal temperatures for breeding and for embryonic development (see table 7.3).

The geographically most widely distributed frog of these four species is *Rana pipiens*. The population of this species from the northern part of the range differs from southern populations in the same way that the northern species differs from southern species. Thus, *Rana pipiens* exists in races that have physiological differences in their response to temperature. This is clearly correlated with the unusually wide geographical range over which this species is distributed (Moore, 1949).

We should now examine more closely natural and experimentally induced changes in temperature tolerance. In large parts of the world the seasonal temperature fluctuations are appreciable, and during one season animals tolerate temperature extremes which in other seasons are fatal. Thus, the bullhead, a catfish (*Ameiurus nebulosus*) has an upper lethal temperature of nearly 36 °C in summer, but in winter this fish will die if the temperature of the water exceeds 28 °C (see fig. 7.12). These changes in temperature tolerance occur in response to the natural climatic change between summer and winter, and therefore present an example of natural acclimatization.

Likewise, the lower lethal temperature changes with the seasons. As an example, summer individuals of the beetle *Pterosticus brevicornis*, from Alaska invariably die if they are frozen (which happens at −6.6 °C). In winter, however, they tolerate cooling to temperatures

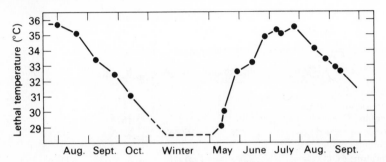

Fig. 7.12. The bullhead, *Ictalurus* (*Ameiurus*) *nebulosus,* has an upper lethal temperature limit that changes with the seasons. Each point indicates the temperature that is
fatal to 50% of the samples for an exposure of 12 hours (Fry, 1947).

below −35 °C and as well as complete freezing (Miller, 1969). The
fact that this beetle actually tolerates freezing is important, for it has
previously been stated that no adult insect is able to tolerate freezing, and that freezing tolerance is limited to larval and pupal forms.
Obviously, this does not hold as a general rule.

Maximum range of tolerance

Since the limits of temperature tolerance are modified by the previous thermal history of an animal, how far can the limit be pushed
by slow and gradual acclimation? Is there an ultimate limit that cannot be exceeded? To answer this question groups of animals are
kept at various constant temperatures for sufficient periods of time
to ensure that they are fully acclimated at these temperatures. They
can then be tested in order to establish the upper and lower limits of
thermal tolerance, and when this has been done for a series of different temperatures, the results can be plotted in a single graph. Fig.
7.13 is such a plot for the goldfish, showing its complete thermal tolerance.

The goldfish has an exceptionally wide temperature tolerance, and
the area enclosed by the tolerance curve is therefore large. The tolerance curve for a fish with a narrower range, such as the chum
salmon (*Oncorhynchus keta*) covers a much smaller area. Like other
salmonids the chum salmon is a typical cold-water fish, and its upper
lethal temperature cannot be extended beyond 24 °C. This is shown
in fig. 7.14, which gives the thermal tolerance range for the chum
salmon and for the brown bullhead, *Ictalurus* (*Ameiurus*) *nebulosus,*
which has a large tolerance range, similar to that of the goldfish.
The total area enclosed by the tolerance curve for the bullhead is
more than twice as large as that for the chum salmon. To quantify

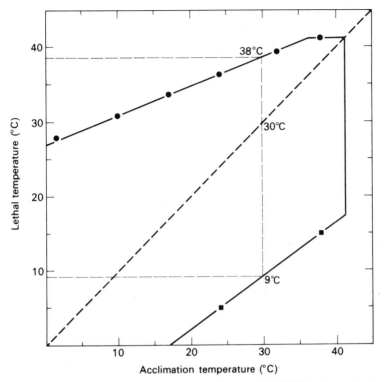

Fig. 7.13. The complete range of thermal tolerance of goldfish is shown by the solid black line. If a goldfish is kept at some temperature on the diagonal broken line until it is fully acclimated, its tolerance to high temperature can be read on the upper solid line, and its tolerance to cold on the lower solid line. For example, a fish kept at 30 °C has an upper lethal limit of 38 °C and a lower lethal limit of 9 °C (Fry, Brett and Clawson, 1942).

the area, we can take as our unit of measurement a small square which corresponds to the division of one degree on each coordinate. The area enclosed by the bullhead curve is 1162 such squares, which properly have the dimension $(°C)^2$, and the curve for the chum salmon encloses 468 squares.

Most other fish for which the thermal tolerance has been established fall between these two extremes. Both the bullhead and the chum salmon tolerate water at freezing temperatures, but many warm-water species will have curves that do not extend down to 0 °C, but are closed at some higher temperature. However, the complete thermal tolerance is known only for a relatively small number of fish, and it seems likely that, on the whole, fresh-water fish may have a wider tolerance than marine fish.

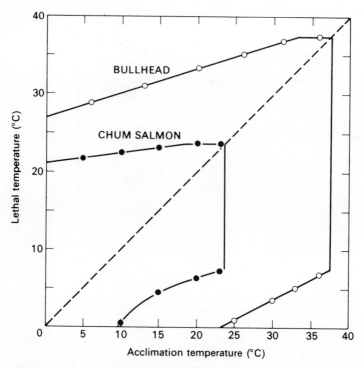

Fig. 7.14. The ultimate thermal tolerance of the chum salmon, a cold-water fish, covers a much smaller range than the bullhead, which has a wide temperature tolerance similar to that of the goldfish (cf. fig. 7.13) (Brett, 1956).

The tolerance of fresh-water fish ranges from 450 $(°C)^2$ for salmon to 1220 for goldfish (Brett, 1956). Of three salt-water fish that have been examined, the silverside has the largest range, 715 $(°C)^2$, a puffer (swellfish) has the smallest, 550 $(°C)^2$, and the winter flounder has a range intermediate between the other two (Hoff and Westman, 1966). (Silverside = *Menidia menidia,* winter flounder = *Pseudopleuronectes americanus,* and puffer = *Spheroides maculatus.*)

More information about the thermal tolerance of fish is much needed, for the limits to thermal adaptation will become increasingly important in connection with problems of thermal pollution and industrial activity, especially power generating plants and atomic reactors.

Rate of acclimation. The time required for a fish to increase its ability to tolerate high temperatures is relatively short. It takes less than 24 hours for a bullhead to adjust fully when it is moved from 20 °C to 28 °C (fig. 7.15). The process of temperature acclimation depends

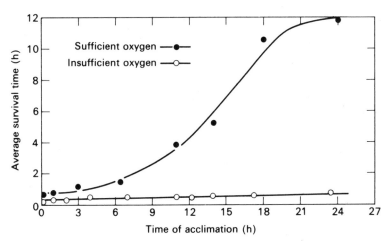

Fig. 7.15. The thermal acclimation of the bullhead takes place within about 24 hours when the water contains adequate oxygen, but proceeds very slowly at low oxygen (Fry, 1947).

on an adequate supply of oxygen. Although the bullhead is tolerant to low oxygen and survives quite well, the process of thermal acclimation seems to be completely impeded in the absence of sufficient oxygen. In general, the acclimation goes faster the higher the temperature. Therefore it seems to be related to metabolic processes, but there is no clear understanding of the mechanisms that are responsible for the modification of the lethal limit, although changes in metabolic enzyme systems seem to have a primary role (Hochachka and Somero, 1973).

When a fish is moved from a higher to a lower temperature, the loss in tolerance to high temperatures and the gain in resistance to low temperature is a slower process. In a minnow (*Pimephales promelas*) the complete adaptation takes between 10 and 20 days when the fish is moved from 24 to 16 °C (Brett, 1956).

We have now seen that the thermal tolerance of a fish depends on the thermal history of the individual, and that oxygen plays a role in the speed with which adaptation to a new temperature regimen takes place. However, other factors are important, such as the age and size of the fish, the quality of the water, and so on. Of particular interest is the finding that goldfish, even if they are kept in the laboratory at completely constant conditions of temperature and diet, show a marked seasonal change in thermal tolerance, with a greater cold resistance in the winter. This difference has been ascribed to the photoperiod (Hoar, 1956). In general, a long photoperiod, as in summer, increases the resistance to heat, while a short photoperiod, as in winter, causes an increase in the tolerance to cold (Roberts, 1964).

Thermal acclimation and metabolic rate

Until now the discussion has been concerned with lethal temperatures and adaptation as a result of the thermal history of an animal. However, adaptation is not restricted to a change in lethal temperature limits, there are other compensating mechanisms as well, and these usually tend to counteract the acute effects of temperature change. Anybody who has been catching fish in winter will have noted that in winter, fish are not noticeably slower and more sluggish than in summer. Laboratory studies confirm the impression that fish, if given time, to a great extent can compensate for temperature changes by appropriate changes in metabolic rate. This is true, not only for fish, but for many other poikilothermic animals as well.

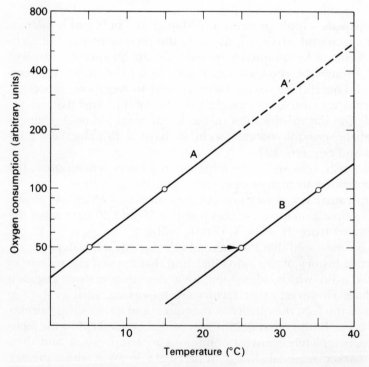

Fig. 7.16. Hypothetical change in metabolic rate (oxygen consumption, on logarithmic ordinate) in a fish acclimated to 5 °C water (A). If the fish were able to compensate fully, acclimation to 25 °C water would make it follow the metabolic curve indicated by (B). If it does not compensate at all, it will follow the broken line marked A'.

Let us assume that a fish, adapted to a winter temperature of 5 °C, has a temperature tolerance range of 0 to 25 °C, and that the Q_{10} throughout this range is 2.0. The metabolic rate of this fish would follow the straight line marked A in fig. 7.16. If we now assume that we acclimate this fish to a summer temperature of 25 °C, its upper lethal limit would be extended, say, to 40 °C. If no metabolic compensation were to take place, the fish would have a metabolic rate extending along the dotted line marked A′. If, on the other hand, the fish compensates fully for the change in temperature, the summer fish after adaptation to 25 °C should have the metabolic rate that it previously had at 5°C. If the Q_{10} remains the same (2.0), the oxygen consumption of the fish would now follow the curve marked B. Thus, the fish would have compensated fully for the change in temperature, and would have returned to its previous metabolic rate.

In addition to the two possibilities just mentioned, no compensation or complete compensation, there are several other long-term types or patterns of adaptation. The various patterns can be generalized as indicated in fig. 7.17. If an animal shows no compensation and continues to consume oxygen at the same rate as caused by an acute effect of a temperature change, it will follow line A (the same as the dashed extension of line A in fig. 7.16). Complete compensation, on the other hand, involves a return to the original oxygen con-

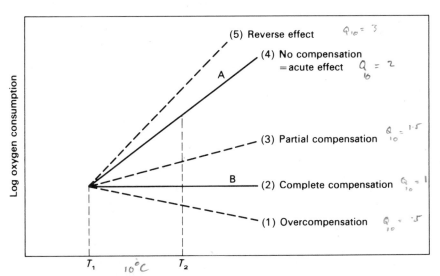

Fig. 7.17. The compensation for a temperature change, from T_1 to T_2, may take any of several courses, as indicated by the various curves. For details, see text (modified from Precht, Christophersen and Hensel, 1955).

sumption after the animal has had time to adapt to a new temperature, and the oxygen consumption will therefore follow line B, showing complete compensation. Any response between these two lines will indicate a partial compensation for the temperature change. The possibility also exists that the animal may overcompensate, as indicated by line (1). A few cases have been reported when a temperature change has a reverse effect, as indicated by line (5), at least within part of its temperature range.

In reality, animals do not often follow these idealized patterns, and their responses are frequently far more complex. As an example, we can examine the response of two aquatic animals, the trout and the European crayfish, to a 10 °C drop in temperature, followed by a 10 °C increase in temperature after full adaptation has taken place (see fig. 7.18).

An example of an animal that shows little or no compensation is the beach flea (*Talorchestia megalophthalma*) from the area of Woods Hole, Massachusetts. The oxygen consumption of this animal in winter is the same as in summer, when measured at the same temperature. There is thus virtually no acclimatization or compensation for the seasonal change in temperature. Accordingly, the winter animals have such a low metabolic rate that they remain inactive, and they can be found immobile in small burrows, hibernating be-

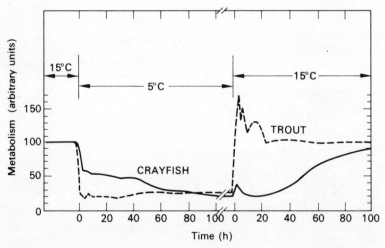

Fig. 7.18. Metabolic response to temperature change of crayfish (solid line) and trout (broken line). Initially the animals were acclimated to 15 °C. When the water temperature was dropped by 10 °C, the trout rapidly achieved a new plateau, while the crayfish reached a plateau more slowly. On raising the temperature to 15 °C again, the trout had an overshoot before returning to the initial level; the crayfish only slowly responded to the temperature rise (Bullock, 1955).

neath the sand. If they are disturbed in winter they respond very sluggishly, but when warmed they soon become as active as in summer (Edwards and Irving, 1943*b*).

In contrast, another crustacean from the same habitat, the sand crab (*Emerita talpoida*) compensates for low winter temperatures by increasing its metabolic rate. At 3 °C, winter animals consume oxygen at a rate four times as high as that of summer animals at the same temperature. Thus, these animals can remain active in winter, but since the sandy beaches of Woods Hole are most inhospitable during winter, the animals move from the intertidal zone out to deeper water (Edwards and Irving, 1943*a*). Thus the sand crab adjusts to the seasonal changes in temperature so that it can remain active at the low winter temperatures, rather than becoming inactive and hibernating. Its lethal temperature also changes with the seasons. Temperatures above 27 °C are lethal in winter, but for summer animals death occurs above 37 °C.

REFERENCES

BOTT, T. L. and BROCK, T. D. (1969). Bacterial growth rates above 90 °C in Yellowstone hot springs. *Science,* **164,** 1411–12.

BRETT, J. R. (1956). Some principles in the thermal requirements of fishes. *Quart. Rev. Biol.,* **31,** 75–87.

BROCK, T. D. (1970). High temperature systems. *Ann. Rev. Ecol. Systm.,* **1,** 191–220.

BROWN, J. H. and FELDMETH, C. R. (1971). Evolution in constant and fluctuating environments: thermal tolerances of desert pupfish (*Cyprinodon*). *Evolution,* **25,** 390–8.

BULLOCK, T. H. (1955). Compensation for temperature in the metabolism and activity of poikilotherms. *Biol. Rev.,* **30,** 311–42.

CARLISLE, D. B. (1968). *Triops* (Entomostraca) eggs killed only by boiling. *Science,* **161,** 279–80.

DEVRIES, A. L. (1970). Freezing resistance in antarctic fishes. In *Antarctic Ecology,* **1** (M. Holdgate, ed.), pp. 320–28, New York: Academic Press.

DEVRIES, A. L., KOMATSU, S. K. and FEENEY, R. E. (1970). Chemical and physical properties of freezing point-depressing glycoproteins from antarctic fishes. *J. Biol. Chem.,* **245,** 2901–8.

EDWARDS, G. A. and IRVING, L. (1943*a*). The influence of temperature and season upon the oxygen consumption of the sand crab, *Emerita talpoida* Say. *J. Cell. Comp. Physiol.,* **21,** 169–82.

EDWARDS, G. A. and IRVING, L. (1943*b*). The influence of season and temperature upon the oxygen consumption of the beach flea, *Talorchestia megalopthalma. J. Cell. Comp. Physiol.,* **21,** 183–9.

FRAENKEL, G. (1968). The heat resistance of intertidal snails at Bimini, Bahamas; Ocean Springs, Mississippi; and Woods Hole, Massachusetts. *Physiol. Zool.*, **41**, 1–13.

FRY, F. E. J. (1947). Effects of the environment on animal activity. *Univ. of Toronto Studies, Biological Ser., no. 55. Publ. Ontario Fish. Res. Lab., no.* **68**, 1–62.

FRY, F. E. J., BRETT, J. R. and CLAWSON, G. H. (1942). Lethal limits of temperature for young goldfish. *Rev. Can. Biol.*, **1**, 50–6.

GORDON, M. S., AMDUR, B. H. and SCHOLANDER, P. F. (1962). Freezing resistance in some northern fishes. *Biol. Bull.*, **122**, 52–6.

X HARGENS, A. R. (1972). Freezing resistance in polar fishes. *Science*, **176**, 184–6.

HART, J. S. (1952). Geographic variations of some physiological and morphological characters in certain freshwater fish. *Univ. Toronto Studies, Biol. ser. no. 60. Publ. of the Ontario Fish. Res. Lab.,* no. **72**, 1–79.

HINTON, H. E. (1960). A fly larva that tolerates dehydration and temperatures of − 270 °C to + 102 °C. *Nature, Lond.*, **188**, 336–7.

HOAR, W. S. (1956). Photoperiodism and thermal resistance of goldfish. *Nature, Lond.*, **178**, 364–5.

HOCHACHKA, P. W. and SOMERO, G. N. (1973). *Strategies of Biochemical Adaptation*, Philadelphia: W. B. Saunders Co., 358 pp.

HOFF, J. G. and WESTMAN, J. R. (1966). The temperature tolerances of three species of marine fishes. *J. Mar. Res.* **24**, 131–140.

KANWISHER, J. (1959). Histology and metabolism of frozen intertidal animals. *Biol. Bull.*, **116**, 258–64.

KOMATSU, S. K., DEVRIES, A. L. and FEENEY, R. E. (1970). Studies of the structure of freezing point-depressing glycoproteins from an antarctic fish. *J. Biol. Chem.*, **245**, 2909–13.

LIN, Y., DUMAN, J. G. and DEVRIES, A. L. (1972). Studies on the structure and activity of low molecular weight glycoproteins from an antarctic fish. *Biochem. Biophys. Res. Commun.*, **46**, 87–92.

LOWE, C. H., LARDNER, P. J. and HALPERN, E. A. (1971). Supercooling in reptiles and other vertebrates. *Comp. Biochem. Physiol.*, **39A**, 125–35.

MARZUSCH, K. (1952). Untersuchungen über die Temperatur abhängigkeit von Lebensprozessen bei Insekten unter Besonderer Berücksichtigung Winterschlafender Kartoffelkäfer. *Zeits. vergl. Physiol.*, **34**, 75–92.

X MILLER, L. K. (1969). Freezing tolerance in an adult insect. *Science*, **166**, 105–6.

MOORE, J. A. (1939). Temperature tolerance and rates of development in the eggs of Amphibia. *Ecology*, **20**, 459–78.

MOORE, J. A. (1949). Geographic variation of adaptive characters in *Rana pipiens* Schreber. *Evolution*, **3**, 1–24.

PITKOW, R. B. (1960). Cold death in the guppy. *Biol. Bull.*, **119**, 231–45.

PRECHT, H., CHRISTOPHERSEN, J. and HENSEL, H. (1955). *Temperatur und Leben*, Berlin: Springer-Verlag, 514 pp.

PROSSER, C. L. (1973). *Comparative Animal Physiology*, 3rd edn, Philadelphia: W. B. Saunders Co., 966 pp.

ROBERTS, J. L. (1964). Metabolic responses of fresh-water sunfish to seasonal photoperiods and temperatures. *Helgol. Wiss. Meeresunter.*,**9**, 459–73.

SALT, R. W. (1959). Role of glycerol in the cold-hardening of *Bracon cephi* (Gahan). *Can. J. Zool.*, **37**, 59–69.

SALT, R. W. (1966). Relation between time of freezing and temperature in supercooled larvae of *Cephus cinctus* Nort. *Can. J. Zool.*, **44**, 947–52.

SCHOLANDER, P. F., FLAGG, W., HOCK, R. J. and IRVING, L. (1953). Studies on the physiology of frozen plants and animals in the arctic. *J. Cell. Comp. Physiol.*, **42**, *Suppl.* **1**, 1–56.

SCHOLANDER, P. F., FLAGG, W., WALTERS, V. and IRVING, L. (1953). Climatic adaptation in arctic and tropical poikilotherms. *Physiol. Zool.*, **26**, 67–92.

SCHOLANDER, P. F., VAN DAM, L., KANWISHER, J. W., HAMMEL, H. T. and GORDON, M. S. (1957). Supercooling and osmoregulation in arctic fish. *J. Cell. Comp. Physiol.*, **49**, 5–24.

SHIER, W. T., LIN, Y. and DEVRIES, A. L. (1972). Structure and mode of action of glycoproteins from an antarctic fish. *Biochim. Biophys. Acta*, **263**, 406–13.

SOMERO, G. N. and DEVRIES, A. L. (1967). Temperature tolerance of some antarctic fishes. *Science*, **156**, 257–8.

SØMME, L. (1964). Effects of glycerol on cold-hardiness in insects. *Can. J. Zool.*, **42**, 87–101.

SØMME, L. (1967). The effect of temperature and anoxia on haemolymph composition and supercooling in three overwintering insects. *J. Insect Physiol.*, **13**, 805–14.

WEISER, C. J. (1970). Cold resistance and injury in woody plants. *Science*, **169**, 1269–78.

8
Temperature regulation

The preceding chapter was concerned with the effect on physiological processes of the temperature of the organism, and it was stated that the temperature of many animals is virtually the same as that of the surroundings. It would seem that great benefit could be derived from independence of this environmental variable, and in this chapter we shall be concerned with animals that keep their temperature more or less constant, independent of the environment. What birds and mammals can achieve in this regard is much more effective than that which other animals can do, and it is reasonable to treat them first. This will set the stage for a discussion of other animals that to some extent can control their temperature independently of the environment.

First of all, it is necessary to clarify what we mean by body temperature, a concept which by no means is simple. Secondly, for the body to maintain a given temperature it is necessary to balance heat transfer so that the loss and the gain of heat are equal. To understand these two processes it is necessary to be familiar with the simple physics of heat transfer.

In a cold environment heat balance can be achieved by manipulating the heat loss and/or the heat gain (heat production). Most mammals and birds do this very successfully, but some mammals and a few birds appear to give up and permit their temperature to drop precipitously, they go into torpor or hibernation. We shall see that they have not fully abandoned temperature regulation, on the contrary, hibernation is a well-regulated and controlled physiological state.

In a hot environment the problems of maintaining the body temperature are reversed, the animal must keep the body temperature from rising and is often compelled to cool itself by evaporation of water.

Birds and mammals are so successful at regulating their body temperature that they live through most of their lives with body temperature fluctuations of no more than a few degrees. However, some other animals are also amazingly adept at keeping their temperature different from that of the environment. This applies to terrestrial vertebrates (e.g. lizards), to many insects, particularly highly active ones, and amazingly, even to some fish which maintain parts of their body at temperatures approaching those of 'warm-blooded' animals.

BODY TEMPERATURE OF BIRDS AND MAMMALS

What is 'body temperature'?

The heat produced by an animal must be transported to the surface before it can be transferred to the environment. Therefore, the surface of the organism must be at a lower temperature than the inner parts, for if the temperature were the same throughout, no heat could be transferred. The conclusion is that the temperature of an organism cannot, of necessity, be uniform throughout.

If we examine where in the mammalian body heat production takes place, we find that some parts produce more heat than others. In man the organs of the chest and the abdomen, although they make up less than 6% of the body weight, produce 56% of the total heat (see table 8.1). If we include the brain, which in man is large and has a high heat production, we have accounted for 72%, or more than two-thirds, of the total heat production.

We can thus consider that the body consists of a core where most of the heat production takes place, and a much larger shell which includes skin and muscles and produces only a small fraction of the total body heat. During exercise, the situation is different, for the total metabolic rate may increase ten-fold or more. Most of this increase occurs in the muscles (including the diaphragm and other respiratory muscles).

During exercise then, for the internal temperature to remain constant, more than ten times as much heat as was produced at rest must be transported to the surface of the organism.

Temperature distribution in the body. The inner, or core temperature remains reasonably constant, but this does not mean that the temperature throughout the core is uniform. Organs which have a high rate of heat production may be warmer than others, but they are

Table 8.1. *Heat production in the major organs of a man at rest* (*body weight, 65 kg, heat production, 1866 kcal per day = 1296 cal per min*). *The main internal organs weigh about 5 kg but account for 72% of the total heat production.* (Data from Aschoff, Günther and Kramer, 1971)

	ORGAN WEIGHT			HEAT PRODUCTION	
	(kg)		(% of body wt)	(cal min^{-1})	(% of total)
Kidneys	0.290		0.45	100	7.7
Heart	0.290		0.45	139	10.7
Lungs	0.600	5.03	0.9	57	4.4
Brain	1.350		2.1	208	16.0
Splanchnic organs *	2.500		3.8	435	33.6
Skin	5.000		7.8	24	1.9
Muscle	27.000	59.97	41.5	203	15.7
Other	27.970		42.0	130	10.0
	65.000			1296	100.0

Braces: organ weight (kg) for Kidneys–Splanchnic = 5.03, for Skin–Other = 59.97; (% of body wt) for Kidneys–Splanchnic = 7.7, for Skin–Other = 91.3; (% of total) heat for Kidneys–Splanchnic = 72.4, for Skin–Other = 27.6.

* Abdominal organs, not including kidneys

cooled by the blood, i.e. the venous blood which leaves these organs is warmer than the arterial blood. The temperature differences in the core may be as much as 0.5 °C from one site to another. We therefore cannot speak about a single core temperature, but as a practical measure the deep rectal temperature is often used as a representative measure.

The surface temperature of a man who is in heat balance is always lower than the core temperature. This means that the arterial blood which flows to the shell loses heat and returns as colder venous blood. This is, of course, how most of the heat produced in the core is brought to the surface, or, in other words, how the core organs are cooled. Depending on the circumstances (external temperatures and the need for heat loss), the surface temperature varies a great deal. Also, the underlying tissues, including a large part of the muscle mass, can take on temperatures considerably below the core temperature (see fig. 8.1).

What is the mean temperature of the body? We can calculate a mean body temperature from multiple measurements at various sites, provided that we have a weighting procedure to apply to the various measurements. However, mean body temperature calculated

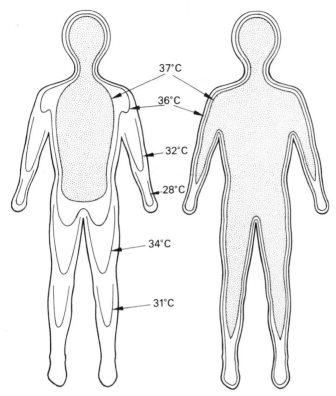

Fig. 8.1. Temperature distribution in the body of a man at room temperatures of 20 °C (left) and at 35 °C (right). The isotherms, which indicate sites of equal temperature, show that, at 35 °C room temperature, a core temperature of 37 °C (shaded area) extends into the legs and arms. At 20 °C room temperature, the temperature gradients in the shell extend throughout the legs and arms, and the core temperature is restricted to the trunk and head (after Aschoff and Wever, 1958).

from multiple measurements is physiologically relatively meaning-less. We have already seen that the core temperature does not repre-sent the heat status of the whole body, that the temperature of the shell can vary widely, and that the depth in the body to which the shell temperature extends changes drastically. A change in the tem-perature of the shell means that the total heat content of the body changes although the core temperature may remain constant. If a man moves from a room temperature of 35 °C to 20 °C, the drop in shell temperature may involve a heat loss of 200 kcal from the shell. The magnitude of this much heat becomes clear when we realize that it corresponds roughly to three hours' resting metabolism of the man.

'Normal' body temperature of birds and mammals

Daily fluctuations in core temperature. The core temperature of man, and that of other mammals and birds, undergoes regular daily fluctuations. Within 24 hours these fluctuations are usually between 1 and 2 °C. Diurnal animals show a temperature peak during the day and a minimum at night; nocturnal animals show the reverse pattern. However, these daily cycles are not caused directly by the alternating periods of activity and rest, for they continue even if the organism is at complete rest.

The daily pattern of the body temperature of many mammals and birds consistently follows the light cycle. When the towhee (*Pipilo aberti*) is kept in 12 hours of darkness and 12 hours of light, its body temperature follows a cycle synchronized with the light cycle. At a room temperature of 23 °C, the core temperature at night is about 39 °C and when the light comes on, it rapidly rises to nearly 42 °C (see fig. 8.2). At a different room temperature, at 5 °C for example, the core temperature follows the same day and night cycle, although it is, on the average, about half a degree higher than the corresponding core temperature measured at 23 °C.

Similar temperature cycles have been recorded in a variety of other birds from several different families. Diurnal forms always have the highest temperature during the day and nocturnal species at night. One such example of a nocturnal bird is the flightless kiwi

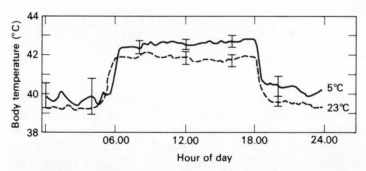

Fig. 8.2. When the towhee (a finch, *Pipilo aberti*) is kept at a constant room temperature of 23 °C, its body temperature varies with the light cycle. When the lights come on at 0600 hours, the body temperature rises by nearly 3 °C, to drop again when the lights go off at 1800 hours. If the room temperature is reduced to 5 °C, the body temperature cycle is similar, but at a slightly higher level (from Dawson, W. 1954. Temperature Regulation and Water Requirements of the Brown and Abert Towhees, Pipilo Fuscus and Pipilo Aberti. University of California Publications in Zoology, **59**, p. 89, Figure 6. Originally published by University of California Press; reprinted by permission of the Regents of University of California).

(*Apteryx*) from New Zealand. Its core temperature is lower than in most other birds, with a daytime mean of 36.9 °C. At night, which is the normal period of activity for the kiwi, the temperature increases to a mean of 38.4 °C, i.e. at night the temperature is about 1.5 °C higher than in the daytime (Farner, 1956).

Both the temperature cycle and the corresponding cycle in metabolic rate can be reversed by reversing the periods of light and dark. This shows that the cycle is governed by the illumination (Dawson, 1954). However, the temperature variations continue on nearly the same timing, even if all variations in the light are removed and the animals are kept in continuous uniform light. Since the regular cycle continues without any external cues, it must be inherent in the organism, that is, it must be a truly endogenous cycle. The cycle is self-sustained in the sense that it continues for days, weeks, and even months, although the timing usually slowly drifts away from an exact 24-hour cycle. To make a measurement of body temperature really meaningful it is therefore desirable to know details of activity, the time of day, and the usual temperature cycle of the animal.

If we disregard variations that amount to a couple of degrees, we see that the usual or 'normal' core temperature is almost uniform within each of the major groups of warm-blooded vertebrates, but that there are characteristic differences between the major groups (table 8.2).

Making allowance for the fact that it is difficult to establish what is 'normal' body temperature for a given animal, and disregarding variations due to external conditions and activity, we can, as a rule of thumb say that most birds maintain their body temperature at 40±2 °C, eutherian mammals at 38±2 °C, marsupial mammals at 36±2 °C, and monotremes at 31±2 °C. Small birds may have somewhat higher body temperatures than large birds (McNab, 1966), but in mammals there is no clear relationship between body size and body temperature (Morrison and Ryser, 1952). For marsupials the available information is insufficient to say whether there is any relationship between body size and body temperature.

The fact that the so-called 'primitive' groups, insectivores, marsupials, and especially monotremes, consistently have a low body temperature raises some very interesting evolutionary problems. These groups are considered to be very ancient, and presumably have had more time than more recent groups to evolve towards a high body temperature, if this is 'desirable' or advantageous. Or have they remained in their more 'primitive' stage because they have lacked the capacity to evolve in this direction? They have certainly

Table 8.2. *Approximate 'normal' core temperature of some major groups of mammals and birds. The lethal temperatures are based on observations made under a wide variety of conditions. Nevertheless, there is rather consistently an approximately 6 °C interval between the 'normal' core and the lethal temperature for the same animal*

	Approximate normal range of core temp. (°C)	Approximate lethal core temp. (°C)
Monotremes (echidna)	30–31 (*a*)	37 (*a*)
Marsupials	35–36 (*b*)	40–41 (*e*)
Insectivore (hedgehog)	34–36	41 (*f*)
Man	37	43
Eutherian mammals	36–38 (*c*)	42–44 (*g*)
Bird (kiwi)	38 (*d*)	
Birds, non-passerine	39–40 (*b*)	46 (*h*)
Birds, passerine	40–41 (*b*)	47 (*i, j*)

(*a*) Schmidt-Nielsen, Dawson and Crawford (1966).   (*f*) Shkolnik, unpublished.
(*b*) Dawson and Hulbert (1970).   (*g*) Adolph (1947).
(*c*) Morrison and Ryser (1952).   (*h*) Robinson and Lee (1946).
(*d*) Farner (1956).   (*i*) Calder (1964).
(*e*) Robinson and Morrison (1957).   (*j*) Dawson (1954).

been successful as witnessed by their ability to survive for so long. The fact is that we do not fully understand the advantages of any given body temperature. In any event, it would be a mistake to interpret a low body temperature as a sign of a 'primitive' and thus inadequate temperature regulation. It has been said that the egg-laying echidna is half-way to being a cold-blooded animal and is thus unable to regulate its body temperature adequately. The fact is that the echidna is an excellent temperature regulator which can maintain its core temperature over a wide range of ambient temperatures down to freezing or below, although it has a poor tolerance to high temperature.

The approximate lethal body temperature for the various groups of warm-blooded vertebrates is given in the last column of table 8.2. It seems that the lethal temperature is regularly at roughly 6 degrees above the normal core temperature. Thus, the echnidna dies when its body temperature reaches 37 °C, which is a normal temperature for mammals and well below the normal temperature for birds. It seems that the margin of safety remains uniformly at about 6 °C

from group to group, but we do not know why the lethal temperature is at such a constant level relative to the normal core temperature.

Do arctic birds and mammals maintain body temperatures within
the same range as species from warmer climates?

To answer this question the body temperature of a number of
Alaskan birds and mammals was measured while the animals were
exposed to a wide range of low air temperatures. The birds studied
belonged to 30 different species and ranged in weight from 0.01 to 2
kg. In air temperatures from + 20 °C down to − 30 °C, their mean
body temperature was 41.1 °C, which is within the normal range for
birds from moderate or tropical climates.

Arctic mammals of 22 species, weighing from 0.1 to 1000 kg, were
exposed to temperatures down to − 50 °C, and one species (the white
fox) even to −80 °C. The individuals of all species maintained their
body temperatures within normal mammalian limits. The mean for
all the observed species was 38.6 °C, which is about 0.5 °C higher
than the previously reported mean temperature for a large number
of mammals from temperate regions, but because of the limited
sample size, a difference of 0.5 °C is insignificant (Irving and Krog,
1954).

We can therefore conclude that arctic birds and mammals maintain body temperatures characteristic of their groups, although they
live in some of the coldest areas on earth.

In contrast to the large amount of information on body temperature in captive and laboratory animals, there are few long-term studies on animals under natural conditions. With the development of
methods to measure temperature with transducers that transmit information by radio (telemetry), it is possible to make recordings
from unrestrained and completely undisturbed animals. By surgically implanting such telemetric units, the temperature of a single
animal has been followed for as long as one year. When this technique was used on sheep that were grazing under field conditions,
the daily range of the deep body temperature was less than 1 °C
whether it was sun, rain, or storm. The total temperature range observed during a whole year from summer to the most severe winter
was 1.9 °C, from 37.9 to 39.8 °C (Bligh, Ingram, Keynes and Robinson, 1965).

The ability to regulate body temperature at such constant levels
requires an extraordinary capacity of the physiological regulatory
mechanisms. Before we discuss these mechanisms we will need an elementary knowledge of the physics of heat and heat transfer.

TEMPERATURE, HEAT, AND HEAT TRANSFER

In the preceding section we were concerned with temperature, and the concept of heat was mentioned only incidentally. It is important to understand the difference between these two physical quantities, and that the measurement of temperature does not necessarily give any information about heat.

Temperature is usually measured in degrees Celcius (°C), although in physical chemistry and thermodynamics we use absolute temperature expressed in degrees Kelvin (K).*

In biology heat is usually measured in calories, and one calorie (cal) is defined as the amount of heat needed to raise the temperature of 1 gram of water by one degree C. (The calorie is not a part of the International System of Units (SI) but its use is so common and ingrained that it will remain in use for some time to come. For conversion to SI units one calorie = 4.190 joule (J).)

To heat one gram of water from room temperature (25 °C) to the boiling point (100 °C) requires 75 cal of heat. To raise the temperature of 100 grams of water by 75 °C requires 7500 cal (7.5 kcal). In both cases the initial and final temperatures are the same, and temperatures alone therefore give no information about the amount of heat added. If we know the amount of water, however, we can calculate the amount of heat from the temperature change, for, by definition, the heat needed to raise the temperature of one gram of water by one degree C is one calorie. The amount of heat needed to warm 1 gram of a substance by 1 degree is known as the specific heat capacity of that substance.† The specific heat capacity of water is 1.0 cal g^{-1} $°C^{-1}$, which compared to other substances is very high. The specific heat capacity of rubber is 0.5, of wood 0.4, and of most metals 0.1 or less. The specific heat capacity of air is 0.24 cal g^{-1} $°C^{-1}$, and since the density of air is 1.2 g $liter^{-1}$ (at 20 °C), the heat capacity of 1 liter air is 0.3 cal.

The amount of heat needed to increase the temperature of the animal body is slightly less than the amount needed to heat the same weight of water. The mean specific heat capacity of the mammalian body is about 0.8. Thus, to increase the temperature of a 1000 g mammal by one degree requires about 800 cal. The exact value for the specific heat capacity of the animal body varies somewhat. For

---

* Absolute zero is at − 273.15 °C. Celcius temperature (T_C) relates to absolute temperature (T_K) as follows:

$$T_K = T_C + 273.15.$$

† It is recommended that the word 'specific' before the name of a physical quantity be restricted to the meaning 'divided by mass' (Council of the Royal Society, 1971).

example, the specific heat capacity of mouse bodies was found to vary between 0.78 and 0.85 with a mean of 0.824 (Hart, 1951). The bulk of the body is water, which has a specific heat capacity of 1.0, and the other components, proteins, bone, and fat, tend to reduce this value. The amount of fat is particularly important, for its amount can vary within wide limits, and its specific heat capacity is only about 0.5. For many purposes it is sufficient to use a mean value for the specific heat capacity of the animal body of 0.8, for in order to determine a change in heat content, we must know the mean body temperature, and as we have seen, it is very difficult to determine this variable with accuracy.

Physics of heat transfer

For a body to maintain a constant temperature there is one absolute requirement, heat loss must exactly equal heat gain. For an animal to maintain a constant temperature this means that heat must be lost from the body at the same rate that it is being produced by metabolic activity. As we have seen, metabolic heat production can easily increase more than ten fold with activity, and unless heat loss is increased in the same proportion, body temperature will rise rapidly. Furthermore, the conditions for heat loss vary tremendously with external factors such as air temperature, wind, etc. An understanding of the physiological mechanisms involved in the regulation of heat production and heat loss require an elementary knowledge of the physics of heat transfer.

Whenever physical materials are at different temperatures, heat will flow from a region of higher temperature to one of lower temperature. This transfer of heat takes place by *conduction* and by *radiation*. A body cannot lose heat by conduction or radiation unless its environment, or some part of it, is at a lower temperature than the surface of the body. There is, however, a third way to remove heat, the *evaporation* of water. These three ways of heat transfer, (1) conduction,* (2) radiation, and (3) evaporation, are the only means available for the removal of the heat produced in the metabolic activity of living organisms.

---

* Transfer of heat between a surface and a fluid (gas or liquid) in contact with it takes place by conduction. Mass movement in the fluid, termed *convection*, contributes to renewal of fluid in the boundary layer and thus complicates the conductive heat transfer (see page 308).

Conduction

Conduction of heat takes place between physical bodies that are in contact with each other, whether they are solids, liquids, or gases. Conduction of heat consists of a direct transfer of the kinetic energy of molecular motion, and it always occurs from a region of higher temperature to one of lower temperature.

Assume that we have a uniform conductor in which we keep one end warm and the other cold (see fig. 8.3). The rate of heat transfer by conduction, $\dot{Q}$, can now be expressed as

$$\dot{Q} = k\, A\, \frac{T_2 - T_1}{l},$$

where k is the thermal conductivity of the conductor, A is the area through which heat flows (normal to the direction of heat flow), and T_2 and T_1 are the temperatures at two points separated by the distance l. The fraction $(T_2 - T_1)/l$ is known as the temperature gradient, and stands for the temperature difference per unit distance along the conductor.

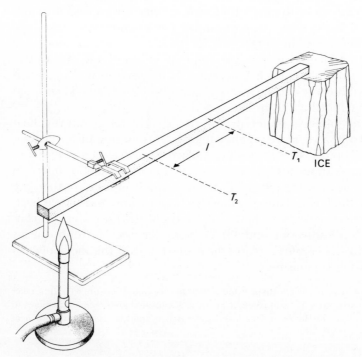

Fig. 8.3. Heat flow in a uniform conductor depends on its cross-section, the temperature gradient, and the material from which it is made.

This expression for heat flow in a conductor can be stated in simple, intuitively obvious terms. Heat flow increases with the thermal conductivity (k) of the conducting material, it increases with increasing cross-sectional area (A) of the conductor, and with an increased temperature difference between two points, T_2 and T_1. Increasing the distance, l, between two given temperatures, T_2 and T_1 (if these remain unchanged), decreases the amount of heat flow.

The thermal conductivity coefficient k, is an expression for how easily heat flows in a given material. Values for the thermal conductivity of some common materials are given in table 8.3. As we know, metals are excellent conductors and have high conductivity coefficients. Glass and wood are poorer conductors; water and human tissues have slightly lower conductivities again, but of a similar order of magnitude. The similarity between the thermal conductivity of human tissue and water is, of course, due to the fact that most tissues consist of roughly two-thirds to three-quarters water. Air and animal fur have very low thermal conductivities, which means that their insulation value (resistance to heat flow) is high. The main reason that the thermal conductivity of fur is low is the large amount of air trapped between the hairs. Other materials that enclose a high proportion of air, such as felt, woolen fabrics, down, etc., are also poor conductors or excellent insulators.

The simple equation for heat conduction given above unfortu-

Table 8.3. *Thermal conductivity coefficients, k, for a variety of common materials.* (Data from Hammel, 1955; Hensel and Bock, 1955; and Weast, 1969)

	k (cal s^{-1} cm^{-1} °C^{-1})
Silver	0.97
Copper	0.92
Aluminum	0.50
Steel	0.11
Glass	0.0025
Soil, dry	0.0008
Rubber	0.0004
Wood	0.0003
Water	0.0014
Human tissue	0.0011
Air	0.000057
Animal fur	0.000091

nately applies only when heat flows through a plane object such as a wall. Most animal surfaces are curved, and this makes the conduction equation considerably more complex. For practical purposes, we can consider the heat flow as dependent on temperature gradients and area, but if we wish to make a quantitatively satisfactory analysis of the heat transfer in an animal, it is necessary to apply a more rigorous treatment of the physics of heat transfer.

The transfer of heat in fluids is almost invariably accelerated by the process of *convection,* which refers to mass movement of the fluid. Assume that a cold fluid is in contact with a warm solid surface. Heat flows into the fluid by conduction and the fluid adjacent to the surface becomes warmer. If the fluid is in motion, the warm fluid adjacent to the solid surface is replaced by cold fluid, and the heat loss from the solid surface is therefore speeded up. Mass flow, or convection, in the fluid thus facilitates heat loss from the solid, although the transfer process between solid and fluid remains one of conduction.

Convection in a fluid may be caused by temperature differences or by external mechanical force. Heating or cooling of a fluid usually causes changes in its density, and this in turn causes mass flow. Thus, if a warm solid surface is in contact with a cold fluid, the heated fluid expands and therefore rises, being replaced by cool fluid. In this case the mass flow, or convection, is caused by the temperature difference, and is called *free* or *natural convection.* This term applies also if the wall is colder than the fluid and the fluid adjacent to the wall becomes denser and sinks.* Free convection, of course, can take place both in air and in water, and contributes substantially to the rate of heat loss from living organisms.

Motion in the fluid can also be caused by external forces, such as wind, water currents, or an electric fan. Convection caused by external forces, as opposed to density changes, is referred to as *forced convection.*

Since convection depends on mass transfer in fluids, the process is governed by the rather complex laws of fluid dynamics, which include such variables as the viscosity and density of the fluids, in addition to their thermal conductivity. Convective heat loss depends not only on the area of the exposed surface. Variables such as the curvature and the orientation of the surface give rise to rather complex mathematical expressions, which cause great difficulties in analyzing

---

* Water has a higher density at +4 °C than at the freezing point. This is of great importance in bodies of fresh water, but in most physiological situations the anomalous density of water is of no significance. Sea water does not exhibit the anomalous density properties at near-freezing temperature (see also p. 279).

the heat transfer from an animal. As we shall see below, however, we can find practical means for analyzing the heat transfer between an animal and the environment which circumvent the need for an exact analysis of the physical processes involved (see p. 313 and 316ff).

Radiation

Heat transfer by radiation takes place in the absence of direct contact between objects. All physical objects at a temperature above absolute zero emit electromagnetic radiation. The intensity and the wavelength of this radiation depend on the temperature of the radiating surface (and its *emissivity*, see p. 310). All objects also receive radiation from their surroundings. Electromagnetic radiation passes freely through a vacuum, and for our purposes atmospheric air can be regarded as fully transparent to radiation.

The *intensity* of radiation from an object is proportional to the fourth power of the absolute temperature of the surface. This is expressed by the Stefan-Boltzmann law for heat radiation flux

$$\dot{Q}_R \propto \sigma\, T^4,$$

where T is the absolute temperature of the radiating surface (in degrees K), and σ is Stefan-Boltzmann's constant (1.376×10^{-12} cal s^{-1} cm^{-2} K^{-4} or 5.67×10^{-8} W m^{-2} K^{-4}). Since the amount of heat or energy radiated increases with the fourth power of the absolute temperature, the emission increases very rapidly indeed with the surface temperature.

The *wavelength* of the emitted radiation depends on the surface temperature, and the hotter the surface, the shorter is the emitted radiation. As the surface temperature of a heated object increases, the radiation therefore includes shorter and shorter wavelengths. Thus, the radiation from a heated piece of iron will just barely begin to include visible red light when its temperature is about 1000 K. If heated further, it emits shorter wavelengths, i.e. more visible radiation is included. Therefore, the visible color shifts from red to yellow to white, as the temperature increases. The sun's radiation, which has its peak in the visible part of the spectrum, corresponds to a surface temperature of about 6000 K, and includes an appreciable amount of radiation in the near ultraviolet as well. Objects that are close to physiological temperatures emit most of the radiation in the middle infrared. For example, the infrared radiation from the living human skin ($T = c$. 300 K) has its peak at about 10 000 nm. Since visible light is between 450 and 700 nm, the radiation from human skin includes no visible light. (The visible

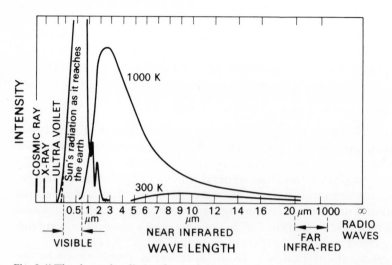

Fig. 8.4. The thermal radiation from a body depends on its surface temperature, both in regard to the spectral distribution of the radiation and its intensity. The higher the surface temperature, the shorter is the wavelength and the higher is its intensity. This figure shows the spectral distribution of the thermal radiation from the sun (6000 K), a red hot stove (1000 K), and the human body (300 K). [Modified from Hardy, J. D.: *J. Clin. Investigation,* **13,** 1934. (Hardy, 1949).]

light from the skin which we perceive is, of course, only reflected light, and in darkness we see no light from the skin.)

Fig. 8.4 shows how the wavelength of emitted radiation as well as its intensity changes with the temperature of the radiating surface, as described in the preceding two paragraphs.

The next concept to consider is that of *emissivity.* The simplest way to approach this concept is to consider first the *absorptivity* of a surface for the radiation which falls on it. A black body, by definition, absorbs radiation completely in all wavelengths and reflects nothing. The absorptivity of a completely black body is therefore 100%. (Although we often think of 'black' in relation to visible light, the physical concept applies to all wavelengths.)

In contrast to a black body, the surface which reflects all radiation is a perfect reflector, and its absorptivity is zero. This condition is approached by a highly polished metallic surface, for example, a silver mirror. Since incident radiation is either absorbed or reflected, absorptivity and reflectivity *must* add up to 1 (or 100%). If 30% of incident radiation is reflected, 70% is absorbed, and so on.

Usually, the absorptivity (and thus reflectivity) is different in different wavelengths of incident radiation. In visible light we recognize this as the color of objects, so that an object which we perceive as yellow reflects mostly yellow light while other components of the visible light are absorbed. In the middle infrared, which is of the greatest interest in connection with heat radiation at physiological temperatures, most surfaces are black bodies. The human skin, for example, absorbs virtually 100% of incoming infrared radiation, and thus is a black body in these wavelengths, irrespective of whether it is light or dark in the visible light.

The concept of surface absorptivity should now be clear, and we can return to the emissivity. The two are numerically equal, a fact that can be intuitively understood from the following example. Consider an object suspended in a vacuum within a hollow sphere of uniform wall temperature. The object within the sphere will receive radiation from the wall of the sphere, part of which is absorbed, and part of which is reflected. Likewise, the object radiates to the spherical surface, and part of this radiation is absorbed while part is reflected. (The reflected portion in turn is either intercepted by the suspended object or reaches another point of the sphere.) When the system is in equilibrium, the object has attained the temperature of the surrounding spherical surface, and the absorbed and emitted radiation from the object are now precisely equal. If this were not so, the object would not be at the temperature of the sphere. This would be a physical impossibility, for otherwise we could tap energy from the system and have the makings of a *perpetuum mobile*. As a practical example, take a highly reflective metal coffee pot that has a reflectivity of 95%, i.e. an absorptivity of only 5%. Its emissivity, therefore, is also 5%, i.e. the polished coffee pot loses heat by radiation very slowly.

Let us next consider that the object within the sphere is nearly black and has an absorptivity of 99%. This object obviously will attain the same temperature as a highly reflective silver object similarly suspended, although the latter may have an absorptivity of only 1%. For each object the absorbed and the emitted radiation at equilibrium are precisely equal. The final temperature will be the same for both, but the highly reflective object will need longer to reach it.

To repeat this relationship, an object of high absorptivity also has high emissivity. A body which is a perfect 'black body' in a given wavelength is a perfect emitter in the same wavelength.

Human skin, animal fur, and all sorts of other non-metallic surfaces have high absorptivities in the middle infrared range of the spectrum, between 5000 and 10 000 nm. For all practical purposes human skin is 'black' in this range, and there is no difference between heavily pigmented and unpigmented skin. The difference which we perceive is in the visible region of the spectrum, but since no radiation is emitted in this range, the difference in pigmentation does not influence radiation from human skin.

As a consequence of the high absorptivity in the middle infrared, emissivity is also close to 100% in this region. Consequently, heat loss by radiation from pigmented and unpigmented skins are about equal, both radiate as virtually black bodies. The same is true of radiation from animal fur, which likewise is independent of the color in the visible part of the spectrum.

A lack of appreciation of this simple physical fact has led to some ill-conceived speculation in regard to animal coloration. For example, it has been suggested that black-colored animals lose heat by radiation faster than white-colored ones. Since the emission of radiation in the infrared has no relation to visible coloration, a difference in emissivity in the infrared can only be established by direct measurement in this range of the spectrum. No such differences have been found.

Skin and fur color may, however, be important to the heat absorbed from solar radiation, which has its peak intensity in the visible range. About half of the energy carried in solar radiation falls within what we call the visible light (see fig. 8.4), and it is important to the heat balance whether this light is absorbed or reflected. When exposed to direct solar radiation, then, dark-colored skin or fur absorbs more of the incident energy than light-colored ones.

Net heat transfer by radiation. If two surfaces are in radiation exchange, each emits radiation according to *Stefan-Boltzmann's law,* and the net radiation transfer ($\dot{Q}_R$) between them will be

$$\dot{Q}_R = \sigma \ \epsilon_1 \ \epsilon_2 \ (T_1{}^4 - T_2{}^4) \ A$$

in which σ is Stefan-Boltzmann's constant, ϵ_1 and ϵ_2 are emissivities of the two surfaces, T_1 and T_2 are their absolute temperatures, and A is an expression for the effective radiating area. If the environment is a uniform sphere, A is a simple expression of the integrated 'visible' surface in the direction of radiation. If the environment is non-uniform, however, and in particular if it includes a point source of heat (such as the sun), the integration of the surfaces of exchange becomes more complex. In this regard, situations in nature are extremely complex, and to describe in exact terms the total heat transfer is very difficult. Nevertheless, once the elementary physics of radiation exchange is understood in principle, we can readily avoid some erroneous conclusions such as those relating to the role of surface pigmentation.

A practical simplification. Although radiation heat transfer changes with the fourth power of the absolute temperature, we may use a simplified expression, provided that the temperature difference between the surfaces is not too great. Within a temperature range of about 20 °C or so, the error of not using the rigorous Stefan-Boltzmann equation can often be disregarded, and, as an approximation, we can regard the radiation heat exchange as being proportional to the difference in temperature between the two surfaces. For small temperature differences, the error is relatively insignificant, but it becomes increasingly important the greater the temperature difference.

We will now use this simplification and regard the radiation heat exchange as proportional to the temperature difference. The rate of heat loss from a warm-blooded animal in cool surroundings consists of conduction and radiation heat loss (for the moment disregarding evaporation). Since both can be considered proportional to the temperature difference $(T_2 - T_1)$, their sum will also be proportional to $(T_2 - T_1)$, or

$$\dot{Q} = C \; (T_2 - T_1)$$

in which all the constants that enter into the heat exchange equation have been combined to a simple proportionality factor, C.

We shall later return to the application of this simplified equation in the discussion of heat loss from warm-blooded animals in the cold.

Evaporation

The evaporation of water requires a great deal of heat. To transfer one gram of water at room temperature to water vapor at the same temperature requires 584 cal. This is an amazingly large amount of heat, for when we consider that it takes 100 cal to heat one gram of water from the freezing point to the boiling point, we see that it takes more than five times again as much heat to change the liquid water into water vapor at the same temperature.

The amount of heat required to achieve the phase change from liquid water to vapor is known as the heat of vaporization (H_v). The heat of vaporization changes slightly with the temperature at which the evaporation takes place; thus, at 0 °C the H_v is 595 cal per gram water, at 22 °C it is 584 cal per gram, and at the boiling point, 100 °C, it is 539 cal per gram. In physiology it is customary to use the figure 580 cal per gram, which is an approximation of the value for vaporization of water at the skin temperature of a sweating man, about 35 °C.

The measurement of the heat loss by vaporization of water has one great convenience, it suffices to know the amount of water that has been vaporized. When a man is exposed to hot surroundings, he cools himself by evaporation of sweat from the general body surface, but a dog evaporates most of the water from the respiratory tract by panting. The amount of heat transferred per gram water is, of course, the same in both cases, i.e. we do not have to know the exact location or the area of the evaporating surface.

The respiratory air of mammals and other air-breathing vertebrates is exhaled saturated with water vapor, and therefore there is normally a considerable evaporation of water from the respiratory tract, even in the absence of heat stress. This evaporation must, of course, be included in any consideration of the total heat balance of an animal, which is the subject we shall now discuss.

HEAT BALANCE

We have repeatedly emphasized that for the body temperature to remain constant, heat loss must equal heat gain. The body temperature does not always remain constant, however. Assume that heat loss does not quite equal the metabolic heat production, but is slightly lower. The body temperature inevitably rises. This means that part of the metabolic heat remains in the body, rather than being lost, and the increase in body temperature thus represents a storage of heat. If the mean body temperature decreases, which happens when heat loss exceeds heat production, we can regard the excess heat loss as heat removed from storage. The amount of heat stored depends on the change in mean body temperature, the mass of the body, and the specific heat capacity of the tissues (which for mammals and birds usually is assumed to be 0.83).

The heat exchange between the body and the environment takes place by the three means described above, conduction (including convection), radiation, and evaporation. Usually, each of these represents a heat loss from the body, but this is not always so. When the air temperature exceeds the body surface temperature, heat flow by conduction will be to, not from, the body. When there is a strong radiation from external sources, the net radiation flux may also be towards the body. Evaporation is nearly always a negative entity, but under unusual circumstances it could be reversed; this happens, for

example, when a cold body comes in contact with moist warm air.*
We can enter these various variables into a simple equation as
follows

$$H_{tot} = \pm H_c \pm H_r \pm H_e \pm H_s$$

in which H_{tot} = metabolic heat production (always positive)
  H_c = convective and conductive heat exchange
      (+ for net loss)
  H_r = net radiation heat exchange (+ for net loss)
  H_e = evaporative heat loss (+ for net loss)
  H_s = storage of heat in the body
      (+ for net heat gain by body) †

The three components of heat exchange, conduction, radiation,
and evaporation depend on external factors among which the most
important single factor is temperature. It is obvious that heat losses
increase when the external temperature falls. If, on the other hand,
external temperature rises, the heat losses decrease, and if external
temperature exceeds body surface temperature, both conduction
and radiation heat exchange may be from the environment to the
organism. The total heat gain is then the sum of heat gain from metabolism plus heat gain from the environment. This situation still
permits the maintenance of a constant body temperature
(storage = 0), provided that evaporation is increased sufficiently to
dissipate the entire heat load.

The physiological responses in cold and in heat differ in many respects, and it is convenient to treat the two conditions separately. We

* When a normal human, with a skin surface temperature of 35 °C, enters a Turkish
steam bath in which the air is nearly saturated and above 40 °C, there is an immediate condensation of water on his skin. In this case the evaporation heat exchange is
in the reverse direction of the usual. In the sauna bath, however, the air is usually
dry, and the visible moisture seen on the skin shortly after entering the sauna is due
to sweat.

† It is important to note that no work term appears in this equation. For heat balance,
the equation is correct as it stands. If, however, oxygen consumption is used for the
determination of metabolic rate and for the calculation of heat production, any external work must be included. External work does not appear as heat, although it
appears as part of the oxygen consumption. This is important when the external
work is a substantial fraction of the total metabolic rate. As an example, a flying bird
uses probably 20–25% of its metabolic rate to impart acceleration to the air
through which it flies. (In the end this external mechanical work degenerates into
heat in a trail of decaying air-vortices.) For a flying bird, therefore, only 75–80% of
the oxygen consumption represents heat released within the body, and only this
part enters into the heat balance equation as metabolic heat (H_{tot}) to be dissipated
through the means to the right of the equality sign.

shall first discuss temperature regulation in cold surroundings, and afterwards discuss temperature regulation in the heat.

Temperature regulation in the cold – keeping warm

To maintain constant body temperature we must satisfy the steady state condition in which the rate of metabolic heat production ($\dot{H}$) equals the rate of heat loss ($\dot{Q}$). For the moment we will consider that the heat loss (in the cold) takes place only through conduction (including convection) and through radiation, and that heat loss through evaporation can be disregarded.* To describe the heat loss we will now use the simplified equation which relates heat loss ($\dot{Q}$) to ambient temperature as developed in the preceding section (page 313).

$$\dot{H} = \dot{Q} = C \ (T_b - T_a).$$

This equation simply says that the rate of metabolic heat production equals the rate of heat loss, which in turn is proportional to the temperature difference between the body and the environment ($T_b - T_a$). The term C is a conductance term which will be discussed later.

What can an animal do to maintain the steady state? Among the terms in the above equation, there is very little an animal can do about the ambient temperature, T_a, short of moving to a different environment. To adjust to an unfavorable T_a an animal therefore has only the three remaining terms that could possibly be adjusted, the heat production ($\dot{H}$), the conductance term (C), or the body temperature (T_b). Since we are concerned with the maintenance of the body temperature, T_b, we are left with adjustments to be made either in heat production, $\dot{H}$, or in the conductance term, C. (The third alternative, a change in body temperature, does occur in hibernation, a subject which will be discussed later.)

Increase in heat production

Although heat production (metabolic rate) cannot be turned down below a certain minimum level, increased metabolic rate permits a wide range of adjustments. The major ways in which heat production is increased are through (1) muscular activity and exercise, (2) involuntary muscle contractions (shivering), and (3) so-called

---

* At moderate to low temperature, evaporation takes place primarily from the respiratory tract, at a rate corresponding to a few per cent of the metabolic heat production.

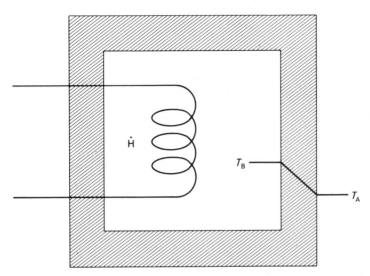

Fig. 8.5. A heated insulated box is maintained at a given temperature, T_b, by a heater supplying heat at the rate $\dot{H}$. If the outside temperature, T_a, is lowered, T_b can be maintained constant by increasing the rate of heat input in proportion to the temperature difference, $T_b - T_a$. This is a model of thermoregulation in a mammal.

non-shivering themogenesis. The last term refers to an increased metabolic rate which takes place without noticeable muscle contractions.

To clarify the role of increased heat production, let us consider a simple physical model of an animal (fig. 8.5). An insulated box is equipped with an electric coil which delivers heat at the rate $\dot{H}$. The temperature inside the box, T_b, will therefore be higher than the ambient temperature, T_a. If we lower the ambient temperature, T_a, the temperature within the box can only be maintained by increasing the rate of heat input $\dot{H}$. The increase in $\dot{H}$ must be proportional to the increase in temperature difference, as long as the insulation (or conductance) in the wall remains unchanged.

This is precisely the situation for an animal exposed to cold. The lower the ambient temperature, the greater the increase in metabolic rate needed to stay warm. In fig. 8.6 we see that below a certain ambient temperature, called the lower critical temperature (T_{lc}), the metabolic rate increases linearly with decreasing temperature. Above the critical temperature the heat production, which cannot be turned down lower than the resting metabolic rate, remains constant as temperature is increased.

If we want to compare a variety of different animals, this can conveniently be done if we assign to their normal resting metabolic rate

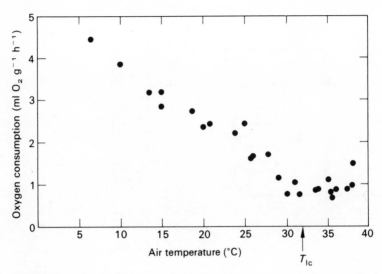

Fig. 8.6. The oxygen consumption of the pigmy possum (*Cercaertus nanus*) at various ambient temperatures. Below a certain point, the lower critical temperature (T_{lc}), the oxygen consumption increases linearly with decreasing ambient temperature (from Bartholomew, G. A. and Hudson, J. W. 1962. *Physiol. Zool.*, **35**, 97, Fig. 2. © University of Chicago Press).

the value 100%. This has been done in fig. 8.7, where we can see that most tropical mammals have a critical temperature between +20 and +30 °C; below this temperature their metabolic rate increases rapidly. Arctic mammals, on the other hand, have much lower critical temperatures, and a well-insulated animal such as an arctic fox does not increase the metabolic rate significantly until the air temperature is below − 40 °C.

The slopes of the lines in fig. 8.7 can be regarded as an expression of the conductance term *C*, in the equation we have been using.* As a rule, tropical mammals have high conductances, and arctic mammals low conductances (high insulation values). As a consequence, a tropical mammal, due to its high conductance (low insulation) must increase its metabolism drastically for even a moderate temperature drop. For example a monkey with a body temperature of 38 °C and a lower critical temperature of 28 °C must double its metabolic rate for a further 10 degree drop in the temperature, to 18 °C. An arctic ground squirrel, on the other hand, which has a lower critical temperature at about 0 °C, will not need a doubling of the metabolic rate until the temperature has dropped to nearly −40 °C. The arctic fox,

* When the data have been normalized by assigning to the resting heat production the value of 100%, the slope does not indicate conductance in absolute units.

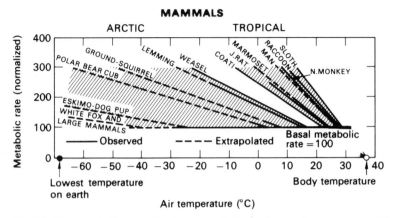

Fig. 8.7. The metabolic rate of various animals in relation to air temperature. The normal resting metabolic rate for each animal, in the absence of cold stress, is given the value 100%. Any increase at lower temperature is expressed in relation to this normalized value, thus making it possible to compare widely differing animals (from Scholander, P. F. et al 1950. *Biol. Bull.*, **99**, 254, Fig. 10).

with its lower critical temperature at−40 °C, should be able to sustain the lowest temperatures measured in arctic climates (−70 °C) with less than a 50% increase in its metabolic rate.

The difference between arctic and tropical animals is very clear in fig. 8.7. The width of the thermoneutral zone is much greater in the arctic animal and the metabolic response to cold is less pronounced than in tropical mammals. The way the graph is drawn, however, is in some ways misleading. Assume that we have two mammals of equal size and equal conductance value, but one has a metabolic rate in the thermoneutral zone which is twice as high as the other. Instead of normalizing the metabolic rate at 100%, we plot their actual metabolic rates (fig. 8.8). The lower critical temperature will be different for the two animals, but once we are below this temperature, the two animals must expend the same number of calories to keep warm. It is therefore obvious that the location of the lower critical temperature (the width of the thermoneutral zone) by itself is not sufficient information as to how well adapted an animal is to cold conditions. Information about the level of the resting metabolic rate, or better, the conductance value, is also necessary in order to evaluate the relationship between energy metabolism and heat regulation.

Conduction, insulation, and fur thickness

Conductance, in the sense it has been used above, is a measure of the heat flow from the animal to the surroundings. The term in-

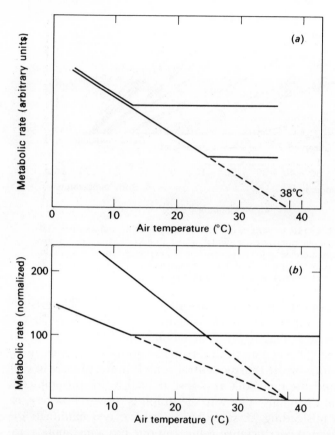

Fig. 8.8. If two animals have the same body temperature and the same conductance, but have different resting metabolic rates, they will also have different lower critical temperatures (*a*). If their metabolic rates are normalized to 100% this procedure will give the incorrect appearance that they have different conductance values (*b*).

cludes the flow of heat from deeper parts of the body to the skin surface and from the skin surface through the fur to the environment. When conductance is low, it means that the insulation value is high. In fact, insulation is the reciprocal value of conduction.

Our discussion is concerned with animals in the cold, where the fur is a major barrier to heat flow. The insulation values of the fur from various animals differ a great deal. Some data, plotted relative to the thickness of the fur, are shown in fig. 8.9. As expected, the insulation value increases with the thickness of the fur, and reaches a maximum for some of the larger animals which have a thick fur, such as the white fox. Among the smaller animals there is a clear correlation between fur thickness (and insulation) and the size of the

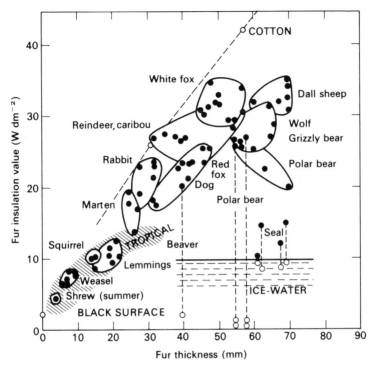

Fig. 8.9. The insulation value of animal fur is related to the fur thickness. Since small mammals of necessity carry relatively short fur, they have poor insulation compared to larger animals. Values for transfer of heat to air indicated by black dots, values for fur submerged in water by open circles. Insulation values of fur (ordinate) measured in air or water at 0 °C with the inside of the fur maintained at 37 °C. The sloping dotted line represents the insulation values of cotton of various thicknesses (from Scholander, P. F. et al 1950. *Biol. Bull.*, **99**, 230, Fig. 3).

animal. Small animals must have relatively short and light fur, or the animal could not move about. This is particularly true for the smallest mammals, small rodents and shrews. Because of their relatively poor insulation, these animals must either take advantage of microclimates (e.g. by living in burrows), or escape from the problem of keeping warm by hibernating.

The polar bear is interesting, for its open and coarse fur provides a poor insulation, relative to its thickness. What is more important, when polar bear skin is submerged in ice water, most of the insulation value of the fur is lost, and heat transfer is 20 to 25 times faster than in air. If the water is agitated (as it would be when the polar bear is swimming) the heat loss is increased even more, some 50 times. This is because water penetrates to the skin surface, dislodging all air from the fur. When in water, the polar bear is helped by subcutaneous blubber which affords insulation (see p. 324).

The skin of the seal has a relatively thin fur, but a heavy layer of blubber affords a considerable amount of insulation. Therefore, the difference in insulation value for seal skin in air and in water is not very great (see fig. 8.9). In air, seal skin with the blubber attached has only a slightly better insulation value than the skin of lemmings, i.e. 60–70 mm of mostly blubber insulates about as well as 20 mm of fur. When the seal skin is submerged in water, the insulation value is reduced, but not as drastically as the polar bear skin.

It is well known that the thickness of animal fur changes with the seasons, and that winter fur is heavier and presumably affords better insulation than the thinner summer fur. The seasonal changes that have been measured are greatest in large animals and relatively insignificant in small rodents. The black bear, for example, in summer loses 52% of the insulation value of the winter fur; i.e. the summer pelt affords less than half the insulation of the winter pelt. The smallest seasonal change observed in a sub-arctic mammal was a 12% reduction of the winter value in the muskrat (Hart, 1956).

Conductance in birds

As we have seen, the conductance value for mammals in the cold requires that the metabolic rate increases in proportion to the imposed cold stress. The straight lines describing the metabolic rate at low temperature extrapolate to body temperature, showing that mammals fit our assumed model (fig. 8.5, p. 317).

For birds the situation is not so simple. Some birds conform to the mammalian pattern with metabolic curves that extrapolate to body temperature. Other birds, however, deviate from this pattern. For example, the metabolic heat production of pigeons and road runners does not increase as much as expected at low ambient temperature (see fig. 8.10). For these birds the metabolic line extrapolates not to the body temperature of about 40 °C, but to a much higher temperature, well above 50 °C.

There is only one possible explanation for such a slope, the bird does not adhere to the simple equation $H = C\ (T_b - T_a)$, where C remains constant at low temperature. One way to explain the observed regression line is to assume that the conductance value, C, gradually decreases with falling ambient temperature, T_a. It is not well understood how the birds can achieve this reduction in conductance below the lower critical temperature. The other way to obtain a change in slope, to permit body temperature to decrease, is not in

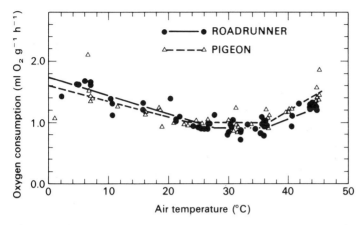

Fig. 8.10. The oxygen consumption of the pigeon and the roadrunner increases at low temperature, but the regression line does not extrapolate to the body temperature of the animals as it commonly does in mammals (from Calder, W. A. and Schmidt-Nielsen, K. 1967. *Am. J. Physiol.,* **213**, 884, Fig. 1).

accord with the observed maintenance of deep body temperature. One possible way to decrease conductance (increase insulation) is to raise the feathers and to draw the feet up into the feathers, thus making the body into a round feather ball. Another possibility is that the peripheral tissues undergo an appreciable temperature drop while the core temperature is maintained constant. A drop in shell temperature and/or an increase in shell thickness has several effects, the circulation to the shell is decreased, the heat production in the shell is decreased, the thickness of the shell is increased, and the volume of the core is reduced. Each of these changes contributes towards a decrease in conductance and a smaller-than-expected increase in metabolic rate, and still permit the maintenance of core temperature (although in a smaller than usual core volume).

Mammals in the sea

Many seals and whales live and swim in the near-freezing water of the arctic and antarctic seas. Not only are the temperatures low, but water has a high thermal conductance and a high heat capacity, and the thermal loss to water is therefore much higher than in air of the same temperature. The heat conductivity coefficient for water is about 25 times as high as for air, but since convection, both free and forced, plays a great role, the cooling power of water may under some circumstances be even greater, perhaps 50 or even as high as 100 times as great as for air.

What can seals and whales do about their heat balance in water with such cooling power? Evidently they manage quite well, for both seals and whales are far more numerous in cold water than in the tropics. Since there is nothing they can do about the water temperature (short of moving to warmer seas), they have a limited choice. They could either (1) live with a lowered body temperature, (2) increase the metabolic rate to compensate for the heat loss, or (3) increase the body insulation to reduce the heat loss.

With regard to body temperature, seals and whales are similar to other warm-blooded animals, and their usual temperature is around 36 to 38 °C (Irving, 1969). We must therefore look at the other possibilities for an explanation of their ability to live in ice-water.

The metabolic rate has been measured in several species of seals and in some dolphins (porpoises) (but not in any of the large whales which are rather unmanageable as experimental animals). Most of these animals have resting metabolic rates somewhat higher than expected from their body size alone, roughly about twice as high or so (Irving, 1969). In the harp seal (*Phoca groenlandica*), an arctic species, the metabolic rate remained the same in water all the way down to the freezing point; in other words, even the coldest water does not cause the heat loss to increase enough to require an increased heat production. The lower critical temperature for the harp seal in water is thus below freezing and has not been determined (Irving and Hart, 1957).

The third possibility, an effective insulation, is obviously the primary solution to the problem. Both seals and whales have a thick layer of subcutaneous blubber which affords the major insulation. Measurements of the skin temperature support this concept, for the temperature at the skin surface is nearly identical to that of the water (see fig. 8.11). If the surface temperature is nearly the same as that of the water, very little heat can be transferred to the water. The temperature gradient is sustained by the blubber, and at a depth of some 50 mm (roughly the thickness of the blubber), the temperature is nearly at body core temperature.

In arctic land mammals the temperature gradients are different, the surface temperature of the body skin under the fur is regularly within a few degrees of core temperature (Irving and Krog, 1955), and most of the insulation therefore resides outside the skin surface. The polar bear, which has a fur that loses most of its insulation value in water, also has a substantial layer of blubber which seems to be of major importance when the polar bear swims in ice-water. Without the layer of blubber, the semi-aquatic way of life would seem impossible (Øritsland, 1970).

If seals and whales are so well insulated, how do they manage to

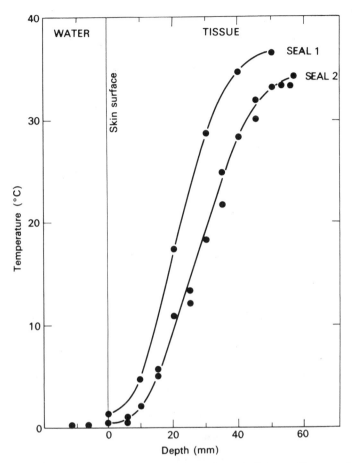

Fig. 8.11. The temperature of the skin surface of a living seal immersed in ice-water is nearly identical to that of the water. Most of the insulation is provided by the thick layer of blubber (after Irving and Hart, 1957).

avoid overheating when the water is warmer, or when the metabolic rate is increased during fast swimming? If a seal is removed from water and placed in air, its skin temperature increases considerably (see fig. 8.12). The higher skin temperature is necessary for dissipation of heat to the air, due to the reduced cooling power of air relative to water. The increase in skin temperature is due to an increased blood flow through the blubber to the superficial layer of the skin, which is well supplied with blood vessels. This system of cutaneous blood vessels permits a precise regulation of the amount of heat that reaches the skin surface and thus is lost to the environment.

We can now conclude that the main difference in the insulation

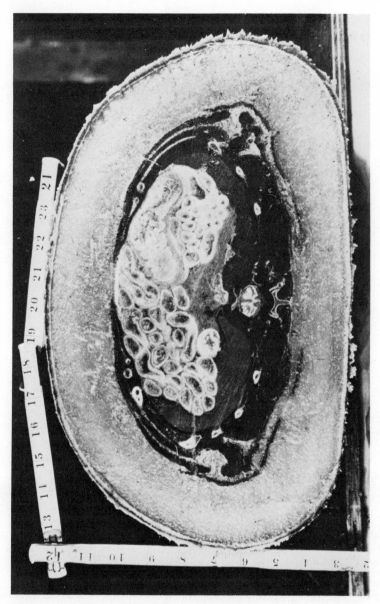

Plate 7. *Seal blubber*. This cross-section of a frozen seal shows the thick layer of blubber. Of the total area in the photo, 58% is blubber and the remaining 42% is muscle, bone, and visceral organs (P. F. Scholander, University of California, San Diego; from *Biological Bulletin*, 1950, **99**, 234).

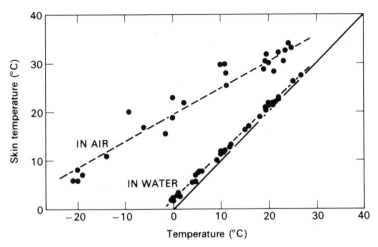

Fig. 8.12. The skin surface temperature of seals in air and in water. Fully drawn line indicates equality of skin and environmental temperature (after Hart and Irving, 1959).

between aquatic and terrestrial mammals is that the insulator of the aquatic mammal (the blubber) is located internally to the surface of heat dissipation. Therefore the blood can bypass the insulator, and heat dissipation during heavy exertion or in warm water can be independent of the insulator. Terrestrial animals, in contrast, have the insulator located externally to the skin surface. Therefore they cannot modulate the heat loss from the general skin surface to any great extent, and must seek other avenues of heat loss when they need to dissipate excess heat to the environment (fig. 8.13).

We must remember that furred arctic animals cannot afford to insulate all parts of the body equally well, they also need surfaces from which heat can be dissipated, especially during exertion. Since the main part of the body is well insulated for maximum heat retention, they need thinly covered skin areas on the feet, face, and other peripheral areas, from which heat loss can take place when the demand for heat dissipation increases.

Seals and whales have flippers and flukes which lack blubber and are poorly insulated. These appendages are well supplied with blood vessels and receive a rich blood supply. This means that these relatively thin structures with their large surfaces can lose substantial amounts of heat and aid in heat dissipation, but how is it possible to avoid excessive heat loss from these structures when heat needs to be conserved? If the blood which returns to the body core from these appendages were at near-freezing temperature, the deeper parts of the body would rapidly be cooled.

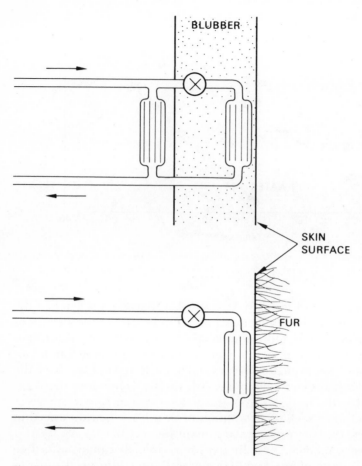

Fig. 8.13. Diagram to show that the insulation afforded by blubber can be by-passed when the need for heat dissipation increases. Fur, in contrast, is located outside the skin surface and its insulation value cannot be drastically changed by a by-pass.

Excessive heat loss from the blood in the flippers is prevented by the special structure of the blood vessels, which are arranged in such a way that they function as heat exchangers. In the whale flipper each artery is completely surrounded by veins (see fig. 8.14), and as warm arterial blood flows into the flipper, it is cooled by the cold venous blood that surrounds it on all sides. The arterial blood therefore reaches the periphery pre-cooled and loses little heat to the water. The heat has been transferred to the venous blood, which thus is pre-warmed before it re-enters the body. If the heat exchange is efficient, the venous blood can reach nearly arterial temperatures, and thus contribute virtually no cooling to the core. This type of

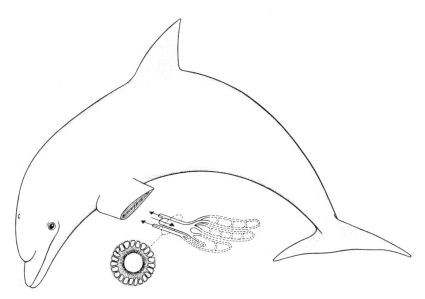

Fig. 8.14. In the flippers and fluke of dolphins (porpoises) each artery is surrounded by several veins. This arrangement permits the venous blood to be warmed by heat transfer from the arterial blood before it re-enters the body (from Schmidt-Nielsen, K. 1970. *Animal Physiology*, 3rd ed. © Prentice Hall, Inc., Englewood Cliffs, New Jersey. Drawing by David E. Murrish).

heat exchanger is known as a *countercurrent heat exchanger* because the blood flows in the opposite direction in the two streams.

A diagram of a countercurrent heat exchanger may help explain how it works (fig. 8.15). Assume that a copper pipe carries water at 40 °C into a coil suspended in an ice bath, and that the water returns in a second pipe located in contact with the first so that heat is easily conducted between them. As the water leaves the coil, it will be very nearly at the temperature of the ice bath, 0 °C. As this water flows adjacent to the warmer pipe leading into the coil, it picks up heat while the incoming water gets cooled. After some time a steady state of temperatures will be reached, which may be as indicated in the diagram, depending on the conditions for heat exchange between the two pipes and their length.

For animals swimming in warm water, heat dissipation presents a greater problem than heat conservation. The anatomical arrangement of the heat exchanger is such that an increased blood supply and an increase in arterial blood pressure will cause the diameter of the central artery to increase, and this in turn causes the surrounding veins to be compressed and collapse. The venous blood must now return in alternate veins, which are located closer to the surface

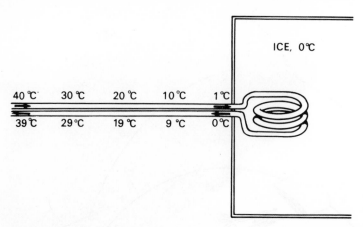

Fig. 8.15. Model of a countercurrent heat exchanger. In this case heat is conducted from the incoming water to the outflowing water so that in the steady-state condition the outflowing water is pre-warmed to within 1 °C of the incoming water. For explanation see text.

of the flipper. Since the heat exchanger is now bypassed, the arterial blood loses heat to the water, and the venous blood returns to the core without being rewarmed, thus cooling the core. In this way the circulatory system of the flipper can function both in heat conservation and in heat dissipation.

It is interesting that such heat exchangers are found in a number of other animals. For example, sea cows, which live in tropical and sub-tropical water, have similar heat exchangers in their appendages. This might seem unnecessary, for they live in warm water, but sea cows move slowly and have relatively low metabolic rates for their size, and thus would have difficulties unless they could reduce the heat loss from the appendages.

Heat exchangers in the limbs are not restricted to aquatic animals. Even in the limbs of man some heat exchange takes place between the main arteries and the adjacent larger veins, located deep within the tissues. In cold surroundings most of the venous blood from the limbs returns in these deep veins, but in warm surroundings much of the venous blood returns in superficial veins under the skin and is thus diverted from heat exchange with the artery (Aschoff and Wever, 1959). In birds, heat exchange in the legs is very important, especially for birds that stand or swim in water. Unless the blood flowing to these thin-skinned peripheral surfaces went through a heat exchanger, the heat loss would be very great indeed. However, due to the heat exchanger that heat loss is minimal. A gull placed with its feet in ice-water for two hours lost only 1.5% of its metabolic heat production from the feet (Scholander, 1955), i.e. a quite insignificant heat loss.

Even some tropical animals have vascular heat exchangers in their limbs. In the sloth the entire artery to the arm splits up into several dozen relatively thin parallel arteries which are intermingled with a similar number of veins. It may seem superfluous for a tropical animal to have such heat exchangers, but on a rainy and windy night the heat loss from an animal sleeping in the crown of a tree can be very great indeed. The value of the heat exchanger in the sloth has been clearly demonstrated by immersing a limb in ice-water. The blood temperature below the heat exchanger decreased drastically, but in the arm above the heat exchanger the temperature of the venous blood was nearly at core temperature (Scholander and Krog,

⋯ ⋯dles in the arm of the

Walker, 1969). However, ⋯ ⋯
upper arm, and the muscles responsible for the grasp are in the lower arm. The fact that the blood runs through a bundle of arteries which join again into a single vessel cannot have any imaginable influence on the muscles in the lower arm. The only difference these muscles could see would be a slightly decreased blood pressure which certainly could not improve the strength of the grip.

Temperature regulation in the heat – keeping cool

We have seen that mammals and birds must increase their heat production in order to stay warm at ambient temperatures below a certain point, the lower critical temperature. We will now be concerned with what happens above this critical temperature.

Above the critical temperature the metabolic heat production (resting metabolic rate) remains constant through a range usually known as the thermoneutral range (see fig. 8.6, page 318). As before, for the body temperature to remain constant, metabolic heat must be lost at the same rate that it is produced. If we return to the now familiar equation for heat balance, $\dot{H} = C \, (T_b - T_a)$, we will see that if $\dot{H}$ and T_b remain constant and T_a changes, the conductance term, C, must also change.

The conductance term refers to the total heat flow from the organism, and it can be increased in several ways, including an increased circulation to the skin so that heat from the core is moved more rapidly to the surface. Also, stretching out and exposing increased surface areas, especially naked or thinly furred areas, contributes to increasing the overall conductance.

Let us return to the basic equation for heat balance

$$H_{tot} = \pm H_c \pm H_r \pm H_e \pm H_s.$$

As the ambient temperature increases, the conditions for heat loss by conduction and convection, H_c, and radiation, H_r, become increasingly unfavorable. As the metabolic heat production, H_{tot}, remains unchanged (it might even increase slightly), the heat balance equation must increasingly emphasize the evaporation and storage terms (H_e and H_s).

In the discussions of heat regulation in the cold we disregarded heat removed by evaporation because it made up a fairly small fraction of the total heat exchange. At high temperature evaporation is the key item in the heat balance. Assume that we have increased the temperature of the environment to equal the body temperature of an animal. If the temperature of the environment equals the body temperature, there can be no loss of heat by conduction, and the net radiation flux approaches zero. If we now wish to maintain a constant body temperature, i.e. keep the storage term H_s at zero, the consequence is that the entire metabolic heat production, H_{tot}, must be removed by evaporation of water.

This prediction has been tested on a number of mammals and birds. When they are kept in air that equals their body temperature, those that are able to keep their body temperature from rising (not all species can) evaporate an amount of water which is equivalent to their metabolic heat production. An example is shown in fig. 8.16, which shows that the desert jackrabbit (actually a hare) follows the prediction.

The ambient temperature can, of course, increase further and exceed the body temperature, as frequently happens in deserts. In this event the body receives heat from the environment, by conduction from the hot air, by radiation from the heated ground surface, and in particular by radiation from the sun. Under these circumstances evaporation must be used to dissipate the sum of metabolic heat production and the heat gain from the environment. The possibility of using heat storage is not a good solution, for as we saw before, animals have only a limited tolerance to an increased body temperature. Nevertheless, as we shall see later, even a moderate increase in body temperature can indeed be important for a desert animal.

In order to evaluate the importance of the heat load from the en-

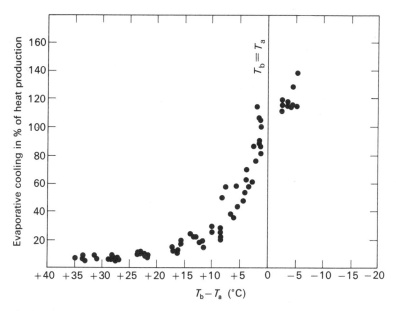

Fig. 8.16. The cooling by evaporation, expressed in per cent of the simultaneous metabolic heat production, increases as the difference between body temperature (T_b) and air temperature (T_a) decreases. As this difference approaches zero (body and air temperature equal), the entire heat production (100%) is dissipated by evaporation (after Dawson and Schmidt-Nielsen, 1966).

vironment, let us examine how much water is used for evaporation (sweating) by a man in a hot desert. His total resting metabolic rate will be about 70 kcal h^{-1}, and this amount of heat, if dissipated entirely by evaporation, requires 120 ml, or 0.12 liter water. We also know that a man, exposed to the sun on a hot desert day, may sweat at a rate of between 1 and 1.5 liter per hour. To make our calculations simple, let us say 1.32 liter h^{-1}. If there is no change in his body temperature, and 0.12 liter is used to dissipate the metabolic heat, the excess of 1.2 liter used must be due to the heat gain from the environment. Thus the sum of conduction from the hot air, radiation from the hot ground, and radiation from the sun, must require the evaporation of 1.2 liter of water; i.e. the heat load from the environment in this example is exactly ten times as high as the metabolic heat production of the man at rest. If the body temperature were to increase by, say 1 °C, the storage of heat would be about 60 kcal, or the equivalent of 0.1 liter water evaporated, i.e. a fairly insignificant amount.

We now have a good idea of how important the environmental heat load can be when the temperature gradients are reversed, and it is worth considering how this influences animals. If a man were to continue evaporating at the indicated rate, and had no water to drink, he would be near death from dehydration at the end of a single hot day in the desert. Yet the deserts of the world have a rich and varied animal life, although in most deserts there is no drinking water available. This requires further discussion.

The importance of body size

If we put a large block of ice outside in the hot sun, and a small piece next to it, the small piece will melt away long before the large block. If a big rock and a small pebble are placed on the ground in the sun, the pebble will be hot long before the big rock. The reason is that a small object has a much larger surface, relative to its volume. If the big rock or block of ice is broken into many small pieces, numerous new surfaces, previously unexposed, will be in contact with the warm air and receive solar radiation. Because of the many new exposed surfaces the total mass of rock will heat up more rapidly and the crushed ice will melt faster.

The relation between surface and volume of similar-shaped objects is simple. If a given cube is cut into smaller cubes with each side one-tenth of that of the larger cube, the total surface of all the small cubes will be ten times that of the original cube. If the linear dimension of the small cube is one-hundredth of the larger one, the aggregate surface is 100 times the original, and so on. This rule holds for any other similar-shaped objects. Thus, any small body has a surface area which, relative to its volume, increases as the linear dimension decreases. Small and large mammals have sufficiently similar shapes that their body surface area is a regular function of the body volume and the same rule applies.

Let us return to the animal in hot surroundings. Its total heat gain consists of two components, heat gain from the environment and metabolic heat gain. The heat gain from the environment by conduction and convection, as well as the radiation from the ground and from the sun, are surface processes, and the total environmental heat load therefore is directly related to the surface area. In chapter 6 we saw that the metabolic heat production of mammals is not quite proportional to body surface, but sufficiently close that, as an approximation, we can assume this to be so. Therefore the total heat load, the sum of metabolic and environmental heat gain, is roughly proportional to the body surface area. This puts the small animal, with its larger relative surface, in a far more unfavorable position with regard to heat load than a large animal.

If a man in the desert must sweat at a rate of one liter per hour

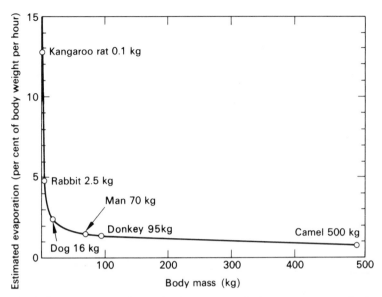

Fig. 8.17. For a mammal to maintain a constant body temperature under hot desert conditions, water must be evaporated in proportion to the heat load. Because of the larger relative surface area of small animals, the heat load, and therefore the estimated evaporation in relation to the body size, increases rapidly in the small animal. The curve is calculated on the assumption that heat load is proportional to body surface (Schmidt-Nielsen, 1964).

(which equals about 0.60 liter per meter^2 body surface per hour), we can use the surface relationship to make an estimate or prediction of how much water other animals should evaporate under similar desert conditions in order to dissipate the heat load. We thus obtain a theoretical curve which predicts the amount of water needed to keep cool (fig. 8.17). The amount is an exponential function of body weight, and on logarithmic coordinates the curve will be a straight line. However, fig. 8.17 is plotted on a linear ordinate in order to emphasize the exponential increase in water required if a small animal were to depend on evaporation to keep cool in the desert. Many small rodents weigh between 10 and 100 g and would have to evaporate water at a rate of 15 to 30% of their body weight per hour. Since a water loss of between 10 and 20% is fatal for mammals, such water loss is impossible. Obviously, because of their small size alone, desert rodents must evade the heat, which they do by retreating to their underground burrows during the day.

Large body size: the camel. Fig. 8.17 shows that large animals, such as camels, derive a substantial advantage simply from being big. The benefit, however, does not increase much as the body size increases

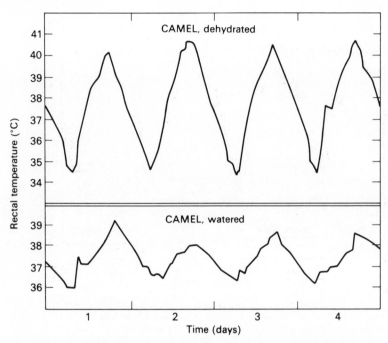

Fig. 8.18. The daily temperature fluctuations in a well-watered camel is about 2 °C per day. When the camel is deprived of drinking water the daily fluctuation may increase to as much as 7 °C. This has a great influence on the use of water for temperature regulation (see text) (from Schmidt-Nielsen, K. 1963. *Harvey Lecture Series,* Academic Press, **58**, p. 622, Fig. 175).

further. Theoretically, if a camel in other respects were like a man, its body size would make it evaporate water at about half the rate of a man. The camel, however, is not a man, and does not use this much water for heat regulation. As we shall see the camel uses a combination of several approaches which help reduce the heat gain from the environment and thus the use of water.

Let us first consider what a camel can do about two of the variables in the heat balance equation, evaporation (H_e) and storage of heat (H_s). Storage of heat is reflected in an increase in body temperature. The temperature of a normal camel which is watered every day and is fully hydrated, varies by less than 2 °C, between about 36 and 38 °C. When the camel is deprived of drinking water, however, the daily temperature fluctuations become much greater (see fig. 8.18) (Schmidt-Nielsen, Schmidt-Nielsen, Jarnum and Houpt, 1957). The morning temperature may be as low as 34 °C, and the highest temperature in the late afternoon may be nearly 41 °C. This large increase in body temperature during the day constitutes a storage of

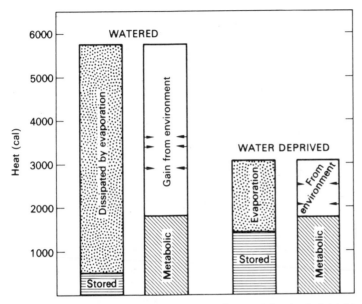

Fig. 8.19. The change in body temperature of a camel deprived of drinking water greatly affects the heat gain from the environment, and thus in turn the amount of water used for heat regulation (dotted area) (from Schmidt-Nielsen, K. 1957. *Am. J. Physiol.*, **188**, 108, Fig. 4).

heat. For a camel that weighs 500 kg, the amount of heat stored due to a 7 °C temperature rise corresponds to 2900 kcal of heat, which equals a saving of five liters of water. In the cool night, the stored heat can be unloaded by conduction and radiation without use of water.

The increase in body temperature during the day has a further advantage, beyond that of storing heat. When the body temperature is increased, the temperature gradient from the hot environment to the body (which determines the amount of heat gain from the environment) is reduced. This reduction in environmental heat gain is at least as important as heat storage with regard to water savings, and is reflected in the reduced amount of water used. This is shown in fig. 8.19 which shows the heat balance of a small adult camel (260 kg), studied in the Sahara desert. In the fully hydrated animal the evaporation during the ten hottest hours of the day was 9.1 liters, corresponding to the dissipation of 5300 kcal heat. In the same camel when deprived of drinking water, the evaporation was reduced to 2.8 liters in ten hours, corresponding to the dissipation of 1600 kcal heat. Thus, the use of water in the dehydrated camel was reduced to less than one-third of that in the fully hydrated animal.

The main reason that the camel in the desert uses substantially less water than a man (about 0.28 liter h^{-1} in the camel, against more than 1 liter h^{-1} for a man who weighs about one-quarter as much) is the higher body temperature in the camel, which means both storage of heat and a reduced environmental heat gain. In contrast, man maintains his body temperature nearly constant at about 37 °C, and thus does not store appreciable amounts of heat. A man also maintains much steeper gradients between the environment and his body surface, for the mean skin surface temperature of a sweating man is about 35 °C. One further important item is that the camel has a thick fur with a high insulation value. This imposes a heavy insulating layer between the body and the source of the heat, and thus reduces the heat gain from the environment. The simplest way of testing the importance of this fur is to shear a camel. In one such experiment, the water expenditure of a camel, under otherwise comparable conditions, was increased by about 50% when the animal was shorn.

We have now seen some of the major reasons for a lower-than-predicted (from body size alone) water loss in the camel, (1) heat is stored because of an increase in body temperature, (2) the increase in body temperature decreases the heat flow from the environment, and finally, (3) the fur is a substantial barrier to heat gain from the environment. The camel has one further advantage over man, it can tolerate a greater degree of water depletion of the body. While a man is close to the fatal limit when he has lost water amounting to between 10 and 12% of the body weight, the camel can tolerate about twice as great water depletion without apparent harm. As a result, the camel can go without drinking for perhaps six to eight days under desert conditions which would be fatal to man in a single day without water. In the end, however, a camel must drink to replenish its body water, and when water is available, it may drink more than a third of its body weight. The immense capacity for drinking has given rise to the legend that a camel, before a long desert journey, will fill its water reserves; in reality, the camel, like other mammals, drinks to replenish lost water and to restore the normal water content of the body, and there is no evidence that it over-drinks in anticipation of future needs.

Small body size: the ground squirrel. We saw earlier that because of their small size and large relative surface, small rodents should be unable to remain above ground and active during the day. Seemingly, the small desert ground squirrels of the North American deserts defy this conclusion. These small rodents, which weigh from less than one hundred to a few hundred grams, are often seen outside their burrows in the daytime, they move around quickly, dashing from one place to another, they frequently disappear in a hole, but soon appear again, often within minutes. During the hottest mid-summer day they are less active in the middle of the day, but are often seen during the morning and later afternoon hours.

Usually, there is no free water available in their environment. How is it possible for ground squirrels to remain active during the day? Apparently, heat storage plays a great role, but not on a diurnal cycle, as for the camel. When they are outside their burrows on a hot day, the ground squirrels may heat up very rapidly. They cannot tolerate any higher body temperature than other mammals and die if heated to 43 °C. However, a temperature of 42.4 °C can be tolerated without apparent ill effects. As an active ground squirrel has become heated outside, it returns to the relatively cool burrow where it rapidly cools off, helped by its large relative surface. By moving in and out of the cool burrow the ground squirrel can rapidly pick up heat and rapidly unload it again, and in this way avoid using water for evaporation. This permits it to move about and be active during the day.

Evaporation – sweating or panting

We know from our own experience that humans sweat to increase cooling by evaporation. Dogs, in contrast, have few sweat glands, and they cool primarily by panting, a very rapid, shallow breathing which increases evaporation from the upper respiratory tract. Some animals use a third method for increasing evaporation, they spread saliva over their fur and lick their limbs, thus achieving cooling by evaporation.

The amount of heat needed to evaporate one gram of water, 580 cal per gram water, is, of course, the same irrespective of the source of the water and where it is evaporated. Let us see if there could be other differences that might make one method of evaporation or the other more or less advantageous for a particular animal. We will therefore examine the characteristic features of each method.

In man, sweating carries virtually the entire burden of increased evaporation. Evaporation from the general body surface is an effective way of cooling, and in the absence of fur the water can evaporate readily. However, several furred mammals, in particular those of large body size, such as cattle, large antelopes, and camels, also depend mainly on sweating for evaporation. The camel in particular has a thick fur, but evaporation is not much impeded because in a desert atmosphere the air is very dry and the sweat evaporates so rapidly that some observers have reported that camels do not sweat at all. In contrast, a number of smaller ungulates, sheep, goats, and many small gazelles, evaporate mostly by panting. Those carnivores that have been studied also pant. The third method, salivation and licking, is common in a fairly large number of Australian marsupials, including the large kangaroos, and can also be found in some rodents, including the ordinary laboratory rat. The salivation-licking method is not very effective, and seems to be used primarily as an emergency measure when the body temperature threatens to approach a lethal level.

In addition to being less effective than sweating and panting, salivation and licking is less widespread, and we shall not discuss it further. Instead, we shall compare sweating and panting and see what possible advantages and disadvantages each method may have.

Birds, in contrast to mammals, have no sweat glands, and they increase evaporation either by panting, or by a rapid oscillation of the thin floor of the mouth and upper part of the throat, a mechanism known as gular flutter. Either mechanism seems to be an effective means of cooling, for birds may use one or the other or often a combination of both (Bartholomew, Lasiewski and Crawford, 1968).

An obvious difference between panting and sweating is that the panting animal provides its own air flow over the moist surfaces, thus facilitating evaporation. In this regard the sweating animal is not so well off. Another difference is that the sweat (at least in man) contains considerable amounts of salt, and a heavily sweating human may lose so much salt in the sweat that he becomes salt deficient. This is the reason for the recommendation that we should increase the intake of sodium chloride when sweating is excessive in very hot situations. Panting animals, in contrast, do not lose any of the electrolytes which are secreted from glands in the nose and the mouth (unless the saliva actually drips to the ground) and are thus better off in this regard as well.

Panting has two obvious disadvantages, one is that increased ventilation easily causes an excessive loss of carbon dioxide from the lungs which can result in severe alkalosis, the other is that increased ventilation requires muscular work which in turn increases the heat production and thus adds to the heat load. The tendency to develop alkalosis can in part be counteracted by shifting to a more shallow respiration (smaller tidal volume) at an increased frequency, so that the increased ventilation takes place mostly in the dead space of the upper respiratory tract. Nevertheless, heavily panting animals regularly become severely alkalotic, and thus they do not utilize fully the possibility of restricting the ventilation to the dead space (Hales and Bligh, 1969; Hales and Findlay, 1968).

The increased work of breathing during panting would be a considerable disadvantage, were it not for the interesting fact that the muscular work, and thus heat production, can be greatly reduced by taking advantage of the elastic properties of the respiratory system. Observations show that when a dog begins to pant, its respiration tends to shift rather suddenly from a frequency of about 30–40 respirations per minute to a relatively constant high level of about 300 to 400. A dog subjected to a moderate heat load does not pant at intermediate frequencies; instead, it pants for brief periods at the high frequency, alternating with periods of normal slow respiration.

The meaning of this becomes clear when we realize that the entire respiratory system is elastic and has a natural frequency of oscillation, like other elastic bodies. That is, on inhalation, much of the muscular work goes into stretching elastic elements, which on exhalation bounce back again, like a tennis ball bouncing. To keep the respiratory system oscillating at its natural frequency (the resonant frequency) requires only a small muscular effort. As a consequence, the heat production of the respiratory muscles is small, adding only little to the heat load (Crawford, 1962). It has been estimated that if panting were to take place without the benefit of a resonant elastic system, the increased muscular effort of breathing at the high frequency of panting would generate more heat than the total heat that can be dissipated by panting. Apparently, birds can also pant at a resonant frequency, thus achieving the same benefits as mammals (Crawford and Kampe, 1971).

Although we recognize some characteristic functional differences between panting and sweating, we do not have a clear picture of why one method or the other is used. Many animals use both methods, and this may indicate that both methods have valuable characteristics that we do not yet fully appreciate.

The greatest advantage of panting may be that an animal under sudden heat stress, such as a fast running African gazelle pursued by a predator, can reach a high body core temperature and yet keep the brain, the most heat sensitive organ, at a lower temperature. This may, at first glance, seem impossible, for the brain is supplied with arterial blood at a high flow rate. It is achieved as follows. In gazelles and other ungulates, most of the blood to the brain flows in the external carotid artery, which at the base of the skull divides up into hundreds of small arteries, which then rejoin before passing into the brain. These small arteries lie in a large sinus of venous blood which comes from the walls of the nasal passages where it has been cooled. The blood that flows through these small arteries is therefore cooled before it enters the skull, and as a result the brain temperature may be two or three degrees lower than the blood in the carotid artery and the body core.

Such temperature differences have been measured in the small East African gazelle, Thomson's gazelle, which weighs about 15 to 20 kg. When it ran for five minutes at a speed of 40 km h^{-1}, a rapid build-up of body heat caused the arterial blood temperature to increase from a normal 39 to 44 °C. The brain temperature did not even reach 41 °C, which is a safe level (Taylor and Lyman, 1972). It would be difficult to design a cooling system that can keep the entire body from heating up when an animal runs in hot surroundings at a speed that may require a 40-fold increase in metabolic rate, but by selective cooling of the blood to the brain the most serious hazard of overheating is avoided.

Selective cooling of blood to the brain has been observed in a number of domestic ungulates as well, and it may be a fairly common mechanism in animals that pant. However, it is unlikely that any such mechanism exists in man, for man doesn't pant and cooling takes place over the entire surface of the body.

It has been known for many years that some reptiles, when they are exposed to a severe heat stress, increase their respiration frequency and breathe with an open mouth, a situation reminiscent of panting. It has been difficult to establish with certainty that this increased respiration, which does increase respiratory evaporation, plays any major role in heat regulation. The amount of water evaporated does not suffice to keep the animals substantially cooler than the environment, and the mechanism therefore seems to be no more than a rudiment of the panting found in birds and mammals. However, in view of the ability of mammals to cool the brain selectively, the panting of lizards can better be understood.

In the desert lizard, the chuckawalla (*Sauromalus obesus*), panting is of marginal importance in the overall heat balance of the animal. When the ambient temperature is kept at 45 °C, the body is barely cooler (44.1 °C), but the brain remains at 42.3 °C, or nearly 3 °C below the ambient temperature. If a chuckawalla is moved from a room at 15 to 45 °C, the brain and cloacal temperatures initially increase rapidly, but as these temperatures reach 41 to 42 °C, the brain temperature stabilizes at this level, while the cloacal temperature continues to rise until it nearly reaches the ambient temperature (see fig. 8.20).

In the chuckawalla the carotid arteries run very close to the surface of the pharynx, so close that they are visible through the open

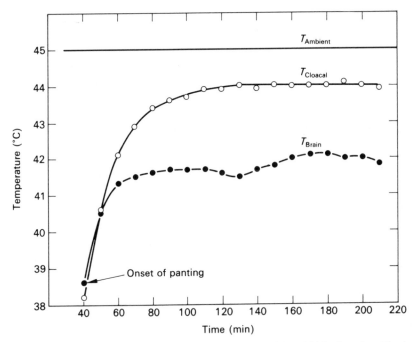

Fig. 8.20. When the lizard *Sauromalus* is moved from 15 °C to 45 °C, cloacal and brain temperatures increase rapidly. As these temperatures reach about 41 °C, they separate and the brain remains about 2 °C below the cloacal temperature and 3 °C below the air temperature (from Crawford Jr., E. C. 1972. Brain and Body Temperatures in a Panting Lizard. *Science,* **177,** 431–33, Fig. 1. © American Association for the Advancement of Science).

mouth of the animal. As the arteries pass right under the moist surfaces where evaporation takes place, the blood is cooled before it enters the brain. This is again an example of an animal that is able to maintain different parts of the body at different, well-regulated temperatures.

TORPOR AND HIBERNATION

Maintaining the body temperature in the cold at a cost of a several-fold increase in metabolic rate is expensive. Small animals have high metabolic rates to begin with, and a further increase may be too expensive when food is scarce or unavailable. The easy way out, and the only logical solution, is to give up the struggle to keep warm and let the body temperature drop. This not only eliminates the increased cost of keeping warm, but cold tissues use less fuel and the energy reserves last longer. This, in essence, is what hibernation is all about.*

Many mammals and a few birds hibernate regularly each winter. This means that the body temperature drops almost to the level of the surroundings, metabolic rate, heart rate, respiration, and many other functions are greatly reduced, and the animal is torpid and shows little response to external stimuli such as noise or being touched. With active life virtually suspended and the metabolic rate greatly reduced, the animal can survive a long winter. Before entering into a period of hibernation, most hibernators also become very fat, i.e. they deposit large fuel reserves. Without the burden of keeping warm, the reserves can last for extended periods of unfavorable conditions.

Most animals that hibernate are of small body size. This makes sense, for their high metabolic rate requires a high food intake. Thus many rodents, hamsters, pocket mice, dormice, etc. hibernate. Insect eaters at high latitudes can find little food in winter, and bats and insectivores (e.g. hedgehogs) could not possibly manage without hibernation. Hibernators are also found among Australian marsupials, such as the pigmy possum (Bartholomew and Hudson, 1962). Hibernating birds include hummingbirds, the smallest of all birds, the insect eating swifts, and some mouse birds (*Colius*, an African genus).

It is not easy to give a completely satisfactory definition of the term hibernation. In physiology the term refers to a torpid condition with a substantial drop in metabolic rate. Thus, bears may sleep during much of the winter, but most physiologists will say that they are not true hibernators, for their body temperature drops only a few degrees and they show only a moderate drop in metabolic rate and other physiological functions, the females often give birth to the cubs during the winter, etc. In other words, the bear does not conform to what a physiologist customarily considers true hibernation.

Another term, *estivation* (Latin *aestas* = summer), refers to inactivity during the summer, and is even less well defined. It may be applied to snails that become dormant and inactive in response to drought, or it may refer to ground squirrels that during the hottest months disappear into their burrows and remain inactive. The Columbia ground squirrel begins to estivate in the hot month of August, but then it remains inactive throughout the autumn and winter and does not appear again until the following May. Does it estivate or hiber-

---

* The word hibernation (from Latin *hiberna* = winter), is also used to designate an over-wintering, inactive stage of poikilothermic animals such as insects or snails.

nate, and when does one condition change into the other? The fact is that no clear physiological distinction can be made between the two states. Furthermore, many hibernators, bats for example, may undergo a daily period of torpor during which the body temperature and metabolic rate fall. Their physiological state is then quite similar to hibernation, although it lasts only for hours instead of weeks or months.

Since we lack precise definitions and there are no sharp lines between these different states, we will treat torpor and hibernation as one coherent physiological phenomenon.

It was earlier believed that hibernation and torpidity are due to a failure of temperature regulation in the cold, and that it is an expression of some kind of 'primitive' condition or poor physiological control. It is now quite clear, however, that hibernation is not due to an inadequate temperature regulation; it is a well-regulated physiological state, and the superficial similarity between a hibernating mammal and a cold-blooded animal (say, a lizard or frog) is misleading. We shall find that hibernation can in no way be considered a physiological 'failure.'

Body temperature and oxygen consumption

Bats are among the animals that may have daily periods of torpor and may also hibernate for longer periods. If exposed to low ambient temperature, bats may respond in one of two different characteristic ways. This is illustrated in fig. 8.21, which shows measured body temperatures of two North American bat species. At temperatures below about 30 °C these bats can be either torpid and have a body temperature within a degree or two of the air temperature, or they can be metabolically active with a normal body temperature between 32 and 36 °C. Even at air temperatures below 10 °C these bats may be either active or torpid.

The two bats are of similar size. The larger, the big brown bat, which weighs about 16 g, ranges over most of the United States, and at least in the northern part of its range it hibernates. The smaller species, the Mexican free-tailed bat, weighs about 10 g, it inhabits the hot, dry southwestern United States and Mexico, and in winter it migrates southwards and is not known to hibernate.

The response to low temperature of these two species is amazingly similar, although in nature one hibernates and the other one does not. The oxygen consumption of those individuals which maintain a high body temperature at low air temperature shows the typical homeothermic pattern. As ambient temperature decreases, the cost

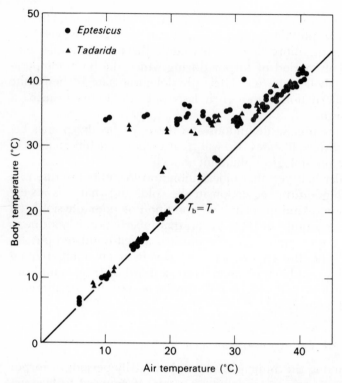

Fig. 8.21. The body temperature of two species of North American bats. At low air temperature these bats may either be active with normal body temperatures about 34 °C, or torpid with the body temperature near the air temperature (from Herreid, C. F. and Schmidt-Nielsen, K. 1966. *Am. J. Physiol.*, **211**, 1108–12, Fig. 1).

of keeping warm is reflected in the increased oxygen consumption, which is at a minimum at about 35 °C (fig. 8.22). Those individuals that become torpid, however, have a very low metabolic rate. At 15 °C, for example, the difference in oxygen consumption between an active and a torpid bat is about 40-fold. The torpid bat thus uses fuel at a rate only one-fortieth of the active bat, and its fat reserves would last 40 times as long.

The amount of energy that can be saved by going into torpor has been more carefully analyzed by Tucker (1965*a*; 1965*b*; 1966). The oxygen consumption during torpor is easily determined, but in addition, the energy cost required to re-warm the animal at the end of a period of torpor must be included in the cost. Tucker studied the California pocket mouse (*Perognathus californicus*), and found that for a short period of torpor the energy cost of reheating was a substantial part of the total cost.

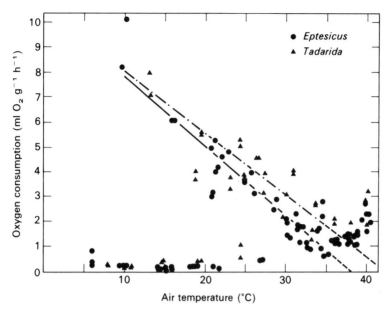

Fig. 8.22. The oxygen consumption of active bats increases with decreasing air temperature, while the oxygen consumption of torpid bats drops to a small fraction of the active rate (same bat species as in fig. 8.21) (from Herreid, C. F. and Schmidt-Nielsen, K. 1966. *Am. J. Physiol.*, **211**, 1108–12, Fig. 2).

The pocket mouse weighs about 20 g and readily becomes torpid at any ambient temperature between 15 and 32 °C. Below 15 °C the normal torpid state is disrupted and the animals usually cannot arouse on their own; in other words, if cooled below 15 °C they are unable to produce enough metabolic heat to get the re-warming cycle started. Pocket mice enter torpor much more readily if their food ration is restricted. With a gradual restriction of the food supply, longer and longer periods are spent in torpor. In this way the pocket mice can maintain their body weight on a food ration which is about one-third of what they normally consume when they remain active. A characteristic curve for the oxygen consumption of a pocket mouse on a restricted food ration is given in fig. 8.23. This particular individual received 1.5 g seed per day, and on this ration it was torpid for about nine hours out of every twenty-four. Although the food ration was less than half of what the animal would normally need if it remained active, it maintained a constant body weight.

When an animal goes into torpor, what happens first? Does heat loss first increase, the body temperature therefore drop, and the metabolic rate as a consequence decrease, thus accelerating the temper-

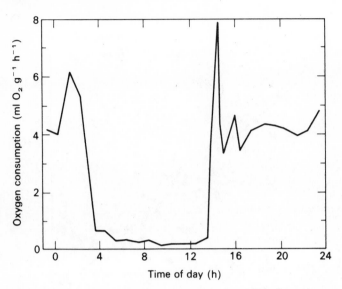

Fig. 8.23. Oxygen consumption over a 24-hour period of a pocket mouse kept on a restricted food ration of 1.5 g seeds per day. During 9 hours the animal was torpid and had a very low rate of oxygen consumption followed by a peak as it returned to the active state. Air temperature was 15 °C (from Tucker, V. A. 1965. *J. Cell. Comp. Physiol.*, **65**, 396, Fig. 3).

ature drop? Or does the animal decrease its metabolic rate, the body temperature therefore begin to drop, the temperature drop causing a further decrease in heat production, and so on?

With information about heat production and heat loss during all stages of the torpor cycle, it is possible to answer this question. The conclusion is that entry into torpor can result simply from a cessation of any thermoregulatory increase in metabolism at air temperatures below the lower critical temperature (about 32.5 °C in the pocket mouse). At any temperature below this point, a decrease in metabolic rate to the thermoneutral resting level alone will, without change in conductance, lead to a decrease in body temperature. In other words, there is no need for a special mechanism to increase the heat loss. As the body temperature begins to fall, metabolism decreases and the pocket mouse slides into torpor with a further drop in heat production and body temperature.

Arousal can take place at any temperature above 15 °C if the animal changes to maximum heat production at that temperature, which is about 10 to 15 times the minimum oxygen consumption at the same temperature. Arousal is thus an active process which requires a considerable expenditure of energy for a considerable period until the body temperature has reached the normal.

Let us ask a simple question. If a pocket mouse goes into torpor at 15 °C and immediately arouses again, would it save any energy, or does re-warming cost too much? The decline in body temperature takes about two hours, and the total oxygen consumption during this period is 0.7 ml O_2 g^{-1}. If arousal follows immediately, normal body temperature is reached in 0.9 h at a cost of 5.8 ml O_2 g^{-1}. Entering into torpor and then immediately arousing therefore costs a total of 6.5 ml O_2 g^{-1} in 2.9 h. The cost of maintaining normal body temperature for the same period of time at 15 °C ambient temperature would be 11.9 ml O_2 g^{-1}. This period of torpor thus consumes only 55% of the energy required to keep warm for the same period of time (Tucker, 1965b).

Remaining in torpor for longer periods, for example ten hours, is even more favorable for the animal. We must now add to the previously calculated cost of entry and arousal the cost of 7.1 h at 15 °C, which is 1.2 ml O_2 g^{-1}, giving a total for a ten-hour cycle, including entry and arousal, of 7.7 ml O_2 g^{-1}. This is less than 20% of the cost of maintaining normal body temperature for ten hours, which at 15 °C would be over 40 ml O_2 g^{-1}.

We can see that in all these cases the cost of arousal is the major part of the cost of the torpor cycle. For the ten-hour period of torpor the cost of arousal alone is 75% of the total energy expenditure for the period.

The widespread occurrence of torpidity as a response to unfavorable conditions suggests that this ability is a very fundamental trait. Animals that can enter torpor include the egg-laying echidna (the spiny anteater, *Tachyglossus aculeatus*), which is only remotely related to modern mammals. The echidna is normally an excellent temperature regulator in the cold and can maintain its normal body temperature at freezing temperatures (Schmidt-Nielsen *et al.*, 1966). However, when it is without food and kept at 5 °C, it readily becomes torpid. The heart rate decreases from about 70 to 7 beats per minute, and the body temperature remains about +5.5 °C. In this state the oxygen consumption of the echidna is about 0.03 ml O_2 g^{-1} h^{-1}, roughly one-tenth of its normal resting oxygen consumption (Augee and Ealey, 1968).

Control mechanisms

Torpor and hibernation are phenomena that are under accurate physiological control. We have just seen how the duration of the daily torpor of a pocket mouse is adjusted to the food ration and the need for energy savings. Obviously, not only entry into torpor must be controlled, but the duration of the cycle and thus the arousal process as well, must be accurately regulated.

In nature the beginning of the hibernation cycle is usually associated with the time of the year, but it is not necessarily induced by low temperature or lack of adequate food. The yearly cycle of hibernation is influenced by the duration of the daily light cycle and is also associated with endocrine cycles. For example it is often impossible to induce even a good hibernator to begin a cycle of torpor during early summer and especially during the reproductive period.

The torpid animal with its low body temperature seems quite passive, it cannot perform coordinated movements, and it hardly responds to sensory stimuli. Superficially it resembles a cold-blooded animal that has been chilled. All of us have experienced how our fingers become numb when the hands are cold; this is because nerve conduction ceases at temperatures below some 10 to 15 °C, and we sense nothing. (The reason that we can still move our cold fingers is that the appropriate muscles are located above the hand in the lower arm.)

If the nerves of hibernators were to become inoperative at lower temperatures, their nervous system could not remain coordinated at, say, 5 °C. Yet respiration and many other functions continue and, although at a lower rate, in a well-coordinated fashion. If the ambient temperature decreases towards or below freezing, some hibernators will be killed. Others, however, respond in one of two ways, leaving no doubt about the integrity of their central nervous system. Either they arouse and return to the fully active condition, or they resist the decrease in body temperature by a regulated increase in heat production, keeping the body temperature at some low level, say + 5 °C. Such a well-regulated heat production does indeed require a well-coordinated central nervous system. For example, the European hedgehog maintains its body temperature at +5 to +6 °C as the ambient temperature decreases to below freezing. This prevents freezing damage, and at the same time it saves fuel, for it eliminates the need of going through an expensive complete arousal and the cost of then maintaining the high body temperature of the active state. For a hibernator that may be exposed to freezing temperatures several times during winter, such repeated arousals might prove too expensive; maintaining a low body temperature, just sufficient to keep from freezing, is far more economical.

A similar well-regulated state of torpor has been observed in the West Indian hummingbird, *Eulampis jugularis*, which like other hummingbirds readily becomes torpid. The temperature of *Eulampis*, when torpid, approaches air temperature, but if the air tempera-

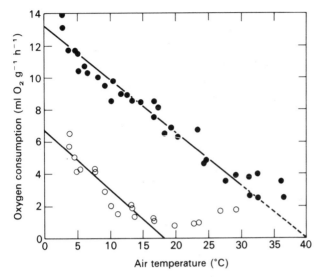

Fig. 8.24. Oxygen consumption of the tropical hummingbird *Eulampis*. In the active *Eulampis* (solid circles) oxygen consumption increases linearly with decreasing temperature. In the torpid *Eulampis* (open circles) the oxygen consumption drops to a low level, but if the air temperature goes below 18 °C, the animal produces more heat and maintains its body temperature at 18–20 °C without arousing from torpor (from Hainsworth, F. R. and Wolf, L. April 1970. Regulation of Oxygen Consumption and Body Temperature During Torpor in a Hummingbird, *Eulampis Jugularis. Science,* **168,** 368–69, Fig. 1. © American Association for the Advancement of Science).

ture falls below 18 °C, *Eulampis* resists a further fall and maintains its body temperature constant at 18 to 20 °C. Below this point the heat production must therefore be increased (see fig. 8.24), and the increase must be linearly related to the drop in the ambient temperature.

Eulampis is particularly interesting because it is a clear case of torpidity in a tropical warm-blooded animal, and this shows that torpidity is not restricted to animals from cold climates. Also, the well-regulated metabolic rate at two different levels of body temperature certainly shows that torpidity in no way is a 'failure' of the process of themoregulation. Another interesting aspect of *Eulampis* is that the overall thermal conductance (the slope of the metabolic regression line) is the same in the torpid and in the normal non-torpid hummingbird.

Arousal

The re-warming during arousal, as we have seen, is metabolically the most expensive part of the torpor cycle. The arousing animal displays violent shivering and muscle contractions, and apparently uses fuel at a maximal rate. However, not only the muscles but also a particular kind of adipose tissue, known as 'brown fat', is important in the re-warming process.

In addition to the ordinary fat, or white fat, many mammals have smaller or larger patches of brown fat, which differ from white fat both in color and in metabolic characteristics. Brown fat has a particularly high content of cytochrome and can consume oxygen at a high rate, while white fat is metabolically rather inactive. Brown fat is particularly prominent in hibernators, and it has long been inferred that there is a direct connection between hibernation and the amount of brown fat. The precise connection remained obscure, however, until biochemical studies revealed that brown fat is capable of a high rate of oxygen consumption, and thus heat production. During arousal the temperature of the brown fat, particularly in some large masses located between the shoulder blades, is among the highest in any part of the organism.

It is characteristic of an arousing hibernator that the temperature is not the same throughout the body, the re-warming is far from uniform. Fig. 8.25 shows temperatures recorded separately in the cheek pouch and in the rectum of an arousing hamster. The an-

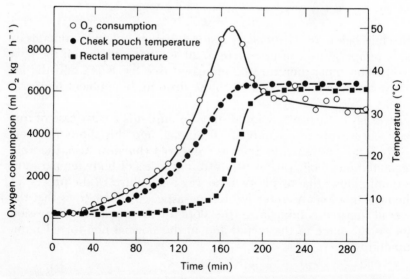

Fig. 8.25. The temperature increase in a hamster arousing from hibernation proceeds more rapidly in the anterior part of the body (cheek pouch temperature) and more slowly in the posterior part (rectal temperature). Open circles show the oxygen consumption during arousal (from Lyman, C. P. 1948. *J. Exp. Zool.*, **109**, 74, Fig. 4).

terior part of the body, which contains vital organs such as heart and brain, warms much more rapidly than the posterior part. Re-warming of the heart is not only essential but must be an initial step, for the heart is needed to provide circulation and oxygen for all the other organs. The major masses of brown fat are also located in the anterior part of the body. The records show that not until the anterior portion of the body has reached near-normal temperatures does the posterior part enter into the re-heating process. At this moment, when the entire body rapidly attains normal temperature, there is a peak in the oxygen consumption, which afterwards declines and settles at a normal level.

The uneven temperature increase in the various parts of the body leads to an inevitable conclusion: during the early stages of arousal the blood flow is directed almost exclusively to the vital organs in the anterior part of the body, and not until these have been re-warmed does circulation increase substantially in the posterior parts as well.

The blood flow to the various organs can be followed by means of radioactive tracers, and such studies have shown that the skeletal muscles in the anterior part of arousing animals receive more than 16 times as much blood as in an awake, non-hibernating individual. This confirms that the muscles are highly involved in the increased heat production. During early arousal the muscles in the hind part of the animal receive only one-tenth as much blood. The brown fat, however, receives even more blood than the most active muscles, thus implicating this tissue in heat production during arousal. As could be expected, the digestive tract, in particular the small intestine, is among the tissues that receive the least blood during arousal (Johansen, 1961).

It is perplexing that brown adipose tissue, which is so important for mammalian hibernators, appears to be absent from a number of birds that regularly hibernate, such as hummingbirds, swifts, and nighthawks (Johnston, 1971).

BODY TEMPERATURE IN 'COLD-BLOODED' ANIMALS

Birds and mammals are traditionally known as warm-blooded, the term cold-blooded being used for all other animals. However poor the term is, it is well established and convenient, and we will use it, for no better single term is available which is both complete and accurate.

Most cold-blooded animals are more or less at the mercy of their environment in regard to the body temperature they attain. However, some so-called 'cold-blooded' animals can and indeed do stay warmer than the medium in which they live, whether it be air or water. What they can achieve in this regard is governed by simple physical principles although the solutions at times seem quite ingenious.

Let us return to the heat balance equation on page 315 and rearrange the terms with only the term for heat storage (H_s) on the left side. On the right side is the total heat production (H_{tot}), heat exchange by radiation (H_r), by conduction and convection (H_c), and by evaporation (H_e). If the body temperature is higher than the surrounding medium, the terms for conduction (H_c) and evaporation (H_e) will signify heat losses and be negative. The equation can therefore be written

$$H_s = H_{tot} \pm H_r - H_c - H_e.$$

If the goal is to increase the body temperature, we must see what the animal can do to maximize heat storage, H_s.

Aquatic animals

For an aquatic animal the situation is simple, for in water there is no evaporation, and no significant radiation source (infrared radiation is rapidly absorbed in water). This leaves the equation: $H_s = H_{tot} - H_c$, and only two parameters to be manipulated in order to increase heat storage, either total heat production to be increased, or conductive heat loss to be minimized.

Because of the high thermal conductivity and high heat capacity of water, small animals lose heat rapidly and have no chance of attaining a body temperature very different from the medium. Even if they have a high level of heat production (metabolic rate) and could increase it further, the oxygen consumption must be increased. Here is where the heat problem comes in. A high rate of oxygen uptake requires a large gill surface and blood for oxygen transport. As the blood flows through the gills, it will inevitably be cooled to water temperature. The gill membrane, which must be thin enough to permit passage of oxygen, provides virtually no barrier to heat loss. The blood will therefore be cooled to water temperature, and it will be impossible for the animal to attain a high body temperature unless a heat exchanger is placed between the gills and tissues.

This is the solution used by some large, fast-swimming fish to achieve independent control of the temperature in limited parts of the body. In this way tuna and some sharks maintain a high temperature in their swimming muscles, independently of the water in which they swim.

The heat exchangers which supply the swimming muscles of the tuna are, in principle, similar to the counter current heat exchangers in the whale flipper, but anatomically they are somewhat differently arranged. In the ordinary fish the swimming muscles are supplied with blood from the large dorsal aorta that runs along the vertebral column and sends branches out to the periphery. In the tuna the pattern is different. The blood vessels that supply the dark red muscles (which the tuna uses for steady, fast swimming) run along the side of the fish just under the skin. From these major vessels come a large number of parallel fine blood vessels which form a slab in which arteries are densely interspersed between veins running in the opposite direction (fig. 8.26). This puts the cold end of the heat exchanger at the surface of the fish and the warm end deep in the muscles. The arterial blood from the gills is at water temperature, and as this blood runs in the fine arteries between the veins, it picks up heat from the venous blood which comes from the muscles. Heat exchange is facilitated by the small diameter of the vessels, about 0.1 mm. When the venous blood has reached the larger veins under the skin, it has lost its heat, which is returned to the muscle via the arterial blood. As a result, the tuna can maintain muscle temperatures as much as 14 °C above the water in which it swims.

The advantage of keeping the swimming muscles warm is that high temperature increases their power output.* A high power output gives the tuna a high swimming speed relatively independent of water temperature, and this in turn makes it possible to pursue successfully prey which otherwise swims too fast to be caught, such as pelagic fish (e.g. mackerel) or squid.

Several of the large sharks, but by no means all of them, have similar heat exchangers which permit their muscles to be kept considerably warmer than the water in which they swim (Carey and Teal, 1969; Carey, Teal, Kanwisher and Lawson, 1971).

The tuna enjoys an additional advantage from the substantial temperature differences between various parts of the body. Separate heat exchangers permit the digestive organs and the liver to be kept at a high temperature. The high power output of the muscles requires a high rate of fuel supply, and this puts a premium on a

---

* The force exerted by a contracting muscle is relatively independent of temperature, and the work performed in a single contraction (force × distance) is therefore also temperature independent. However, at higher temperatures the muscle contracts faster, and since the number of contractions per unit time increases, the power output (work per unit time) increases accordingly.

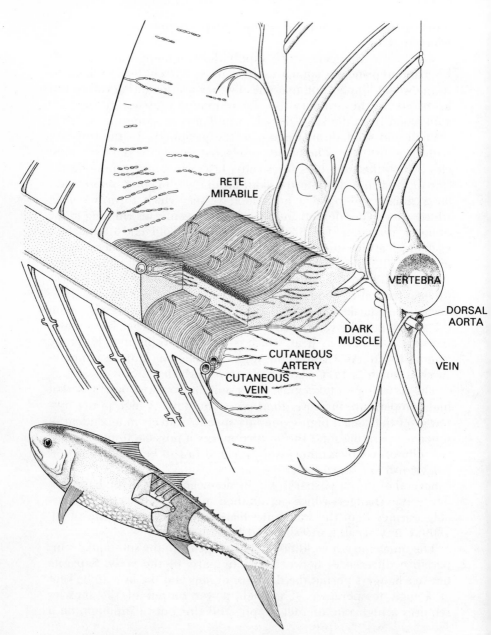

Fig. 8.26. The blood supply to the swimming muscle of the tuna comes through a heat exchanger which is arranged so that these muscles retain a high temperature although the arterial blood is at water temperature (Carey and Teal, 1966).

rapid rate of digestion, which is most readily achieved by a high temperature in the digestive tract.

Let us now compare the warm-bodied fish with marine mammals and consider the main differences and similarities between them. Seals and whales have heavily insulated surface (blubber), heat loss from the extremities is reduced by heat exchangers, and most importantly, they are air breathers whose blood escapes cooling to water temperature at the respiratory surface. The warm-bodied fish, in contrast, would gain no advantage from a better surface insulation, for all the arterial blood coming from the gills is already at water temperature. They do achieve independent temperature control in limited parts of the body by strategic location of efficient heat exchangers which help retain locally produced heat.

Terrestrial animals

The heat balance of terrestrial animals includes all the terms listed in the heat balance equation on p. 354. To raise the body temperature, i.e. to increase heat storage, it is important to reduce evaporation and heat loss by conduction, and to maximize heat gain by radiation and metabolic heat production. The only commonly available radiation source is the sun, and in the absence of solar radiation, metabolic heat production is the only way to increase the heat gain. Of course, external and internal sources of heat, i.e. sun and metabolism, can be used simultaneously, but it is more economical to use an external source rather than body fuel.

Solar radiation is used especially by insects and reptiles. To increase the amount of radiant energy absorbed, these animals depend both on their color and on their orientation relative to the sun. Many reptiles can change their color by dispersion or contraction of black pigments in their skin. Since about half of the solar radiation energy is in the visible light, a dark skin substantially increases the amount of solar energy that is absorbed rather than reflected. The absorption in the near infrared is not much affected by color change, for the animal surface is already close to being 'black' in this range of the spectrum.

The other way of increasing heat gain from solar radiation is to increase the exposed area. This is done by orienting the body at right angles to the direction of the sun's rays, and by spreading the legs and flattening the body. In this way lizards can attain temperatures much higher than the surrounding air. When a suitable temperature has been reached, further heating is avoided by lightening the skin color and changing the orientation towards a posture more

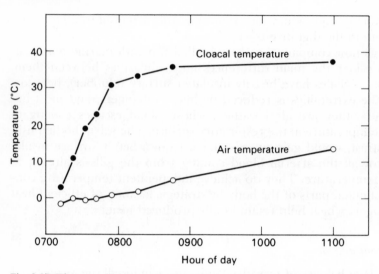

Fig. 8.27. The peruvian mountain lizard *Liolaemus* comes out in the morning while the air temperature is still below freezing, and, by exposing itself to the sun, it rapidly heats up to a body temperature that permits full activity (data from Pearson, 1954.) (After Schmidt-Nielsen, 1964.)

parallel to the sun's rays, and eventually by moving into shade. The temperature of the substratum is also important, for a cool lizard can place itself in close contact with a warm rock and thereby increase the heat gain.

The degree of heating that a lizard can attain by heating up in the sun can be spectacular. It has been reported from the mountains of Peru, at an altitude of 4000 meters where the air temperature is low even in summer, that the lizard *Liolaemus* remains active and comes out at temperatures below freezing. One lizard which weighed 108 g was tethered in the sun in the morning while the air temperature was still at −2 °C. Within the next hour the air temperature increased to +1.5 °C, but the cloacal temperature of the lizard had increased from +2.5 °C to 33 °C (fig. 8.27). The lizard actually had a body temperature of more than 30 degrees above the air, which would permit it to be active at the near-freezing air temperature. Once the lizard is warm, the body temperature remains more or less constant around 35 °C while the air temperature slowly continues to rise. This is achieved by reducing the amount of absorbed solar energy by changing posture, so that the lizard maintains its 'preferred' body temperature, also known as the *eccritic temperature.*

It is, of course, important for the animal to heat up rapidly and become fully active as fast as possible. On the other hand, when the

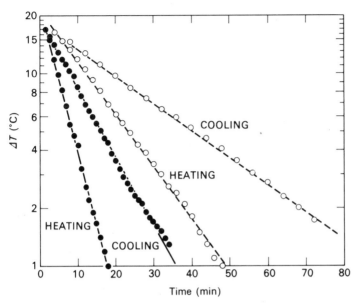

Fig. 8.28. Heating and cooling rates of the Galapagos marine iguana (body weight 652 g) when immersed in water (filled circles) or in air (open circles). The *initial* temperature difference between animal and medium (ΔT) was, in all cases, 20 °C (Bartholomew and Lasiewski, 1965).

animal moves out of the sun, it would be an advantage to retard the inevitable cooling. Can the lizard do anything to increase the rate of heating and decrease the rate of cooling? The heating of the body can be accelerated by increasing the circulation to the heated skin area, but what about the cooling?

This question has been studied in several species of lizards. It is of particular importance to the Galapagos marine iguana, which feeds in water where the temperature is 22 to 27 °C. When it is out of water the animal sits on the rocks near the surf where it can warm itself in the sun. When it enters the water to feed on the seaweed, its temperature begins to fall toward that of the water, which is below the eccritic temperature. It would be advantageous for the marine iguana if it could reduce its rate of cooling while in water, and thus extend the period during which it can feed actively and move fast enough to escape predators. Conversely, after leaving the sea, it would be advantageous to heat up as rapidly as possible.

A comparison of the rates of heating and cooling for the marine iguana shows that heating rates are about twice as high as cooling rates, both in air and in water (fig. 8.28). When a cold animal is immersed in water which is 20 degrees above body temperature, the heating is twice as fast as the cooling when a warm animal is immersed in water 20 degrees below the body temperature. In air both rates are lower, but again, heating is about twice as fast as cooling at the same temperature differential.

The different rates of heating and cooling can best be explained by changes in circulation, especially in the skin. During heating the heart rates are high and increase with body temperature, but in an animal being cooled the heart rate drops off to half of what it otherwise is at the same body temperature. This ability to regulate the rates of heating and cooling through adjustment of the circulation is an obvious advantage for an animal that seeks its food by periodically entering cool water.

Not only lizards but other reptiles habitually bask in the sun. Both tortoises and snakes, and even the most aquatic reptiles, crocodiles, can often be seen sunning themselves at the edge of the water. How important this is for heating and temperature regulation is unknown for most of them.

Internal heat production, i.e. metabolic rate, can also be used to attain a high body temperature. It has been reported that some female snakes do this while incubating their eggs. Observations on this were made in the New York Zoological Park where a 2.7 meter long female python started incubating a clutch of 23 eggs. It coiled itself around the eggs, and temperatures were measured with thermocouples placed between the tightly appressed coils of the snake. As the air temperature was lowered from 33 °C to 25 °C, the animal increased its rate of oxygen consumption and kept its body temperature at 4 and 5 °C above the air (fig. 8.29). The increase in oxygen consumption seemed to be caused by strong contractions of the body musculature, reminiscent of the shivering of mammals. At 25.5 °C the oxygen consumption of the brooding snake was 9.3 times higher than during the non-brooding period. When the temperature was decreased further to 21 °C, the animal seemed unable to increase its metabolic rate further, for its temperature fell and there was a drop in the oxygen consumption. After 30 days of incubation the animal had decreased in weight from 14.3 kg to 10.3 kg, a decrease of nearly 30%, supposedly due to the fuel consumption needed to keep warm (Hutchison, Dowling and Vinegar, 1966).

Flying insects

Most insects become increasingly sluggish and are unable to fly at low temperatures. Some insects, however, can warm up their flight muscles and be active in quite cold air. The flight muscles are located

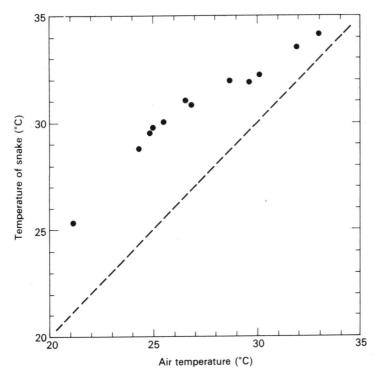

Fig. 8.29. A brooding female python while incubating a clutch of eggs keeps her body temperature above that of the surroundings by strong muscular contractions reminiscent of shivering in mammals (Vinegar, Hutchison and Dowling, 1970).

in the thorax, and can produce large amounts of heat by a process similar to shivering in vertebrates. Such heating of the muscles before take-off occurs mainly in large insects such as locusts, large moths, butterflies, and bumblebees, and also in wasps and bees which are strong and rapid fliers.

One of the most carefully studied insects in this regard is the sphinx moth, *Manduca sexta* (whose larva is the tobacco hornworm). The adult sphinx moths weigh between 2 and 3 g and feed on nectar while hovering in front of a flower, much like a hummingbird. Since they are nocturnal, they often fly at quite low air temperatures and flight would be impossible unless they could heat themselves. Hovering flight requires a muscle temperature of at least 35 °C, but nevertheless, sphinx moths can fly and feed when the air temperature is as low as 10 °C. This is possible through pre-flight heating of the thorax. The lower the initial temperature, the lower is the rate of heating (fig. 8.30). During the warm-up the animal seems to shiver

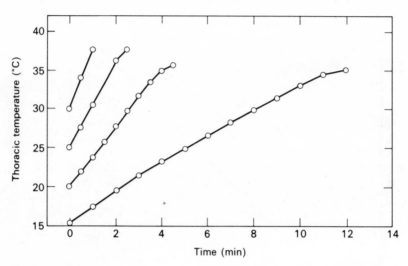

Fig. 8.30. The increase in thorax temperature of the sphinx moth (*Manduca sexta*) proceeds more rapidly when the initial temperature of the animal is higher (after Heinrich and Bartholomew, 1971).

as the wings vibrate lightly. The heat is generated by contraction of the wing muscles, and the reason that the wings do not flap is that the muscles for upstroke and downstroke contract simultaneously, rather than alternately as in flight. We might expect that the warm-up should progress faster and faster as the temperature of the moth increases, for more heat can be produced at higher temperatures. However, as the insect becomes warmer, the rate of heat loss to the air also increases, and for this reason the temperature curves in fig. 8.30 appear as more or less straight lines.

The thorax of the moth is covered by long, furry scales. This helps retain the generated heat in the thorax during the warm-up. In free flight the sphinx moth maintains its thoracic temperature at about 40–42 °C over a wide range of air temperatures, but due to the high rate of heat production in the flight muscles it must be able to increase the rate of heat loss to keep the temperature from rising even higher. This is achieved through an increased circulation of blood between the thorax and the abdomen. Heat loss from the abdomen is facilitated by its larger surface and a smaller amount of surface insulation. If we tie off the large dorsal blood vessel leading from the heart (located in the abdomen) so that the flight muscles cannot be cooled, the moth rapidly overheats and cannot continue to fly even at an air temperature of 23 °C. The cause could, of course, be that the blood supply to the muscles is necessary for supplying fuel. However, a simple experiment can answer this objection. If the scales are removed from the thorax of a moth with ligated circulation, the heat loss is sufficiently improved to enable the animal to fly again (Heinrich, 1970).

Honeybees regulate not only their own individual temperature, but that of the entire colony as well. In summer the temperature of a bee-hive is maintained at about 35 °C and fluctuates very little. This provides the colony with an optimum temperature for development of the brood. Rapid reproduction permits a fast build-up of the population during the most favorable season, and an optimal temperature for the brood is therefore ecologically highly desirable. If the bee-hive temperature rises above the regulated level, the bees spread droplets of water over the comb, and evaporation is aided by air currents created by a large number of bees fanning their wings. Below 30 °C bees do not usually consume any water, but at higher temperatures their water intake increases enormously as water is used for cooling (Free and Spencer-Booth, 1958).

At low outside temperatures the bees cluster together and maintain a temperature within the cluster of about 20 to 30 °C. The temperature at the center of the cluster is the highest, and it appears that the bees at the periphery force their way into the center where it is warmer, thus exposing others to the outside. The core temperature of the cluster is regulated between 18 and 32 °C at outside air temperatures ranging as low as − 17 °C. In fact, there is a tendency for the cluster temperature to be higher at the very lowest ambient temperatures. For each 5 °C drop in the outside air temperature, the cluster temperature increases by about 1 °C (Southwick and Mugaas, 1971). In this way a cluster of bees behaves much like a single large organism which maintains a well-regulated high temperature at external temperature conditions which would rapidly kill individual bees.

REFERENCES

ADOLPH, E. F. (1947). Tolerance to heat and dehydration in several species of mammals. *Am. J. Physiol.*, **151**, 564–75.

ASCHOFF, J. and WEVER, R. (1958). Kern und Schale im Wärmehaushalt des Menschen. *Die Naturwissenschaften*, **45**, 477–85.

ASCHOFF, J. and WEVER, R. (1959). Wärmeaustausch mit Hilfe des Kreislaufes. *Deutsche Medizinische Wochenschrift*, **84**, 1509–17.

ASCHOFF, J., GÜNTHER, B. and KRAMER, K. (1971). *Energiehaushalt und Temperaturregulation*, München: Urban and Schwarzenberg, 196 pp.

AUGEE, M. L. and EALEY, E.H.M. (1968). Torpor in the echidna, *Tachyglossus aculeatus*. *J. Mammal.*, **49**, 446–54.

BARTHOLOMEW, G. A. and HUDSON, J. W. (1962). Hibernation, estivation, temperature regulation, evaporative water loss, and heart rate of the pigmy possum, *Cercaertus nanus. Physiol. Zool.,* **35,** 94–107.

BARTHOLOMEW, G. A. and LASIEWSKI, R. C. (1965). Heating and cooling rates, heart rate and simulated diving in the Galapagos marine iguana. *Comp. Biochem. Physiol.* **16,** 575–82.

BARTHOLOMEW, G. A., LASIEWSKI, R. C. and CRAWFORD, E. C., JR. (1968). Patterns of panting and gular flutter in cormorants, pelicans, owls, and doves. *Condor,* **70,** 31–4.

BLIGH, J., INGRAM, D. L., KEYNES, R. D. and ROBINSON, S. G. (1965). The deep body temperature of an unrestrained Welsh mountain sheep recorded by a radiotelemetric technique during a 12-month period. *J. Physiol.,* **176,** 136–44.

BUETTNER-JANUSCH, J. (1966). *Origins of Man,* New York: John Wiley & Sons, Inc., 674 pp.

CALDER, W. A. (1964). Gaseous metabolism and water relations of the zebra finch, *Taeniopygia castanotis. Physiol. Zool.,* **37,** 400–13.

CALDER, W. A. and SCHMIDT-NIELSEN, K. (1967). Temperature regulation and evaporation in the pigeon and the roadrunner. *Am. J. Physiol.,* **213,** 883–9.

CAREY, F. G. and TEAL, J. M. (1966). Heat conservation in tuna fish muscle. *Proc. Nat. Acad. Sci.,* **56,** 1464–9.

CAREY, F. G. and TEAL, J. M. (1969). Mako and porbeagle: warm-bodied sharks. *Comp. Biochem. Physiol.,* **28,** 199–204.

CAREY, F. G., TEAL, J. M., KANWISHER, J. W. and LAWSON, K. D. (1971). Warm-bodied fish. *Am. Zool.,* **11,** 137–45.

CRAWFORD, E. C., JR. (1962). Mechanical aspects of panting in dogs. *J. Appl. Physiol.,* **17,** 249–51.

CRAWFORD, E. C., JR. (1972). Brain and body temperatures in a panting lizard. *Science,* **177,** 431–3.

CRAWFORD, E. C., JR. and KAMPE, G. (1971). Resonant panting in pigeons. *Comp. Biochem. Physiol.,* **40A,** 549–52.

COUNCIL OF THE ROYAL SOCIETY (1971). *Quantities, Units, and Symbols.* London: The Royal Society. 48 pp.

DAWSON, W. R. (1954). Temperature regulation and water requirements of the brown and abert towhees, *Pipilo fuscus* and *Pipilo aberti. Univ. Calif. Publ. Zool.,* **59,** 81–124.

DAWSON, T. J. and HULBERT, A. J. (1970). Standard metabolism, body temperature, and surface areas of Australian marsupials. *Am. J. Physiol.,* **218,** 1233–8.

DAWSON, T. and SCHMIDT-NIELSEN, K. (1966). Effect of thermal conductance on water economy in the antelope jack rabbit, *Lepus alleni. J. Cell. Physiol.,* **67,** 463–72.

FARNER, D. S. (1956). The body temperatures of the North Island kiwis. *Emu,* **56,** 199–206.

FREE, J. B. and SPENCER-BOOTH, Y. (1958). Observations on the temperature regulation and food consumption of honeybees (*Apis mellifera*). *J. Exp. Biol.,* **35,** 930–7.

HAINSWORTH, F. R. and WOLF, L. L. (1970). Regulation of oxygen consumption and body temperature during torpor in a hummingbird, *Eulampis jugularis*. *Science,* **168,** 368–9.

HALES. J. R. S. and BLIGH, J. (1969). Respiratory responses of the conscious dog to severe heat stress. *Experientia,* **25,** 818–19.

HALES, J. R. S. and FINDLAY, J. D. (1968). Respiration of the ox: normal values and the effects of exposure to hot environments. *Respir. Physiol.,* **4,** 333–52.

HAMMEL, H. T. (1955). Thermal properties of fur. *Am. J. Physiol.,* **182,** 369–76.

HARDY, J. D. (1949). Heat transfer. In *Physiology of Heat Regulation and the Science of Clothing* (L. H. Newburgh, ed.), pp. 78–108, Philadelphia: W. B. Saunders Co.

HART, J. S. (1951). Calorimetric determination of average body temperature of small mammals and its variation with environmental conditions. *Can. J. Zool.,* **29,** 224–33.

HART, J. S. (1956). Seasonal changes in insulation of the fur. *Can. J. Zool.,* **34,** 53–7.

HART, J. S. and IRVING, L. (1959). The energetics of harbor seals in air and in water with special consideration of seasonal changes. *Can. J. Zool.* **37,** 447–57.

HEINRICH, B. (1970). Thoracic temperature stabilization by blood circulation in a free-flying moth. *Science,* **168,** 580–2.

HEINRICH, B. and BARTHOLOMEW, G. A. (1971). An analysis of pre-flight warm-up in the sphinx moth, *Manduca sexta. J. Exp. Biol.,* **55,** 223–39.

HENSEL, H. and BOCK, K. D. (1955). Durchblutung und Warmeleitfähigkeit des menschlichen Muskels. *Pflügers Archiv,* **260,** 361–7.

HERREID, C. F. II and SCHMIDT-NIELSEN, K. (1966). Oxygen consumption, temperature, and water loss in bats from different environments. *Am. J. Physiol.,* **211,** 1108–12.

HUTCHISON, V. H., DOWLING, H. G. and VINEGAR, A. (1966). Thermoregulation in a brooding female Indian python, *Python molurus bivittatus. Science,* **151,** 694–6.

IRVING, L. (1969). Temperature regulation in marine mammals. In *The Biology of Marine Mammals* (H. T. Andersen, ed.), pp. 147–74, New York: Academic Press.

IRVING, L. and HART, J. S. (1957). The metabolism and insulation of seals as bare-skinned mammals in cold water. *Can. J. Zool.,* **35,** 497–511.

IRVING, L. and KROG, J. (1954). Body temperatures of arctic and subarctic birds an mammals. *J. Appl. Physiol.,* **6,** 667–80.

IRVING, L. and KROG, J. (1955). Temperature of skin in the arctic as a regulator of heat. *J. Appl. Physiol.,* **7,** 355–64.

JOHANSEN, K. (1961). Distribution of blood in the arousing hibernator. *Acta Physiol. Scand.,* **52,** 379–86.

JOHNSTON, D. W. (1971). The absence of brown adipose tissue in birds. *Comp. Biochem. Physiol.,* **40A,** 1107–8.

LYMAN, C. P. (1948). The oxygen consumption and temperature regulation of hibernating hamsters. *J. Exp. Zool.,* **109,** 55–78.

MCNAB, B. K. (1966). An analysis of the body temperatures of birds. *Condor,* **68,** 47–55.

MORRISON, P. R. and RYSER, F. A. (1952). Weight and body temperature in mammals. *Science,* **116,** 231–2.

ØRITSLAND, N. A. (1970). Temperature regulation of the polar bear (*Thalarctos maritimus*). *Comp. Biochem. Physiol.,* **37,** 225–33.

PEARSON, O. P. (1954). Habits of the lizard *Liolaemus multiformis multiformis* at high altitudes in Southern Peru. *Copeia, 1954*(2), 111–16.

ROBINSON, K. W. and LEE, D.H.K. (1946). Animal behaviour and heat regulation in hot atmospheres. *Univ. Queensland Papers, Dept. of Physiol.,* **1,** 1–8.

ROBINSON, K. W. and MORRISON, P. R. (1957). The reaction to hot atmospheres of various species of Australian marsupial and placental animals. *J. Cell. Comp. Physiol.,* **49,** 455–78.

SCHMIDT-NIELSEN, K. (1963). Osmotic regulation in higher vertebrates. *Harvey Lectures,* Ser. **58,** 53–93.

SCHMIDT-NIELSEN, K. (1964). *Desert Animals. Physiological Problems of Heat and Water,* New York: Oxford Univ. Press, 277 pp.

SCHMIDT-NIELSEN, K., DAWSON, T. J. and CRAWFORD, E. C., JR. (1966). Temperature regulation in the echidna (*Tachyglossus aculeatus*). *J. Cell. Physiol.,* **67,** 63–72.

SCHMIDT-NIELSEN, K., SCHMIDT-NIELSEN, B., JARNUM, S. A. and HOUPT, T. R. (1957). Body temperature of the camel and its relation to water encomy. *Am. J. Physiol.,* **188,** 103–12.

SCHOLANDER, P. F. (1955). Evolution of climatic adaptation in homeotherms. *Evolution,* **9,** 15–26.

SCHOLANDER, P. F. and KROG, J. (1957). Countercurrent heat exchange and vascular bundles in sloths. *J. Appl. Physiol.,* **10,** 405–11.

SCHOLANDER, P. F. and SCHEVILL, W. E. (1955). Counter-current vascular heat exchange in the fins of whales. *J. Appl. Physiol.,* **8,** 279–82.

SCHOLANDER, P. F., WALTERS, V., HOCK, R. and IRVING, L. (1950). Body insulation of some arctic and tropical mammals and birds. *Biol. Bull.,* **99,** 225–36.

SCHOLANDER, P. F., HOCK, R., WALTERS, V., JOHNSON, F. and IRVING, L. (1950). Heat regulation in some arctic and tropical mammals and birds. *Biol. Bull.,* **99,** 237–58.

SOUTHWICK, E. E. and MUGAAS, J. N. (1971). A hypothetical homeotherm: the honeybee hive. *Comp. Biochem. Physiol.,* **40A,** 935–44.

SUCKLING, J. A., SUCKLING, E. E. and WALKER, A. (1969). Suggested function of the vascular bundles in the limbs of *Perodicticus potto. Nature, Lond.,* **221,** 379–80.

TAYLOR, C. R. and LYMAN, C. P. (1972). Heat storage in running antelopes: independence of brain and body temperatures. *Am. J. Physiol.,* **222,** 114–17.

TUCKER, V. A. (1965*a*). Oxygen consumption, thermal conductance, and torpor in the California pocket mouse *Perognathus californicus. J. Cell. Comp. Physiol.,* **65,** 393–403.

TUCKER, V. A. (1965*b*). The relation between the torpor cycle and heat exchange in the California pocket mouse *Perognathus californicus. J. Cell. Comp. Physiol.,* **65,** 405–14.

TUCKER, V. A. (1966). Diurnal torpor and its relation to food consumption and weight changes in the California pocket mouse *Perognathus californicus. Ecology,* **47,** 245–52.

VINEGAR, A., HUTCHISON, V. H., and DOWLING, H. G. (1970). Metabolism, energetics, and thermoregulation during brooding of snakes of the genus *Python* (Reptilia, Boidae). *Zoologica,* **55,** 19–50.

WEAST, R. C. (1969). *Handbook of Chemistry and Physics,* 50th edn, Cleveland, Ohio: Chemical Rubber Co.

USEFUL REFERENCE MATERIAL FOR PART III

FISHER, K. C., DAWE, A. R., LYMAN, C. P., SCHÖNBAUM, E. and SOUTH, F. E., JR. (eds) (1967). *Mammalian Hibernation*, vol. III, Edinburgh, Scotland: Oliver & Boyd, 535 pp.

GATES, D. M. (1962) *Energy Exchange in the Biosphere*, New York: Harper & Row, 151 pp.

HARDY, J. D. GAGGE, A. P. and STOLWIJK, J.A.J. (eds) (1970). *Physiological and Behavioral Temperature Regulation*, Springfield, Ill.: Charles C. Thomas Publ., 944 pp.

MALOIY, G.M.O. (ed.) (1972). *Comparative Physiology of Desert Animals. Symp. Zool. Soc. Lond.*, **31**, London: Academic Press, 413 pp.

NEWBURGH, L. H. (ed.) (1968). *Physiology of Heat Regulation and the Science of Clothing* (reprint edn), Darien, Conn.: Hafner Publ. Co.

SCHMIDT-NIELSEN, K. (1964). *Desert Animals. Physiological Problems of Heat and Water*, New York: Oxford University Press, 277 pp.

WHITTOW, G. C. (ed.) (1970). *Comparative Physiology of Thermoregulation*, vol. I, *Invertebrates and Nonmammalian Vertebrates*, New York: Academic Press, 333 pp.

WHITTOW, G. C. (ed.) (1971). *Comparative Physiology of Thermoregulation*. vol. II, *Mammals*, New York: Academic Press, 410 pp.

WHITTOW, G. C. (ed.) (1973). *Comparative Physiology of Thermoregulation*. vol. III, *Special Aspects of Thermoregulation*, New York: Academic Press, 278 pp.

IV

WATER

9
Water and osmotic regulation

In a very crude way the living organism can be described as an aqueous solution contained within the body surface. Both the volume of the organism and the concentration of solutes normally remain constant within rather narrow limits. Substantial deviation from the usual conditions are harmful or fatal, i.e. an animal should remain in a steady state with regard to water and solutes.

The problems of keeping water and solute concentrations constant vary with the environment, for an animal meets entirely different osmotic problems in sea water, in fresh water, and on land. It is therefore convenient to treat the environments separately, analyze the main physiological problems in each of them, and survey the solutions arrived at by different animals.

In this chapter we will first be concerned with aquatic animals and their problems, and afterwards take up terrestrial animals for discussion.

AQUATIC ANIMALS

The aquatic environment

Before discussing the physiological problems peculiar to an environment, it is helpful to be familiar with its most important physical and chemical characteristics.

More than two-thirds (71%) of the earth's surface is covered with water. Most of this is ocean, the total fresh water in lakes and rivers makes up less than 1% of the area and 0.01% of the volume of sea water (Hutchinson, 1957; 1967; Sverdrup, Johnson and Fleming, 1942). On land, life exists in a thin film on, just below, and just above the surface; in water, organisms live not only along the solid bottom but extend throughout the water masses to the greatest depths of the oceans, in excess of 10 000 meters.

All water contains dissolved substances, salts, gases, minor amounts of organic compounds, various pollutants, etc. and furthermore, the temperature of the water is of the greatest physiological importance. In this chapter we are concerned primarily with dissolved salts; the role of dissolved gases was discussed in chapter 1 and temperature in chapters 7 and 8. A brief summary of the properties of solutions, osmotic pressure, and related matters is found in appendix 5.

Sea water contains about 3.5% salt, i.e. one liter sea water contains 35 g salt.* The major ions are sodium and chloride, with magnesium, sulfate, and calcium present in substantial amounts (see table 9.1). The total salt concentration varies somewhat with the geographical location, the Mediterranean, for example, has a salt content of nearly 4% because the high evaporation is not balanced by an equal inflow of fresh water from rivers. In other areas, particularly in coastal regions, the salt content is somewhat lower than in the open ocean, but the relative amounts of the dissolved ions remain very nearly constant in the proportions given in table 9.1.

Table 9.1. *Composition of sea water. In addition to the ions listed, sea water contains small amounts of virtually all elements found on earth.* (Potts and Parry, 1964)

	In 1 liter sea water		In 1 kg H_2O	
ION	(mmol)	(g)	(mmol)	(g)
Sodium	470.2	10.813	475.4	10.933
Magnesium	53.57	1.303	54.17	1.317
Calcium	10.23	0.410	10.34	0.414
Potassium	9.96	0.389	10.07	0.394
Chloride	548.3	19.440	554.4	19.658
Sulfate	28.25	2.713	28.56	2.744
Bicarbonate	2.34	0.143	2.37	0.145

Fresh water, in contrast to sea water, has a highly variable content of solutes. Minute amounts of salts are present already in rainwater, but its composition is greatly modified as the water runs over and through the surface of the earth. The salts in rainwater are derived from the sea; droplets of ocean spray evaporate and salt particles are

* It is not feasible to determine the exact amount of dissolved salts by evaporating the water and weighing the residue. The reason is that the drying conditions influence both the amount of water included as water of crystallization and the weight of the bicarbonate and carbonate salts.

carried by air currents, often far inland, and are deposited with the rain. If the water runs over hard, insoluble rock such as granite, it dissolves little additional material and is called 'soft'. If, on the other hand, it percolates over and through porous limestone, it can dissolve relatively large amounts of calcium salts and is called 'hard'. The total salt content in fresh water may vary from less than 0.1 mmol liter^{-1} to more than 10 mmol liter^{-1}, and the relative amounts of various ions can vary over a tremendous range. This is of physiological significance, especially if magnesium and sulfate are the major ions. The composition of various types of 'hard' and 'soft' waters are given in table 9.2.

Table 9.2. *Typical composition of soft water, hard water, and inland saline water, given in mmol per kg H$_2$O and listed in same order as in table 9.1.* (Recalculated from Livingstone, 1963)

	Soft lake water (a)	River water (b)	Hard river water (c)	Saline water (d)	Dead Sea (e)
Sodium	0.17	0.39	6.13	640	840
Magnesium	0.15	0.21	0.66	6	2302
Calcium	0.22	0.52	5.01	32	583
Potassium	—	0.04	0.11	16	152
Chloride	0.03	0.23	13.44	630	6662
Sulfate	0.09	0.21	1.40	54	8.4
Bicarbonate	0.43	1.11	1.39	3	trace

(a) Lake Nipissing, Ontario.
(b) Mean composition of North American rivers.
(c) Tuscarawas River, Ohio.
(d) Bad Water, Death Valley, California.
(e) Dead Sea, Israel. This water also contains 118 mmol per kg H$_2$O of bromide.

Some inland water has a very high salt content. The Great Salt Lake in Utah is saturated with NaCl, which crystallizes out on the shore. The Dead Sea in Israel is likewise saturated, but the predominant ions are magnesium and chloride with calcium sulfate crystallizing out. In the Great Salt Lake no fish can live, but some animals thrive, the brine shrimp (*Artemia*), for example. The saturated salt solution of the Dead Sea has a different composition, it harbors no higher animal or plant life and only micro-organisms survive. Some springs may have a high and unusual salt content, but such habitats, although interesting, are of minor importance compared to the larger bodies of fresh and sea water.

Brackish water occurs in coastal regions where sea water is mixed with fresh water. At the mouth of large rivers fresh water dilutes the ocean water for a considerable distance, and if the tides are large the estuary at the mouth may extend far up the river. In this area the salinity varies rapidly with the tides, often from nearly fresh to nearly undiluted sea water. In a larger enclosed area, such as the Baltic Sea, the situation is different, the salinity at the west coast of Sweden is some 3%, and gradually declines to less than 0.5% in the northernmost part of the Baltic. In this large area there is a relatively stable geographical gradient in the salinity with an almost imperceptible change from brackish to fresh water. It is difficult to say exactly where sea water becomes brackish, and at the other extreme, where a very dilute brackish water for practical purposes is fresh. As a commonly accepted definition we can say that brackish water refers to salinities between 3.0% and 0.05%. Brackish water is physiologically extremely important, it forms a barrier to the distribution of many marine animals on one hand as well as fresh-water animals on the other, and it also forms an interesting transition between marine and fresh-water habitats. In geographical extent, however, brackish water covers less than 1% of the earth's surface.

Definitions. Some aquatic animals can tolerate wide variations in the salt concentration of the water in which they live, they are called *euryhaline* animals (Greek *eurys* = wide, broad, and *halos* = salt). Other animals have a limited tolerance to variations in the concentration of the medium, they are called *stenohaline* (Greek *stenos* = narrow, close). A marine animal that can penetrate into brackish water and survive is euryhaline, an extremely euryhaline animal may even be able to tolerate shorter or longer periods in fresh water. The term euryhaline is used also for fresh-water animals that can withstand considerable increases in the salt content of the water. A stenohaline organism, whether marine or fresh water, on the other hand, can tolerate only a narrow range of changes in the concentration of the water.

There is no sharp separation between euryhaline and stenohaline animals, and there is no commonly accepted definition which places a given animal in one or the other of these groups.

Most marine invertebrates have body fluids with the same osmotic pressure as the sea water, they are *iso-osmotic* or *isosmotic* with the medium (Greek *isos* = equal). When there is a change in the concentration of the medium, an animal may respond in one of two ways. The osmotic concentration of the body fluids of the animal may change with that of the medium, the animal thus remaining isosmotic with the medium. Such an animal is an *osmoconformer*. If, on the other hand, the animal can maintain or regulate its osmotic concentration in spite of external concentration changes, the animal is called an *osmoregulator*. For example, a marine crab that maintains a high concentration in its body fluids after it has been moved to dilute brackish water is a typical osmoregulator.

The concentrations of the various individual solutes in the body fluids of an animal usually differ substantially from those in the medium, even if the animal is isosmotic with the medium. The differences are usually carefully regulated, a subject dealt with under the term *ionic regulation*. Some degree of ionic regulation seems to occur in all living organisms, both in osmoregulators and in osmoconformers.

Fresh-water animals have body fluids which are osmotically more concentrated than the medium, these animals are *hyperosmotic*. If an animal has a lower osmotic concentration than the medium, such as a marine teleost fish, it is said to be *hypo-osmotic* or *hyposmotic*.

The term *isotonic* is used in a different sense. We say that a living cell is isotonic with a given solution if the cell neither swells nor shrinks in the solution. For example, if mammalian red blood cells are suspended in a sodium chloride solution of 150 mmol liter^{-1} (about 0.9%), the cells retain their size, shape, and volume. If, on the other hand, they are suspended in an isosmotic solution of urea (0.3 mol liter^{-1}), they rapidly swell and burst. The urea solution, although isosmotic, is not isotonic. The solute, urea, rapidly penetrates the red cell membrane so that the urea concentration inside and outside the cell is equal, but the electrolytes do not move out of the cell, which behaves as if it were suspended in distilled water. Because of the osmotic concentration differences, water flows into the cell, which swells and bursts. While *isosmotic* is defined in terms of physical chemistry, *isotonic* is a descriptive word based on the behavior of cells in a given solution.

Invertebrates

The marine environment

Most marine invertebrates have an osmotic concentration in their body fluids which equals that of the surrounding sea water, they are *osmoconformers*. From the osmotic viewpoint, this eliminates one major physiological difficulty, they do not have to cope with the problem of osmotic movement of *water*.

Although as a rule marine invertebrates are osmoconformers, this does not mean that their body fluids have a *solute* composition identical to sea water. On the contrary, there are characteristic differences which the animals must maintain, and this requires extensive regulation of ionic concentrations.

The concentration of the most important ions in the blood of some invertebrates is given in table 9.3 (Potts and Parry, 1964). Some of the concentrations are similar to the concentrations in sea water, but others differ substantially. For example, several invertebrates have magnesium present in the same concentration as in sea water, but others have a much lower concentration of this ion. The same holds true for sulfate, which in some animals is similar to and in others substantially lower than in sea water. Such differences can only be maintained if the body surface, including the thin surface membrane of the gills, is relatively impermeable to the ion in question. Some amounts of these ions will enter anyway, for no animal is completely impermeable and any ingested food will also contain some solutes. Therefore the animals must have a mechanism for elimination of some ions, while others are maintained at a higher level than in the water. The regulated elimination of solutes is one major function of excretory organs, such as the kidney.

Several of the animals in table 9.3 regulate their concentration of sulfate at less than half of that in sea water. Such great differences obviously indicate both exclusion and elimination of this ion, in other words, an active ion regulation. If the concentration in the animal differs only slightly from that in sea water, however, it is

Table 9.3. *Concentration of common ions (mmol per kg water) in some marine animals* (Potts and Parry, 1964)

	Na	Mg	Ca	K	Cl	SO₄	Protein (g liter⁻¹)
Sea water	478.3	54.5	10.5	10.1	558.4	28.8	—
Jellyfish (*Aurelia*)	474	53.0	10.0	10.7	580	15.8	0.7
Polychaete (*Aphrodite*)	476	54.6	10.5	10.5	557	26.5	0.2
Sea-urchin (*Echinus*)	474	53.5	10.6	10.1	557	28.7	0.3
Mussel (*Mytilus*)	474	52.6	11.9	12.0	553	28.9	1.6
Squid (*Loligo*)	456	55.4	10.6	22.2	578	8.1	150
Isopod (*Ligia*)	566	20.2	34.9	13.3	629	4.0	—
Crab (*Maia*)	488	44.1	13.6	12.4	554	14.5	—
Shore crab (*Carcinus*)	531	19.5	13.3	12.3	557	16.5	60
Norwegian lobster (*Nephrops*)	541	9.3	11.9	7.8	552	19.8	33
Hagfish (*Myxine*)	537	18.0	5.9	9.1	542	6.3	67

more difficult to say whether this is due to a regulation. In particular, proteins have a considerable influence on the distribution of the ions across a semi-permeable membrane (known as the Donnan effect, see appendix 5). A difference in ion concentrations therefore does not necessarily indicate an active regulation of the ion in question.

It is difficult to evaluate accurately the role of proteins and the Donnan effect on the various ions, but this has been circumvented by the British investigator J. D. Robertson (1957) in a simple way. The sample is put in a semi-permeable cellophane bag which is then placed in sea water. Since salts and water can pass through the cellophane but proteins cannot, the ion concentrations inside the bag will, at equilibrium, differ somewhat from those in the sea water, the differences being caused by the Donnan effect of the proteins. This procedure is called *dialysis*. The concentrations now found inside the bag serve as a baseline, and the concentration of each ion as found in the living animal is expressed as a percentage of this baseline. If the observed value differs appreciably from the concentration reached passively by dialysis, it must be due to an active regulation of this particular ion. Some results of such experiments are listed in table 9.4.

Table 9.4. *Ionic regulation in some marine invertebrates. Concentrations found in plasma or coelomic fluid expressed as percentage of concentration in body fluid dialyzed against sea water.* (Robertson, 1957)

		Na (%)	Mg (%)	Ca (%)	K (%)	Cl (%)	SO₄ (%)
Coelenterates	*Aurelia aurita*	99	97	96	106	104	47
Echinoderms	*Marthasterias glacialis*	100	98	101	111	101	100
Tunicates	*Salpa maxima*	100	95	96	113	102	65
Annelids	*Arenicola marina*	100	100	100	104	100	92
Sipunculids	*Phascolosoma vulgare*	104	69	104	110	99	91
Arthropods	*Maia squinado*	100	81	122	125	102	66
	Dromia vulgaris	97	99	84	120	103	53
	Carcinus maenas	110	34	108	118	104	61
	Pachygrapsus marmoratus *	94	24	92	95	87	46
	Nephrops norvegicus	113	17	124	77	99	69
Molluscs	*Pecten maximus*	100	97	103	130	100	97
	Neptunea antiqua	101	101	102	114	101	98
	Sepia officinalis	93	98	91	205	105	22

* This grapsoid crab is the only animal in the table which is hypo-osmotic to sea water (ionic concentration 86% that of sea water).

An examination of the table reveals that echinoderms show no significant ionic regulation of any ion. The coelenterate *Aurelia,* a jelly fish, shows little or no regulation for most of the ions, with the exception of sulfate which is kept considerably below the concentration in sea water. In this animal the low sulfate is directly related to a problem of buoyancy, exclusion of the heavy sulfate ion reduces the density of the jellyfish which thus is kept from sinking (see page 539).

Among the arthropods in table 9.4 we find the interesting feature that the magnesium level in the plasma is low in those that have the ability to move quickly. The crab *Pachygrapsus* and the lobster-like *Nephrops* are active and fast moving animals; the spider crab *Maia,* on the other hand, is slow moving and has a high magnesium concentration. Magnesium is an anesthetic which depresses neuromuscular transmission, and one might hastily conclude that a high magnesium concentration is related to a low level of activity of these crustaceans. The cuttlefish *Sepia,* on the other hand, moves quickly and swims well, but has a magnesium concentration as high as in the clam *Pecten,* thus making us suspicious of any cause-and-effect correlation between activity and magnesium concentration.

Fresh water and brackish water

If we transfer a variety of marine animals to somewhat dilute sea water, say, 80% of the usual strength, most of them are likely to survive. When, after some time, we examine their body fluids, it turns out that they have adjusted to the dilution and have established new and lower concentrations of the ions in their body fluids. The osmoconformers, starfish and oysters, for example, will have the same osmotic concentration as the dilute medium, although the concentration of individual ions still differs from the diluted sea water. The osmoregulators, on the other hand, resist the dilution more or less successfully and remain hyperosmotic. Thus, marine animals that penetrate into brackish water can be of two types, either passive osmoconformers or active osmoregulators. An osmoconformer such as the oyster may tolerate considerable dilution; it can also to some extent resist the effects of periodic dilution of the water in an estuary by keeping its shells closed. In the long run, however, the active regulators can better resist the fluctuations in the environment.

Let us see how the active regulators manage in dilute sea water. The relationship between the concentrations in the body fluids and the water is shown in fig. 9.1 for a number of osmoregulators. Even the good regulators meet certain limitations. An example is the Eu-

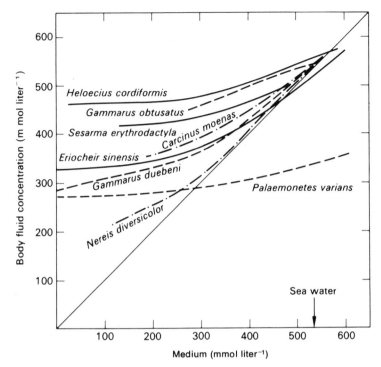

Fig. 9.1. Relation between the concentration of body fluids and medium in various brackish water animals. Full-strength sea water is indicated by an arrow. Diagonal line indicates equal concentrations in body fluid and medium (Beadle, 1943).

ropean shore crab, *Carcinus*, which will not survive in brackish water more dilute than about one-third of normal sea water.*

Another crab, the Chinese mitten crab *Eriocheir*, can tolerate far greater dilutions, in fact, it is able to penetrate into fresh water. To reproduce, the mitten crab must return to the sea, and it therefore cannot establish itself permanently and carry out its complete life cycle in fresh water.

In principle, fresh-water animals behave osmotically in a way similar to the successful osmoregulators in brackish water, but there are great differences in the concentrations at which they maintain their body fluids. A crustacean such as the crayfish *Potamobius* maintains an osmotic concentration of about 500 mosm liter^{-1}, but the fresh-water clam *Anodonta* maintains less than one-fifth of this, only about

* The limit of dilution tolerated by the shore crab, and by many other brackish water animals, varies with the geographical locale. A shore crab from the North Sea is far less tolerant to dilute water than a shore crab from the Baltic Sea, where the salt concentration normally is much lower than in the open sea.

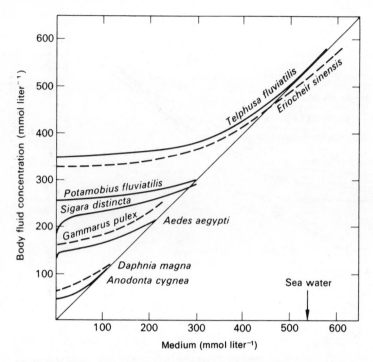

Fig. 9.2. Relation between the concentration of body fluids and medium in various fresh-water animals. Full-strength sea water is indicated by an arrow. Diagonal line indicates equal concentrations in body fluid and medium (Beadle, 1943).

80 mosm liter^{-1} (see fig. 9.2). *Anodonta* still is hyperosmotic, and no fresh-water animal is known that permits its concentration to decrease to the same as the fresh water in which it lives.

Most of the major animal phyla have representatives both in the sea and in fresh water, although the number of species in the sea is much, much greater. No echinoderms are found in fresh water, and likewise, among molluscs the entire class Cephalopoda (octopus and squid) is absent in fresh water.

Mechanism of osmoregulation

An animal that remains hyperosmotic to the medium encounters two physiological problems, solutes tend to be lost to the dilute medium, and water tends to flow into the animal because of the higher osmotic concentration inside. These problems could be reduced in magnitude by making all surfaces highly impermeable, but no completely impermeable animal is known; at least the respiratory surfaces must be thin enough to permit diffusion of gases. Water that enters the animal osmotically can be eliminated as urine, but since urine cannot be made to consist of pure water only, the solute loss through the surface is increased by the urinary loss.

How can the animal compensate for the solute loss? It might be possible to obtain the necessary ions in food, and this might suffice for a highly impermeable animal. However, many fresh-water animals are not particularly impermeable, and yet, even when starved they maintain their normal high concentrations. This is achieved by a direct uptake of ions from the medium.

The simplest way to demonstrate such uptake is to keep an animal, for example a crayfish, in running or frequently changed distilled water. This will gradually deplete the salts in the crayfish so that its concentration is reduced from, say, 500 mosm liter^{-1} to 450, which it tolerates without difficulty. If the crayfish is now returned to ordinary fresh water, its blood concentration increases again, although the fresh water is a hundred times more dilute than the blood (say 5 mosm liter^{-1}).

Since, in this experiment, the crayfish has removed ions from a dilute solution and moved them to a much higher concentration in the blood, the ions have been moved against a concentration gradient and the transport is an 'active' transport.*

The organs responsible for active ion uptake are not always known. It has been assumed that the general body surface of some animals is the organ of uptake, but this is mostly a conclusion based on lack of anatomical differentiation and any obvious recognizable organs that could be tested for this role. In other animals, notably crustaceans and insects, it is probable that the general body covering does not participate in the active transport. There is conclusive evidence that the crustacean gill is an organ of active ion transport, and likewise that certain appendages of aquatic insect larvae, notably the 'anal gills', are organs of ionic regulation. There is no conflict between osmoregulation and respiration, the same organ can serve both functions. This is the case of the gills of fresh-water crabs and crayfish, but the 'anal gills' of insect larvae probably have no respiratory function and are exclusively organs of osmoregulation.

Active transport requires *energy*, and it would be interesting to know the increase in energy requirement as an animal moves to more dilute water. If the shore crab *Carcinus*, which is quite tolerant of brackish water, is moved from sea water to gradually more di-

---

* *Active transport* is defined as a transport against an electrochemical gradient, and as such is an energy-requiring process. It is often referred to as an 'up-hill' transport, as opposed to passive or 'down-hill' diffusion along the concentration gradient.

luted solutions, its rate of oxygen consumption increases appreciably. When placed in sea water of one-quarter full strength, the oxygen consumption of *Carcinus* is increased by about 50%, and it could easily be concluded that this increase is due to the work of active ion transport necessary for the crab to maintain itself in the dilute water. In contrast, the mitten crab *Eriocheir* has the same metabolic rate in sea water, in brackish water, and in fresh water. The mitten crab therefore does not seem to spend measurable extra energy to maintain itself, not even in fresh water. This difference could be explained if the mitten crab were more impermeable than the shore crab, but there is no striking difference in the permeability of the two animals. The approach of determining the change in oxygen consumption therefore does not give reliable information about the energy cost of osmoregulation (Potts and Parry, 1964).

The energy required for osmoregulation can, however, be calculated from thermodynamic considerations. From information about an animal's permeability, its surface area, the concentration in its blood and in the medium, and the urine concentration, Potts calculated the minimum energy requirement for osmoregulation (Potts, 1954). The permeability of the animal determines the water influx, and since the water must be eliminated by the excretory organs, the rate of urine flow is thus determined. Many fresh-water animals have an extremely dilute urine, thus minimizing the loss of solutes, but since osmotic work is also required to make the urine dilute, relative to the body fluids from which it is formed, this item should be added to the cost of ion uptake from the external medium. In some animals in fresh water, the mitten crab *Eriocheir* is an example, the urine has the same osmotic concentration as the blood. In this case no osmotic work is required to form the urine, but on the other hand, a larger amount of solute is lost in the urine, thus increasing the amount that must be taken from the water by active transport, thus also increasing the energy requirements.

The minimum work required for osmoregulation has been calculated for several animals about which the necessary information is available for permeability, blood concentration, and urine concentration. The results indicate that the total energy requirement for osmoregulation is a small fraction of the normal oxygen consumption of the animal. For the crayfish *Potamobius*, the requirement is approximately 0.3% of its total metabolic rate, for the mitten crab *Eriocheir* in fresh water about 0.5% of the total metabolic rate, and for the fresh-water clam *Anodonta*, 1.2% of the total metabolic rate (Potts, 1954). The energy required for osmoregulation is therefore such a small fraction of the total metabolic rate that it would be difficult to measure on a living animal, for the metabolic rate normally fluctuates much more than a few per cent. We must assume that the much greater increases in metabolic rate observed as animals are moved to dilute media are not directly caused by the increased energy requirement for osmoregulation.

Saline habitats, hyporegulation

Two of the animals in fig. 9.1, the shrimps *Palaemonetes* and *Leander*, differ in one important way from all the others. In normal, full strength sea water they are hypotonic, i.e. their body fluids are osmotically more dilute than the medium, and this must require active osmoregulation. Hyporegulation is very uncommon for a marine invertebrate, and it is generally assumed that these two shrimps belong to a group that is at home in fresh water, and that they have secondarily invaded the sea while maintaining a lower concentration than common in sea water.

Some water is far more concentrated than sea water, and in such extreme environments hyporegulation is of greater importance. The best known example is probably the brine shrimp, *Artemia*, which is found in tremendous numbers in many salt lakes, and in coastal evaporation ponds where salt is obtained by evaporation of sea water.

While the brine shrimp cannot survive in fresh water, it can adapt to media which vary from about one-tenth sea water to crystallizing brine, which contains about 300 g salt liter^{-1}. In dilute sea water, *Artemia* is hypertonic to its medium and behaves like a brackish water organism. At higher concentrations, *Artemia* is an excellent hyporegulator which in concentrated brine still maintains an osmotic pressure of the body fluid of not much more than one-tenth of the medium (see fig. 9.3). The brine shrimp maintains its low osmotic concentration, not by being exquisitely impermeable to water and ions, but by active regulation. It continuously swallows the medium, and the osmotic pressure of the gut fluid is appreciably greater than that of the hemolymph (blood). However, although the osmotic pressure of the gut fluid remains greater than the hemolymph, the concentrations of sodium and chloride in the gut fluid are considerably below those in the hemolymph (Croghan, 1958*b*). Sodium and chloride must therefore be removed from the gut by active uptake, and to eliminate these ions from the body, so that their blood concentration is kept low, excretion must take place elsewhere. It is probable that the epithelium of the gills has the major role in this process.

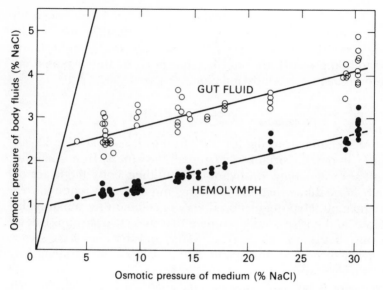

Fig. 9.3. Osmotic pressure of body fluids of brine shrimp in highly concentrated NaCl solutions. The steep line running through the origin indicates equal concentrations in body fluid and medium (Croghan, 1958*a*).

Hyporegulation is, on the whole, the exception rather than the rule among invertebrates. As we now turn to the vertebrates we shall find that osmotic hyporegulation is much more widespread, although still not universally present, for there are other ways as well to solve osmotic problems.

Aquatic vertebrates

This section will deal primarily with fish and amphibians. Nobody would deny that whales or sea turtles are aquatic vertebrates, but we will treat them in a different context. They are descended from terrestrial ancestors and are air breathers, and it is more convenient to discuss them as terrestrial animals living in an environment where no fresh water is available.

The major strategies used by aquatic vertebrates will be evident from an examination of table 9.5. The table lists examples of both marine and fresh-water vertebrates (indicated by M and F, respectively). The marine representatives fall into two distinct groups, their osmotic concentrations (last column) are either the same as or slightly above sea water (hagfish, elasmobranchs, *Latimeria*, and crab-eating frog), or in a range of about one-third of the sea water (lamprey, teleosts). The former group should have no major problem of water balance, but those that remain distinctly hyposmotic live in

constant danger of losing water to the osmotically more concentrated medium. The osmotic problems and the means to solve them
thus differ drastically among marine vertebrates. Fresh-water vertebrates, on the other hand form a uniform group with concentrations of about one-quarter to one-third of sea water, thus they are
hyperosmotic to the medium and in principle similar to fresh-water
invertebrates.

Table 9.5. *Concentrations of major solutes (in mmol per liter) in blood
plasma of aquatic vertebrates*

	Habitat *	Na	K	Urea **	Osmotic conc. (mosm liter^{-1})
Sea water		~450	10	0	~1000
Cyclostomes					
Hagfish (*Myxine*) (*a*)	M	549	11		1152
Lamprey (*Petromyzon*) (*b*)	M	—	—		317
Lamprey (*Lampetra*) (*a*)	F	120	3	< 1	270
Elasmobranchs					
Ray (*Raja*) (*a*)	M	289	4	444 ***	1050
Dogfish *Squalus* (*a*)	M	287	5	354 ***	1000
Fresh water ray	F	150	6	< 1	308
(*Potamotrygon*) (*c*)					
Coelacanth					
Latimeria (*a, d*)	M	181	—	355 ***	1181
Teleosts					
Goldfish (*Carassius*) (*a*)	F	115	4		259
Toadfish (*Opsanus*) (*a*)	M	160	5		392
Eel (*Anguilla*) (*a*)	F	155	3		323
	M	177	3		371
Salmon (*Salmo*) (*a*)	F	181	2		340
	M	212	3		400
Amphibia					
Frog (*Rana*) (*e*)	F	92	3	~ 1	
Crab-eating frog	M	252	14	350	830 ****
(*R. cancrivora*) (*f*)					

* M = marine, F = fresh water.
** When no value is listed for urea, the concentration is of the order of 1 mmol
liter^{-1} and osmotically insignificant.
*** Also includes trimethylamine oxide (TMAO).
**** Values for frogs kept in a medium of about 800 mosm liter^{-1}, or four-fifths
the concentration of normal sea water.

(*a*) (Bentley, 1971).	(*d*) (Lutz and Robertson, 1971).
(*b*) (Robertson, 1954).	(*e*) (Mayer, 1969).
(*c*) (Thorson, Cowan	(*f*) (Gordon, Schmidt-Nielsen and
and Watson, 1967).	Kelly, 1961).

Cyclostomes

The cyclostomes are eel-shaped fish and considered the most primitive of all living vertebrates. They lack a bony skeleton, paired fins, and jaws (they are grouped in the class Agnatha, jawless vertebrates).

There are two groups of cyclostomes, lampreys and hagfish. Lampreys live both in the sea and in fresh water. The hagfish, on the other hand, are strictly marine and stenohaline. Interestingly, lampreys and hagfish have employed two different solutions to the problem of life in the sea. The hagfish are the only true vertebrates which have body fluids with a salt concentration similar to that of sea water, in fact, the normal sodium concentration in hagfish blood slightly exceeds that in the medium. Nevertheless, the hagfish have pronounced ionic regulation, but in being isosmotic and having high salt concentrations, they behave osmotically like invertebrates.

With the exception of hagfish, all marine vertebrates maintain salt concentrations in their body fluids at a fraction of that in the medium. This situation has been used in favor of the argument that the vertebrates originally evolved in fresh water, and only later invaded the sea. Cyclostomes show many similarities to the ancestral forms of modern vertebrates and an understanding of their anatomy has been of great importance to the interpretation of fossil records and to the understanding of the early evolution of vertebrates. The fact that hagfish deviate from the general vertebrate pattern of low salt concentrations means that evolutionary theories of a fresh water origin of all vertebrates is not supported by physiological evidence, a low salt concentration is not a universal vertebrate trait. However, present-day physiological characteristics cannot be used in evolutionary arguments, for, on the whole, physiological adaptation takes place more easily than morphological change. Anatomical structure and fossil records therefore remain more important to evolutionary hypotheses than physiological evidence.

The other group of cyclostomes, the lampreys, live in both fresh water and in the sea, but even the sea lamprey (*Petromyzon marinus*) is anadromic * and ascends rivers to breed in fresh water.

The lampreys, whether fresh-water or marine, have an osmotic concentration of about one-quarter to one-third of sea water. Their

---

* A fish that ascends from the sea to spawn in fresh water is called *anadromic*. Shad and salmon are well-known examples. The word comes from Greek *ana*, up, and *dramein*, to run. *Catadromic* (Greek *cata* = down) refers to living in fresh water and descending to the sea. The common eel is catadromic, it grows to adult size in fresh water and descends to the sea to breed.

main osmotic problem is similar to that of teleost fish, whether marine or fresh-water species, problems which will be discussed in detail below (page 390ff).

Elasmobranchs

The elasmobranch fishes, sharks and rays, are almost without exception marine. They have solved the osmotic problem of life in the sea in a very interesting way. Like most vertebrates they maintain salt concentrations in their body fluids of roughly one-third that of sea water, but they still maintain osmotic equilibrium. This is achieved by addition to the body fluids of large amounts of organic compounds, primarily urea, so that the total osmotic concentration of their blood equals or slightly exceeds that of sea water (see table 9.5, p. 385).

In addition to urea, an osmotically important organic compound in elasmobranch blood is trimethylamineoxide (TMAO).

$$O = C \diagdown_{NH_2}^{NH_2} \qquad\qquad H_3C - \overset{\overset{\displaystyle CH_3}{|}}{\underset{\underset{\displaystyle CH_3}{|}}{N}} = O$$

Urea   TMAO

Urea is the end product of protein metabolism in mammals and some other vertebrates; it is excreted by the mammalian kidney, but in contrast, the shark kidney actively reabsorbs this compound so that it is retained in the blood. TMAO is a compound found in many marine organisms, but its origin and metabolism are still poorly understood. Whether TMAO is obtained by the sharks through the food chain or is produced in the body remains uncertain.

The blood urea concentration in marine elasmobranchs is of the order of more than 100 times as high as that in mammals, and such concentrations could not be tolerated by other vertebrates. In the elasmobranchs urea is a normal component of all body fluids, and the tissues cannot function normally in the absence of a high urea concentration. The isolated heart of a shark can continue to contract normally for hours if it is perfused with a saline solution of ionic composition similar to the blood, if urea is also present in a high concentration. If the urea is removed, however, the heart rapidly deteriorates and stops beating.

Although the elasmobranchs have solved the osmotic problem of life in the sea by being isosmotic, they still have extensive ionic regulation. The sodium concentration, for example, is maintained at about half of that in sea water. This means that sodium tends to diffuse from the medium into the shark, primarily through the thin gill epithelium, and furthermore some sodium will always be ingested with the food. Since the sodium concentration tends to increase but must be kept down, excess sodium must be eliminated. Part of the sodium excretion is handled by the kidney, but a special gland, the *rectal gland,* is probably more important. This small gland opens via a duct into the posterior part of the intestine, the rectum. The gland secretes a fluid with high sodium and chloride concentrations, in fact somewhat higher than in the sea water. For example, in sharks that were maintained in sea water with a sodium concentration of 440 mmol liter^{-1}, the secretion from the rectal gland contained from 500 to 560 mmol Na liter^{-1} (Burger and Hess, 1960).

The function of the rectal gland, however, does not fully explain the salt elimination in elasmobranchs. If the rectal gland is surgically removed in the spiny dogfish (*Squalus acanthias*), these sharks are still able to maintain their plasma ion concentration at the usual level of about half that in sea water. Since the gills are slightly permeable to salts, the blood concentrations should gradually increase unless some other means of excretion is available. It is likely that the kidney plays a major role in this excretion, but whether or not the gill is also a site for active transport of ions out from the elasmobranch blood is not known.

The fact that elasmobranchs are nearly in osmotic equilibrium with sea water eliminates the problem of a severe osmotic water loss (which for marine teleost fish is very important). Elasmobranchs therefore do not need to drink, and thus they avoid the high sodium intake which would be associated with the drinking of sea water. It is an interesting fact, however, that elasmobranch blood is usually slightly more concentrated than sea water. This causes a slight osmotic inflow of water via the gills because of the higher concentration inside. In this way the elasmobranch slowly gains water osmotically, and this water is used for the formation of urine and for the secretion from the rectal gland. Since the excess osmotic concentration is due to urea, the retention of urea can be considered as a rather elegant solution to the otherwise difficult osmotic problem of maintaining a low salt concentration while living in the sea.

Fresh-water elasmobranchs. The overwhelming number of elasmobranchs belong in the sea, but a few enter rivers and lakes, and some may belong permanently in fresh water. Even those elasmobranchs which are thought of as typically marine include some species with a remarkable tolerance to low salinity in the external medium. In several parts of the world both sharks and rays enter rivers and apparently thrive in fresh water. A well-known case is the existence in Lake Nicaragua of the shark *Carcharhinus leucas*. It was previously assumed that this shark was land-locked, but recent evidence indicates that the shark in Lake Nicaragua is morphologically indistinguishable from the marine form, and that it is able to move in free communication with the sea (Thorson, Watson and Cowan, 1966).

Four elasmobranch species which are found in the Perak River in Malyasia probably do not live permanently in fresh water but enter regularly from the sea. Their blood concentrations are lower than in strictly marine forms, and in particular, the urea is reduced to less than a third, although it remains far above the normal level for other vertebrates. The low level of solutes in the blood reduces the problems of osmotic regulation, for the osmotic inflow of water will be diminished, and lower salt concentrations are easier to maintain. The reduced osmotic inflow of water gives less water to be eliminated by the kidney. Since the urine inevitably contains some solutes, a low urine flow in turn reduces the urinary salt losses. It is, of course, difficult to say whether the lowered blood concentration is a primary adjustment or merely a passive result of the increased inflow of water and concomitant urinary losses (Smith, 1931).

One elasmobranch, the Amazon sting ray *Potamotrygon*, is permanently established in fresh water. It is common in the Amazon and Orinoco drainage systems up to more than 4000 km from the sea. It does not survive in sea water, even when the transfer takes place through a gradual concentration increase (Pang, Griffith and Kahn, 1972). The average composition of its blood (table 9.6) shows com-

Table 9.6. *The blood serum of the Amazon sting ray has solute concentrations similar to a teleost fish. Although the ray is an elasmobranch, urea is virtually absent from its body fluids.* (Thorson *et al.*, 1967)

Sodium	150	mmol liter^{-1}
Potassium	5.9	mmol liter^{-1}
Calcium	3.6	mmol liter^{-1}
Magnesium	1.8	mmol liter^{-1}
Chloride	149	mmol liter^{-1}
Urea	0.5	mmol liter^{-1}
TMAO	0	mmol liter^{-1}
Osmolality	308	mosm liter^{-1}
Protein	18	g liter^{-1}

plete adaptation to fresh water with a low blood urea concentration, similar to fresh water teleosts.

The most striking feature is the low urea concentration, which is in fact even lower than in most mammals. It is therefore clear that the retention of urea is not a universal requirement for elasmobranchs, an interesting physiological observation, for again it shows that physiological function is far more changeable than most anatomical structures. In other words, evolutionary arguments cannot be reliably based on physiological similarities or differences.

The coelacanth

Until the year 1938 it was believed that the group of fishes known as the Crossopterygii had been extinct for more than 75 million years, for they disappeared completely from the later fossil record. Their evolutionary position is far away from modern fish, close to the lung-fish, and in the ancestry of amphibians. In 1938 a living specimen of the coelancanth *Latimeria* was caught off the coast of South-East Africa, causing a world-wide scientific sensation. It was a large specimen, more than a meter and a half long and weighing over 50 kg, but it was poorly preserved and no detailed information on its anatomy was available.

After intensive search several additional specimens of *Latimeria* have been caught near Madagascar, and although no living specimen has been kept long enough to perform physiological experimentation, it is known that the coelacanth has solved its problems of osmoregulation in the same way as elasmobranchs. The data entered in table 9.5 (p. 385) were obtained on a frozen specimen of *Latimeria*, and the high urea concentration places it physiologically with the elasmobranchs. Additional analyses have confirmed the high urea content, and also shown that TMAO is high in blood (>100 mmol liter^{-1}) and muscle (>200 mmol liter^{-1}) (Lutz and Robertson, 1971). The figures given for sodium concentration in the plasma may have to be adjusted upwards, for freezing and thawing causes exchange of sodium and potassium between blood plasma and the red cells, thus lowering the value for sodium in the plasma and increasing the potassium value (which indeed was found to be abnormally high, 51 mmol liter^{-1}) (Pickford and Grant, 1967).

Teleost fish

Teleost fish maintain their osmotic concentration at about one-quarter to one-third that of sea water (see table 9.5, p. 385). On the whole, marine and fresh water fish are within the same range, although marine fish tend to have somewhat higher blood concentrations. Some fish can tolerate a wide range of salinities and move between sea, brackish, and fresh water. Such moves are often associated with the life cycle; the salmon, for example, reproduces in fresh water, migrates to the sea, and after reaching maturity returns to fresh water to spawn. The life cycle of the common eel is the reverse, it breeds in the sea, the larvae drift with the currents and reach coastal areas where they ascend into fresh water, and when maturity approaches the eel returns to the sea to reproduce. The change from one environment to the other requires profound changes in the osmoregulatory processes.

Marine fish are hyposmotic and in constant danger of losing body water to the more concentrated sea water, for their body surface, in particular the large gill surface, is somewhat permeable to water. They must somehow compensate for the inevitable osmotic loss of water, and this they do by drinking sea water.

Although drinking restores the water content, large amounts of salts are also ingested, and are absorbed from the intestinal tract together with the water. Thus, the salt concentration in the body increases, and the problem now becomes one of eliminating the excess salt. To achieve a net gain in water from the ingestion of sea water, the salts must be excreted in a higher concentration than in the water taken in. The teleost kidney cannot serve this purpose, for it cannot produce a urine which is more concentrated than the blood.

Some other organ must therefore eliminate the excess salt. This is done by the gills, which thus have a dual function, participation in osmotic regulation as well as in gas exchange. The secretion of salt across the gill epithelium must be an active transport, for it takes place from a lower concentration in the blood to a higher concentration in the surrounding medium.

The main aspects of osmotic regulation in marine teleosts are summarized in fig. 9.4. The top diagram shows the movements of water; water is lost osmotically across the gill membrane and also in the urine. To compensate for the losses, the fish drinks sea water, absorbing both water and salts from the intestine. The lower diagram shows the movement of salts, which are taken in by mouth as the fish drinks from the surrounding sea water. The double arrow at the gills indicates the elimination, by active transport, of sodium and chloride. The excretion of sodium and chloride in the urine is of minor importance because teleost urine is usually more dilute than the body fluids. However, the kidney plays a major role in the excre-

MARINE TELEOSTS

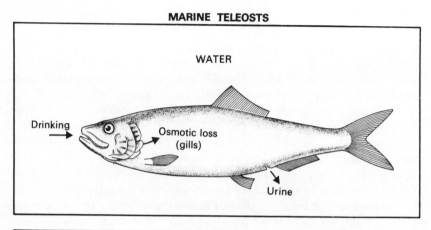

WATER

Drinking →

Osmotic loss (gills)

Urine

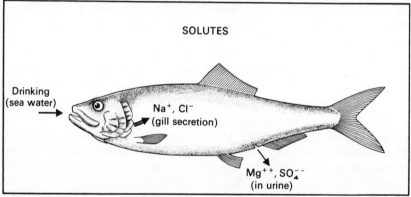

SOLUTES

Drinking (sea water) →

Na^+, Cl^- (gill secretion)

Mg^{++}, SO_4^{--} (in urine)

Fig. 9.4. A marine teleost is osmotically more dilute than the water in which it lives. Because of the higher osmotic concentration in the medium the fish constantly loses water, primarily across the thin gill membranes. Additional water is lost in the urine. To compensate for the water loss, the marine teleost drinks substantial amounts of sea water. Of the ingested salts, sodium and chloride are absorbed in the intestine and eliminated via the gills by active transport (indicated by double arrow in the lower diagram), while magnesium and sulfate are excreted by the kidney.

tion of divalent ions, magnesium and sulfate, which make up roughly one-tenth of the salts in sea water. These ions are not eliminated by the gills, which seem to transport only sodium and chloride.

Although marine fish drink sea water, measurements of the drinking rates have shown that only a smaller part of the total sodium intake is due to drinking, and the larger part of the sodium influx takes place elsewhere, presumably because the gills are somewhat permeable. Whether the general body surface or the gills are responsible, it is certain that fish adapted to sea water are relatively permeable to ions, and that fresh-water adapted fish are relatively impermeable (Montais and Maetz, 1965).

The killifish (*Fundulus heteroclitus*), which readily adapts to both fresh and sea water has been used to study the changes in permeability to sodium and chloride that take place during adaptation to various concentrations. The permeability decreases within a few minutes of transfer to fresh water, but the increase in permeability on return to sea water takes many hours (Potts and Evans, 1967). The advantage of a low permeability to ions in fresh water is obvious, but it is difficult to see the advantage of a high permeability in sea water. Marine fish must expend work to maintain their osmotic steady state in sea water, and it seems that a low permeability would reduce this work. It takes the fish several hours to return to the higher permeability in sea water, and we can only wonder why they do not permanently retain the low permeability which seems to be within their physiological capacity.

It is unlikely that the general gill epithelium participates in the ion transport, which is probably carried out by some large cells known as 'chloride cells'. Until recently it was uncertain whether the chloride ion is actively transported and sodium follows passively, or the sodium ion is actively transported and chloride follows passively. The cells were named 'chloride cells' without any definite knowledge of their function (Keys and Willmer, 1932). However, it now appears that the name was quite appropriate, for eels kept in sea water transport the chloride ion actively (Maetz and Capanini, 1966). The potential difference across the gill surface indicates active chloride transport, but sodium is not always in passive equilibrium and may be actively transported as well (House, 1963).

Fresh-water teleosts. The osmotic conditions for fish in fresh water resemble those for fresh-water invertebrates. The osmotic concentration in the blood, roughly in the range of 300 mosm liter^{-1}, is much higher than in the surrounding fresh water.

The main events in the osmoregulation of fresh-water teleosts are outlined in fig. 9.5. The major problem is the osmotic water inflow. The gills are important in this gain because of their large surface and relatively high permeability, while the skin is less important. Excess water is excreted as urine, which is very dilute and may be produced in quantities up to one-third of the body weight per day. Although the urine contains no more than perhaps 2 to 10 mmol liter^{-1} of solutes, the large urine volume nevertheless causes a sub-

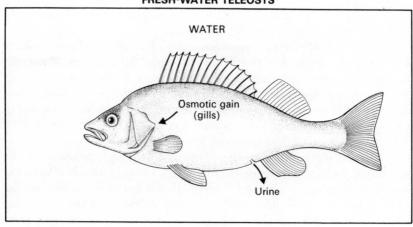

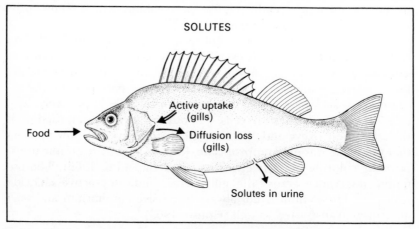

Fig. 9.5. A fresh-water teleost is osmotically more concentrated than the medium, and therefore suffers a steady osmotic influx of water, mainly through the gills (top diagram). The excess water is eliminated as urine. Loss of solutes through the gills and in the urine is compensated for, primarily through active uptake in the gills (double arrow, bottom diagram).

stantial loss of solutes, which must be replaced. The gills are also slightly permeable to ions, and this loss must likewise be covered by ion uptake.

Some solutes are taken in with the food, but the main intake is by active transport in the gills. This has been shown by placing a fish in a divided chamber, in which the head and the remainder of the body can be studied separately (fig. 9.6) In such experiments active uptake of ions takes place only in the front chamber. It is therefore

Rubber membrane

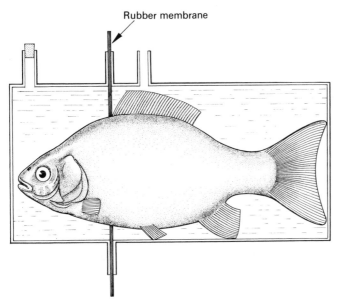

Fig. 9.6. By placing a fish in a chamber where the front and hind parts are separated by a rubber membrane, it can be shown that the skin does not participate in the active uptake of ions (which is carried out by the gills) (after Krogh, 1937).

concluded that the skin does not participate in active absorption, and we assume that only the gills are responsible.

Most teleost fish have a limited ability to move between fresh water and the sea, they are relatively stenohaline. We have already mentioned, however, that some fish such as lampreys, salmon, and eel move between fresh and salt water as part of their normal life cycle (Koch, 1968). Such moves between fresh and sea water expose the fish to cataclysmic changes in the demands on their osmoregulatory mechanisms.

If an eel is moved from fresh to sea water, the osmotic loss of water reaches 4% of the body weight in ten hours (Keys, 1933). If the eel is prevented from drinking sea water by placing an inflated balloon in the esophagus, it continues to lose water and dies from dehydration within a few days. If it can drink, however, it soon begins to swallow sea water, the weight loss subsides, and a steady state is reached within one or two days. If the eel is transferred in the opposite direction, from sea water to fresh water, there is an initial gain in weight, but as urine formation increases, a steady state is reached, again within one or two days.

When the eel is moved between fresh and sea water, not only does the osmotic flow of water change direction, but to achieve a steady state and compensate for the solute gain or loss, the active ion transport in the gills must change direction. How this change takes place is unknown, although we assume that endocrine mechanisms are involved. Likewise, it is not known whether different cell populations are responsible for the transport in the two directions, one or the other being activated as need arises. The other possibility is that the polarity of the transport mechanism in all participating cells can be reversed on demand. This remains an unsolved problem.

With our present knowledge it seems unlikely that the direction of transport in the individual cell can be reversed. Of the many organs and cell types that participate in active transport of some sort, none is known that with certainty can carry out such reversals. The frog skin, which is analogous to the fish gill in performing active uptake from the dilute solutions of fresh water, does not seem able to reverse the direction of the transport in the one species that is known to live in sea water, the crab-eating frog (see page 398).

Amphibians

Most amphibians are aquatic or semi-aquatic. The eggs are laid in water, and the larvae are gill-breathing aquatic animals. At the time of metamorphosis, many amphibians (but not all) change to respiration by lungs. Some salamanders retain the gills and remain completely aquatic as adults; most frogs, on the other hand, become more terrestrial although usually tied to the vicinity of water or moist habitats.

With regard to osmotic regulation, amphibians are quite similar to teleost fish. Virtually all amphibians are fresh-water animals, and in the adult the skin serves as the main organ of osmoregulation. When the animals are in water, there is an osmotic inflow of water, which is excreted again as a highly dilute urine. There is, however, a certain loss of solutes, both in the urine and through the skin. This loss is balanced by active uptake of salt from the highly dilute medium. The transport mechanism is located in the skin of the adult, and the frog skin has become a well-known model for studies of active ion transport.

Pieces of frog skin can be readily removed and used as a membrane separating two chambers which can be filled with fluid of various concentrations. By analyzing the changes in the two chambers, the transport function by the skin can be studied (see fig. 9.7). Such isolated pieces of skin survive for many hours. This apparatus for the study of active transport processes was originally devised by Ussing and is known as an Ussing cell.

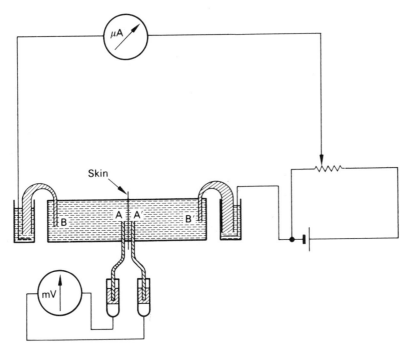

Fog. 9.7. Apparatus used for determining the sodium transport in an isolated piece of frog skin. The skin separates two chambers containing Ringer solution, and sodium transport in the skin will produce a potential or voltage across the skin. If an external current is now applied in a direction opposite to the potential produced by sodium transport, this current will, when it is adjusted to give a zero voltage across the skin, be a direct measure of the sodium transport in the skin (Ussing and Zerahn, 1951). A and A' = agar bridges to connect solutions to calomel electrodes; B and B' = agar bridges to provide electrical connections to the outside voltage source.

When frog skin in the Ussing cell separates two salt solutions * of the same composition, a potential difference of about 50 mV is rapidly established between the inside and the outside of the skin. The inside is positive, and it is therefore postulated that the potential is due to an inward active transport of sodium ion. When the potential difference has been established, the chloride ion will pass through the skin by diffusion, accelerated by the electric field. An enormous amount of evidence has accumulated to indicate that this interpretation is correct, and the fact that the transport is active is clear from

* The salt solution should have the same pH and osmotic concentration as the blood, and have approximately the same concentrations of the major ions, Na^+, K^+, Ca^{++}, and Cl^-. Such a balanced salt solution is called a Ringer solution, named after a British physiologist who discovered that survival of the isolated frog heart depended on balanced proportions of these ions.

the developed potential and from the fact that metabolic inhibitors (cyanide, for example) prevent the formation of a potential and also inhibit the transport.

By applying an external potential across the skin, equal to the sodium potential but in the opposite direction, the potential across the skin can be reduced to zero. The current needed to maintain the potential at zero must equal the current generated by sodium transport in the skin. The current, called the short-circuit current, is therefore a direct measure of the inward transport of sodium. This method has, in similar form, become an extremely valuable tool in measuring active transport of ions in many other active transport systems as well.

A salt-water frog. Frogs and salamanders are usually restricted to fresh water and die within a few hours if placed in sea water. There is one exception, however, the crab-eating frog (*Rana cancrivora*) of South-East Asia. This small, ordinary-looking frog lives in coastal mangrove swamps where it swims and seeks its food in full-strength sea water.

If a frog is to maintain the characteristic relatively low salt concentrations of vertebrates, it has two possible avenues for solving the problems. If it were to use the strategy of marine teleost fish, it would have to counteract the osmotic water loss through the skin, and also compensate for the inward diffusion of salt through the skin. The other possible strategy is that of marine elasmobranchs, which retain urea and have body fluids in osmotic equilibrium with the medium, thus eliminating the problem of osmotic water loss. The salt-water frog has employed the elasmobranch strategy of adding large amounts of urea to the body fluids, which may contain as much as 480 mmol urea liter^{-1} (Gordon *et al.*, 1961).

In retrospect it seems reasonable that the salt-water frog would use this solution. Amphibian skin is relatively permeable to water, and it is therefore simpler to maintain the same osmotic concentration as that of the medium and thus eliminate the problem of osmotic water loss. If this were to be achieved solely by increasing the internal salt concentrations, the frog would need a salt tolerance unique among vertebrates (with the sole exception of the hagfish). If it were to use the teleost strategy and remain hyposmotic, the salt balance would be further impaired by the need to drink from the medium.

A crab-eating frog placed in sea water is not completely isosmotic with the medium, like sharks it remains slightly hyperosmotic. The result is a slow osmotic influx of water, which is desirable because it provides the water required for formation of urine. This is certainly more advantageous than obtaining water by drinking sea water, which would inevitably increase the salt intake.

In the crab-eating frog, as in the elasmobranch, the urea is an important osmotic constituent and not merely an excretory product. In addition to its osmotic importance, urea is necessary for normal muscle contraction, which rapidly deteriorates in the absence of urea (Thesleff and Schmidt-Nielsen, 1962). Since urea is essential for the normal life of the animal, it is desirable that it should be retained and not be excreted in the urine. In sharks, urea is retained by an active reabsorption in the kidney tubules (see p. 479). In the crab-eating frog, however, urea retention is achieved primarily by a reduction in urine volume when the frog is in sea water. It seems that urea is not actively reabsorbed, for the urea concentration in the urine consistently remains slightly above that in the plasma (Schmidt-Nielsen and Lee, 1962).

The tadpoles of the crab-eating frog have an even greater tolerance for high salinities than the adults. Their pattern of osmotic regulation, however, is similar to that of teleost fish, and thus differs from the adult frogs which adhere to the elasmobranch pattern (Gordon and Tucker, 1965).

Although both the tadpole and the adult crab-eating frog are highly tolerant to sea water, this frog is not independent of fresh water, for both fertilization of the eggs and metamorphosis to the adult depend on a relatively low salt concentration in the water. Because of frequent torrential rains in the tropics, temporary freshwater pools readily form near the shore, and spawning can therefore take place in dilute water. Although the tadpole is highly tolerant to salt, metamorphosis is delayed for as long as the salinity remains high, and the frog goes through this critical stage only after dilution by heavy rain.

Although the crab-eating frog depends on fresh water for reproduction, its tolerance to sea water permits the exploitation of a rich tropical coastal environment which is closed to all other amphibians.

Some atypical frogs, at home in very dry habitats and highly resistant to water loss by evaporation, have recently been studied in Africa and South America. Their unusual physiological characteristics are described in further detail on page 405–6.

TERRESTRIAL ANIMALS

The greatest physiological advantage of terrestrial life is the easy access to oxygen, the greatest physiological threat to life on land is the danger of dehydration. Successful, large-scale evolution of terrestrial life has taken place only in two animal phyla, arthropods and vertebrates, which live and thrive in some of the driest and the hottest habitats found anywhere. In addition, some snails thrive on land and are truly terrestrial, some even live in deserts.

The threat of dehydration is evidently a serious barrier to terrestrial life, for many 'terrestrial' animals, except in the phyla just mentioned, depend on the selection of a suitably moist habitat, and are terrestrial only in the technical sense of the word. An earthworm, for example, is highly dependent on a moist environment and soon succumbs to desiccation if exposed to the open atmosphere for any length of time. It has little resistance to water loss, depends on behavior, and seeks out microhabitats where air and surroundings are humid. Other examples are frogs and snails, and we can group these together as 'moist-skinned' animals which have a high rate of evaporation. However, before we discuss more animals, we must know something about those physical factors that influence evaporation.

Evaporation. We know that evaporation from a free water surface increases with temperature, and also that evaporation is faster in a dry atmosphere than under humid conditions. In order to understand the evaporation from animal surfaces, it is necessary to have a more precise concept of the physical laws that govern the rate of transfer of water from the liquid to the gas phase.

The water vapor pressure over a free water surface increases rapidly with temperature, as shown in fig. 9.8. As a rule-of-thumb we can remember that at mammalian body temperature (38 °C) the water vapor pressure (50 mm Hg) is about twice as high as at a room temperature of 25 °C (24 mm Hg), and more than ten times as high as the vapor pressure at the freezing point (4.6 mm Hg).

If the air already contains some water vapor, the difference between the vapor pressure at the surface of liquid water and in the air will be less, and the driving force for water vapor to diffuse from the saturated boundary layer at the water surface into the air will be correspondingly reduced. As a rough approximation to the measure of this driving force, many investigators have used the *saturation deficit*. The saturation deficit is expressed as the difference in the vapor pressure over a free water surface at the temperature in question, and the water vapor pressure in the air. If the relative humidity of the air is, say, 50%, the saturation deficit increases with temperatures as shown by the difference between the two curves in fig. 9.8.

The 'saturation deficit rule' has been very useful in many ecological studies, both of plants and animals, but it is theoretically inadequate, for the saturation deficit is not the only physical factor that

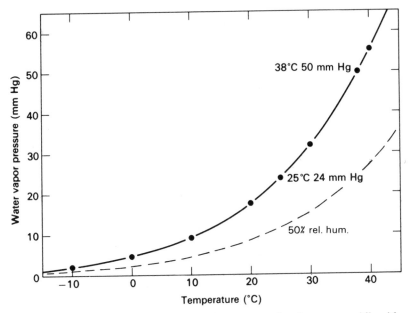

Fig. 9.8. The water vapor pressure over a free water surface increases rapidly with temperature. At the body temperature of mammals it is roughly twice as high as at room temperature, and at higher temperature the rise is increasingly steep.

determines the rate of evaporation (Ramsay, 1935; Edney, 1957). First of all, the rate of diffusion of water vapor into the air increases with temperature. Another physical factor that is important under some circumstances is the fact that the surface from which evaporation takes place loses heat and therefore attains a lower temperature than the remainder of the system. This difference in temperature will depend on the rate of evaporation as well as the transfer of heat from other parts of the system, a parameter that is often difficult to measure.

When it comes to the evaporation from living organisms, whether under natural or experimental conditions, the extent of air movement is one of the most important variables. If there is wind, and in nature the air is virtually never completely still, the air layer close to the surface is rapidly renewed, and evaporation increases. This we feel as a cooling effect when moist skin is exposed to the moving air from wind or a fan. The cooling of a surface, due to evaporation, also causes a change in the density of the adjoining air, thus causing convection currents. The extent of such convection differs between a horizontal and a vertical surface, and as a consequence, under other-wise identical physical conditions the evaporation from a moist surface varies with the orientation of the surface. Finally, evaporation also depends on the curvature of a surface.

Because of these complexities, it is difficult to describe accurately the variables that govern the rate of evaporation. The saturation deficit remains the most practical approximation we have, and when we express a certain rate of evaporation relative to the saturation deficit, it gives a reasonably good basis for comparisons. We should only realize that results on animals of various sizes and shapes obtained under seemingly similar conditions do not give an accurate basis for comparison, and, in particular, cannot be directly compared to the evaporation from the horizontal surface of free liquid water.

In the following sections we will first discuss some moist-skinned terrestrial animals that depend heavily on water in their environment, and then turn to insects and terrestrial vertebrates.

Moist-skinned animals

Consider the two extreme cases. From a moist animal surface the evaporation is high, and the rate of water loss is determined primarily by the transfer of water vapor into the surrounding air (diffusion aided by convection). In the opposite case (for example the dry insect cuticle), the greatest resistance to evaporation is in the surface itself, and any change in the permeability of the barrier greatly affects the rate of evaporation. In this case changes in the saturation deficit, convection, etc. are of minor importance and in some cases may even be disregarded. The two cases can be described as (1) a *vapor-limited system* where the resistance to evaporation lies in water transport in the air, and (2) a *membrane-limited system* in which the membrane is the major barrier to water evaporation (Beament, 1961).

An *earthworm* kept in dry air rapidly loses weight due to evaporation. The rate of water loss (see table 9.7) is quite high, and in dry air the earthworm soon dies. If a partially dehydrated worm is placed in a U-shaped tube, covered with water but with the mouth and anus above the water surface, water is absorbed by the worm. Thus, the skin is readily permeable to water in both directions. A more careful study will show that the earthworm behaves much like a typical fresh-water animal. If it is placed in salt solutions of various concentrations, its body fluids remain hypertonic to the medium (fig. 9.9). The urine, however, remains hypotonic to the body fluids, and also in this regard the earthworm resembles a typical fresh-water animal (Ramsay, 1949).

These osmotic relations show that, with regard to water balance,

Table 9.7. *Evaporation of water from the body surface of various animals at room temperature. The data indicate orders of magnitude and exact figures will vary with experimental conditions* (from Schmidt-Nielsen, 1969)

	Evaporation * (μg)
Earth worm	400
Frog	300
Salamander	600
Garden snail, active	870
Garden snail, inactive	39
Man (not sweating)	48
Rat	46
Iguana lizard	10
Mealworm	6

* All data refer to μg water evaporated per cm^2 body surface per hour per mm Hg saturation deficit.

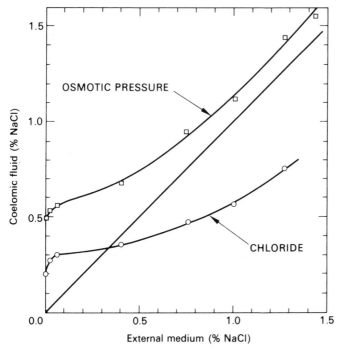

Fig. 9.9. The earthworm when placed in sodium chloride solutions of various concentrations behaves like a typical fresh-water osmoregulator. It remains hyperosmotic to the medium at all concentrations, and more so at the lower concentrations (Ramsay, 1949).

the earthworm is really more of a fresh-water animal than a truly terrestrial animal. It tunnels in the soil where the air is fully saturated, and it is in contact with soil particles which are covered with a thin film of free water. An earthworm is unable to live in completely dry soil where it rapidly becomes dehydrated and dies.

Frogs and other amphibians. A frog, even when picked up on land, has a moist and cool skin. The rate of evaporation from the skin is similar to that of the earthworm (see table 9.7), and the two-fold difference between frog and salamander is, for reasons that were explained above, not a meaningful difference. For both, the rate of evaporation is of the same magnitude as evaporation from a free water surface, and the amphibian skin does not seem to present any significant barrier to evaporation. Accordingly, those adult amphibians that are terrestrial and air-breathing usually live near water and in humid habitats where evaporation is low. When they enter water, they behave osmotically like the typical fresh-water animal.

Knowing that amphibians are moist-skinned and tied to the vicinity of free water makes it seem a paradox that some live successfully in deserts. Several species of frogs are found in the dry arid interior of Australia. These animals retreat to burrows several feet deep in the ground, where they estivate during the long periods of drought. Since reproduction takes place in water, they can breed only after rainfall, which gives an abundant but temporary supply of water. When it rains these frogs reappear from their burrows, restore their water content, deposit their eggs which develop into tadpoles at exceptional speed, and metamorphose before the pools dry up. When the frogs enter estivation again, their urinary bladders are filled with dilute urine, which is of considerable importance in their water balance. Some can store as much as 30% of their gross body weight as urine in the urinary bladder. The urine is extremely dilute and has an osmotic concentration corresponding to less than 0.1% NaCl. This urine is the main water reserve and is gradually depleted during estivation (Ruibal, 1962).

It has been reported that the Australian aborigines use desert frogs (*Chiroleptes*) as a source of drinking water because these frogs are so distended with dilute urine that they 'resemble a knobbly tennis ball' (Buxton, 1923). Once the animal has used up the water reserve in the bladder, further dehydration results in increased body concentrations, and the animal then undergoes progressive dehydration of blood and tissues.

The situation for frogs that remain deep in their burrows during dry periods is similar to that described for lung-fish, which also estivate in the ground during drought. When a lake dries out, the lung-fish wiggle into the mud, where they remain in a dry cocoon with a breathing channel to the rock-hard surface. They may remain alive in this condition for several years, and to keep their minimal metabolic rate going, they gradually use up their body proteins. No water is available for the excretion of the urea which is formed in protein metabolism, and the concentration of urea in their body fluids may rise as high as 500 mmol liter^{-1} (Smith, 1959).

One important question is whether the rate of evaporation from the skin of desert amphibians is reduced, compared to other amphibians. When determinations are made under comparable conditions, including temperature and atmospheric humidity, the differences in the rates of evaporation seem to be minor and not sufficient to explain the great variations in habitat. It has been claimed that frogs lose water more rapidly than toads, which have a thicker and more cornified skin, but this does not appear to be true (Bentley, 1966). Therefore, resistance to cutaneous evaporation does not seem to be a major factor in the adaptation of amphibians to terrestrial life.

There is, however, one spectacular exception, the South African frog *Chiromantis* loses water by evaporation at a rate which is a small fraction of that in other frogs; in fact, its rate of water loss is similar to that of reptiles (fig. 9.10). *Chiromantis* also has an additional physiological characteristic completely unlike other frogs, instead of excreting urea as other amphibians do, it excretes uric acid, which is a characteristic of reptiles. This will be discussed in further detail in connection with nitrogen excretion (see p. 475–90).

The finding of 'reptilian' physiological characteristics in a typical amphibian illustrates the point we have made before. Physiological characteristics as a rule are much more plastic and adaptable to the environment than are morphological characters, and are therefore not suitable for establishing evolutionary relationships.

Snails. Snails and slugs are moist-skinned, and evaporation from their surface is high (see table 9.7). Naked slugs depend on the humidity of their habitat, and they are active mostly after rain and at night when the relative humidity is high. Otherwise they withdraw to and remain in microhabitats where the humidity is favorably high. Snails, in contrast, carry a water-impermeable shell into which they can withdraw.

It has been suggested that the *mucus* which covers a snail forms a barrier to evaporation, but this is difficult to understand. When snail mucus is removed from the animal, water evaporates from the iso-

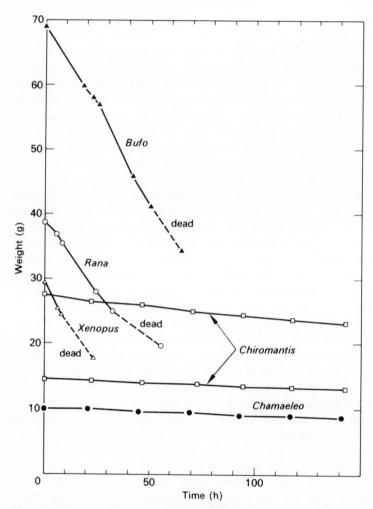

Fig. 9.10. When kept at 25 °C and 20 – 30% relative humidity the frog *Chiromantis* loses water by evaporation at a very low rate, similar to that of a reptile (*Chamaeleo*). After six days in air its weight loss is only a few grams. Other amphibians lose weight rapidly. A toad (*Bufo*) loses over 40% of its body weight and dies within two days, an ordinary frog (*Rana*) dies in a day, and the aquatic *Xenopus* in less than 10 hours (after Loveridge, 1970).

lated mucus at practically the same rate as from a free water surface, until the mucus is completely dried out. Nevertheless, this question is not entirely clear, for an undisturbed snail evaporates water more slowly than it does after the surface layer of mucus is disturbed, and new, fresh mucus is extruded from the mucus glands in the skin (Machin, 1964).

Terrestrial snails that are inactive or estivate have an excellent barrier to water loss. They cover the opening to the shell with a membrane, an epiphragm, which usually consists of dry mucus, but in some cases also contains a large amount of crystalline calcium carbonate.

The extremely low rate of water loss from an inactive snail in the shell and covered with an epiphragm permits some species to survive in hot, dry desert areas. One such snail, *Sphincterochila*, is found in the deserts of the Near East. Withdrawn into the shell and dormant, this snail can be found on the desert surface in mid-summer, fully exposed to sun and heat. They become active after rains, which are concentrated in the winter months, from November to March, when they feed and reproduce. During the long dry season they remain dormant. The water loss from a dormant *Sphincterochila* is less than 0.5 mg per day, and since one animal contains about 1.5 g water, it can survive for several years without becoming severely dehydrated. In fact, the snail itself (not including the calcareous shell), contains over 80% water. This high water content remains the same, even during the hottest and driest part of the year. This shows that these snails do not gradually deplete their water resources but have solved the water problem by having a uniquely low water loss while remaining dormant (Schmidt-Nielsen, Taylor and Shkolnik, 1971).

Arthropods

The most successful terrestrial animals are found among the arthropods, which comprise insects, crustaceans, and several other groups of animals characterized by a jointed, rigid exoskeleton. Insects are by far the most numerous animals in regard to number of species, for nearly one million different species have been described. Insects and arachnids (spiders, ticks, mites, scorpions, etc.) are primarily terrestrial, and they are highly adapted to life on land and respiration in air. Only a small number have secondarily invaded fresh water, and almost no marine forms are found in these two classes. Two additional classes of arthropods are terrestrial, the millipedes (Diplopoda) and the centipedes (Chilopoda). In contrast, most crustaceans are aquatic, as are the Xiphosura, represented by the 'living fossil' the horseshoe crab (*Limulus*).

Crustaceans

Although most crustaceans are aquatic, an appreciable number, including some decapods (crabs and crayfish), are terrestrial, even to the extent that they will drown if kept submerged. However, they are mostly restricted to moist habitats and the vicinity of water. On the other hand, there are some isopods, notably sowbugs (or pill-bugs) and woodlice, which are completely terrestrial, and a few even live in hot dry deserts.

Crabs are common in the intertidal zone, and in the tropics many live permanently above the high tide mark. They depend on a moist habitat, however, and often have burrows where a pool of water remains at the bottom. In the more terrestrial species emancipation from water is nearly complete, but they do return to water for spawning.

Apparently, invasion of land has taken place both through the tidal zone and from fresh water.

The semi-terrestrial and terrestrial crabs are generally excellent osmoregulators. Depending on conditions and need they can usually carry out effectively both hyper-regulation and hyporegulation. An interesting aspect of some land crabs (*Cardiosoma* and *Gecarcinus*) is that they are able to take up water from damp sand or a moist sub-stratum, even in the absence of visible amounts of free water. In this way they can survive for many months without direct access to water (Gross, Lasiewski, Dennis and Rudy, 1966; Bliss, 1966).

The ghost crab, *Ocypode quadrata,* which is often seen scurrying over sandy beaches, requires frequent immersion in sea water. Simi-larly, the well known fiddler crabs, *Uca,* which are found along the coasts of all tropical seas and into temperate areas, dig deep burrows into the beach near or above high tide mark but depend on frequent return to water.

The common terrestrial isopods known as pillbugs, sowbugs, or woodlice are often found in humid habitats, well hidden from ex-posure. They remain hidden during the day and move around at night when the relative humidity is higher. These crustaceans are completely independent of free water for reproduction, and are thus truly terrestrial.

Although woodlice live mostly in microhabitats with high humid-ity, they are sometimes exposed to drier air. Then they lose water by evaporation, and this water must be replaced. Normally woodlice feed on moist decaying plant material, which is probably their nor-mal source of water. It has been found, however, that several iso-pods can take up free water by drinking and also through the anus. When kept on a moist porous slab of plaster of Paris, they can even absorb water from this surface (Edney, 1954).

As compared to insects, the cuticle of isopods is relatively perme-able to water. The main reason for this difference seems to be that the cuticle of insects is covered by a thin layer of wax which greatly reduces water loss, but all attempts at demonstrating a wax layer in isopods have been unsuccessful (Edney, 1954).

It may seem strange to find a land crustacean in a desert, but the woodlouse, *Hemilepistus,* is quite common in desert areas. These animals still breathe by gills, and they survive by digging narrow vertical holes about 30 cm deep, where they spend the hot part of the day. Temperature and humidity measurements show that their hideout remains much cooler than the desert surface, and most importantly, the relative humidity is high, around some 95%. The cuticle of *Hemilepistus* is somewhat less permeable to water than that of other woodlice, but their major adaptation to life in the desert seems to be a question of behavior (Edney, 1956).

Insects and arachnids

If numbers are used to judge success, insects are undoubtedly the most successful animals on earth. There are more species of insects than all other animals combined, and furthermore, many insect species are extremely numerous. They live in almost every conceivable habitat on land and in fresh water, only the polar regions and the ocean seem to be relatively free from them. Dryness, desert, and complete absence of free water seems to be no barrier to these animals. A clothesmoth that thrives on a woolen garment, or a flour beetle that completes its life cycle from egg to adult in dry flour, lives, grows, and thrives virtually without any free water in its environment. Nevertheless, as the adult insect consists of more than two-thirds water, these animals must have extra-ordinary capacities to retain water and reduce losses.

Since there are considerable similarities between insects and arachnids, these two classes will be discussed together in this chapter.

Water balance, gains and losses. For an organism to remain in water balance, it is necessary that, over a period of time, all loss of water is balanced by an equal gain of water. The components in the water balance are:

WATER LOSS	WATER GAIN
Evaporation	Drinking
(*a*) body surface	Uptake via body surface
(*b*) respiratory organs	(*a*) from water
Feces	(*b*) from air
Urine	Water in food
Other (specialized secretions)	Oxidation water (= metabolic water)

We should realize that the problem of maintaining balance may work either way, there may be an excess of water or a shortage of water, and the physiological mechanisms must therefore be able to cope with both situations. For example, a fresh-water insect may swallow large amounts of water with its food, and there is also the osmotic inflow of water. The excess water must be eliminated, and this task is normally carried out by the renal organ or kidney, which must then produce a large volume of dilute urine. A terrestrial insect living in a dry habitat, on the other hand, has a very limited water intake, and all losses must be reduced to such a level that their sum in the long run does not exceed the total intake.

Any organism can tolerate some variation in its water content, and some will be more tolerant than others. Mammals for example, can usually withstand losing 10% of their body water, although they will then be in rather poor condition, while a loss of 15 or 20% is probably fatal to most. Many lower organisms can withstand greater losses, some frogs, for example, can withstand the loss of 40% of their body water, but very few animals can tolerate losing half of their body water.*

When the water content of an animal is expressed as per cent of its body weight, the results are not always easy to interpret. For example, the fat content of animals varies a great deal, and fatty tissues have a low water content, about 10% or less. An animal with a large fat storage therefore will have a lower overall per cent water content in the body than a lean animal in which muscles and internal organs have exactly the same degree of hydration. If the water content is expressed in relation to the fat-free tissue mass, the figures are usually much less variable. The disadvantage is that this requires a fat determination and therefore becomes much more complicated.

The custom of expressing water content as per cent of the initial body weight tends to be misleading in another way that is best explained by an example. If an insect which initially contains 75% water, loses exactly half of all water in its body, the animal now contains 60% water. An examination of fig. 9.11 shows that the arith-

---

* Two methods are available for the determination of total body water. The water content can be obtained by killing the animal, drying the body completely (e.g. in a drying oven), and determining the weight loss. If such a destructive method is used, some individuals are used to establish a normal or baseline value, and others are used for experimentation. A non-destructive method is the isotope dilution technique. A small volume of water labelled with an isotope (deuterium or tritium) is injected into the animal. After the labelled water is evenly distributed in all body water, a sample of blood or any other body fluid is withdrawn and the concentration of the label determined. It is now simple to calculate the water volume into which the isotope label was distributed. The method permits repeated determinations on the same individual animal.

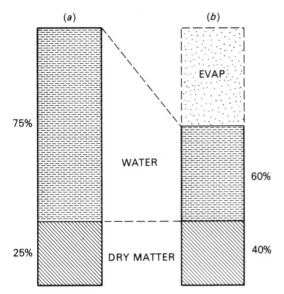

Fig. 9.11. Diagram to illustrate that it can be deceptively misleading to express body water content in per cent of body weight. If an animal originally has a water content of 75% of its body weight (*a*), and loses exactly half of its body water, the water content of the dehydrated animal will now be 60% of the body weight (*b*). Thus a water loss that in reality is very serious gives the appearance of not being very great.

metic is correct. At first glance, a reduction in water content from 75% to 60% does not seem excessive, although in reality half of all body water was lost. The sometimes deceptive effect of percentage values can be avoided by expressing results in absolute units. If the insect in the above example weighed 100 mg, it would initially contain 75 mg water and after dehydration 37.5 mg water, and it is immediately apparent that half of its water was lost.

Water loss, evaporation. Evaporation takes place both from the respiratory organs and from the general body surface, but it is often experimentally difficult to determine these two variables separately. The respiratory organs of insects and vertebrates differ profoundly in structure and function. The lining of the vertebrate lung is always moist, the respiratory air is saturated with water vapor, and the water loss from the respiratory tract is substantial. The respiratory organs of insects, on the other hand, consist of tubes which are lined with chitin, and only the finest branches of these are relatively permeable to water. Nevertheless, for insects the water loss from the respiratory system is important, and the possible means to reduce this loss were discussed under insect respiration (p. 71ff).

The general body surface of insects is covered by a hard, dry cuticle, and it is deceptively easy to assume that this cuticle would be completely impermeable to water. This is not so, however. There are enormous differences in evaporation between species that normally live in moist environments and others that live in completely dry surroundings (table 9.8). The evaporation from the aquatic larvae of the marsh fly, when in air, is of the same order of magnitude as evaporation from a moist-skinned animal (see table 9.7, p. 403). This means that, essentially, the cuticle is no barrier to evaporation. On the other hand, in insects that live in very dry habitats the permeability of the cuticle can be lower than in any other animals about which we have information.

Table 9.8. *Evaporation from the body surface of insects and arachnids. Water loss from respiratory organs is not included* (from Schmidt-Nielsen, 1969)

		Evaporation * (µg)
Marsh fly larvae	*Bibio*	900
Cockroach	*Periplaneta*	49
Desert locust	*Schistocerca*	22
Tse-tse fly	*Glossina*	13
Mealworm	*Tenebrio*	6
Flour mite	*Acarus*	2
Tick	*Dermacentor*	0.8

* All data refer to µg water evaporated per cm² body surface per hour per mm Hg saturation deficit.

It is helpful to realize that the insect cuticle is a highly complex organ which consists of several layers. It is of particular interest that the hard chitin in itself is not particularly impermeable to water, the resistance to evaporation resides in a thin covering layer of wax. If this layer is scratched by abrasives, for example alumina particles, the evaporation increases greatly and insects that can otherwise live in very dry material, such as stored grain, succumb from dehydration (Wigglesworth, 1945). This is of some practical interest, for abrasives have been used as non-toxic insecticides which can later be separated from the grain with relative ease due to differences in density. Not only sharp particles, but also materials such as dry clay dust have a similar effect, presumably because, likewise, they disturb the structural arrangement of the wax layer.

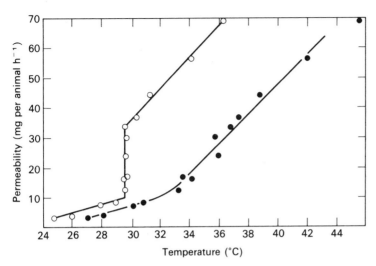

Fig. 9.12. The water loss from the cuticle of a cockroach increases with air temperature (black circles). If the data are plotted against the insect surface temperature (open circles) a sharp transition point appears at about 30 °C, which corresponds to a transition point of the surface wax. Ordinate shows water loss from the animal in mg H_2O per hour (after Beament, 1958).

The evaporation from insect cuticle increases with rising temperatures (fig. 9.12). The increase is far greater than the increase in water vapor pressure and seems related to physical changes in the wax layer. If the permeability is plotted against surface temperature, rather than air temperature, a sharp transition point is observed in evaporation. This transition point differs from species to species. The surface waxes have been isolated and their physical characteristics show a transition point at the same temperature as the observed transition in water loss (Beament, 1959).

It seems that a wax layer is not a universal prerequisite for successful terrestrial life of all arthropods. It was mentioned above that many isopods are adapted to terrestrial life, and that some even live in deserts. No wax has been found on their cuticle, and their evaporation is directly related to the saturation deficit in the atmosphere and is otherwise temperature independent. In insects, in contrast, the evaporation changes with temperature even if the vapor pressure deficit is kept constant.

Water loss in feces and urine. Insects eliminate the feces and the urine through the same opening, the anus. The urine is formed by the Malpighian tubules, which open into the posterior part of the gut. Liquid urine, as well as fecal material from the intestine, enter the rectum where water reabsorption takes place. This matter will be discussed in greater detail in chapter 10.

Storage excretion. If excretory products, instead of being eliminated in the urine, could be withheld in the body, no water would be expended for their excretion. Since uric acid is a highly insoluble compound, its retention in the body is in fact a feasible approach to the problem of 'excretion', and the deposition of uric acid in various parts of the insect body seems to be a regular feature. For example, in various species of cockroaches, as much as 10% of the total dry weight of the body may be uric acid.

Many insects store uric acid in the fat body, in others it may be found in the cuticle. Uric acid that is deposited in the cuticle is probably never mobilized again. In the fat body, however, the situation may be different. It is possible that this uric acid represents a storage or depot from which nitrogen could be mobilized during periods of nitrogen deprivation and used for metabolic purposes. So far evidence for such a function is inadequate.

'Active' transport of water. An insect that lives in very dry surroundings, e.g. a mealworm, withdraws water from the rectal contents until the fecal pellets are extremely dry. This withdrawal of water has the appearance of being an active transport of water, for water is moved from a high osmotic concentration in the rectum to a lower concentration in the blood. However, because of the complexity of the rectum and its surrounding structures, alternate hypotheses are possible. In general, most 'up-hill' transport of water in animal systems can be explained by a primary transport of a solute, with water following passively due to osmotic forces, acting as explained by the *three-compartment theory* proposed by Curran (1960).

The Curran hypothesis can be understood by reference to the artificial model in fig. 9.13, which shows a cylinder divided into three compartments. Compartments A and B are separated by a cellophane membrane which is readily permeable to water, but not to sugar. Compartments B and C are separated by a porous glass disc (a sheet of filter paper would do if given suitable mechanical support). Let us initially fill compartment A with 0.1 molar sucrose solution, compartment B with 0.5 molar sucrose, and compartment C with distilled water. The concentrated solution in B will immediately draw water osmotically from A and C, but if the liquid in B is kept from expanding, it will flow through the porous disc into compartment C. The over-all result is therefore a net movement of water from A to C, although the osmotic concentration in A is higher than

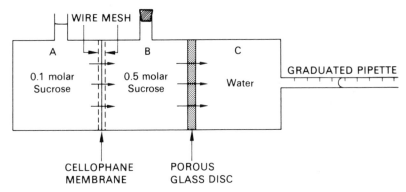

Fig. 9.13. The Curran model to demonstrate net transfer of water against an osmotic gradient between chambers A and C. During experiments the exit tube from chamber B is closed. For details, see text (after Curran and MacIntosh, 1962).

in C. The movement of water against the over-all osmotic gradient obviously cannot continue forever, for the concentration differences will gradually disappear and the system run down. However, if somehow we can maintain a high solute concentration in compartment B, the water movement will continue.

Movement of water through a variety of biological membranes can be explained according to the Curran model. The diagram in fig. 9.14 represents a typical epithelial cell, which has deep infoldings of the membrane on the side facing the tissue fluid. Active transport of sodium establishes a high osmotic concentration within the lumen of the infolding. Then, because of the osmotic forces, water diffuses into the lumen, and the increased hydrostatic pressure causes a bulk flow of liquid through the opening into the tissue fluid, resulting in an over-all movement of water from the outside. This water transport, which can be against an over-all osmotic gradient from the outside of the cell to the tissue fluid, depends on the active transport of sodium as the primary event in the process.

Water transport as explained by this mechanism may be a common way of moving water. Long, narrow, intercellular or intracellular spaces are characteristic of many epithelia, and available experimental evidence strongly supports the hypothesis (Diamond and Bossert, 1967; Diamond, 1962).

Water reabsorption from the insect rectum has been studied in cockroaches, and the results conform to the three-compartment hypothesis. Samples of fluid, as small as 0.1 nl (0.0001 mm³) have been withdrawn from the rectal pads. This fluid was found to be more concentrated than the fluid in the rectal lumen, but less concentrated than the blood. It thus corresponds to compartment B in

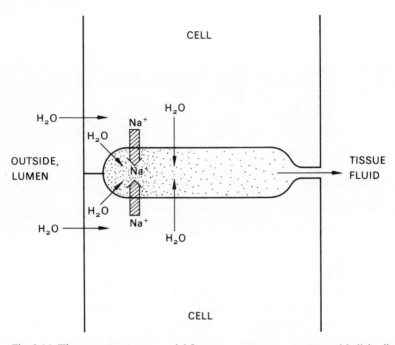

Fig. 9.14. Three-compartment model for water transport across an epithelial cell. Heavy arrows, Na⁺ active transport. Dotted area corresponds to compartment B in fig. 9.3.

the Curran model (see fig. 9.13) (Wall, Oschman and Schmidt-Nielsen, 1970; Wall and Oschman, 1970).

Water gain, drinking. The most obvious form of water intake is drinking of free water. Such water is, of course, available to all fresh-water insects and to insects under many other circumstances such as when there is dew or rain. For most insects, however, the situation is that free water is available only intermittently and at irregular intervals, and many live in dry habitats where no free water is available at any time. For these, water must be obtained from elsewhere.

Water in the food. Insects that eat plant material may obtain large amounts of water in the food, for fresh vegetable material has a high water content. Succulent fruits, leaves, etc. may contain over 90% water, but it should be realized that even the driest plant material contains some free water. Dry grains and seeds, flour, wool and other seemingly completely dry substances on which insects are able to subsist, grow, and reproduce, may contain around 5 to 10% free water.

When water is plentiful in the food, the question is one of eliminating the excess. This is the normal function of the kidney, which then produces a dilute urine in large quantities. Under dry conditions, in contrast, it becomes a question of producing as little urine as possible, i.e. eliminate the excretory products with a minimum of water and produce a urine as concentrated as the renal mechanism is able to achieve.

Oxidation water. For animals living under dry conditions, the most significant item on the 'gain' side is water formed in the combustion of organic materials. We could properly call it *oxidation water*, although the term 'metabolic water' is also common.

Everybody knows that water is formed when organic materials burn; we can see water condense on the outside of a cold pot placed over a gas flame, or water dripping out of the exhaust pipe of a car on a cold morning. The oxidation reaction for glucose is as follows:

$$C_6H_{12}O_6 + 6\ O_2 \rightarrow 6\ CO_2 + 6\ H_2O$$
$$(180\ g) + (192\ g) \rightarrow (264\ g) + (108\ g)$$

In the over-all reaction water can be regarded as formed by oxidation of the hydrogen. In the oxidation of food, the amount of water will depend on the amount of hydrogen present in the foodstuff in question. From the above equation it can be readily calculated that one gram glucose yields 0.60 gram water, and no amount of physiological maneuvering could be used to obtain more water than that indicated by the equation.

In the metabolism of other carbohydrates, such as polysaccharides and starch, slightly less water is formed because of their lower hydrogen content. The oxidation of starch, for example, will be as follows:

$$(C_6\ H_{10}\ O_5)_n + n\ O_2 \rightarrow 6\ n\ CO_2 + 5\ n\ H_2O.$$

In this case, the amount of water formed in oxidation is 0.56 gram water per gram starch metabolized.

Oxidation of fat gives larger amounts of oxidation water than obtained from carbohydrates, about 1.07 gram water per gram fat, a figure that varies slightly with the composition of the fat and the degree of saturation of the fatty acids. Thus, about twice as much water is formed in fat oxidation as in the oxidation of starch. In some respects, however, this figure is misleading, for fat also gives more energy per gram (9.4 kcal against 4.2 for starch). For a given metabolic rate, an animal therefore uses less than half the amount of fat, with a corresponding reduction in the yield of oxidation water.

The amount of oxidation water formed, relative to a given metabolic rate, is therefore slightly more favorable for starch than for fat. This is shown in the last column of table 9.9, which lists the amount of water formed in relation to the energy value of the food.

Table 9.9. *Amounts of water formed in the oxidation of various foodstuffs.* (King, 1957; Schmidt-Nielsen, 1964)

	grams water formed per gram food	Metabolic heat value, (kcal g^{-1})	g water formed per kcal
Starch	0.56	4.2	0.13
Fat	1.07	9.4	0.11
Protein (urea excretion)	0.39	4.3	0.09
Protein (uric acid excretion)	0.50	4.4	0.11

Protein metabolism is somewhat more complex, because the nitrogen contained in the protein yields excretory products that contain some hydrogen, this hydrogen, therefore, is being excreted and not oxidized to water. The amount of oxidation water depends on the nature of the end product of protein metabolism. If it is urea, the amount of oxidation water formed will be 0.39 gram water per gram protein; if the end product is uric acid, which is common in insects, the amount of oxidation water formed will be higher. Urea (CH_4ON_2) contains two hydrogen atoms per nitrogen atom, while uric acid ($C_5H_4O_3N_4$) contains only one hydrogen per nitrogen atom, that is, only half as much. This increases the yield of oxidation water to 0.50 gram water per gram protein when uric acid is the end product. To give an accurate account of the amount of oxidation water, we must therefore know, not only the exact amounts and composition of the foodstuffs oxidized, but also the nature of the metabolic end product of protein metabolism. This is of importance only if a very accurate account is needed, for the differences in oxidation water formed for the various foodstuffs are relatively small (table 9.9, last column).

Water uptake via body surface. The uptake of water through the body surface of aquatic insects is of the same nature as uptake in other fresh water animals, the higher osmotic concentration of solutes in the body fluids causes an osmotic inflow of water. For aquatic insects the problem is to eliminate excess water and, as was mentioned above, this problem is usually handled by the kidney or equivalent excretory organ.

In terrestrial insects, however, the situation is different; they are often in water shortage. A very interesting phenomenon has been observed in some terrestrial insects and arachnids, they are able to absorb water vapor directly from atmospheric air. When first reported, this observation was met with considerable doubt because it is extremely difficult to suggest any mechanism that can take up water vapor directly. The reports have been confirmed by several competent investigators, and a number of well-documented cases are listed in table 9.10. This list also contains information about the lowest relative humidity at which each animal can carry out such absorption of water vapor.

Table 9.10. *Arthropods in which water absorption from atmospheric air has been demonstrated. The last column gives the limiting relative humidity below which a net gain in water is no longer possible.* (Schmidt-Nielsen, 1969)

		Limit r.h. (%)
Tick	*Ornithodorus*	94
Tick	*Ixodes*	92
Mealworm	*Tenebrio*	90
Mite	*Echinolaelaps*	90
Desert roach	*Arenivago*	83
Grasshopper	*Chortophaga*	82
Flour mite	*Acarus*	70
Flea	*Xenopsylla*	50
Firebrat	*Thermobia*	45

The characteristics of the uptake mechanism can best be described with the aid of an example, and we will use the desert roach *Arenivaga*. *Arenivaga* takes up water from the atmosphere only after it has been partly dehydrated. If it has been placed in a very dry atmosphere until, say, 10% of the body weight has been lost, uptake of water begins when the animal is moved to any relative humidity above 83%. The uptake continues until the animal is fully rehydrated and then ceases, i.e. the absorption process is accurately regulated according to the need for water. We saw that the atmospheric humidity below which uptake cannot take place, differs from species to species (table 9.10); this species-specific limit is not influenced by temperature. Therefore, the relative humidity rather than the vapor pressure deficit is limiting to the uptake, i.e. the process appears to be similar to a hygroscopic effect. This is difficult to understand, and any hypothesis ought to explain this point adequately. One of the hypotheses which have been considered for the uptake of water vapor is that temperature microgradients within the animal lead to condensation of water vapor. To achieve condensation from an atmosphere of 90% relative humidity requires gradients in excess of 2 °C, which is rather unthinkable when experiments are done under constant temperature conditions. For 50% relative humidity, roughly the lowest humidity from which uptake has been reported in any species, the air temperature would have to be lowered from 25 to 14 °C to achieve condensation. Obviously, in experiments under controlled constant temperature conditions, condensation due to such temperature differences is out of the question (Edney, 1966).

The anatomical location for the uptake of water vapor is difficult to establish. It might seem simple to determine whether the respiratory system is involved, for it is relatively easy to seal the openings of the tracheae with wax. In some insects such sealing stops water absorption, but in the tick *Ixodes* uptake still continues after the spiracles have been sealed. In those animals where uptake ceases, this can be explained as a secondary effect of sealing the respiratory system, for this prevents oxygen uptake, and anoxia as such also stops water vapor uptake. It is possible that the rectum, which we know is capable of absorbing water from the feces, is the site of the water vapor uptake.

Terrestrial vertebrates

Reptiles. There are four major orders of living reptiles. Of these the crocodilians are always associated with water. The other three orders, snakes, lizards, and tortoises, are considered well adapted to dry habitats, but they also have some representatives that are aquatic or semi-aquatic. All the aquatic reptiles have lungs and are air breathers, and are obviously descended from terrestrial stock.

The skin of reptiles is dry and scaly, and it has been assumed that it is impermeable to water. Let us therefore examine the cutaneous evaporation from a number of reptiles (table 9.11). To make a comparison easier we will use the same units as we used before (see tables 9.7 and 9.8). We can now see that the evaporation from the skin of a dry-habitat reptile is only a small fraction of that in an aquatic reptile's, and even in the aquatic reptiles (when kept in air) the evaporation is one magnitude lower than in moist-skinned animals such as frogs (see table 9.7).

Table 9.11. *Evaporation from the body surface of reptiles at 23–25 °C* (from Schmidt-Nielsen, 1969)

		Evaporation * (μg)
Caiman	*Caiman*	65
Water snake	*Natrix*	41
Pond turtle	*Pseudemys*	24
Box turtle	*Terrapene*	11
Iguana	*Iguana*	10
Gopher snake	*Pituophis*	9
Chuckawalla	*Sauromalus*	3
Desert tortoise	*Gopherus*	3

* All data refer to μg water evaporated per cm^2 body surface per hour per mm Hg saturation deficit.

We would perhaps expect that the evaporation from the moist respiratory tract would be much greater than the evaporation from the dry reptilian skin, but this is not so. The contribution made by the skin to the total evaporation always exceeds the respiratory evaporation by a factor of two or more (fig. 9.15). When the water snake *Natrix* is kept in air, the skin contributes nearly 90% of the total evaporation. The relationships are similar for turtles and lizards, the skin remains more important than the respiratory tract in the water loss. Even in the chuckawalla, a desert lizard, two-thirds of the total evaporation is from the skin and only one-third from the respiratory tract.

There is a close correlation between evaporation and habitat. The drier the normal habitat, the lower the rate of evaporation. The total evaporation, combining that from body surface and respiratory tract, from a desert rattlesnake is less than half a per cent of its body weight per day, and the snake could probably survive for two or three months at this rate. If it were to remain in an underground burrow or tunnel where the humidity is higher, it could undoubtedly last even longer.

In addition to water lost through evaporation, water is also needed for urine formation. Reptiles excrete mainly uric acid as an end product of protein metabolism, and as this compound is highly insoluble it requires only small amounts of water for excretion. The relationship between nitrogen excretion and water metabolism will be discussed in greater detail in the next chapter, which deals with excretion and excretory organs.

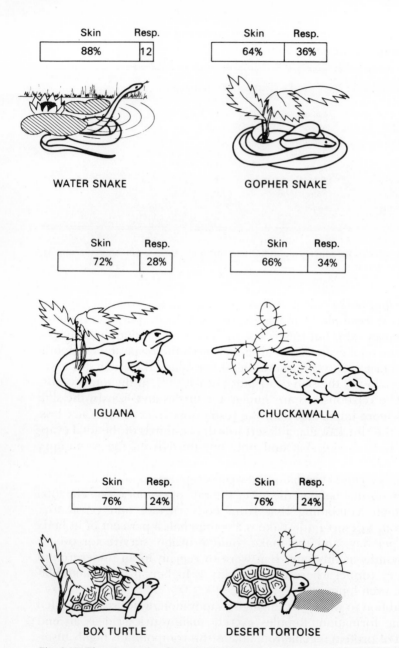

Fig. 9.15. The evaporation from the dry skin of reptiles exceeds the evaporation from the respiratory tract, ranging in these animals from 66 to 88% of the total evaporation. The total evaporation (skin and respiration combined) is habitat related, and is more than 10 times as high in a water snake as in a desert lizard. All observations in dry air at 23 °C (Schmidt-Nielsen, 1969).

Marine reptiles have a special problem because the water they live in and much of their food contain large amounts of salt. The particular problems of excretion that arise from this will be discussed later in this chapter.

The gain in water which is necessary to balance the losses are, of course, the same for reptiles as for other animals (see list on page 409). There is no indication that reptiles have the ability to obtain water by absorption from the atmosphere, as some insects can. Therefore, when drinking water is unavailable, their total water intake must be derived from food and from oxidation water.

Birds and mammals. Until now we have discussed water balance without considering problems of temperature regulation, but some animals, in particular birds and mammals, use water to keep cool in hot surroundings. Man and some other mammals sweat, dogs and many other mammals and birds pant, and the increased evaporation cools the animal. This aspect of water expenditure was discussed in the preceding chapter, and we shall therefore not discuss it further here.

Mammals the size of rodents are convenient for a discussion of the basic aspects of water balance, for they do not pant and lack skin glands in sufficient numbers to be of importance in heat regulation. This permits us to examine the basic, unavoidable components of their water expenditure, evaporation and the losses in urine and feces, without a large and highly variable amount used for heat regulation. We can then see how well the available sources of water cover the needs.

In most deserts free water is available only on the rare occasions of rain, yet birds and small rodents are found in many deserts where it may rain less than once a year. These animals must obtain all their water from the food, for dew seems to play a minimal role, at least for higher vertebrates. Many animals obtain more than sufficient water from green leaves, stems, fruits, roots and tubers, and carnivorous animals obtain much water from the body fluids of their prey, which may contain from 50 to 80% water.

Some desert rodents depend on such moist food, but others live primarily on dry seeds and other dry plant material, and their intake of free water is minimal. The kangaroo rats and pocket mice which are abundant in the North American deserts are well-known examples of this type. They can live indefinitely on dry food, and yet never drink. An examination of their water balance will therefore explain the mechanisms through which a mammal can live and remain in water balance seemingly without any water intake.

First of all, a kangaroo rat is not an exceptionally 'dry' animal, its body contains as much water as other mammals (about 66%). Even when a kangaroo rat has lived on a diet of only dry barley or oats for weeks or months, its water content remains the same. It maintains its body weight, or may even gain in weight. This shows that the animal remains in water balance on the dry food, in other words, it means that water loss does not exceed the gain.

An account of the over-all water metabolism of a kangaroo rat is given in table 9.12. The account refers to a period during which a kangaroo rat would eat and metabolize 100 grams of barley. For a kangaroo rat that weighs about 35 g, this amount of food might be consumed in about one month. The exact time is of no consequence, however, for the balance would look exactly the same if the animal had a higher metabolic rate, and the same amount of food were consumed in, say, two weeks.

Table 9.12. *Over-all water metabolism, balancing gains and losses, during a period of time in which a kangaroo rat consumes and metabolizes 100 g barley (this would usually be about 4 weeks).** (Schmidt-Nielsen, 1964)

WATER GAINS	(ml)	WATER LOSSES	(ml)
Oxidation water	54.0	Urine	13.5
Absorbed water		Feces	2.6
(at 20% r.h.)	6.0	Evaporation	43.9
Total water gain	60.0	Total water loss	60.0

* Air temperature 25 °C, 20% r.h.

On the gain side we find that 54 grams of water are formed in the oxidation of the food. This figure can be calculated from an analysis of the composition of the grain. By using the figures for oxidation water formed in the metabolism of starch, protein, and fat (as given in table 9.9) we arrive at the listed figure. The grain also contains a small amount of free water, the exact amount depending on the humidity of the air. At 20% relative humidity it will be 6 grams of water in 100 grams of barley, giving a total amount on the gain side of 60 g. This has to suffice for all the needs.

On the loss side we find that nearly one-quarter of the available water goes for urine formation. In amount, the most important excretory product is urea, formed from the protein in the grain. This excretory product must be eliminated (there is no evidence for

Plate 8. *Kangaroo rat (Dipodomys spectabilis)*. This rodent is common in the deserts of North America. It does not drink and subsists mostly on seeds and other dry plant material. Through economical use of water for urine formation and evaporation from the respiratory tract, it manages on the water formed in metabolic oxidation processes as its main water source (K. Schmidt-Nielsen, Duke University).

'storage excretion' as was described for insects), and the more concentrated the urine, the less water is used. The kangaroo rat has a remarkable renal concentrating ability, which far exceeds that of most non-desert mammals (see table 10.2, p. 468). If we know the amount of protein in 100 grams of barley, we can calculate how much urea must be excreted, and thus in turn how much water is needed. Likewise, the amount of water lost in the feces can be determined by collecting the feces and determining their water content.

The most important avenue for water loss is evaporation, most of this being from the respiratory tract. In table 9.12, water gains and losses just balance. If the atmospheric humidity is lower, the water gain will be slightly reduced because less free water is absorbed in the grain, and at the same time, in the drier air more water is evaporated from the respiratory tract. In a very dry atmosphere, therefore, the kangaroo rat cannot maintain water balance. In more humid air, on the other hand, above 20% relative humidity, less water evaporates from the respiratory tract because the inhaled air contains more water, and the kangaroo rat now readily maintains water balance. In nature, kangaroo rats spend much time in their underground burrows where the air humidity is somewhat higher than in the outside desert atmosphere, and this aids in their water balance.

The amount of evaporation from the respiratory tract depends on how much air is brought into the lungs (the ventilation volume) and on the fact that the exhaled air is always saturated with water vapor. The amount of water already present in inhaled air determines how much additional water is needed to saturate the air, for there is no evidence that any mammal exhales air at less than full saturation. Ventilation volume is determined by the rate of oxygen consumption (which in turn determines how much oxidation water is available). Mammals in general remove about 5 ml O_2 from 100 ml alveolar air before it is exhaled. If the oxygen extraction could be increased, the volume of respired air could be reduced (for a given oxygen consumption), and the amount of water evaporated would be correspondingly reduced. This avenue for reducing the respiratory evaporation apparently has not been used by any mammal. To extract more oxygen in the lung would require that the hemoglobin of the blood should have a higher affinity for oxygen, but kangaroo rat blood does not differ from the blood of other rodents in this regard. An increased oxygen extraction would also lead to an increase in carbon dioxide in the blood which would be reflected in the acid–base balance, but this system shows no such changes from the usual mammalian pattern.

However, an important reduction in respiratory evaporation is achieved by exhalation of air at a lower temperature than the body core. Although the lung air is at core temperature and saturated, it is cooled as it passes out through the nose. The mechanism is very simple (see fig. 9.16); during inhalation the walls of the passageways lose heat to the air flowing over them, the wall temperature decreases, and because of evaporation it may fall to below the temperature of the inhaled air. On exhalation, as warm air from the lungs passes over the cool surfaces, the air is cooled and water condenses on the walls. The extent of cooling, i.e. the final temperature of the exhaled air, varies with the temperature and humidity of the inhaled air.

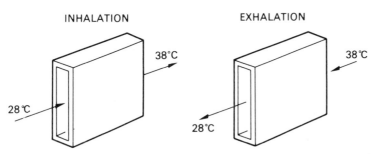

Fig. 9.16. Model of heat exchange in the nasal passageways. Ambient air is 28 °C and saturated, body temperature is 38 °C. As inhaled air flows through the passageways (left) it gains heat and water vapor, and is saturated and at 38 °C before it reaches the lungs. On exhalation (right) the air flows over the cool walls, gives up heat, and water recondenses. As heat and water exchange approaches completion, exhaled air approaches 28 °C, saturated (Schmidt-Nielsen, 1972).

How important is the cooling for the water balance? That varies, of course, with the conditions of the air the kangaroo rat breathes, but let us use a reasonable example with the air at 30 °C and 25% r. h. This air when inhaled into the lungs becomes heated to body temperature, 38 °C, and is saturated with water vapor so that it contains 46 mg H_2O per liter. On exhalation this air is cooled to 27 °C, and although still saturated, it contains only half as much water, i.e. half of the water is retained in the nasal passageways and is used to humidify the next inhaled breath. But note that the exhaled air is saturated and always contains more water vapor than inhaled air; in other words, the nose is not a system to wring water out of the air, and there is always a net loss of water from the respiratory tract.

Heat exchange in the nasal passages is not unique for kangaroo rats; it occurs in all animals. It is an inevitable result of heat ex-

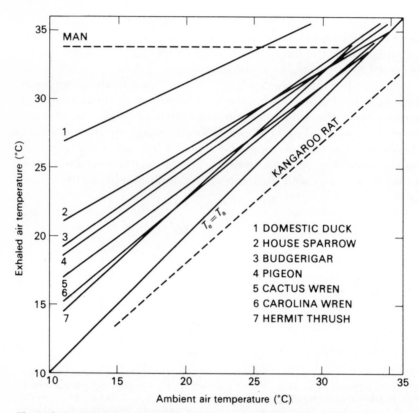

Fig. 9.17. Temperature of exhaled air in seven species of birds, measured in dry air at temperatures between 12 and 30 °C. The body temperature of all the birds was between 40 and 41 °C. Two of the birds, cactus wren and budgerigar (parakeet) are desert species, but there is no difference in the extent of cooling (and thus water recovery) between these and birds from other habitats. The less effective cooling of exhaled air in the duck comes with the larger size of its respiratory passageways (Schmidt-Nielsen, Hainsworth and Murrish, 1970).

change between air and the nasal walls, and this heat exchange is more complete when the passageways are narrow and the surface is large, as it is in small rodents. In birds, which have shorter and wider nasal passages, the cooling is less effective, as shown in fig. 9.17. Kangaroo rats, in dry air, exhale air at temperatures below the inhaled air, while birds, although they do cool the air appreciably, have higher exhaled-air temperatures. Man, with his much wider nasal passages, has a very incomplete heat exchange, and the exhaled air is commonly only a few degrees below body temperature. In man and other large animals, therefore, heat exchange in the nose and the associated water conservation are of lesser importance, and larger amounts of water are lost in the exhaled air.

Marine air-breathing vertebrates

The higher vertebrates, reptiles, birds, and mammals, are typically terrestrial, and, as we have seen, some are at home in the most arid deserts of the world. However, several lines of terrestrial vertebrates have secondarily invaded the sea, and in doing so they have remained air breathers. Therefore, in regard to problems of water and salt, they are still essentially 'terrestrial' animals and, compared to a fish, they are physiologically isolated from the surrounding sea water. In contrast to fish, which have gills and skin relatively permeable to water, the higher marine vertebrates have lungs and thus escape the osmotic problem of an intimate contact with the sea water over a large gill surface. The higher marine vertebrates differ physiologically from their terrestrial relatives primarily in that they have only sea water to drink, and that much of their food has a high salt content.

The sea contains enough water; the problem is that all this water contains about 35 g salt liter^{-1} and has an osmotic concentration of some 1000 mosmolar. Much of the food also has a high salt content, although its ionic composition often differs substantially from sea water. Plants and invertebrate animals, when eaten, present roughly the same osmotic problem as the drinking of sea water. Marine teleost fish, as we have seen, contain much less salt, and animals that subsist primarily on fish have therefore less of a salt problem than those that feed on plants or invertebrates. In any event, all of them must excrete the end products of protein metabolism, which, in mammals, is mainly urea and in birds and reptiles is uric acid.

If a vertebrate animal drinks sea water, the salts are absorbed and the concentration of salt in the body fluids will increase. Unless the salts are eliminated with a smaller volume of water than that which was taken in, there can be no net gain of water. In other words, the salts must be excreted in a solution at least as concentrated as sea water, otherwise the body will become more and more dehydrated. The reptilian kidney cannot produce urine which is more concentrated than the body fluids, and the bird kidney can usually produce urine no more than twice as concentrated as the blood. Since their body fluids are about 300–400 mosm liter^{-1} the urine cannot reach the concentration of sea water (1000 mosm liter^{-1}), i.e. the kidney does not have a sufficient concentrating ability to permit these animals to drink sea water or eat food with a high salt content, and if they do, they must have other mechanisms for salt excretion.

Marine reptiles

Three orders of reptiles, turtles, lizards, and snakes, have marine representatives. Some of the sea snakes are completely independent of land, even in their reproduction, for they bear live young and never leave the sea and would in fact be quite helpless on land. Sea turtles spend most of their lives in the open ocean, but they return to sandy tropical beaches for reproduction. Only the female turtles go on land to lay their eggs, the males never set foot on land after they as hatchlings enter the sea. The marine lizards are more tied to land. An example is the Galapagos marine iguana, *Amblyrhynchus cristatus*. It lives in the surf of the Galapagos Islands where it climbs on the rocks and feeds exclusively on seaweed. The fourth living order of reptiles, the crocodiles, probably have no truly marine representative. The saltwater crocodile, *Crocodylus porosus,* is primarily estuarine in its habits, it subsists mainly on fish and probably cannot survive indefinitely in a truly marine environment.

The excretion of excess salt, which the reptilian kidney is unable to handle, is carried out by glands in the head which are called salt-excreting glands, or simply salt glands. The salt glands produce a highly concentrated fluid which contains primarily sodium and chloride in concentrations substantially higher than in sea water. The glands do not function continually, as the kidney does, they secrete only intermittently in response to a salt load which increases the plasma salt concentrations. Similar salt-secreting glands are found in marine birds, in which they have been studied in greater detail.

In the marine lizards the salt glands empty their secretion into the anterior portion of the nasal cavity, and a ridge keeps the fluid from draining back and being swallowed. A sudden exhalation occasionally forces the liquid as a fine spray of droplets out through the nostrils. The Galapagos iguana eats only seaweed, which has a salt concentration similar to that of sea water. It is therefore a necessity for this animal to have a mechanism for excretion of salts in a high concentration (Schmidt-Nielsen and Fänge, 1958).

The marine turtles, whether plant eating or carnivorous, have a large salt-excreting gland in the orbit of the eye. The duct from the gland opens into the posterior corner of the orbit, and a turtle which has been salt loaded therefore cries salty tears. (The tears of man, which everybody knows have a salty taste, are isosmotic with the blood plasma. These tear glands therefore play no particular role in salt excretion.)

The sea snakes, which also excrete a salty fluid in response to a salt load, have salt glands which open into the oral cavity, from which the secreted liquid is expelled (Dunson, 1968). The sea snakes are close relatives of the cobras and are highly venomous, and physiological work on their salt metabolism therefore has certain exciting aspects which have tended to slow down their study.

Although the marine reptiles have the physiological mechanism necessary to eliminate salt in a highly concentrated fluid, the question of whether many of them normally do drink sea water in appreciable quantities remains unresolved.

Marine birds

Many birds are marine, but most of them live on and above, rather than in the ocean. Many are coastal, but some are truly pelagic. The young albatross that is hatched on a Pacific island spends three or four years over the open ocean before it returns to the breeding grounds. The penguin, the most highly adapted marine bird, has lost its power of flight. It is an excellent swimmer and is well advanced in the evolution towards a fully aquatic life. Nevertheless, in the physiological sense it has remained essentially a terrestrial air-breathing animal which reproduces on land. The saying that the emperor penguin does not even breed on land is a play on words, for it hatches its egg while standing on the ice during the long, dark antarctic winter.

All marine birds have a paired nasal salt gland which, through a duct, connects with the nasal cavity (Fänge, Schmidt-Nielsen and

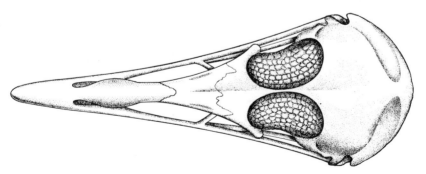

Fig. 9.18. The salt-secreting glands of the sea gull are located on top of the head, above the eye, in shallow depressions in the bone (Schmidt-Nielsen, 1963).

Osaki, 1958). A nasal gland is found also in most terrestrial birds, although in these it is very small while in marine birds it is without exception a large gland. Most frequently it is located on top of the skull, above the orbit of the eye, in shallow depressions in the bone (see fig. 9.18). In birds that regularly eat a diet high in salt or receive salt solutions to drink, the glands increase in size and become even larger than normal (Schmidt-Nielsen and Kim, 1964).

The salt glands are usually inactive and start secreting only in response to an osmotic stress, for example, when sea water or salty food is ingested. Otherwise the glands remain at rest, and in this respect they differ from the kidney which produces urine continuously. The secreted fluid has a simple composition, it contains mostly sodium and chloride in rather constant concentrations. Also in this regard the salt gland differs drastically from the kidney, which changes the concentrations and the relative proportions of the secreted components over a very wide range. Urine also contains a wide variety of organic compounds, while the salt gland secretion has no more than a trace amount of non-electrolytes (Schmidt-Nielsen, 1960).

Although the salt concentration in the secreted fluid is consistently high, there are characteristic species differences which are related to the normal ecology of the birds and their feeding habits. For example, the cormorant, which is a coastal bird and a fish-eater, secretes a fluid with a relatively low salt content, about 500 to 600

Table 9.13. *Concentration of sodium in the nasal secretion of different species of birds. The chloride concentration in a given sample is nearly identical to the sodium concentration, and other ions are found only in small amounts.* (Schmidt-Nielsen, 1960)

SPECIES	Concentration of sodium (meq liter^{-1})
Duck, mallard	400 – 600
Cormorant	500 – 600
Skimmer, black	550 – 700
Pelican, brown	600 – 750
Gull, herring	600 – 800
Gull, black-backed	700 – 900
Penguin, Humboldt's	725 – 850
Guillemot	750 – 850
Albatross, blackfooted	800 – 900
Petrel, Leach's	900 – 1100

mmol Na liter^{-1} (see table 9.13). The herring gull, which eats more invertebrate food and consequently ingests more salt, has a concentration of 600 to 800 mmol Na liter^{-1}. The petrel, a highly oceanic bird that feeds on planktonic crustaceans, has sodium concentrations in the nasal fluid up to 1100 mmol liter^{-1}.

The salt gland has an extraordinary capacity to excrete salt. In an experiment in which a gull received nearly one-tenth of its body weight of sea water (which would correspond to seven liters for a man) the entire salt load was eliminated in about three hours. The details of this experiment are recorded in table 9.14. The bird weighed 1420 grams and was given 134 ml sea water by stomach tube. During the following three hours the combined volume of excreted liquid equalled that ingested. The amount of sodium excreted by the nasal gland was, however, about ten times as high as the amount appearing in the cloacal discharge (urine mixed with some fecal material), with less than half (56.3 ml) coming from the salt gland and the remainder from the cloaca (75.2 ml).

Table 9.14. *Nasal and cloacal excretion by a black-backed gull during 175 min following the ingestion of sea water in an amount of nearly one-tenth of its body weight.* (Schmidt-Nielsen, 1960)

	NASAL EXCRETION			CLOACAL EXCRETION		
Time (min)	Vol. (ml)	Sodium conc. (mmol liter^{-1})	Sodium amount (mmol)	Vol. (ml)	Sodium conc. (mmol liter^{-1})	Sodium amount (mmol)
15	2.2	798	1.7	5.8	38	28
40	10.9	756	8.2	14.6	71	1.04
70	14.2	780	11.1	25.0	80	2.00
100	16.1	776	12.5	12.5	61	0.76
130	6.8	799	5.4	6.2	33	0.21
160	4.1	800	3.3	7.3	10	0.07
175	2.0	780	1.5	3.8	12	0.05
Total	56.3		43.7	75.2		4.41

An examination of table 9.14 shows that the volume of nasal secretion was highest during the second hour, after which it gradually tapered off. The concentration of the nasal secretion was quite constant, in spite of changes in flow rate. The volume of cloacal discharge (mostly urine) exceeded in volume the nasal secretion, but its salt concentration was quite low. The bird kidney can produce urine with a maximum sodium concentration of about 300 mmol liter^{-1}, but in this experiment the urine concentration was only a small fraction of this. Also, the urine sodium concentration varied about eightfold during the experiment, by no means unusual for the kidney, but it is in sharp contrast to the very constant concentration in the nasal secretion.

The volume of fluid secreted by the salt gland is remarkably high, in particular in view of the exceptional osmotic work performed in producing the concentrated salt solution. The flow from the salt glands, estimated per kilogram body weight, is about twice as high as the urine production of a man in maximum water diuresis (see table 9.15). Calculated per gram gland, the difference is even more striking. One gram of the salt gland can produce 0.6 ml fluid per minute, whereas the kidney of man in maximum water diuresis produces only $^1/_{20}$ as much. For the human kidney, when it produces a concentrated urine (which is still only half as concentrated as the secretion from the salt gland) the flow rate may decline to only 1% of that in water diuresis. This makes it evident that the salt gland is one of the most effective ion transport systems known.

Table 9.15. *Comparison of secretion rates of the salt gland of the herring gull with that of the human kidney during maximum water diuresis*

	Salt gland	Kidney
Secretion rate per kg body weight (ml min^{-1})	0.5	0.24
Secretion rate per gram gland (ml min^{-1})	0.6	0.03

The fluid secreted from the salt gland of marine birds always contains mostly sodium and very little potassium. The ratio between these two ions remains at about 30 : 1, and if the amount of potassium in the diet of gulls is increased, the Na^+ : K^+ ratio does not change much (Schmidt-Nielsen, 1965). Sea turtles and sea snakes have similarly high Na^+ : K^+ ratios, but in the marine iguana we find a lower ratio and the relative amount of potassium excreted by the salt gland is higher (see table 9.16). This is easy to understand; the diet of the marine iguana is primarily marine algae, and in general plants contain large quantities of potassium. The marine iguana therefore lives on a diet with a relatively high potassium content, and its need for potassium excretion is correspondingly high.

A typical land lizard from a dry habitat, the false iguana (*Ctenosaura*), has a nasal secretion with a very high potassium concentration while the sodium concentration is low, giving a Na^+ : K^+ ratio of 0.15. Another difference in the salt gland secretion of land reptiles is

Table 9.16. *The concentrations of sodium and potassium in the nasal salt gland secretion. The Na⁺:K⁺ ratio is high in the marine species and low in terrestrial plant-eaters*

	Na⁺ (mmol liter⁻¹)	K⁺ (mmol liter⁻¹)	Na⁺:K⁺ ratio
Sea water	470	10	47
Herring gull (*Larus*) (*a*)	718	24	30
Sea turtle (*Lepidochelys*) (*b*)	713	29	25
Sea snake (*Pelamis*) (*b*)	607	28	24
Marine iguana (*Amblyrhynchus*) (*b*)	1434	235	6.7
False iguana (*Ctenosaura*) (*c*)	78	527	0.15

(*a*) Schmidt-Nielsen, 1960; (*b*) Dunson, 1969; (*c*) Templeton, 1967.

that the composition of the secreted fluid changes with the nature of the load. If animals are given a sodium load, the secretion changes in favor of this ion, if they are given a potassium load, the secretion contains mainly potassium. Thus, in terrestrial reptiles the composition of the secreted fluid is regulated according to need, while the salt glands of marine birds and reptiles appear to be organs highly specialized for the specific excretion of sodium and chloride.

Marine mammals

Three orders of mammals, seals, whales, and sea cows, are exclusively marine in the sense that they spend practically their entire lives in the sea. Seals return briefly to land to bear and nurse their young, but whales and sea cows (manatee and dugong) even bear their young in water.

The food these mammals eat varies a great deal in salt content. Seals and whales are carnivorous and feed on fish, certain large invertebrates, and marine plankton organisms. Those that feed on fish, as many seals and some whales do, obtain food with a rather low salt content, less than 1%, but with a relatively high protein content. Some seals, the crab-eating seal of the Antarctic and the walrus (which feeds on clams and other bottom organisms), live on food organisms which are isosmotic with sea water. The baleen whales feed on crustacean plankton organisms which have the high salt content characteristic of marine invertebrates. Also, some sea water may be incidentally ingested with the food, and this would add further to the salt load. Dugongs and manatees are herbivorous and feed on plants which are in osmotic equilibrium with sea water and thus also have a high salt intake.

Do marine mammals have some physiological mechanism that corresponds to the salt-secreting glands of birds and reptiles? This should not be necessary, for the kidneys of whales and seals can produce a urine more concentrated than sea water. The highest chloride concentration reported in whale urine is 820 mmol liter^{-1} (Krogh, 1939). This is well above the concentration of sea water (about 535 mmol Cl liter^{-1}), and a whale that takes food with a high salt content, or even sea water, should be able to eliminate the salts without difficulty. It is well known that sea water is toxic to man, and that a castaway at sea only hastens the dehydration processes if he drinks sea water. The effect of drinking sea water on the water balance for a man is compared to that of a whale in table 9.17. A whale could drink one liter of sea water and have a net gain of about one-third liter of pure water after the salts are excreted. The kidney of man is less powerful, the maximum urine concentration is below that in sea water, and if he drinks one liter of sea water, he inevitably ends up with a net water loss of about one-third liter and is worse off than if he didn't drink at all. His dehydration is further aggravated by the large amount of magnesium and sulfate in sea water, which acts as a laxative and causes diarrhea, thus increasing the water loss.

There is still inadequate information about whether seals and whales ingest any appreciable amounts of sea water, either by drinking or incidentally with the feed. There is good evidence, however, that seals do not have to resort to the drinking of sea water. California sea lions (*Zalophus californianus*) have been kept in captivity and given nothing but fish to eat, and even after 45 days without access

Table 9.17. *Effect on the water balance of ingesting one liter of sea water in man and in whale.*

	Sea water consumed		Urine produced		Water balance
	Vol. (ml)	Cl conc. (mmol liter^{-1})	Vol. (ml)	Cl conc. (mmol liter^{-1})	Gain or loss (ml)
Man	1000	535	1350	400	− 350
Whale	1000	535	650	820	+ 350

to water, these animals were perfectly normal and in positive water balance (Pilson, 1970). It is interesting to note that if sea lions eat invertebrate animals, such as squid, they probably manage equally well. Although the blood of the squid is in osmotic equilibrium with sea water and has a high salt content, the osmotically active substances within the cells of the squid are only in part salts, for organic compounds are important tissue constituents and make up a substantial fraction of the total osmotic concentration. Thus, squid and many other invertebrates actually impose a lower salt load than we would expect merely from the osmotic concentration of their body fluids.

Mammals have in their water balance an item that does not apply to birds and reptiles, the female nurses her young and large quantities of water are required for production of milk. One way of reducing this loss of water would be to produce a more concentrated milk. It has long been known that seal and whale milk have a very high fat content, and also a higher protein content than cow's milk (table 9.18). This has usually been interpreted as a necessity for the rapid growth of the young, and particularly as a means to transfer a large amount of fat to be deposited as blubber and serve as insulation.

Table 9.18. *Composition of mammalian milk (g per 100 g milk)*

	H_2O	Fat	Protein	Carbo-hydrate	kcal per g H_2O
Cow (a)	87.3	3.7	3.3	4.8	0.8
Human (a)	87.6	3.8	1.2	7.0	0.8
Harp seal (b)	45.3	42.7	10.5	0	9.9
Weddell seal, mean (c)	43.6	42.2	(14.1) *	—	10.5
extreme (c)	27.2	57.9	(19.5) *	—	23

* This figure refers to total non-fat solids, but as seal milk is virtually carbohydrate free, most of it is protein.

(a) Kon and Cowie, 1961; (b) Sivertsen, 1935; (c) Kooyman and Drabek, 1968.

It seems that the high fat content of seal milk can also be viewed in light of the limited water resources of the mother. In the Weddell seal, which has been better studied than other seals, the fat content of the milk gradually increased during the lactation period, while the water content decreased correspondingly. The highest fat content found in Weddell seal milk is 57.9% (which is nearly twice as much fat as contained in whipping cream), with a water content in the same sample of only 27.2% (which contrasts to a water content in ordinary lean meat of about 65%). Indeed, seals provide nutrients for their young with a minimal expenditure of water. For each gram of water used, seal milk transfers nourishment more than ten times as effectively as land mammals (table 9.18, last column). It is in accord with these observations that female Weddell seals that nurse their young do not suffer dehydration when observed in captivity during the lactation period (Kooyman and Drabek, 1968).

REFERENCES

BEADLE, L. C. (1943). Osmotic regulation and the faunas of inland waters. *Biol. Rev.*, **18**, 172–83.

BEAMENT, J.W.L. (1958). The effect of temperature on the waterproofing mechanism of an insect. *J. Exp. Biol.*, **35**, 494–519.

BEAMENT, J.W.L. (1959). The waterproofing mechanism of arthropods. I. The effect of temperature on cuticle permeability in terrestrial insects and ticks. *J. Exp. Biol.*, **36**, 391–422.

BEAMENT, J.W.L. (1961). The water relations of insect cuticle. *Biol. Rev.*, **36**, 281–320.

BENTLEY, P. J. (1966). Adaptations of amphibia to arid environments. *Science*, **152**, 619–23.

BENTLEY, P. J. (1971). *Endocrines and Osmoregulation. A Comparative Account of the Regulation of Water and Salt in Vertebrates*, New York: Springer-Verlag, 300 pp.

BLISS, D. E. (1966). Water balance in the land crab, *Gecarcinus lateralis*, during the intermolt cycle. *Am. Zool.*, **6**, 197–212.

BURGER, J. W. and HESS, W. N. (1960). Function of the rectal gland in the spiny dogfish. *Science*, **131**, 670–71.

BUXTON, P. A. (1923). *Animal Life in Deserts. A Study of the Fauna in Relation to the Environment*, London: Arnold, 176 pp.

CURRAN, P. F. (1960). Na, Cl, and water transport by rat ileum *in vitro. J. Gen. Physiol.*, **43**, 1137–48.

CURRAN, P. F. and MACINTOSH, J. R. (1962). A model system for biological water transport. *Nature, Lond.*, **193**, 347–8.

CROGHAN, P. C. (1958a). The osmotic and ionic regulation of *Artemia salina* (L.). *J. Exp. Biol.*, **35**, 219–33.

CROGHAN, P. C. (1958b). The mechanism of osmotic regulation in *Artemia salina* (L.): The physiology of the gut. *J. Exp. Biol.*, **35**, 243–9.

DIAMOND, J. M. (1962). The mechanism of water transport by the gall-bladder. *J. Physiol.*, **161**, 503–527.

DIAMOND, J. M. and BOSSERT, W. H. (1967). Standing-gradient osmotic flow. *J. Gen. Physiol.*, **50**, 2061–83.

DUNSON, W. A. (1968). Salt gland secretion in the pelagic sea snake *Pelamis. Am. J. Physiol.*, **215**, 1512–17.

DUNSON, W. A. (1969). Electrolyte excretion by the salt gland of the Galapagos marine iguana. *Am. J. Physiol.*, **216**, 995–1002.

EDNEY, E. B. (1954). Woodlice and the land habitat. *Biol. Rev.*, **29**, 185–219.

EDNEY, E. B. (1956). The micro-climate in which woodlice live. *Proc. 10th Int. Congr. Entomol.*, **2**, 709–12.

EDNEY, E. B. (1957). *The Water Relations of Terrestrial Arthropods*, Cambridge, England: Cambridge University Press, 109 pp.

EDNEY, E. B. (1966). Absorption of water vapour from unsaturated air by *Arenivago* sp. (Polyphagidae, Dictyoptera). *Comp. Biochem. Physiol.*, **19**, 387–408.

FÄNGE, R., SCHMIDT-NIELSEN, K. and OSAKI, H. (1958). The salt gland of the herring gull. *Biol. Bull.*, **115**, 162–71.

GORDON, M. S., SCHMIDT-NIELSEN, K. and KELLY, H. M. (1961). Osmotic regulation in the crab-eating frog (*Rana cancrivora*). *J. Exp. Biol.*, **38**, 659–78.

GORDON, M. S. and TUCKER, V. A. (1965). Osmotic regulation in the tadpoles of the crab-eating frog (*Rana cancrivora*). *J. Exp. Biol.*, **42**, 437–45.

GROSS, W. J., LASIEWSKI, R. C., DENNIS, M. and RUDY, P., JR. (1966). Salt and water balance in selected crabs of Madagascar. *Comp. Biochem. Physiol.*, **17**, 641–60.

HOUSE, C. R. (1963). Osmotic regulation in the brackish water teleost *Blennius pholis*. *J. Exp. Biol.*, **40**, 87–104.

HUTCHINSON, G. E. (1957). *A Treatise on Limnology*, vol. I, *Geography, Physics, and Chemistry*, New York: John Wiley & Sons, Inc., 1015 pp.

HUTCHINSON, G. E. (1967). *A Treatise on Limnology*, vol. II, *Introduction to Lake Biology and the Limnoplankton*, New York: John Wiley & Sons, Inc., 1115 pp.

KEYS, A. B. (1933). The mechanism of adaptation to varying salinity in the common eel and the general problem of osmotic regulation in fishes. *Proc. Roy. Soc. Lond. B.*, **112**, 184–99.

KEYS, A. B. and WILLMER, E. N. (1932). 'Chloride secreting cells' in the gills of fishes, with special reference to the common eel. *J. Physiol.*, **76**, 368–78.

KING, J. R. (1957). Comments on the theory of indirect calorimetry as applied to birds. *Northwest Science*, **31**, 155–69.

KOCH, H. J. A. (1968). Migration. In *Perspectives in Endocrinology. Hormones in the Lives of Lower Vertebrates* (E.J.W. Barrington and C.B. Jørgensen, eds), pp. 305–49, London: Academic Press.

KON, S. K. and COWIE, A. T. (1961). *Milk: The Mammary Gland and Its Secretion*, vol. II, New York: Academic Press, 423 pp.

KOOYMAN, G. L. and DRABEK, C. M. (1968). Observations on milk, blood, and urine constituents of the Weddell seal. *Physiol. Zool.*, **41**, 187–94.

KROGH, A. (1939). *Osmotic Regulation in Aquatic Animals*, Cambridge, England: Cambridge University Press, 242 pp.

LIVINGSTONE, D. A. (1963). Chemical composition of rivers and lakes. In *Data of Geochemistry*, 6th edn, (M. Fleischer, ed.), Geological Survey Professional Paper 440, Chapter G, 64 pp. Washington, D.C.: U.S. Govt. Printing Ofc.

LOVERIDGE, J. P. (1970). Observations on nitrogenous excretion and water relations of *Chiromantis xerampelina* (Amphibia, Anura). *Arnoldia,* **5,** 1–6.

LUTZ, P. L. and ROBERTSON, J. D. (1971). Osmotic constituents of the coelacanth *Latimeria chalumnae* Smith. *Biol. Bull.,* **141,** 553–60.

MACHIN, J. (1964). The evaporation of water from *Helix aspersa.* I. The nature of the evaporating surface. *J. Exp. Biol.,* **41,** 759–69.

MAETZ, J. and CAMPANINI, G. (1966). Potentiels transépithéliaux de la branchie d'anguille *in vivo* en eau douce et en eau de mer. *J. Physiol. (Paris),* **58,** 248.

MAYER, N. (1969). Adaptation de *Rana esculenta* à des milieux variés. Etude speciale de l'excrétion rénale de l'eau et des électrolytes au cours des changements de milieux. *Comp. Biochem. Physiol.,* **29,** 27–50.

MOTAIS, R. and MAETZ, J. (1965). Comparaison des échanges de sodium chez un téléostéen euryhalin (le flet) et un téléostéen sténohalin (le serran) en eau de mer. Importance rélative du tube digestif et de la branchie dans ces échanges. *C.R. Acad. Sci. Paris,* **261,** 532–5.

PANG, P.K.T., GRIFFITH, R. W., and KAHN, N. (1972). Electrolyte regulation in the fresh water stingrays (Potamotrygonidae). *Fed. Proc.,* **31,** 344.

PICKFORD, G. E. and GRANT, F. B. (1967). Serum osmolality in the coelacanth, *Latimeria chalumnae:* urea retention and ion regulation. *Science,* **155,** 568–70.

PILSON, M.E.Q. (1970). Water balance in sea lions. *Physiol. Zool.,* **43,** 257–69.

POTTS, W.T.W. (1954). The energetics of osmotic regulation in brackish- and fresh-water animals. *J. Exp. Biol.,* **31,** 618–30.

POTTS, W.T.W. and EVANS, D. H. (1967). Sodium and chloride balance in the killifish *Fundulus heteroclitus. Biol. Bull.,* **133,** 411–25.

POTTS, W.T.W. and PARRY, G. (1964). *Osmotic and Ionic Regulation in Animals,* Oxford, England: Pergamon Press, 423 pp.

RAMSAY, J. A. (1935). Methods of measuring the evaporation of water from animals. *J. Exp. Biol.,* **12,** 355–72.

RAMSAY, J. A. (1949). The osmotic relations of the earthworm. *J. Exp. Biol.,* **26,** 46–56.

ROBERTSON, J. D. (1954). The chemical composition of the blood of some aquatic chordates, including members of the Tunicata, Cyclostomata and Osteichthyes. *J. Exp. Biol.,* **31,** 424–42.

ROBERTSON, J. D. (1957). Osmotic and ionic regulation in aquatic invertebrates. In *Recent Advances in Invertebrate Physiology,* pp. 229–46. Eugene, Ore.: University of Oregon.

RUIBAL, R. (1962). The adaptive value of bladder water in the toad, *Bufo cognatus. Physiol. Zool.,* **35,** 218–23.

SCHMIDT-NIELSEN, K. (1960). The salt-secreting gland of marine birds. *Circulation,* **21,** 955–67.

SCHMIDT-NIELSEN, K. (1963). Osmotic regulation in higher vertebrates. *Harvey Lectures,* Ser. **58,** 53–95.

SCHMIDT-NIELSEN, K. (1964). *Desert Animals. Physiological Problems of Heat and Water,* Oxford, England: Clarendon Press, 277 pp.

SCHMIDT-NIELSEN, K. (1965). Physiology of salt glands. In: *Funktionelle und morphologische Organisation der Zelle. Sekretion und Exkretion,* pp. 269–88, Berlin: Springer-Verlag.

SCHMIDT-NIELSEN, K. (1969). The neglected interface: the biology of water as a liquid gas system. *Quart. Rev. Biophys.,* **2,** 283–304.

SCHMIDT-NIELSEN, K. (1972). *How Animals Work,* Cambridge, England: Cambridge University Press, 114 pp.

SCHMIDT-NIELSEN, K. and FÄNGE, R. (1958). Salt glands in marine reptiles. *Nature, Lond.,* **182,** 783–5.

SCHMIDT-NIELSEN, K., HAINSWORTH, F. R. and MURRISH, D. E. (1970). Countercurrent heat exchange in the respiratory passages: effect on water and heat balance. *Respir. Physiol.,* **9,** 263–76.

SCHMIDT-NIELSEN, K. and KIM, Y. T. (1964). The effect of salt intake on the size and function of the salt gland of ducks. *Auk,* **81,** 160–72.

SCHMIDT-NIELSEN, K. and LEE, P. (1962). Kidney function in the crab-eating frog *(Rana cancrivora). J. Exp. Biol.,* **39,** 167–77.

SCHMIDT-NIELSEN, K., TAYLOR, C. R. and SHKOLNIK, A. (1971). Desert snails: problems of heat, water and food. *J. Exp. Biol.,* **55,** 385–98.

SIVERTSEN, E. (1935). Über die chemische Zusammensetzung von Robbenmilch. *Magazin for Naturvidenskaberne,* **75,** 183–5.

SMITH, H. (1931). The absorption and excretion of water and salts by the elasmobranch fishes. I. Fresh water elasmobranchs. *Am. J. Physiol.,* **98,** 279–95.

SMITH, H. W. (1959). *From Fish to Philosopher,* Summit, N.J.: CIBA Pharmaceutical Products, Inc., 304 pp. Reprinted (1961), Garden City, N.Y.; Doubleday & Co.

SVERDRUP, H. U., JOHNSON, M. W. and FLEMING, R. H. (1942). *The Oceans. Their Physics, Chemistry, and General Biology.* New York: Prentice-Hall, Inc. 1087 pp.

TEMPLETON, J. R. (1967). Nasal salt gland excretion and adjustment to sodium loading in the lizard, *Ctenosaura pectinata. Copeia, 1967;* no. 1: 136–40.

THESLEFF, S. and SCHMIDT-NIELSEN, K. (1962). Osmotic tolerance of the muscles of the crab-eating frog *Rana cancrivora. J. Cell. Comp. Physiol.,* **59,** 31–4.

THORSON, T. B., COWAN, C. M. and WATSON, D. E. (1967). *Potamotrygon* spp.: elasmobranchs with low urea content. *Science,* **158,** 375–7.

THORSON, T. B., WATSON, D. E. and COWAN, C. M. (1966). The status of the freshwater shark of Lake Nicaragua. *Copeia, 1966,* no. 3, 385–402.

USSING, H. H. and ZERAHN, K. (1951). Active transport of sodium as the source of electric current in the short-circuited isolated frog skin. *Acta Physiol. Scand.,* **23,** 110–27.

WALL, B. J. and OSCHMAN, J. L. (1970). Water and solute uptake by rectal pads of *Periplaneta americana. Am. J. Physiol.,* **218,** 1208–15.

WALL, B. J., OSCHMAN, J. L. and SCHMIDT-NIELSEN, B. (1970). Fluid transport: concentration of the intercellular compartment. *Science,* **167,** 1497–8.

WIGGLESWORTH, V. B. (1945). Transpiration through the cuticle of insects. *J. Exp. Biol.,* **21,** 97–114.

10
Excretion

The previous chapter repeatedly mentioned the role of excretory organs in osmoregulation and in the maintenance of a steady-state of water and solutes. This chapter will discuss in greater detail how the excretory organs work.

Excretory organs have a number of functions, all related to the maintenance of a constant internal environment in the organism. The maintenance of a constant composition entails one basic requirement, that any material taken in must be balanced by an equal amount removed. This in turn requires that the excretory paths must have a variable capacity that can be adjusted to remove judiciously controlled amounts of each of a tremendous variety of different substances. The major functions of excretory systems can be listed as follows:

(1) Maintenance of proper concentrations of individual ions, e.g. Na^+, K^+, Cl^-, Ca^{++}, H^+, etc.
(2) Maintenance of proper body volume (water content).
(3) From the preceding follows: maintenance of osmotic concentrations.
(4) Removal of metabolic end products, e.g. urea, uric acid, etc.
(5) Removal of foreign substances and/or their metabolic products.

The first three of these functions were repeatedly brought up in the previous chapter. The fourth concerns metabolic excretory products. One major end product of metabolic activity is carbon dioxide, and the bulk of it is removed in the respiratory organs; most other metabolic end products are removed by the excretory organs, which also remove a wide variety of foreign substances, either unchanged or after some modification that makes them harmless (= detoxification) or more easily excreted.

The main role of excretory organs therefore becomes one of removing from the body, accurately regulated amounts of materials that are in excess, thus helping to maintain a steady-state in response to all those influences that tend to impose a change.

There is a large variety of excretory organs, but there are, in principle, only two basic processes responsible for the formation of the excreted fluid, (*a*) *ultrafiltration,* and (*b*) *active transport.* Ultrafiltration means that pressure forces a fluid through a semi-permeable membrane which withholds protein and similar large molecules, while water and small molecular solutes can pass (see appendix 5). Active transport means transport against a gradient. If it is directed from the animal into the lumen of the excretory organ or organelle we call it an *active secretion.* If the active transport is in the opposite direction, from the lumen back into the animal, we speak about an *active reabsorption.* In those cases where the initial fluid is formed by ultrafiltration, this fluid is modified as it passes through the canals of the excretory organ, certain substances being removed from the filtrate by active reabsorption, while others may be added to the ultrafiltrate by active secretion. Thus active transport is often, if not always, superimposed on the filtration system.

Excretory organs and organelles show a great variety of morphological structure and anatomical location, yet they can be classified into a relatively small number of functional types. Some are generalized, or non-specialized excretory organs and can, in a general sense, be regarded as kidneys and their excretory product as urine. Other excretory organs have more specialized roles in that they carry out one particular function. Some examples will help:

GENERALIZED EXCRETORY ORGANS	SPECIALIZED EXCRETORY ORGANS
Contractile vacuoles of Protozoa	Gills (e.g. crustacea, fish)
Invertebrate nephridial organs	Rectal glands (elasmobranchs)
Malpighian tubule of insects	Salt glands (reptiles, birds)
Vertebrate kidneys	Liver (vertebrates)

The vertebrate kidney is the best known of the generalized excretory organs. It has been studied extensively, and its function is well understood. The contractile vacuole is only a part of a cell (i.e. an organelle) and is so small that studies of the fluid it produces have been extremely difficult, but a few facts about its function are known. The Malpighian tubule of insects has been carefully studied in a few species, and it is quite well understood. The excretory organs of other invertebrates, however, are much less well understood, and we have only approximate ideas about how some of them may work. For most of them we do not even know whether a secretory process or ultrafiltration is the initial process in the formation of the excreted fluid.

Several specialized excretory organs were discussed in the preceding chapter on osmotic regulation. However, no mention was made of the vertebrate liver, whose excretory function is as follows: red blood cells that have reached their normal life span are sequestered from circulation by the liver, broken down, and the porphyrin of the hemoglobin molecule is transformed to compounds known as bile pigments. These are excreted in the bile, discharged into the intestine, and finally eliminated with the feces. Thus, in addition to its many other functions, the liver is an excretory organ specialized for porphyrin excretion. It also plays a major role in the metabolism and detoxification of a wide variety of foreign substances and may have other minor roles in excretion.

ORGANS OF EXCRETION

Contractile vacuoles

Two animal groups have contractile vacuoles in common, protozoans and sponges. It seems that all fresh-water protozoans have contractile vacuoles. Whether contractile vacuoles are present in marine forms is more uncertain, but in at least some marine ciliates they have been demonstrated with certainty. The occurrence of contractile vacuoles in fresh-water sponges, although previously doubted, has been confirmed beyond doubt (Jepps, 1947).

Since fresh-water forms are always hypertonic to the medium in which they live, and their surface is permeable to water, they must continually perform osmotic work. They must continually eliminate excess water and also replace lost solutes, presumably by active uptake of salts from the external medium. Estimates of the water permeability of the large amoeba *Chaos chaos* indicate that the calculated osmotic influx of water is in good agreement with the observed volume of fluid eliminated by the contractile vacuole. This confirms the widely held opinion that the primary function of the contractile vacuole is in osmotic and volume regulation (Lövtrup and Pigon, 1951).

Microscopic observation of the contractile vacuole of fresh-water protozoans reveals continuous cyclic changes. It collects fluid and gradually increases in volume until a critical size is reached. It then suddenly expels its contents to the outside and decreases in size, whereupon it again begins to enlarge.

The lumen of the contractile vacuole of amoebas is surrounded by a single thin membrane. Surrounding this membrane again is a thick layer (0.5–2 μm thick) densely packed with small vesicles, each some

0.02–0.2 μm in diameter. Around this layer of small vesicles is a layer of mitochondria, which presumably provide the energy required for the osmotic work of producing the hypotonic contents of the vacuole. In electron-micrographs it appears that the small vesicles empty into the contractile vacuole by fusion of their membrane (fig. 10.1).

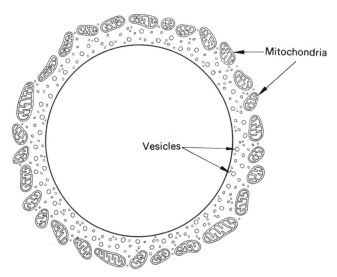

Fig. 10.1. The contractile vacuole of *Amoeba proteus* is enclosed in a membrane and is surrounded by a layer of tiny vesicles which are filled with fluid and appear to empty into the vacuole. Around this structure is a layer of mitochondria which presumably provides the energy for the secretory process (Mercer, 1959).

The role of the contractile vacuole in osmotic regulation has been well demonstrated in the euryhaline amoeba, *Amoeba lacerata*. This amoeba is originally a fresh-water organism, but it has a high salt tolerance and can eventually be adapted to 50% sea water. The rate of emptying of the contractile vacuole of this amoeba, when it is adapted to various salt concentrations, varies inversely with the osmotic concentration of the medium (see fig. 10.2). It appears, therefore, that the contractile vacuole works to bail out water as fast as it enters by osmotic inflow, for as the concentration of the medium increases, the amount of entering water decreases. In a marine habitat, where the inside and outside osmotic concentrations must be assumed to be nearly equal, contractile vacuoles (in those forms where they have been observed) empty at a very low rate. In these cases we must assume that they are not primarily organelles of osmoregulation but carry out other excretory functions as well.

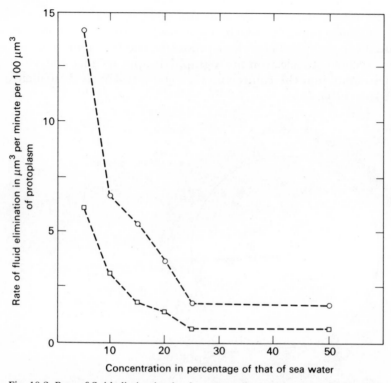

Fig. 10.2. Rate of fluid elimination by the contractile vacuole of *Amoeba lacerata* in relation to the concentration of the medium. The amoebae were tested in the solution in which they were grown (after Hopkins, 1946). Squares, average rates; circles, maximum rates.

If the primary function of the contractile vacuole of fresh-water protozoans is to remove water, the contents of the vacuole should be hypotonic to the remainder of the cells. This is indeed the case. Withdrawal of minute samples of fluid from the contractile vacuole has shown that the osmotic concentration is about one-third of the protoplasm, yet several times as high as the surrounding medium (B. Schmidt-Nielsen and Schrauger, 1963). The contractile vacuole can thus eliminate a hypotonic fluid and serve to remove water. Nevertheless, since the eliminated fluid has a higher osmotic concentration than the medium, solutes are continuously lost and it is necessary to postulate that the amoeba can take up the needed solutes, presumably by active transport directly from the medium.

The question of how the vacuole can increase in volume and yet contain a hypotonic fluid could be explained in different ways, one being active water transport into the vacuole. This is, however, for a number of reasons an unlikely hypothesis. Another alternative is that the vacuole originally contains an isotonic fluid from which osmotically active substances are withdrawn before the fluid is discharged to the surface. This proposal is contrary to the observation that the fluid concentration is hypotonic and relatively constant in composition throughout the growth of the vacuole.

Information about the composition of the vacuolar fluid permits us to propose a mechanism for secretion. Table 10.1 shows that the osmotic concentration of the vacuolar fluid is about half that of the cytoplasm, but more than 25 times as high as that of the surrounding medium. The sodium concentration in the vacuolar fluid is relatively high, in fact, 33 times as high as the cytoplasmic concentration of sodium. The potassium concentration in the vacuolar fluid, on the other hand, is relatively low, and substantially below the potassium concentration in the cytoplasm. The sum of sodium and potassium in the vacuolar fluid is about 25 mmol liter^{-1}, and if the anion is, say chloride, the total osmotic concentration of the fluid (51 mosm liter^{-1}) would be nearly accounted for.

Table 10.1. *Solute concentrations in plasma and in the contractile vacuole of a fresh-water amoeba. Mean volume of the vacuole was about 0.2 nl.* (Riddick, 1968)

	Medium	Cytoplasm	Vacuole	Ratio $\dfrac{\text{vacuole}}{\text{cytoplasm}}$
osmotic (mosm liter^{-1})	<2	117	51	<0.49
Na$^+$(mmol liter^{-1})	0.2	0.60	19.9	33
K$^+$ (mmol liter^{-1})	0.1	31	4.6	0.15

The most likely mechanism for formation of the contractile vacuole is as follows: the small vesicles which surround the contractile vacuole are initially filled with a fluid isotonic with the cytoplasm. The vesicles then pump sodium into the fluid by active transport, and remove potassium, also by active transport, in such a manner that removal of potassium exceeds sodium accumulation. The membrane of the vesicles must be relatively impermeable to water to permit the formation of a vesicle fluid hypotonic to the cytoplasm. If the hypotonic vesicles now fuse and open into the contractile vacuole, as the electron-micrographs indicate, the contractile vacuole becomes a receptacle for fluid produced by the vesicles, the necessary energy for the osmotic work being provided by the layer of mitochondria adjacent to the vesicles. Since sodium is continuously lost through the activity of the contractile vacuole, we must assume that it is replaced by active uptake at the cell surface (Riddick, 1968).

Invertebrate nephridial organs

The function of various excretory organs is easier to discuss if we first classify them into groups or types.* The major types and their distribution in the animal kingdom are approximately as follows:

Contractile vacuoles	Protozoans
	Sponges
None demonstrated	Coelenterates
	Echinoderms
Nephridial organs	
Protonephridium, closed end	Platyhelminthes
	Aschelminthes
Metanephridium, open ended	Annelids
Nephridium	Molluscs
Antennal gland	
(= green gland)	Crustaceans
Malpighian tubules	Insects
Kidneys	Vertebrates

Contractile vacuoles, which occur in fresh-water protozoans and sponges, and at least in some marine species, have already been discussed.

No specific excretory organs have been identified in coelenterates and echinoderms. This is curious, for fresh-water coelenterates are distinctly hypertonic to the medium, and they undoubtedly gain water by osmotic influx. How the excess water is eliminated is not known. Echinoderms, on the other hand, have no problem of osmoregulation, for echinoderms do not occur in fresh water and the marine forms are always isosmotic with sea water.

---

* This functional grouping of the renal organs of invertebrates is widely used, but is not in accord with the morphological grouping of excretory organs of Goodrich (1945), whose classification is based on whether the excretory organ is derived from a tube that develops from the outside and inward (a *nephridium*) or from the inside (the coelom) and outward (a *coelomoduct*). The renal organs of molluscs, arthropods, and vertebrates are derived from coelomoducts, those of other invertebrates (except protozoans, sponges, coelenterates, and echinoderms) are nephridia. Goodrich also recognizes a complex variety of mixed renal organs. The renal organs of insects, the *Malpighian tubules,* arise from the posterior part of the gut and do not fit into the Goodrich classification.

True organs of excretion are found only in those animal phyla that show bilateral symmetry. The most common type, widely distributed among invertebrates, is a simple or branching tube which opens to the outside through a pore (nephridial pore). There are two major types, *protonephridia* whose internal end is closed and terminates blindly, and *metanephridia* which connect to the body cavity through a funnel-like structure called nephridiostome or nephrostome.

Protonephridia occur mainly in animals that lack a true body cavity (coelom). They may have two or more protonephridia, which are often extensively branched. The closed ends terminate in enlarged bulb-like structures with a hollow lumen in which is found one or several long cilia (fig. 10.3). If there is a single cilium, the terminal cell is called a *solenocyte;* if there are numerous (often several dozen) cilia projecting into the lumen, the structure is called a *flame cell*

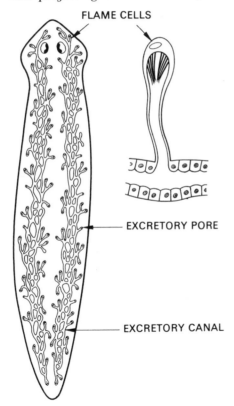

FLAME CELLS

EXCRETORY PORE

EXCRETORY CANAL

Fig. 10.3. The excretory system of a planaria is a widely branched system. The excreted fluid is initially formed at the solenocytes, or flame cells, and then passed down the nephridial ducts and discharged through excretory pores.

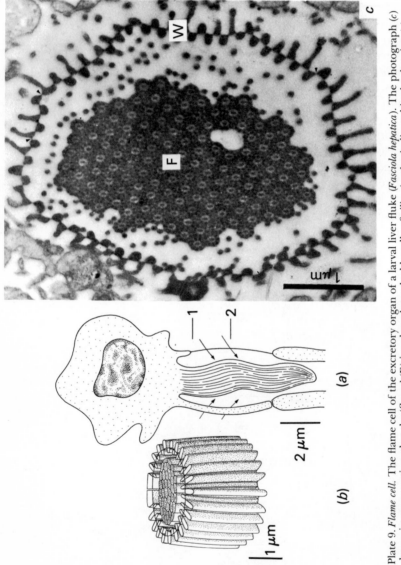

Plate 9. *Flame cell.* The flame cell of the excretory organ of a larval liver fluke (*Fasciola hepatica*). The photograph (*c*) shows, in cross-section, that the 'flame' (F) is a densely packed bundle of cilia. As also indicated in the drawing (*a*) and (*b*), the flame is surrounded by a thin wall (W) of pillar-like rods. The initial urine is presumably formed by ultrafiltration through this wall (G. Kümmel, Freie Universität Berlin).

because the tuft of cilia, as it undulates, has some resemblance to the flickering flame of a candle. There is no information available about functional differences between flame cells and solenocytes.

Metanephridia are characteristically unbranched, and the inner end opens through a funnel into the coelomic cavity. Metanephridia are found only in animals with a coelom, but the reverse is not true, some animals with a coelom have protonephridia, which otherwise are characteristic of acoelomate and pseudocoelomate animals (fig. 10.4).

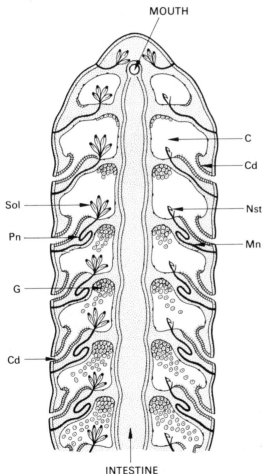

Fig. 10.4. Diagram of the front end of a primitive annelid in longitudinal section, showing relations of nephridia and coelomoducts to segmental coelomic cavities (C). On left side, protonephridia (Pn) with solenocytes (Sol); on right side metanephridia (Mn) with funnelshaped nephrostomes (Nst). The gonadal products (G) empty through the coelomoducts (Cd) at the nephrostome (Goodrich, 1945).

The function of the metanephridum (often called simply nephridium) is reasonably clear, primarily thanks to studies by J. A. Ramsay, who succeeded in removing minute samples of fluid from various parts of the earthworm nephridium. The results corroborate the following view: fluid from the coelom drains into the nephridium through the funnel-shaped nephrostome, and as it passes down through the extensively looped duct, its composition is modified. When originally entering the nephridium the fluid is isotonic, but salt is withdrawn in the terminal parts of the nephridium, and a dilute urine is discharged (Ramsay, 1949). (The importance of a dilute urine for the osmoregulation of the earthworm was discussed on p. 402). The metanephridium therefore functions as a filtration–reabsorption kidney, in which an initial fluid is modified as it passes through a uriniferous tubule.

How the protonephridium functions is more uncertain, for it has been assumed that the closed end would not be suitable for an ultrafiltration process to take place. Furthermore, the closed end is located in tissue between the body cells, rather than in contact with coelomic fluid, as in the case of metanephridia. The function of solenocytes and flame cells is obscure, although it has been suggested that their constant beating could produce sufficient negative pressure to cause an ultrafiltration to take place. Direct evidence for this hypothesis may be difficult to obtain because of the small size of these structures. There is evidence, however, that the protonephridium of a rotifer, *Asplanchna,* functions on the basis of a filtration–reabsorption mechanism. *Asplanchna* has body fluids hypertonic to the medium and produce a dilute urine, and if the animal is moved to a more dilute medium, more dilute urine is formed. This shows at least that the protonephridium is involved in osmoregulation and the excretion of water (Braun, Kümmel and Mangos, 1966).

The most important question in the study of the protonephridium is whether an ultrafiltration takes place. The generally accepted method for demonstrating ultrafiltration is to inject the substance inulin into the body and observe whether it appears in the urine. Inulin (which will be further discussed under the vertebrate kidney) is a soluble polysaccharide with a molecular weight of about 5000. Inulin is not metabolized in the body, and it can appear in the urine only if a filtration process takes place. It is never excreted by a cellular transport or secretion process, for it is inert to all known active transport processes. When injected into *Asplanchna*, inulin does appear in the urine, and this indicates that there is a filtration process. Furthermore, the concentration of inulin in the urine was higher than in the body; this indicates that water was reabsorbed before the urine was discharged to the exterior. The wall of the flame cell of *Asplanchna* is extremely thin and seems suitable for ultrafiltration to take place (Pontin, 1964).

Obviously, it is possible that other protonephridia may function differently, but since it has been shown in *Asplanchna* that there is an initial filtration, the need to separate functionally between the several types of nephridia seems less important.

Molluscs and crustaceans

The more highly organized invertebrates have more elaborate excretory organs which are often called kidneys (see footnote on p. 448). We shall discuss here only some aspects of excretion in molluscs and crustaceans.

The major groups of molluscs are cephalopods (octopus and squid), bivalves (clams, etc.) and gastropods (snails). Octopus and squid are strictly marine, but clams and snails occur in both marine and fresh water, and snails, of course, include terrestrial species as well. Renal function has been studied quite well in representatives of all these groups. In all of them an initial fluid is formed by ultrafiltration of the blood. This fluid contains the same solutes as present in the blood, except the proteins, and in virtually identical concentrations.* The ultrafiltrate therefore contains not only substances to be excreted, but also valuable substances such as glucose and amino acids. The discharge of the ultrafiltrate would be very wasteful, were it not for the fact that these valuable substances are reabsorbed before the urinary fluid is discharged to the outside.

In addition to ultrafiltration followed by selective reabsorption, there is also active secretion of certain substances, which are added to the urinary fluid in specific portions of the kidney. Two compounds that are actively secreted in a wide variety of kidneys (including vertebrate kidneys), are *para-aminohippuric acid* (often abbreviated *PAH*) and the dye *phenol red* (phenolsulfonphthalein). Phenol red has the advantage that, due to its color, its presence is readily observed and its concentration is easily measured. Phenol red and PAH will be mentioned again in the discussion of the vertebrate kidney because of their importance in demonstrating secretory processes.

One aspect of molluscan excretion seems rather peculiar, the function of the two kidneys is not always identical. For exam-

* The small concentration differences between blood and ultrafiltrate are due to the Donnan effect of the blood proteins. See appendix 5.

ple, the following information is about the abalone, *Haliotis.* Both kidneys have an initial ultrafiltration, for inulin appears in the urine from both kidneys, and its concentration is the same in blood, in pericardial fluid, and in the final urine from both kidneys. This indicates that the volume of urine is the same as the filtered volume, and that water is not reabsorbed to any appreciable extent. It is reasonable that a marine animal, which is isotonic with sea water and thus has no major problem of water regulation, does not reabsorb water in its kidney. It must, however, regulate its ionic composition and excrete metabolic products, and in this regard the two kidneys differ in their function, for PAH and phenol red are actively secreted primarily by the right kidney, while glucose reabsorption seems to take place primarily in the left kidney (Harrison, 1962).

The renal organ of crustaceans is the so-called *antennal gland* or the *'green gland'*. This paired gland is located in the head; it consists of an initial sac, a long coiled excretory tubule, and a bladder, which opens into an excretory pore near the base of the antennae. Hence the name, antennal gland.

Urine is formed in the antennal gland by filtration and reabsorption, with tubular secretion added. Ultrafiltration can be demonstrated by injection of inulin, which then appears in the urine. The lobster, a typical marine animal, produces urine with the same inulin concentration as in the blood, i.e. the urine:blood concentration ratio is 1.0, which shows that water is not reabsorbed. PAH and phenol red, however, are found in higher concentrations in the urine than in the blood. This shows that they are secreted into the urine, and that, in addition to the filtration–reabsorption process, secretion is also involved in urine formation (Burger, 1957).

In contrast to the lobster, the common shore crab, *Carcinus*, reabsorbs water from the ultrafiltrate. *Carcinus* is a good osmoregulator and can penetrate into brackish water. The inulin concentration in its urine may exceed that in the blood by several-fold and this must be due to water reabsorption from the initial filtrate. The most likely explanation for the water movement is that there is an active reabsorption of sodium, followed by a passive reabsorption of water. The result is a urine with a lower sodium concentration than in the blood, and this is of obvious value to an animal that penetrates into a dilute environment where it will have problems with maintaining high blood concentrations (Riegel and Lockwood, 1961).

The fresh-water amphipod, *Gammarus pulex,* produces urine which is not only very dilute, but may even be more dilute than the medium (fig. 10.5). A close relative, *G. duebeni,* a brackish-water species which is able to tolerate fresh water, has urine concentrations closer

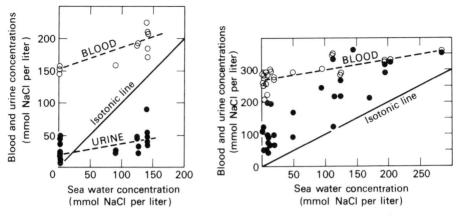

Fig. 10.5. Left: the relation between blood, urine, and medium concentration of the amphipod *Gammarus pulex*. This fresh-water species can form a highly dilute urine. Open circles, blood concentration; solid circles, urine concentration.
Right: the closely related brackish water species, *Gammarus duebeni*, is unable to produce urine more dilute than the medium (Lockwood, 1961).

to the blood concentration. This species is unable to form a very dilute urine, and in fresh water it cannot compete successfully with *G. pulex*.

Insects

For many invertebrate excretory organs, the question of whether the initial urine is formed by ultrafiltration is unresolved. For the excretory organs of the insects, however, the answer seems to be unequivocally negative. This conclusion is supported by the fact that inulin, which is considered the most reliable indicator of whether ultrafiltration takes place, does not appear in insect urine, thus indicating the complete absence of ultrafiltration (Ramsay and Riegel, 1961). How, then, is insect urine formed?

The excretory system of insects consists of tubules, known as *Malpighian tubules*, which may number anywhere from two to several hundred. The Malpighian tubules open into the intestine between the midgut and the hindgut, the other end is blind and, in most insects, lies in the hemocoel (fig. 10.6). Some insects, however, notably beetles that feed on dry substances (e.g. the mealworm, *Tenebrio*), have a particular arrangement associated with remarkable powers of withdrawal of water from the excrement. In these the blind end of the tubule lies closely associated with the rectum, the entire structure being surrounded by a membrane, the perirectal membrane. The

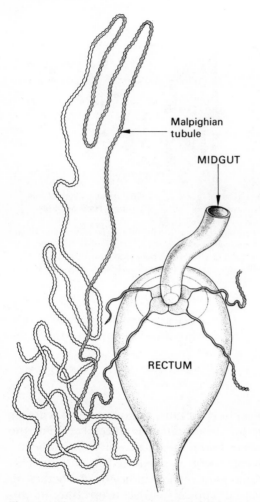

Fig. 10.6. The excretory system of the bug *Rhodnius*. Only one of the four Malpighian tubules is shown in full. The Malpighian tubules have two distinct parts, the upper and lower parts. They empty into the rectum at its junction with the midgut (Wigglesworth, 1931).

space formed by this membrane is filled with fluid (*perirectal fluid*) which surrounds both the Malpighian tubule and the rectal epithelium, but is separated from the general hemolymph.

The Malpighian tubules function as follows: potassium is actively secreted into the lumen of the tubule and water follows passively, due to osmotic forces. As a result a copious amount of a potassium-rich fluid is formed in the tubule, from which it enters the hindgut. In the hindgut solutes and much of the water are reabsorbed, and uric acid (which entered the fluid as water-soluble potassium urate) is precipitated. This facilitates the further withdrawal of water, for the precipitated uric acid does not contribute to the osmotic activity of the rectal contents. What remains in the rectum is eventually deposited as mixed urine and feces.

An insect that lives on fresh vegetation eats food with a high water content and excretes copious amounts of liquid urine. Other insects may live on dry food, they can produce very dry excreta and therefore may lose virtually no water in feces and urine. The mealworm, the larva of the flour beetle (*Tenebrio molitor*), is a good example – it spends its entire life cycle in dry flour and yet has all the water it needs. The mealworm has the characteristic perirectal membrane described above, and it can produce excreta so dry that they absorb water from air with 90% relative humidity (Ramsay, 1964). Let us see how the mechanism of water withdrawal can be explained. The osmotic concentration of the fluid in the perirectal space can be very high, its freezing point depression (ΔT) reaching as much as 8 °C. This is very much greater than the concentration ever is in the hemolymph ($\Delta T = 0.7$ to 1.4 °C). The difference is more pronounced if the mealworms have been kept at particularly low humidity. The rectal complex of the mealworm could function in either of two ways: (1) active transport of water from the lumen of the rectal complex into the hemolymph, or (2) active transport of a solute (presumably potassium chloride) from the hemolymph into the perirectal space, a high concentration in this space in turn accounting for the osmotic withdrawal of water from the rectal lumen (Grimstone, Mullinger and Ramsay, 1968).

The second of these alternatives for water reabsorption is, in principle, the same as Curran's three-compartment theory for fluid transport (see p. 414). The study of cockroaches has given further support for this hypothesis. When cockroaches are provided with water, they excrete a dilute urine, but if deprived of water, they produce dry fecal pellets and their rectal contents are now highly hyperosmotic to the hemolymph. A detailed analysis of osmotic concentrations in the rectal complex led to the mechanism for water reabsorption which is suggested in diagrammatic form in fig. 10.7.

In conclusion, it appears that the excretory system of insects operates without any initial ultrafiltration, and that it is based on a primary secretion of potassium into the Malpighian tubules, followed by passive movement of water, and that water as well as solutes are withdrawn in the hindgut and the rectal complex. Good evidence has been obtained that the movement of water is based on a primary solute transport.

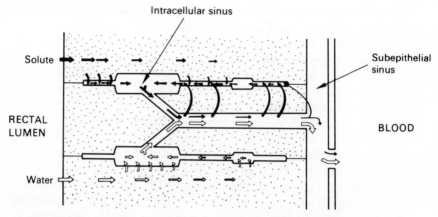

Fig. 10.7. Model of fluid movements in the rectal wall of the roach *Periplaneta americana*. Movement of solute is shown by black arrows, of water by open arrows. They are shown separately for reasons of clarity, although both occur in the same intracellular sinus. Solute is pumped actively into the intracellular sinus, both from the rectal lumen and from the fluid flowing into the subepithelial sinus. The high osmotic concentration in the intracellular sinus causes water to enter osmotically, and as a result there is a bulk flow of water and solutes towards the subepithelial sinus. The system is similar to Curran's three-compartment model (fig. 9.13) (from Oschman, J. L. and Wall, B. J. 1969. *J. Morph.*, **127**, 475–510, Fig. 34).

Vertebrate excretion

The kidneys of all vertebrates, fish, amphibians, reptiles, birds, and mammals, are similar to the extent that they function on the *filtration–reabsorption principle* with tubular secretion added. A few teleost fish differ from this general pattern, they lack the ultrafiltration mechanism and depend entirely on a secretory-type kidney.

What are the advantages and disadvantages of a filtration mechanism? The initial ultrafiltrate contains all the compounds present in the blood, with the exception of substances of large molecular size, such as proteins. Many of the filtered compounds are valuable and should not be lost. Reabsorption mechanisms must therefore be present so that compounds such as glucose, amino acids, vitamins and so on are conserved. A filtration–reabsorption kidney can process large volumes of fluid, and often more than 99% of the filtered volume is reabsorbed and less than 1% is excreted as urine. It might seem more advantageous to design a kidney that functions by tubular secretion only, for this would save a great deal of the work of reabsorption. We know that this solution in the evolutionary sense is feasible, for some teleost fish manage this way, and, as we have seen, the insect kidney also lacks an ultrafiltration process.

There is one unique consequence of ultrafiltration, any substance that has been filtered will remain in the urine, unless it is re-absorbed. An organism that encounters 'new' substances to be ex-creted can, with the aid of a filtration kidney, eliminate these with-out the need to develop a specialized secretory mechanism for each particular new substance it may meet. This gives an organism much greater freedom in exploring new environments, changing food habits, and so on. A secretory kidney is in this sense much more re-strictive. It is in accord with this viewpoint that, among vertebrates, secretory-type kidneys are found primarily in a small number of marine fish that live in a stable and 'conservative' environment.

All vertebrate kidneys can produce urine which is isotonic or hypotonic to the blood, only birds and mammals can also form urine more concentrated than the body fluids. In fresh water a dilute urine serves to bail out excess water while withholding solutes. Ani-mals in sea water cannot use a dilute or isotonic urine to eliminate excess salt, and they have accessory organs for salt excretion (e.g. gills, rectal glands, salt glands). Marine mammals have kidneys with exceptional concentrating ability, and they manage their salt prob-lems by renal excretion. For land mammals the ability to produce a concentrated urine is of great importance in their water balance. The fact that birds and reptiles excrete uric acid (rather than urea) enables them to produce a semi-solid urine which puts only moder-ate demands on the use of water for excretion (see further under Nitrogen Excretion, p. 475ff).

To understand how the kidney functions, we must know its struc-ture (see fig. 10.8). All vertebrate kidneys consist of a large number of units, *nephrons*. A small fish may have only a few dozen nephrons in their kidneys, a large mammal may have several million in each kidney. Each nephron begins with a *Malpighian body,* in which ultra-filtration of the blood plasma takes place. The tubule that leads from each Malpighian body can be divided into two parts, the first or *proximal tubule* in which many solutes such as salt and glucose as well as water are reabsorbed, and a *distal tubule* that continues the process of changing the tubular fluid into urine. The distal tubules which join with other tubules to form collecting ducts, empty the urine into the renal pelvis. From here, the urine passes through the ureters to the urinary bladder, from which at intervals it is dis-charged to the exterior. In the nephron of mammals the proximal and distal segments of the tubule are separated by a characteristic thin segment which forms a loop, like a hairpin, known as *Henle's loop.* A similar structure, although not as well developed, is found in the bird kidney. The loop is the particular structure in the nephron

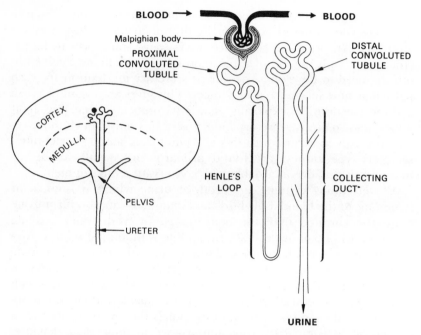

Fig. 10.8. Schematic diagram of a mammalian kidney. The kidney contains a large number, up to several million, of single nephrons. Only one nephron is indicated in this diagram, and is shown enlarged to the right. The outer layer of the kidney, the *cortex,* contains the *Malpighian bodies* and the proximal and the distal convoluted tubules. The capillary network within the Malpighian body is known as the *glomerulus.* The inner portion, the *medulla,* contains *Henle's loops* and *collecting ducts.*

The urine is initially formed by ultrafiltration in the Malpighian bodies, and the filtered fluid is modified and greatly reduced in volume as it passes down the renal tubule and into the collecting ducts, which empty the urine into the *pelvis,* from where it is conveyed to the bladder via the *ureter.*

which is responsible for the formation of a urine which is more concentrated than blood plasma. The loop is missing in fish, amphibians, and reptiles, and these can produce urine no more concentrated than the blood plasma.

How renal function is studied

The methods that are used to study renal function can give a surprising amount of information without any direct experimentation on the kidney itself. From the intact animal, we can obtain information about the amount of fluid formed by ultrafiltration, the amount of fluid reabsorbed in the renal tubule, the process of tubular secretion, and even the rate of blood flow to the kidney. The methods are routinely used, without any undue hazard or discomfort, to study kidney function in man. We should therefore understand the principles involved in such renal studies. Much information about renal function has also been obtained by direct experimentation on the kidney. This is especially true of studies of the mechanism responsible for formation of a concentrated urine. Through so-called micropuncture methods, in which fine pipettes are inserted into various segments of the nephron to withdraw minute samples of fluid, information has been gathered about the function of various parts of the nephron. Such methods require extensive surgery and are not suitable for use on man.

Ultrafiltration, which takes place in the Malpighian body, is caused by the blood pressure, which forces fluid out through the thin walls of the capillaries. Since the wall does not permit proteins to pass, the plasma proteins are withheld and only substances of low molecular weight are filtered with the water. For an ultrafiltration to take place, the blood pressure must exceed the osmotic pressure of the blood proteins, the so-called colloidal osmotic pressure. If the artery to a dog kidney is gradually constricted with a clamp so that the blood pressure in the kidney is decreased, urine production ceases when the blood pressure in the kidney falls to the colloidal osmotic pressure. This was shown many years ago by Starling. When the clamp is gradually loosened again so that the blood pressure in the capillaries is increased, filtration begins again when the blood pressure just exceeds the colloidal osmotic pressure, and the higher the excess pressure, the higher the filtration rate (Starling, 1899).

The Malpighian bodies of the kidney of mammals are located deep within the kidney, but in the amphibian kidney the Malpighian bodies are visible from the surface. The distinguished American renal physiologist, A. N. Richards succeeded in inserting micropipettes into the Malpighian bodies of amphibian kidneys, and he showed that the concentrations of chloride, glucose, urea, and phosphate, as well as the total osmotic pressure in the filtrate was the same in the blood, except for the minor differences which would be caused by the Donnan effect of the proteins. These experiments gave the essential proof that the initial urine is indeed an ultrafiltrate of the blood (Richards, 1935).

The total amount of fluid formed by ultrafiltration can be determined with the aid of the polysaccharide inulin without exposing the kidney or performing any other drastic manipulation. Inulin has a molecular weight of about 5000 and therefore passes through the capillary wall, which only withholds molecules larger than about 70 000 molecular weight. Inulin is not metabolized, and is neither reabsorbed nor secreted by the renal tubule, and all the inulin found in the final urine is therefore derived only from the filtration process. For example, if the inulin concentration in the urine is 100 times as high as in the plasma, it must mean that 99% of the water which was filtered has been reabsorbed, and only 1% of the water remains to form the final urine. By measuring the urine volume, the volume of the initial filtrate can be calculated. Let us say that we measure a urine flow of 1.3 ml min^{-1} in a man; if the inulin concentration in the urine is 100 times that in the plasma, the filtration rate would be 130 ml min^{-1}, which is a normal filtration rate for man.

We can now see that in order to determine the filtration rate in an intact animal we need only to (1) inject a suitable amount of inulin,* (2) collect the urine during a known period of time and measure its volume, and (3) determine the inulin concentration in the urine and in the plasma. It turns out that in man the filtration rate is rather constant, about 130 ml min^{-1}, and variations in urine volume are due primarily to variations in the amount of water reabsorbed. In many other animals, in particular in the lower vertebrates, the filtration rate can vary considerably, and in the frog the filtration rate can, without harm, cease completely for long periods of time.

If the filtration rate in man is 130 ml min^{-1}, or 0.13 liter min^{-1}, it means that in one hour 7.8 liters of filtrate is formed. This is roughly twice the total volume of blood plasma in man, and this volume has been subject to removal of excretory products. On the other hand, if glucose, amino acids, and other important substances in the filtrate were lost, the filtration process would soon deplete the organism of these valuable compounds. The saving process is tubular reabsorption.

One important substance that is reabsorbed in the proximal tubule is glucose, and it can serve as an example to illustrate the general principle of *tubular reabsorption*. Reabsorption of glucose is an active transport, and normally all filtered glucose is reabsorbed. If the amount of glucose in the plasma is increased above the normal level, the reabsorption mechanism is presented with larger than usual amounts of glucose, and if the amount exceeds the capacity of the transport mechanism, not all the glucose is reabsorbed and some remains in the urine. This is what happens to persons suffering from diabetes mellitus; their blood glucose readily increases to higher than normal levels, and when the filtered amount exceeds the capacity for reabsorption, glucose appears in the urine.

---

* Usually, inulin is infused continuously throughout an experiment. In this way the plasma concentration of inulin is maintained constant, rather than undergoing a continuous drop during the observation period.

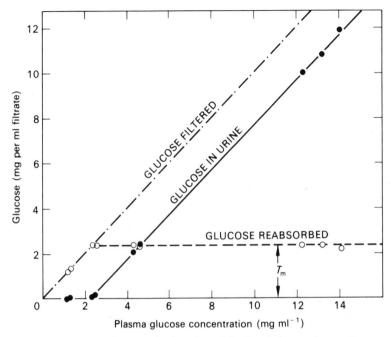

Fig. 10.9. Excretion of glucose in the urine of dog in relation to plasma glucose concentration (fully drawn line). The amount filtered is shown by the diagonal dashed line. Below a certain limit (*c.* 2.3 mg glucose per ml plasma) no glucose appears in the urine; this is because all the filtered glucose has been removed by tubular reabsorption. Above this limit the reabsorption mechanism is fully saturated, and although it continues to work at full capacity, glucose appears in the urine in amounts that increase with further increases in the plasma glucose concentration. The amount reabsorbed is the difference between the filtered and the excreted amount – it is shown by the horizontal curve, which remains constant above the saturation point, and thus represents the tubular maximum for glucose reabsorption (after Shannon and Fisher, 1938).

Let us examine the situation by reference to fig. 10.9. Glucose is continually filtered, and since the filtration rate remains constant, the filtered amount increases linearly with the plasma concentration of glucose, as indicated by the diagonal dashed line. At low plasma glucose levels, all filtered glucose is reabsorbed, and none appears in the urine. This is the normal situation, for the plasma glucose usually remains about 100 mg per 100 ml. If plasma glucose increases, all filtered glucose is still reabsorbed up to the threshold value, about 230 mg per 100 ml, and none is found in the urine. Should the plasma glucose exceed the threshold value, however, reabsorption is incomplete and some glucose will remain in the urine. In other words, the maximum capacity of the reabsorption mechanism has been reached. This amount is often designated the *tubular maximum* (T_m) for glucose. Above the threshold value, larger and larger amounts of glucose will be found in the urine (fig. 10.9). The line for filtered glucose and the line for glucose in the urine run parallel, and this means that the tubular maximum remains constant, as indicated by the constant distance between these two lines.

The tubular maximum for any other reabsorbed substance can be determined in a similar way by increasing the plasma concentration to above the threshold value for that substance, so that the reabsorption mechanism is saturated and the substance begins to appear in the urine.

In addition to filtration and reabsorption, a third process, *tubular secretion,* is important in urine formation. It was mentioned above that phenol red and PAH are secreted by the renal tubules. Since these substances are also filtered, the amount appearing in the urine will be the sum of the filtered and the secreted amounts.

If we inject one of these substances into the blood stream, the total amount secreted will depend on the plasma concentration, as shown in fig. 10.10. The amount filtered increases linearly with the plasma concentration. At low plasma concentrations the amount added by tubular secretion exceeds the filtered amount. However, again we

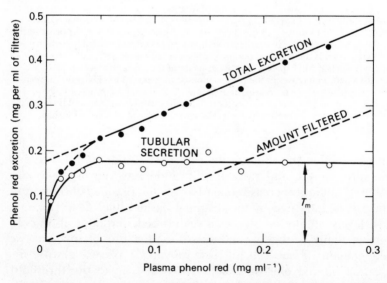

Fig. 10.10. Excretion of phenol red by the bullfrog kidney. The amount filtered (dashed line) increases in proportion to the plasma concentration. In addition, phenol red is also added to the urine by active tubular transport, giving a larger total amount in the urine (fully drawn line). The amount added by tubular transport remains constant at concentrations above 0.05 mg per ml plasma; this indicates that the tubular maximum for phenol red has been reached (Forster, 1940).

find that, at a given plasma concentration, the transport mechanism becomes saturated. The total excretion increases parallel to the filtered amount, and the difference between the total excretion and the filtered amount (the distance between the two parallel lines) gives the amount added by tubular secretion, in other words, this amount is the tubular maximum for phenol red excretion.

Other substances that are foreign to the body are also eliminated by tubular secretion. Often this is the case with phenolic compounds and their detoxification products. One well-known substance that is rapidly eliminated by tubular secretion is penicillin, a disadvantage that is counteracted by repeated injections of large amounts.

The process of tubular secretion can, in a very interesting way, be used to determine how much blood flows to the kidney, i.e. to determine the *renal blood flow*. For this we can use a substance that is excreted by tubular secretion so readily that it is completely removed from the blood that passes through the kidney. One such substance is PAH, which is removed so completely that none is present in the blood that leaves the kidney (unless its tubular maximum is exceeded). The method depends on the fact that the renal artery is the only blood supply to the mammalian kidney, all the blood first flows through the glomerular capillaries, and then on to the capillaries that surround the renal tubules. We now inject some PAH and determine the amount of PAH which appears in the urine in a given time. From the concentration of PAH in the blood plasma, we can then calculate the volume of plasma which has flowed through the kidney in that period of time. In man the renal plasma flow is normally about 0.7 liter min^{-1}. If the hematocrit is 45%, this corresponds to a blood flow to the kidneys of 1.25 liter min^{-1}. This is between one-quarter and one-fifth of the total cardiac output at rest (5–6 liter min^{-1}), showing that the kidneys of mammals receive a surprisingly large blood supply.

The same method cannot be used to determine the arterial blood flow to the kidneys of lower vertebrates, because the kidneys in these animals have a different type of circulation and receive a double blood supply. In addition to receiving blood from the renal artery, these kidneys also receive venous blood from the posterior part of the body (called the renal portal system). The importance of the renal portal circulation is not very clear. It may be related to the fact that, when a renal portal circulation is present, urine production can continue even if glomerular filtration ceases. In the frog, for example, filtration may stop, and phenol red which arrives in the kidney via the renal portal system is still secreted by tubular secretory activity.

The vertebrate kidney

Fish. The kidneys of most teleosts consist of nephrons with typical vertebrate characteristics. Fresh-water fish are hypertonic to the medium, and excess water which enters the body due to the osmotic gradient is eliminated as a hypotonic urine.

Marine teleosts, which suffer a water shortage because they are hypotonic to the medium, have a low rate of urine production. They cannot produce a concentrated urine, and excess salts are excreted by the gill. To make up for the osmotic water loss, they drink sea water, and one major function of the kidney is to excrete the divalent ions, magnesium and sulfate, found in sea water.

Some marine teleosts have kidneys in which the glomerulus is absent. This holds true for the goose fish, *Lophius piscatorius,* the toad fish, *Opsanus tau,* and some pipe fish, *Syngnathus.* Aglomerular marine teleosts have been very important in the study of renal function, for they permit the study of secretory processes in a kidney where no filtration can take place. If these fish are injected with inulin, no inulin appears in the urine. The absence of inulin from the urine is an excellent functional test for the filtration mechanism.

It is peculiar that a few fresh-water pipe fish also seem to lack glomeruli, but no detailed study has been made of these fish and their renal function. A fresh-water fish should have a considerable osmotic inflow of water, and it is difficult to explain how a non-filtering kidney can eliminate excess water. Obviously, these fish deserve further study.

The elasmobranchs have glomerular kidneys, and as we saw earlier, urea is essential for elasmobranchs to maintain their osmotic balance. Urea, being a small molecule (molecular weight $= 60$) is filtered in the glomerulus, but it is reabsorbed by the tubules. This means that there is, in the elasmobranch kidney, a specific mechanism for active transport of urea. Further details of this mechanism are discussed on p. 481.

Amphibians. The first unequivocal evidence for an initial ultrafiltration process in the kidney was obtained on frogs. This is because many of the glomeruli in the amphibian kidney are rather large and located so close to the surface of the kidney that it is possible to insert a fine micropipette and withdraw minute samples for analysis. Another advantage of experimentation on cold-blooded animals is that the kidney functions quite well at room temperature as opposed to the mammalian kidney which must be elaborately maintained at body temperature.

Most amphibians live at or near fresh water. When they are in water there is an osmotic influx of water, and to eliminate the water they produce a large volume of highly dilute urine. The inevitable loss of sodium is compensated for by cutaneous active uptake of sodium from the dilute medium.

In the frog kidney urea is eliminated, not only by glomerular filtration, but by tubular secretion as well (Marshall, 1933). This has certain advantages, for the following reason. When a frog is exposed to dry air, it reduces its urine output (water loss) by reducing the glomerular filtration rate and by increasing the tubular reabsorption of water. However, urea excretion may still remain high. In fact, of the total urea excreted by the kidney, as little as one-seventh may be filtered and the remainder added to the urine by tubular secretion (B. Schmidt-Nielsen and Forster, 1954). This means that the excretion of urea can continue at a high rate even when glomerular filtration has almost ceased. Thus, in contrast to the situation we found in the elasmobranch where urea is actively reabsorbed, in the amphibian kidney urea is actively secreted.

In the chapter on osmoregulation we saw that one amphibian, the crab-eating frog of South-East Asia (*Rana cancrivora*), lives and seeks its food in full-strength sea water. Physiologically this frog resembles elasmobranchs in that it carries a high level of urea in the blood, thus achieving osmotic equilibrium with the surrounding sea water and yet retaining the relatively low salt concentrations characteristic of vertebrates in general. However, the crab-eating frog differs from sharks in the way its kidney handles urea. The kidney of the crab-eating frog does not reabsorb urea, but when the animal is in sea water a very low rate of urine flow combined with a cessation of urea secretion reduces the loss of urea. If the crab-eating frog is transferred to more dilute water, the osmotic influx of water increases, urine volume goes up, and more urea is lost. This reduces the osmotic difference between the frog and the medium, which in turn reduces the osmotic influx of water (Schmidt-Nielsen and Lee, 1962).

Reptiles. The reptilian kidney conforms to the typical vertebrate pattern. It can produce dilute or isotonic urine, but it cannot produce urine more concentrated than the blood plasma. For aquatic freshwater reptiles (for example crocodiles and aquatic turtles) this is adequate, for a dilute urine serves to eliminate excess water. Marine reptiles have the opposite problem, water shortage and an overabundance of salt. Their kidneys cannot handle excess salts, which are eliminated instead by a nasal or an orbital salt gland.

Terrestrial reptiles which live in dry habitats and have limited water supplies excrete their nitrogenous waste in the form of uric acid, which is highly insoluble and therefore precipitates in the urine. As a result, urine can be eliminated as a paste or as a semi-solid pellet, thus requiring very little water. Since terrestrial reptiles have a salt gland, excess salts, particularly Na^+ and K^+ can be eliminated extra-renally. One of the major functions of the kidney, therefore, becomes that of uric acid excretion.

Birds and mammals. Uniquely among vertebrates, birds and mammals can produce urine hyperosmotic to the blood plasma. In birds this ability is not very pronounced; the maximum urine osmotic concentration can reach about two times the plasma concentration. The mammalian kidney, in contrast, can produce urine up to perhaps 25 times the plasma concentration (MacMillen and Lee, 1969). The highest urine concentrations are found in animals from desert habitats, and animals such as beavers, which always have a superabundance of water, have a very moderate concentrating ability (see table 10.2).

The mechanism responsible for formation of a concentrated urine in birds and mammals is based, paradoxically, on a reabsorption of sodium. It would seem that the production of a concentrated urine would require, either reabsorption of water from the tubular fluid, or active secretion of solute into the tubule. Neither is the case. Through an ingenious geometric arrangement of the renal tubule,

Table 10.2. *The maximum concentrating ability of the kidney of various mammals is correlated with the normal habitat of the animal, desert animals having the highest urine concentrations and fresh water animals the lowest*

	Urine max. osm. conc. (osm liter^{-1})	Urine:Plasma conc. ratio
Beaver (*a*)	0.52	2
Pig (*a*)	1.1	3
Man (*b*)	1.4	4
White rat (*b*)	2.9	9
Cat (*b*)	3.1	10
Kangaroo rat (*b*)	5.5	14
Sand rat (*b*)	6.3	17
Hopping mouse (*c*)	9.4	25

(*a*) B. Schmidt-Nielsen and O'Dell (1961); (*b*) K. Schmidt-Nielsen (1964); (*c*) MacMillen and Lee (1967).

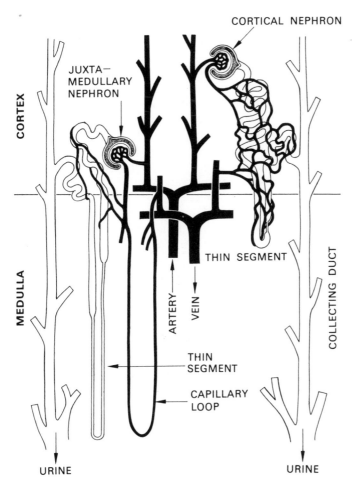

Fig. 10.11. This diagram shows two typical nephrons, one 'long-looped' (located close to the border of the medulla and therefore called juxtamedullary) and one 'short-looped' (or cortical) nephron. The long-looped nephron is paralleled by a loop formed by the blood capillary. The short-looped nephron is surrounded by a capillary network. Most mammalian kidneys contain a mixture of the two types of nephrons, but in some species only one or the other kind is found (after Smith, H. 1951. *The Kidney:* Oxford University Press, Plate II).

the reabsorption of sodium from the tubular contents is used to produce, in the end, a highly concentrated urine. This arrangement is found in birds, and is developed to the highest perfection in those mammals that form the most concentrated urine.

   The concentrating ability of the mammalian kidney is closely related to the length of Henle's loop, which is the hairpin loop formed by the thin segment interposed between the proximal and the distal tubule. Most mammalian kidneys have two types of nephrons (see fig. 10.11). Some nephrons have long loops, others have short ones. Those animals that produce the most highly concentrated urine have only long-looped nephrons with loops so long that they extend into the renal papilla. Others, such as the beaver and the pig, have only short loops. These are the animals that cannot produce urine more than about twice as concentrated as the plasma (B. Schmidt-Nielsen and O'Dell, 1961).

The role of the loop structure in the formation of a concentrated urine has been clarified by the Swiss investigator Wirz and his collaborators, and a number of other investigators in Europe and the USA have added further evidence.

The mechanism can best be understood by reference to the diagram in fig. 10.12. Let us follow the fluid as it flows through the loop. The fluid enters the loop with a sodium concentration equal to that in the plasma, say 150 mmol liter^{-1}. The scheme is now as follows. Sodium is transported actively out of the ascending limb of the loop, and this increases the sodium concentration in the surrounding tissue to, say, 200 mmol liter^{-1}. The descending limb is permeable to sodium, and sodium therefore diffuses in until the fluid inside the limb reaches the same sodium concentration as in the surrounding tissue, 200 mmol liter^{-1} (the ascending limb must be impermeable to water, otherwise water would passively follow the sodium). The fluid which flows into the distal tubule at the top of the ascending limb has had sodium removed from it, and its concentration will therefore be less than that in the plasma. As the fluid which leaves the loop contains less sodium than the fluid that entered, sodium must remain in and around the loop and accumulate in higher and higher concentrations. Gradually, as more sodium is returned to the loop, a very high concentration can be built up at the tip of the loop. The limit is determined primarily by the relative length of the loop; the longer the loop, the higher concentration can be reached at the tip.

Fluid that enters the collecting duct from the distal tubule is relatively dilute; how can this lead to the formation of a highly concentrated urine? The wall of the collecting duct is permeable to water and permits water to pass out. The sodium concentration is high in the surrounding tissues, and this causes an osmotic withdrawal of water, out of the collecting duct. In turn, the concentration of the fluid within the collecting duct increases, thus producing a concentrated urine which passes on to the bladder.

Determinations of the osmotic concentrations in the various parts of the nephron have shown that (1) the glomerular filtrate is iso-

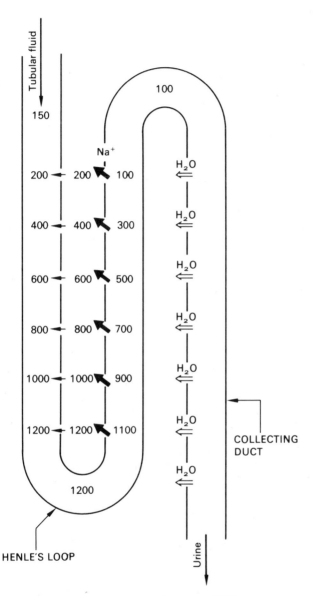

Fig. 10.12. Diagram of the countercurrent multiplier system in the mammalian kidney. The figures refer to sodium concentrations, arbitrarily chosen to indicate the increase in concentration towards the bottom of the loop. Heavy black arrows refer to active sodium transport, thin arrows to passive diffusion of sodium. Open arrows indicate passive movement of water due to osmotic forces. Further explanation in text.

tonic, (2) the tubular fluid at the beginning of the loop is also isotonic, (3) at the tip of the hairpin loop it is highly concentrated, (4) in
the distal convoluted tubule the fluid is hypotonic to the blood, and
(5) the fluid along the collecting duct becomes increasingly concentrated and in the end equals the concentration at the tip of the hairpin loop.

The fluids in the descending and the ascending limbs flow in opposite directions, thus the flow is a countercurrent flow. In any one
place along the loop the concentration difference between the ascending limb and the descending limb is moderate, but along the
length of the loop this moderate concentration difference is additive,
i.e. there is a multiplication effect. The entire system is therefore
called a *countercurrent multiplier system*. The final concentration that
can be reached in this system depends on the length of the loop; the
longer the loop, the higher the urine concentrations that can be
achieved. Therefore, there is a high degree of correlation between
the relative length of the loop and the maximum concentration of
the urine in any given animal species.

The bird kidney differs from the mammalian kidney in that some
nephrons have loops but others have no loops at all. It appears that
the maximum concentrating ability of the bird kidney is related to
the proportion of looped nephrons to the number of nephrons without loops, rather than to the maximum length of the loops, as in
mammals (Poulson, 1965).

Many birds live under relatively dry conditions, eat dry food such
as seeds, and have limited access to water. Thus it is reasonable to
ask why their kidneys cannot match the concentrating ability of
mammals. Perhaps the answer is to be found in the fact that birds
excrete uric acid as an end product of their nitrogen metabolism. If
the bird kidney were to withdraw water to a much greater extent
than it does, the highly insoluble uric acid would precipitate in the
tubules and clog them, unless sufficient liquid were to remain to
carry the precipitate along. Although bird urine contains precipitated uric acid, it must remain relatively liquid in the nephrons and
in the ureter until it enters the cloaca, which is the very last portion
of the intestine where urine and fecal material accumulate until at
intervals, they are emptied through the single exterior opening.
After the urine has entered the cloaca, further water can be withdrawn before the cloacal contents are voided to the exterior.

The cloaca. As the urine of birds and reptiles enters the cloaca, it is
liquid or semi-liquid. If an animal needs to conserve water, however,
the urine is dropped as a semi-solid paste or pellet, in other words,

water has been withdrawn. This reabsorption of water may, at least in part, take place outside the cloaca, for it has been shown that urine from the cloaca may migrate into the lower part of the intestine where both salt and water are reabsorbed (Nechay, Boyarsky and Catacutan-Labay, 1968; Skadhauge, 1967, 1968).

What is the nature of the process of water withdrawal from the cloacal contents? It can be a solute-linked water transport, and, at least in lizards, it can be a passive process, resulting from the colloidal osmotic pressure of the plasma proteins. In the lizard cloaca water is absorbed until the force necessary to withdraw more water corresponds to the colloidal osmotic pressure of the plasma proteins, and then further water absorption ceases. If the osmotic pressure of the plasma proteins is counterbalanced by placing inside the cloaca a protein solution with the same colloidal osmotic concentration as the plasma, water reabsorption is prevented. Thus, it seems that in lizards a passive osmotic removal of water suffices for the production of a semi-dry urine within the cloaca. Nevertheless, it remains possible that an active transport of sodium or other solutes also takes place in the lizard cloaca, thus causing solute-linked water withdrawal (Murrish and Schmidt-Nielsen, 1970).

When salt is withdrawn from the cloaca together with water, we encounter one problem, the animal will have difficulties in eliminating excess salt. However, lizards (and many birds) have the possibility of excreting excess salt via the nasal gland. Thus, lizards that live in a dry habitat can use the kidney primarily for the excretion of uric acid and lose a minimal amount of water, and they can use the nasal gland for the excretion of salts in a highly concentrated solution, again with only a small loss of water (Schmidt-Nielsen, Borut, Lee and Crawford, 1963).

Regulation of urine concentration and volume. If an animal is in caloric balance, its body weight, and thus volume, remains remarkably constant. A major factor in this constancy is the accurate regulation of the rate of urine production. The total volume of urine production depends on the need for water excretion. If there is an excess intake of water, a large volume of dilute urine will be excreted, if there is a shortage of water in the organism, the urine volume decreases and its concentration increases towards the maximum possible.

In man, the urine flow may vary more than 100-fold. When a man is deprived of water, urine production may be as low as 10 ml h^{-1}, but it will not decrease further until the situation becomes critical and renal failure sets in. In the opposite case, when the water intake is high, urine flow may exceed one liter per hour. Some persons may think that by drinking more, urine production could be increased further, but this is not possible. The maximum urine volume is determined by the amount of fluid filtered, which, in man, remains quite constant and does not increase when we drink more. In the proximal tubule about 85% of the filtered volume of water is invariably withdrawn, leaving 15% of the filtered water as the maximum volume of urine that can be excreted. The reason for the reabsorption of water in the proximal tubule is that major solutes such as glucose and sodium chloride are reabsorbed here. As solutes are removed, the tubular content becomes hyposmotic, and due to the osmotic effect water follows passively through the tubular wall. Thus, in the proximal tubule the reabsorption of water is a direct and inevitable result of the reabsorption of glucose and sodium.

The distal mechanism (Henle's loop and the collecting duct) is, as was described above, the structure responsible for formation of a concentrated urine. The concentrating process is controlled by the *antidiuretic hormone* (ADH), now usually called *vasopressin*. In the absence of this hormone, which is secreted from the anterior hypophysis, a large volume of highly dilute urine is produced. Under the influence of vasopressin, more water is reabsorbed, urine volume decreases, and urine concentration increases.

Vasopressin acts by influencing the permeability of the collecting ducts. In its absence the permeability of the collecting duct to water is low. Thus, the dilute fluid which enters the collecting duct from the distal tubule undergoes no further loss of water, even if the osmotic concentration in the surrounding tissues is high. With increasing amounts of vasopressin, the permeability of the collecting ducts increases. Thus, because of the high osmotic concentration in the tissue of the papilla (produced by the countercurrent multiplier mechanism), water is osmotically withdrawn from the fluid within the collecting ducts, leaving a small volume of concentrated urine.

This view of the role of the collecting duct has been verified by introducing filament-thin polyethylene tubes through the tip of the papilla in rats and withdrawing samples from various levels of the collecting duct (Hilger, Klümper and Ullrich, 1958). When vasopressin is present in the blood, the collecting duct urine is in osmotic equilibrium with the fluid in the loop, but in the absence of vasopressin such equilibrium does not exist because of the lowered permeability of the collecting duct (Gottschalk, 1961).

It may seem paradoxical that a hormone which causes the formation of a concentrated urine in fact acts by increasing the permeability of the collecting ducts. Another paradoxical aspect is that the transport mechanism that leads to the formation of a concentrated urine is based on the withdrawal of sodium from the tubular fluid, rather than its addition. In the end, however, it is precisely this reabsorption which, in combination with the high permeability of the collecting ducts under the influence of vasopressin, causes the final urine to reach a maximum concentration.

It is interesting that the effect of vasopressin in increasing the permeability is similar to its effect on frog skin, which is a much used model in studies of salt transport. In the frog skin the sodium transport takes place in the same direction as the sodium pump in the kidney tubule, by returning sodium to the organism. In both systems vasopressin increases the permeability to water. In the mammalian kidney it has a water-saving effect, in the frog skin the increased permeability leads to an increased osmotic influx of water.

NITROGEN EXCRETION

Most of the food that animals eat contains three major nutrient components (1) carbohydrates, (2) fats, and (3) proteins (plus smaller amounts of nucleic acids). When carbohydrates and fats are metabolized they give rise to carbon dioxide and water as the only end products of oxidation. Proteins and nucleic acids also yield carbon dioxide and water, but in addition, the chemically bound nitrogen in these foodstuffs gives rise to the formation of some relatively simple nitrogen-containing excretory products. The three compounds of greatest interest to us are

Ammonia	Urea ·	Uric acid (keto-form)
NH_3	CH_4ON_2	$C_5H_4O_3N_4$

Which one of these is the main excretory product for a particular animal is closely correlated with the environment in which the animal normally lives, and in particular with the availability of water.

When amino acids are metabolized, the amino group (-NH₂) is removed by the process of deamination and forms ammonia (NH₃). The ammonia is excreted unchanged by many animals, in particular by aquatic invertebrates, but by some animals it is synthesized into urea and by others into uric acid before being excreted. The nitrogen compounds contained in nucleic acids are either purines or pyrimidines, and they may be excreted as any of the degradation products listed in table 10.3. The bulk of the excreted nitrogen compounds is derived from protein (amino acids) by deamination and we shall discuss these first.

Table 10.3. *Metabolic end products of the major groups of foodstuffs. Ammonia from protein metabolism may be excreted as such or may be synthesized into other N-containing excretory products; purines from nucleic acids may be excreted as such or as any of a number of degradation products, including ammonia*

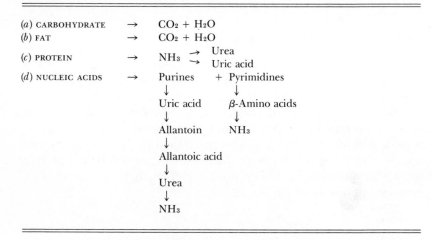

Some other nitrogen-containing products are also excreted, some being derived from the diet, others arising in metabolic pathways. For example, the urine of vertebrates normally contains a small amount of creatinine. This compound is normally formed by the muscles and is of particular interest because it can be used in studies of renal function in a manner similar to inulin, aided by the fact that it is formed in the body at a relatively constant rate. The injection of a foreign substance is therefore not necessary for such studies. Creatinine is apparently derived from the metabolism of the high-energy phosphate compound, phosphocreatine. Another group of nitrogen compounds frequently found in urine is a result of detoxification processes. An example, benzoic acid (present in certain plant foods) when combined with the amino acid glycine, forms the conjugation product hippuric acid (named after the Greek *hippos* = horse, because it was originally discovered in horse urine). Other phenolic compounds are detoxified in a similar manner and are then excreted in the urine. Hippuric acid, and in particular the closely related compound, *para*-aminohippuric acid (PAH), has been important in studies of renal physiology because they are readily secreted by tubular activity and can be used to determine renal blood flow (see p. 465).

One important aspect of nitrogen metabolism is the assumption that nitrogen gas never enters into the metabolism of animals; it has been universally accepted that only nitrogen-fixing bacteria can assimilate molecular nitrogen (N_2) into chemical compounds, and that only de-nitrifying bacteria have metabolic pathways that lead to the formation of molecular nitrogen gas.*

Aquatic animals and the transition to land

Whether the nitrogen from protein metabolism is excreted as ammonia, urea, or uric acid is closely related to the normal habitat of the animal and the availability of water (see table 10.4). Ammonia is a highly toxic compound, even in minute concentrations, and must

Table 10.4 *Major nitrogen excretory products in various animal groups. The role of cleidoic eggs is discussed on p. 483–4*

Animal	Major end product of protein metabolism	Adult habitat	Embryonic environment
Aquatic invertebrates	Ammonia	Aquatic	Aquatic
Teleost fish	Ammonia, some urea	Aquatic	Aquatic
Elasmobranchs	Urea	Aquatic	Aquatic
Crocodiles	Ammonia, some uric acid	Aquatic	
Amphibians, larval	Ammonia	Aquatic	Aquatic
Amphibians, adult	Urea	Semiaquatic	Aquatic
Mammals	Urea	Terrestrial	Aquatic
Turtles	Urea and uric acid	Terrestrial	Cleidoic egg
Insects	Uric acid	Terrestrial	Cleidoic egg
Land gastropods	Uric acid	Terristrial	Cleidoic egg
Lizards	Uric acid	Terrestrial	Cleidoic egg
Snakes	Uric acid	Terrestrial	Cleidoic egg
Birds	Uric acid	Terrestrial	Cleidoic egg

* Recent evidence strongly indicates that this is not true under all circumstances, formation of gaseous nitrogen in mammals on a high-protein diet has been reported by several investigators (e.g. Costa, Ulrich, Kantor and Holland, 1968). These findings are bound to upset some of our present ideas about protein metabolism and nitrogenous end products.

therefore be removed rapidly, either to the exterior or by synthesis into less toxic compounds (urea or uric acid).

Ammonia

Most aquatic invertebrates excrete ammonia as the end product of protein metabolism. Because of its high solubility and small molecular size ammonia diffuses extremely rapidly. It can be lost therefore, to a great extent, through any surface in contact with water and need not be excreted by the kidney. Thus, in teleost fish most of the nitrogen is lost as ammonia from the gills. In the carp and goldfish the gills excrete six to ten times as much nitrogen as the kidneys do, and only 10% of this is urea, the remaining 90% is ammonia (Smith, 1929).

Urea

Urea is easily soluble in water and has a moderately low toxicity. The synthesis of urea in higher animals has been clarified by the British biochemist Hans Krebs, the same man whose name is attached to the well-known Krebs cycle of oxidative energy metabolism (the tricarboxylic acid cycle).

In the synthesis of urea, ammonia and carbon dioxide are condensed with phosphate to form carbamyl phosphate, which enters a synthetic pathway to form citrulline, as shown in fig. 10.13. A second ammonia is added from the amino acid aspartic acid, leading to the formation of the amino acid arginine. In the presence of the enzyme arginase, arginine is decomposed into urea and ornithine. This frees the ornithine for renewed synthesis of citrulline, repeating the entire cycle; the total pathway is therefore known as the *ornithine cycle* for urea synthesis. The presence of arginase in an animal shows that it has the ability to produce urea, and it often indicates that urea is the major nitrogenous excretory product. This is not invariably so, however, for arginase can be present with the remaining pathway absent.

The vertebrates that excrete mainly urea, and possess the ornithine cycle enzymes for urea synthesis, are circled with a line of long dashes in fig. 10.14. Some urea is excreted by teleost fish, and in elasmobranchs, amphibians, and mammals it is the main nitrogenous excretory product. In elasmobranchs (sharks and rays), as well as in the crab-eating frog and in the coelacanth, *Latimeria*, urea is retained and serves a major role in osmotic regulation and is therefore a valuable metabolic product. Urea is filtered in the glomerulus of the elasmobranch kidney, but it should be kept from being lost in the urine

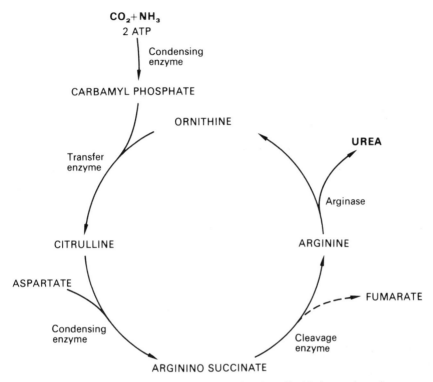

Fig. 10.13. Urea is synthesized from ammonia and carbon dioxide by condensation with the amino acid ornithine. Through several more steps arginine is formed, which, with the aid of the enzyme arginase, splits off urea, forming ornithine which can re-enter the cycle.

and must therefore be recovered by active tubular reabsorption. In amphibians the situation is different, urea is filtered, but in addition a substantial amount is added to the urine by active tubular secretion. Thus, both elasmobranchs and amphibians have active tubular transport of urea, but the transport is in the opposite direction in the two groups. Apparently, the mechanism of the 'pump' is not metabolically identical in the two groups, for some closely related urea derivatives are treated differently by the two animal groups (table 10.5). This serves as an excellent example that a physiological function can arise independently in two groups, and that it does not necessarily have to use an identical mechanism to achieve the same end (in this case the active transport of urea).

The crab-eating frog, which also retains urea for purposes of osmoregulation, does not show evidence of active tubular reabsorption of urea (Schmidt-Nielsen and Lee, 1962). The crab-eating frog has a

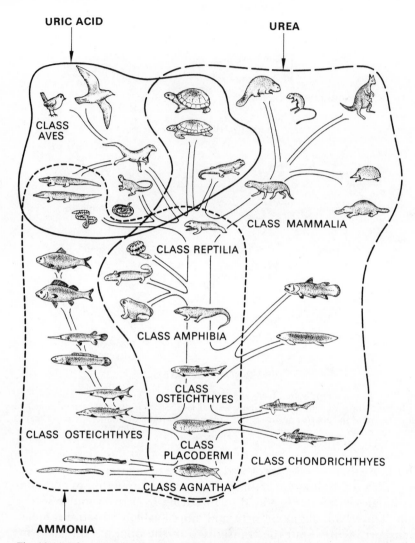

Fig. 10.14. Nitrogen excretion in relation to the phylogeny of vertebrates. The lines enclose groups of animals which use ammonia, urea, and uric acid, respectively, as the major nitrogenous excretory product (B. Schmidt-Nielsen, 1972).

low rate of urine production, and the renal tubules are highly permeable to urea. Urea therefore diffuses from the tubular fluid back into the blood, and appears in the urine in approximately the same concentration as in the blood. Thus, only small amounts of urea are lost in the urine. Since ordinary frogs have an active tubular secretion of urea, the absence of active reabsorption in the crab-eating

Table 10.5. *Urea is actively transported by the kidney tubule of both shark (active reabsorption) and frog (active secretion). Three closely related compounds, however, are not transported alike by the two kinds of animals. This provides evidence that the cellular transport mechanism is not identical in the two kidneys.* (B. Schmidt-Nielsen and Rabinowitz, 1964)

Compound	Shark	Frog
Urea	Active	Active
Methylurea	Active	No
Thiourea	No	Active
Acetamide	Active	No

frog shows that a 'pump' for active transport of a substance is not easily 'turned around', the direction of transport is not readily reversed. The direction of an active transport mechanism seems to be a very conservative physiological function. As we have seen, the frog skin and the mammalian kidney both retain the direction of active transport of sodium from the exterior into the organism. In the mammalian kidney, however, this inward transport of sodium has been utilized in the countercurrent multiplier so that the end result is nevertheless the formation of a concentrated urine.

The common concept of urea excretion in the mammalian kidney is that urea is filtered in the glomerulus and then treated passively by the tubules, although some urea, because of its high diffusibility, re-enters the blood by passive diffusion. There is, however, convincing evidence that the mammalian kidney does not treat urea entirely passively, but rather that there is a well-regulated control of urea excretion.

Amphibian metamorphosis. The tadpoles of frogs and toads excrete mostly ammonia, the adults excrete urea. At metamorphosis there is a well-defined changeover from ammonia to urea excretion in frog (*Rana temporaria*), toad (*Bufo bufo*), newt (*Triturus vulgaris*) and others. The South African frog *Xenopus,* however, which remains aquatic during adult life, continues to excrete ammonia as an adult (table 10.6).

The change to urea excretion at the onset of metamorphosis in the semi-terrestrial amphibians is associated with a marked increase in the activities of all the liver enzymes of the ornithine cycle (Brown, Brown and Cohen, 1959).

It is interesting that adult specimens of the aquatic *Xenopus,* if kept out of water for several weeks, accumulate urea in the blood and tis-

Table 10.6. *Ammonia excretion of the terrestrial toad* Bufo bufo *and of the fully aquatic frog* Xenopus laevis. *The figures give the excretion of free ammonia, expressed in per cent of the total ammonia and urea excretion at various stages of development.* (Munro, 1953)

STAGE	Bufo (%)	Xenopus (%)
No hind limbs		85
Hindlimbs ¾ developed	80	
Hindlimbs functional	85	83
Forelimbs free. Tadpole tailed	50	81
Tail atrophying	36	
Tail-less	20	77
Adult	15	81

sues. Accumulation of urea can also be induced by keeping the animals in 0.9% NaCl solution. If *Xenopus* adults were kept out of water but in moist moss to prevent dehydration, the blood urea concentration increased between 10- and 20-fold, rising to nearly 100 mmol liter^{-1}. When the animals were returned to water, the excess urea was excreted again (Balinsky, Cragg and Baldwin, 1961).

A group of *Xenopus* which was found naturally estivating in the mud near a dried pool, had urea concentrations raised similarly by a factor of 15 to 20. Of the enzymes responsible for urea synthesis, the enzyme carbamyl phosphate synthetase, responsible for the first synthetic step in fig. 10.13 (p. 479) was increased about six-fold, but the activities of the other enzymes of the cycle were unchanged. The synthesis of carbamyl phosphate may be the rate-limiting step in urea synthesis, and an increase in this enzyme is probably responsible for keeping the plasma ammonia low in animals out of water (Balinsky, Choritz, Coe and Schans, 1967).

Lung-fish. The changes in the African lung-fish, *Protopterus*, are exactly analogous to those in amphibians. Normally, when the lung-fish lives in water, it excretes large amounts of ammonia (and some urea), but when it estivates in a cocoon within the dried mud, it channels the entire waste nitrogen into urea, which accumulates in the blood and may reach concentrations as high as 3% (500) mmol liter^{-1}) at the end of three years of estivation (Smith, 1959).

The presence of all five enzymes of the ornithine cycle has been demonstrated in the liver of the African lung-fish (Janssens and Cohen, 1966). The levels of the two enzymes that are rate-limiting in the urea synthesis are similar in the lung-fish to the levels of these same enzymes in the tadpole of the frog *Rana catesbeiana,* and considerably lower than the levels reported for adult frogs. This is consistent with the predominant excretion of ammonia in the lung-fish when it is in water. It has been calculated, however, that the amounts of the ornithine cycle enzymes present in the liver of non-estivating lung-fish are sufficient to account for the accumulation of urea actually observed in estivating lung-fish (Forster and Goldstein, 1966).

In the Australian lung-fish, *Neoceratodus,* the concentrations of the ornithine cycle enzymes are low. This is in accord with the life habits of the Australian lung-fish, which uses its lung only as an accessory respiratory organ and cannot survive in air for any length of time (see discussion of lung-fish respiration, p. 50). The rate of synthesis of urea by liver slices of the Australian lung-fish show that it is only one-hundredth of the rates observed in the African lungfish. This again is consistent with the completely aquatic nature of the Australian lung-fish (Goldstein, Janssens and Forster, 1967).

Uric acid

Excretion of uric acid is dominant in insects, land snails, most reptiles, and in birds. These are typical terrestrial animals, and the formation of uric acid must be considered as a successful adaptation to water conservation in a terrestrial habitat. Since uric acid and its salts are only slightly soluble in water (the solubility of uric acid is about 6 mg per liter water), the withdrawal of water from the urine will cause uric acid and its salts to precipitate. The semi-solid white portion of bird droppings is urine and consists mostly of uric acid; thus very little water is used for excretion of the nitrogenous excretory products in these animals. Some insects have carried the reduction in urine water loss so far that they do not excrete the uric acid at all but deposit it in various parts of the body, mainly in the fat body. In these forms, therefore, no water whatsoever is required for the elimination of nitrogenous end products.

It has been suggested by Joseph Needham that the difference between those higher vertebrates that form urea (mammals and amphibians), and those that form uric acid (reptiles and birds) is primarily correlated with their mode of reproduction. The amphibian egg develops in water, and the mammalian embryo develops in a liquid medium where waste products are transferred to the blood of the mother. The development in reptiles and birds, on the other hand, takes place in a closed egg, a so-called *cleidoic* * egg, where

* Greek *kleistos* = closed, from *kleis* = key.

only gases are exchanged with the environment and all excretory products remain within the egg shell. In the cleidoic egg the embryo has a very limited water supply, and ammonia is, of course, too toxic to be tolerated. If urea were produced, it would remain inside the egg and accumulate in solution. Uric acid, on the other hand, can be precipitated and thus essentially eliminated, as happens when it is deposited as crystals in the allantois, which thus serves as an embryonic urinary bladder.

It has been suggested that birds gain one further advantage by using uric acid as their main excretory product. Since little water is needed for urine formation, uric acid excretion has been said to save weight for flying birds. This argument, however, does not seem very convincing, for birds that have access to water, both fresh-water and marine species, often eliminate large quantities of liquid urine.

Uric acid in reptiles. Lizards and snakes excrete mostly uric acid, many turtles excrete a mixture of uric acid and urea, and crocodiles excrete mainly ammonia (Cragg, Balinsky and Baldwin, 1961). This fits the generalization that nitrogen excretion is closely related to the availability of water in the environment. In crocodiles and alligators ammonia is excreted in the urine, where the principal cation is NH_4^+ and the principle anion is HCO_3^- (Coulson, Hernandez and Baldwin, 1950; Coulson and Hernandez, 1955). It is possible that the presence of these ions in the urine permits an improved retention of sodium and chloride in these fresh-water animals, which, incidentally, also lose very little sodium and chloride in the feces.

There is little doubt that there is a close correlation between the habitat of turtles and their nitrogen excretion. Table 10.7 shows the composition of urine samples from eight species of turtles obtained in the London Zoo. The most aquatic species excrete considerable amounts of ammonia and urea and only traces of uric acid, while the most terrestrial species excrete over half of their nitrogen as uric acid.

There have been conflicting reports as to whether turtles excrete mostly urea or uric acid. The reason is not only the differences between species, but also that different individuals of one species may excrete mainly uric acid, or mainly urea, or a mixture of both (Khalil and Haggag, 1955). One individual may even in the course of time change from one compound to the other. The fact that some precipitated uric acid may remain in the cloaca while the liquid portion of the urine is voided to the exterior, makes it unreliable to determine the amount of uric acid formed by analyzing a single or a few urine samples. An incomplete emptying of the cloaca may give an entirely

Table 10.7. *Partition of nitrogen in the excreta of turtles, given in per cent of total nitrogen excretion. The most aquatic species excretes almost no uric acid, while this compound dominates in the most terrestrial species.* (Moyle, 1949)

SPECIES	HABITAT	Uric acid	Ammonia	Urea	Amino acids	Unaccounted for
Kinosternon subrubrum	Almost wholly aquatic	0.7	24.0	22.9	10.0	40.3
Pelusios derbianus	Almost wholly aquatic	4.5	18.5	24.4	20.6	27.2
Emys orbicularis	Semi-aquatic, feeds on land in marshes	2.5	14.4	47.1	19.7	14.8
Kinixys erosa	Damp places, frequently enters water	4.2	6.1	61.0	13.7	15.2
K. youngii	Drier than above	5.5	6.0	44.0	15.2	26.4
Testudo denticulata	Damp, swampy ground	6.7	6.0	29.1	15.6	32.1
T. graeca	Very dry, almost desert conditions	51.9	4.1	22.3	6.6	4.0
T. elegans	Very dry, almost desert conditions	56.1	6.2	8.5	13.1	12.0

There were small amounts of allantoin, guanine, xanthine, and creatinine, and a variable amount not accounted for.

too low amount of uric acid, and an evacuation which includes precipitate that has accumulated over a period of time will give too large amounts.

The cause of the shift between urea and uric acid in the excretion of the tortoise, *Testudo mauritanica,* seems to be a direct function of temperature and hydration of the animal. Uric acid excretion increases when the water balance is unfavorable, but the mechanism which controls the shift in biochemical activity from urea to uric acid synthesis is unknown and deserves further study (Drilhon and Marcoux, 1942).

An unusual frog. It was mentioned earlier that the African frog, *Chiromantis xerampelina,* loses water from the skin very slowly, at rates comparable to those in reptiles (see p. 405). *Chiromantis* resembles reptiles also in that it excretes mainly uric acid, rather than urea as most adult amphibians normally do. This is a sensational finding, for it challenges our commonly accepted view of nitrogen excretion in amphibians. There is no doubt about the accuracy of the report, for uric acid was determined in *Chiromantis* urine by means of an enzymatic method which is specific for uric acid, and uric acid was found to make up 60 to 75% of the dry weight of the urine (Loveridge, 1970).

A South-American frog, *Phyllomedusa sauvagii,* shows similar reptile-like traits. Its cutaneous water loss is of the same magnitude as from the dry skin of reptiles, and its urine contains large amounts of a semi-solid urate precipitate (Shoemaker, Balding and Ruibal, 1972). Since *Chiromantis* and *Phyllomedusa* live on different continents and belong to two different amphibian families, it appears that their reptile-like physiological characteristics have evolved independently in response to their arid habitat.

Ammonia and renal function

The preceding discussion makes it appear that ammonia is excreted mainly by aquatic animals, but this is not completely true. Ammonia is normally found in the urine of terrestrial animals, where it serves to regulate the pH of the urine. If urine becomes acid due to the excretion of acid metabolic products, ammonia is added to neutralize the excess acid.

Excess acid is normally formed in protein metabolism, for sulfuric acid is the end product of oxidation of the sulfur-containing amino acid cysteine. The more acid the urine is, the more ammonia is added. The ammonia which is used to neutralize an acid urine is produced in the kidney and is derived from the amino acid glutamine. Glutaminase is found in the kidney, and its presence there serves the particular purpose of producing ammonia. Ammonia found in mammalian urine, therefore, has no direct connection with the ammonia produced in the liver by the deamination of amino acids, and in this sense should not be considered as a normal end product of protein metabolism.

Nucleic acids and nitrogen excretion

Nucleic acids contain two groups of nitrogen compounds, purines (adenine and guanine) and pyrimidines (cytosine and thymine). In some animals purines are excreted as uric acid (which is, itself, a purine), in other animals the purine structure is degraded to a number of intermediaries or to ammonia, any one of which may be excreted.

The metabolic degradation and excretion of purines has not been as carefully studied as the metabolism of protein nitrogen. The main features, however, are listed in table 10.8. Birds, terrestrial reptiles, and insects degrade purines to uric acid and excrete this compound. These are the same animals that synthesize uric acid from amino nitrogen; it would obviously be meaningless for an animal on the one hand to synthesize uric acid, and on the other have mechanisms for its degradation. We therefore cannot expect to find any further breakdown of purines in animals in which uric acid is the end product of protein metabolism.

Table 10.8. *Nitrogenous end products of purine metabolism in various animals.* (Keilin, 1959)

Animal	End product of purine metabolism
Birds	Uric acid
Reptiles, terrestrial	Uric adic
Insects	Uric acid
Man, apes, dalmatian dog	Uric acid
Mammals (except those above)	Allantoin
Gastropod molluscs	Allantoin
Diptera	Allantoin
Some teleost fish	Allantoic acid
Amphibia	Urea
Teleosts	Urea
Elasmobranchs	Urea
Lamellibranch molluscs (fresh-water)	Urea
Lamellibranch molluscs (marine)	Ammonia
Most other aquatic invertebrates	Ammonia

Among mammals, man, apes, and the dalmatian dog are in a special group in that they excrete uric acid, while mammals in general excrete allantoin. Allantoin is formed from uric acid in a single step in the presence of the enzyme uricase. This enzyme is absent in man and apes. Due to its low solubility uric acid is at times deposited in the human organism, causing swelling of the joints and the well-known and very painful disease gout. If man had retained the enzyme uricase, gout would be an unknown disease.

While the dalmatian dog excretes uric acid in amounts much higher than other dogs, this is not due to a metabolic defect of the dog. Its liver contains uricase and some allantoin is formed. However, the dalmatian dog has a kidney defect which prevents tubular reabsorption of uric acid (as found in other mammals, including man); uric acid is therefore lost in the urine of the dalmatian dog faster than it can be converted to allantoin by the liver (Yü, Berger, Kupfer and Gutman, 1960). There is considerable evidence that uric acid in the dalmatian dog, in addition to being filtered in the glomerulus, is also excreted by active tubular transport (Keilin, 1959).

The structure of the purines adenine and guanine is very similar to that of uric acid, containing one six-membered and one five-membered ring. The pyrimidines (cytosine and thymine), however, are single six-membered rings which contain two nitrogen atoms. In higher vertebrates the pyrimidines are degraded by opening this ring, forming one molecule of ammonia and one molecule of a β-amino acid. These are then metabolized in the normal metabolic scheme by further deamination.

The most striking feature of nucleic acid metabolism is that the 'higher' animals listed at the top of table 10.8 completely lack the enzymes necessary to degrade the purines. As we go down the list to 'lower' animals we find an increasing complexity in the biochemical and enzymatic systems for the further degradation of the purines. Thus the 'lowest' animals in this case possess the most complete enzyme systems.

Other nitrogen compounds

Amino acids are not a major end product of nitrogen metabolism, but they are found in small amounts in the urine of many animals. It would seem more economical for an animal to deaminate the amino acid, excrete the ammonia in the usual way, and use the resulting organic acid in energy metabolism. Because of the minor role that excretion of amino acids plays, it will not be discussed further here.

Guanine is the major excretory product in spiders. It appears to be synthesized from amino nitrogen, although the complete pathway is unknown. It has been found that some spiders, notably the bird-eating tarantulas, after a meal excrete more than 90% of the total nitrogen as guanine (Peschen, 1939). In the common garden spider, *Epeira diadema,* the identification of guanine has been confirmed by a highly specific enzymatic method (Vajropala, 1935).

Guanine is also found rather widely in a variety of other animals, for example, the silvery sheen in fish scales is due to a deposit of guanine crystals. The garden snail, *Helix,* excretes guanine, but it makes up only about 20% of the total purine excretion, the remainder being uric acid. It is possible that this fraction is derived from nucleic acid metabolism, while the remainder of the nitrogen is derived from protein metabolism.

The recapitulation theory

The nitrogen excretion of the developing chick embryo is said to change with time and go through a series of peaks, first with ammonia as the main end product, then urea, and finally uric acid. This development would *recapitulate* the evolutionary events that terminate in uric acid excretion in birds. Ammonia production in the chick embryo is reported to culminate at four days, urea at nine days, and uric acid at 11 days of incubation (Baldwin, 1949).

A more recent study of the nitrogen excretion in the chick embryo claims to differ sharply from these reported peaks (Clark and Fisher, 1957). All the three major excretory products, ammonia,

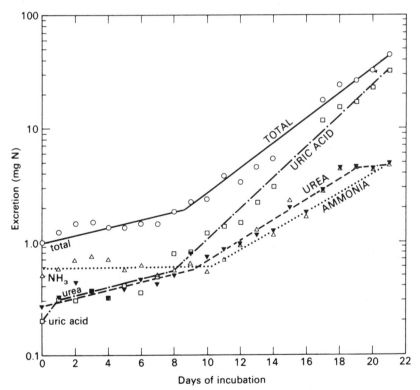

Fig. 10.15. The accumulation of nitrogenous products in the chick embryo during incubation of the egg. Urea (▼) and uric acid (□) are formed from the beginning of incubation; uric acid, however, dominates later. Ammonia (△), also present from the beginning, remains constant for about 10 days, and then increases further throughout the incubation period. (Based on data from Clark and Fischer 1957. *J. Exp. Zool.*, 136).

urea, and uric acid are present and are formed from the beginning of the embryological development. Towards the end of the incubation period uric acid is by far the dominant excretory product. However, the amounts of both urea and ammonia continue to increase throughout the incubation, and at the time of hatching both are present in approximately equal quantities. At the end of incubation the chick has excreted 40 mg nitrogen, of which 23% is divided equally between urea and ammonia, and the remainder is present as uric acid (fig. 10.15).

What is the reason for this conflict in reported results? The older results may have been less accurate because of more primitive analytical techniques, but the main reason is that the results were expressed as the amount of each excretory product relative to the weight of the embryo. Since the embryo continuously increases in weight, and at a rising rate, an artificial peak is created by dividing the amounts of each excretory product by the weight of the embryo. In fact, all three excretory products are present from the beginning and increase in amount throughout the embryological development, but ammonia does not increase appreciably until after the tenth day of incubation. The urea produced by the chick embryo is not synthesized from amino-acid-nitrogen by the ornithine cycle, but from the action of arginase upon arginine (Eakin and Fisher, 1958). Thus, neither ammonia nor urea production in the chick embryo support the claim that embryonic biochemical development recapitulates the evolutionary history of nitrogen excretion.

REFERENCES

BALDWIN, E. (1949). *An Introduction to Comparative Biochemistry,* Cambridge, England: Cambridge University Press, 164 pp.

BALINSKY, J. B., CHORITZ, E. L., COE, C. G. L. and VAN DER SCHANS, G. S. (1967). Amino acid metabolism and urea synthesis in naturally aestivating *Xenopus laevis. Comp. Biochem. Physiol.* **22**, 59–68.

BALINSKY, J. B., CRAGG, M. M. and BALDWIN, E. (1961). The adaptation of amphibian waste nitrogen excretion to dehydration. *Comp. Biochem. Physiol.,* **3**, 236–44.

BARNES, R. D. (1968). *Invertebrate Zoology,* 2nd edn, Philadelphia: W. B. Saunders Co. 743 pp.

BRAUN, G., KÜMMEL, G. and MANGOS, J. A. (1966). Studies on the ultrastructure and function of a primitive excretory organ, the protonephridium of the Rotifer *Asplanchna priodonta. Pflügers Archiv,* **289**, 141–54.

BROWN, G. W., JR., BROWN, W. R. and COHEN, P. P. (1959). Comparative biochemistry of urea synthesis. II. Levels of urea cycle enzymes in metamorphosing *Rana catesbeiana* tadpoles. *J. Biol. Chem.,* **234**, 1775–80.

BURGER, J. W. (1957). The general form of excretion in the lobster, *Homarus. Biol. Bull.,* **113,** 207–23.

CLARK, H. and FISCHER, D. (1957). A reconsideration of nitrogen excretion by the chick embryo. *J. Exp. Zool.,* **136,** 1–15.

COSTA, G., ULLRICH, L., KANTOR, F. and HOLLAND, J. F. (1968). Production of elemental nitrogen by certain mammals including man. *Nature, Lond.,* **218,** 546–51.

COULSON, R. A. and HERNANDEZ, T. (1955). Renal excretion of carbon dioxide and ammonia by the alligator. *Proc. Soc. Exp. Biol. Med.,* **88,** 682–7.

COULSON, R. A., HERNANDEZ, T. and BRAZDA, F. G. (1950). Biochemical studies on the alligator. *Proc. Soc. Exp. Biol. Med.,* **73,** 203–6.

CRAGG, M. M., BALINSKY, J. B. and BALDWIN, E. (1961). A comparative study of nitrogen excretion in some amphibia and reptiles. *Comp. Biochem. Physiol.,* **3,** 227–35.

DRILHON, A. and MARCOUX, F. (1942). Etude biochimique du sang et de l'urine d'un chélonien: *Testudo mauritanica. Bull. Soc. Chim. Biol. Paris,* **24,** 103–7.

EAKIN, R. E. and FISHER, J. R. (1958). Patterns of nitrogen excretion in developing chick embryos. In *The Chemical Basis of Development* (W. D. McElroy and B. Glass, eds), pp. 514–22, Baltimore, Md.: Johns Hopkins Press.

FORSTER, R. P. (1940). A renal clearance analysis of phenol red elimination in the frog. *J. Cell. Comp. Physiol.,* **16,** 113–22.

FORSTER, R. P. and GOLDSTEIN, L. (1966). Urea synthesis in the lungfish: relative importance of purine and ornithine cycle pathways. *Science,* **153,** 1650–2.

GOLDSTEIN, L., JANSSENS, P. A. and FORSTER, R. P. (1967). Lungfish *Neoceratodus forsteri:* activities of ornithine-urea cycle and enzymes. *Science,* **157,** 316–17.

GOODRICH, E. S. (1945). The study of nephridia and genital ducts since 1895. *Quart. J. Micr. Sci.,* **86,** 113–392.

GOTTSCHALK, C. W. (1961). Micropuncture studies of tubular function in the mammalian kidney. *The Physiologist,* **4,** 35–55.

GRIMSTONE, A. V., MULLINGER, A. M. and RAMSAY, J. A. (1968). Further studies on the rectal complex of the mealworm, *Tenebrio molitor,* L. (Coleoptera, Tenebrionidae). *Phil. Trans. Roy. Soc. Lond. B,* **253,** 343–82.

HARRISON, F. M. (1962). Some excretory processes in the abalone, *Haliotis rufescens, J. Exp. Biol.,* **39,** 179–92.

HILGER, H. H., KLÜMPER, J. D. and ULLRICH, K. J. (1958). Wasserrückresorption und Ionentransport durch die Sammelrohrzellen der Säugetierniere (Mikroanalytische Untersuchungen). *Pflügers Archiv,* **267,** 218–37.

HOPKINS, D. L. (1946). The contractile vacuole and the adjustment to changing concentration in fresh-water amoebae. *Biol. Bull.,* **90,** 158–76.

JANSSENS, P. A. and COHEN, P. P. (1966). Ornithine-urea cycle enzymes in the African lungfish, *Protopterus aethiopicus. Science,* **152,** 358–9.

JEPPS, M. W. (1947). Contribution to the study of the sponges. *Proc. Roy. Soc. Lond. B,* **134,** 408–17.

KEILIN, J. (1959). The biological significance of uric acid and guanine excretion. *Biol. Rev.,* **34,** 265–96.

KHALIL, F. and HAGGAG, G. (1955). Ureotelism and uricotelism in tortoises. *J. Exp. Zool.,* **130,** 423–32.

KITCHING, J. A. (1956). Contractile vacuoles of protozoa. *Protoplasmatologia,* 3(D 3a), 1–45.

LOCKWOOD, A. P. M. (1961). The urine of *Gammarus duebeni* and *G. pulex. J. Exp. Biol.,* **38,** 647–58.

LOVERIDGE, J. P. (1970). Observations on nitrogenous excretion and water relations of *Chiromantis xerampelina* (Amphibia, Anura). *Arnoldia,* **5,** 1–6.

LØVTRUP, S. and PIGON, A. (1951). Diffusion and active transport of water in the Amoeba *Chaos chaos* L. *C. R. des Travaux du Lab. Carlsberg,* Sér. Chim., **28,** 1–36.

MAC MILLEN, R. E. and LEE, A. K. (1967). Australian desert mice: Independence of exogenous water. *Science,* **158,** 383–5.

MAC MILLEN, R. E. and LEE, A. K. (1969). Water metabolism of Australian hopping mice. *Comp. Biochem. Physiol.* **28,** 493–514.

MARSHALL, E. K., JR. (1933). The secretion of urea in the frog. *J. Cell. Comp. Physiol.,* **2,** 349–53.

MERCER, E. H. (1959). An electron microscopic study of *Amoeba proteus. Proc. Roy. Soc. Lond. B,* **150,** 216–32.

MOYLE, V. (1949). Nitrogenous excretion in chelonian reptiles. *Biochem. J.,* **44,** 581–4.

MUNRO, A. F. (1953). The ammonia and urea excretion of different species of Amphibia during their development and metamorphosis. *Biochem. J.,* **54,** 29–36.

MURRISH, D. E. and SCHMIDT-NIELSEN, K. (1970). Water transport in the cloaca of lizards: active or passive? *Science,* **170,** 324–6.

NECHAY, B. R., BOYARSKY, S. and CATACUTAN-LABAY, P. (1968). Rapid migration of urine into intestine of chickens. *Comp. Biochem. Physiol.,* **26,** 369–70.

NEEDHAM, J. (1931). *Chemical Embryology.* vol. I, 1–614; vol. II, 615–1254; vol. III, 1255–724, Cambridge, England: Cambridge University Press.

OSCHMAN, J. L. and WALL, B. J. (1969). The structure of the rectal pads of *Periplaneta americana* L. with regard to fluid transport. *J. Morphol.,* **127,** 475–510.

PANG, P.K.G., GRIFFITH, R. W. and KAHN, N. (1972). Electrolyte regulation in the fresh water stingrays (Potomotrygonidae). *Fed. Proc.,* **31,** 344.

PESCHEN, K. E. (1939). Untersuchungen über das Vorkommen und den Stoffwechsel des Guanins im Tierreich. *Zool. Jahrb.,* **59,** 429–62.

PONTIN, R. M. (1964). A comparative account of the protonephridia of *Asplanchna* (Rotifera) with special reference to the flame bulbs. *Proc. Zool. Soc. Lond.,* **142,** 511–25.

POULSON, T. L. (1965). Countercurrent multipliers in avian kidneys. *Science,* **148,** 389–91.

RAMSAY, J. A. (1949). The site of formation of hypotonic urine in the nephridium. *J. Exp. Biol.,* **26,** 65–75.

RAMSAY, J. A. (1964). The rectal complex of the mealworm *Tenebrio molitor,* L. (Coleoptera, Tenebrionidae). *Phil. Trans. Roy. Soc. Lond. B,* **248,** 279–314.

RAMSAY, J. A. and RIEGEL, J. A. (1961). Excretion of inulin by Malpighian tubules. *Nature, Lond.,* **191,** 1115.

RICHARDS, A. N. (1935). Urine formation in the amphibian kidney. *Harvey Lectures,* Ser. **30,** 93–118.

RIDDICK, D. H. (1968). Contractile vacuole in the amoeba, *Pelomyxa carolinensis. Am. J. Physiol.,* **215,** 736–40.

RIEGEL, J. A. and LOCKWOOD, A.P.M. (1961). The role of the antennal gland in the osmotic and ionic regulation of *Carcinus maenas. J. Exp. Biol.,* **38,** 491–9.

SCHMIDT-NIELSEN, B. (1972). Mechanisms of urea excretion by the vertebrate kidney. In *Nitrogen Metabolism and the Environment* (J. W. Campbell and L. Goldstein, eds), pp. 79–103, London: Academic Press.

SCHMIDT-NIELSEN, B. and FORSTER, R. P. (1954). The effect of dehydration and low temperature on renal function in the bullfrog. *J. Cell. Comp. Physiol.,* **44,** 233–46.

SCHMIDT-NIELSEN, B. and O'DELL, R. (1961). Structure and concentrating mechanism in the mammalian kidney. *Am. J. Physiol.,* **200,** 1119–24.

SCHMIDT-NIELSEN, B. and RABINOWITZ, L. (1964). Methylurea and acetamide: active reabsorption by elasmobranch renal tubules. *Science,* **146,** 1587–8.

SCHMIDT-NIELSEN, B. and SCHRAUGER, C. R. (1963). *Amoeba proteus:* studying the contractile vacuole by micropuncture. *Science,* **139,** 606–7.

SCHMIDT-NIELSEN, K. (1964). *Desert Animals. Physiological Problems of Heat and Water,* Oxford, England: Clarendon Press, 277 pp.

SCHMIDT-NIELSEN, K., BORUT, A., LEE, P. and CRAWFORD, E. C., JR. (1963). Nasal salt excretion and the possible function of the cloaca in water conservation. *Science,* **142,** 1300–1.

SCHMIDT-NIELSEN, K. and LEE, P. (1962). Kidney function in the crab-eating frog (*Rana cancrivora*). *J. Exp. Biol.* **39,** 167–77.

SHANNON, J. A. and FISHER, S. (1938). The renal tubular reabsorption of glucose in the normal dog. *Am. J. Physiol.,* **122,** 766–74.

SHOEMAKER, V. H., BALDING, D. and RUIBAL, R. (1972). Uricotelism and low evaporative water loss in a South American frog. *Science,* **175,** 1018–20.

SKADHAUGE, E. (1967). *In vivo* perfusion studies of the cloacal water and electrolyte resorption in the fowl (*Gallus domesticus*). *Comp. Biochem. Physiol.,* **23,** 483–501.

SKADHAUGE, E. (1968). The cloacal storage of urine in the rooster. *Comp. Biochem. Physiol.,* **24,** 7–18.

SMITH, H. W. (1929). The excretion of ammonia and urea by the gills of fish. *J. Biol. Chem.,* **81,** 727–42.

SMITH, H. W. (1951). *The Kidney. Structure and Function in Health and Disease,* New York: Oxford University Press, 1049 pp.

SMITH, H. W. (1959). *From Fish to Philosopher,* Summit, N.J.: CIBA Pharmaceutical Products, Inc., 304 pp. Reprinted (1961), Garden City, N.Y.: Doubleday & Co.

STARLING, E. H. (1899). The glomerular functions of the kidney. *J. Physiol.*, **24,** 317–30.

VAJROPALA, K. (1935). Guanine in the excreta of arachnids. *Nature, Lond.*, **136,** 145.

WIGGLESWORTH, V. B. (1931). The physiology of excretion in a blood-sucking insect, *Rhodnius prolixus* (Hemiptera, Reduviidae). II. Anatomy and histology of the excretory system. *J. Exp. Biol.*, **8,** 428–42.

YÜ, T. F., BERGER, L., KUPFER, S. and GUTMAN, A. B. (1960). Tubular secretion of urate in the dog. *Am. J. Physiol.*, **199,** 1199–1204.

USEFUL REFERENCE MATERIAL FOR PART IV

BENTLEY, P. J. (1971). *Endocrines and Osmoregulation. A Comparative Account of the Regulation of Water and Salt in Vertebrates,* New York: Springer-Verlag, 300 pp.

BRENNER, B. M. *et al.* (1974). Symposium on renal handling of sodium. *Fed. Proc.,* **33,** 13–36.

BURTON, R. F. (1973). The significance of ionic concentrations in the internal media of animals. *Biol. Rev.,* **48,** 195–231.

KEYNES, R. C. (1971). A discussion on active transport of salts and water in living tissues. *Phil. Trans. Roy. Soc. Lond., B,* **262,** 83–342.

KIRSCHNER, L. B. (1967). Comparative physiology: invertebrate excretory organs, *Ann. Rev. Physiol.,* **29,** 169–96.

KIRSCHNER, L. B. *et al.* (1970). Refresher course on ionic regulation in organisms. *Am. Zool.,* **10,** 329–436.

KROGH, A. (1965). *Osmotic Regulation in Aquatic Animals* [republication of 1st edn (1939)], New York: Dover Publ. Inc., 242 pp.

ORLOFF, J., BERLINER, R. W. and GEIGER, S. R. (eds) (1973). *Handbook of Physiology,* sect. 8, *Renal Physiology,* Washington, D.C.: American Physiological Society, 1082 pp.

POTTS, W. T. W. and PARRY, G. (1964). *Osmotic and Ionic Regulation in Animals,* Oxford, England: Pergamon Press, 423 pp.

SMITH, H. W. (1951). *The Kidney. Structure and Function in Health and Disease,* New York: Oxford University Press, 1049 pp.

SMITH, H. W. (1961). *From Fish to Philosopher,* Garden City, N.Y.: Doubleday & Co. Inc.

RIEGEL, J. A. (1972). *Comparative Physiology of Renal Excretion,* Edinburgh, Scotland: Oliver and Boyd, 204 pp.

V

MOVEMENT, INFORMATION
AND INTEGRATION

11
Muscle, movement locomotion

Most animals, even those that remain attached and never move about (e.g. corals and sponges), show a great deal of movement or motility. Let us first realize that movements serve not only in locomotion, i.e. an animal moving from place to place, but also a large number of other purposes. The latter include such varied activities as the pumping of blood in the vascular system, the transport of food through the intestinal tract, and the moving of the external medium, such as pumping air in and out of the lungs or water over the gills.

There is only a limited number of basic mechanisms used in achieving motility, but the variation in how these mechanisms are used is very great indeed. We will therefore concentrate on some basic mechanisms, and avoid detailed descriptions of the wide variety of designs employed by different animals. The three basic mechanisms are *amoeboid, ciliary,* and *muscular* movement.

Amoeboid movement derives its name from the motion of the amoeba, the unicellular organism described in every biology school book. This form of locomotion involves extensive changes in cell shape, flow of cytoplasm, and pseudopodal activity. It is the most poorly understood type of locomotion, although it has received a great deal of attention.

Ciliary locomotion is the characteristic way in which ciliated protozoans such as *Paramecium* move, but cilia are found in all animal phyla and serve a variety of different purposes. For example, the respiratory passageways of air-breathing vertebrates are lined with ciliated cells which move foreign particles that lodge on their surfaces, thus slowly removing these particles. Cilia serve to set up currents or move fluid in internal structures such as the water-vascular system of echinoderms, or in the external medium such as the flow of water over the gills of lamellibranchs. The sperm of virtually all animals are motile, and most move with the aid of a tail, which in principle acts in a fashion similar to a cilium.

The overwhelming majority of animal movements, however, depend on the use of muscle, which throughout the animal kingdom has one characteristic in common, the ability to exert a force by shortening (contraction).

We have already discussed the use of muscle for the pumping of fluids, whether the pumps are peristaltic (intestinal transport of foodstuffs) or valved (the heart). In this chapter we shall be concerned with mechanisms involved in animal locomotion and how it is achieved, the use of muscle being by far the most widespread and most important mechanism. In the chapter on energy metabolism we discussed the energy requirements for locomotion (p. 244ff); at the end of this chapter we shall discuss ways in which some aquatic animals can reduce their use of muscular energy.

AMOEBOID, CILIARY, AND FLAGELLAR LOCOMOTION

Amoeboid movement

What we designate as amoeboid movement is characteristic of some protozoans, slime molds, and vertebrate white blood cells. The movement of these cells is connected with cytoplasmic streaming, change in cell shape, and extension of pseudopodia. These changes are easily observed in a microscope, but the mechanisms involved in achieving the movement are poorly understood.

When an amoeba moves, its cytoplasm flows into newly formed arm-like extensions of the cell, pseudopodia, which gradually extend and enlarge so that the entire cell occupies the space where previously only a small pseudopodium began to form. As the cell moves, new pseudopodia are formed in the direction of movement, while the 'posterior' parts are withdrawn.

Protoplasmic streaming (*cyclosis*) is a very common phenomenon in all sorts of cells and plays an important role in intracellular transport. The molecular mechanism underlying protoplasmic streaming is not known, although various hypotheses postulate changes in viscosity of the protoplasma or sol–gel transformations. There is no doubt that in an amoeba the outermost layer, known as the ectoplasm, is a somewhat stiffer gel-like layer. As a pseudopodium is formed, the more liquid endoplasm streams into it, and new ectoplasm is formed on the surface. In the rear part of the advancing cell the ectoplasmic gel should then be converted to a more liquid endoplasmic sol by a gel–sol transformation.

Such theories may have some merit, but in any event, movement cannot occur without a driving force, and sol–gel transformations by themselves do not provide the necessary forces. It has been suggested that contraction of material at the posterior end of an amoeba drives material forward and forces it into the extending pseudopod. A lowered viscosity in certain parts of the endoplasm would facilitate streaming, and it is possible that a liquefied layer between the ectoplasm and the endoplasm is of importance.

When it comes to the actual driving force and how it is developed, it has been impossible to pinpoint contractile elements. The greatest experimental difficulties arise from the simple fact that we are unable to make direct mechanical measurements of the forces involved, partly because of the small size of amoeboid cells, and partly because of the absence of mechanical structures that can be subject to manipulation and measurement.

One of the most frequently mentioned theories suggests that hydrostatic pressure plays a major role in pseudopod extension. Certain experimental evidence tends to contradict this hypothesis. If one pseudopod of an amoeba is sucked into a capillary connected to a reduced pressure, the extension of other pseudopods is not prevented (Allen, Francis and Zeh, 1971). This presumably indicates that endoplasmic streaming cannot be attributed to pressure gradients along the length of the stream, for if it were, streaming should be reversed under an applied pressure gradient of opposite sign. Nevertheless, the results do not exclude the possibility of local pressure gradients, for in the absence of appropriate forces, no movement could take place. What these forces are and what molecular mechanism is responsible for their generation remains to be solved, and there may, of course, be several mechanisms functioning independently or in combination (Jahn and Bovee, 1971).

Movement by cilia and flagella

Cilia and flagella are organelles which were originally classified as being different, based on differences in beating patterns. Cilia beat asymmetrically with a fast oar-like stroke, followed by a recovery motion in which the bending cilium returns to its original position (see fig. 11.1). Flagella, on the contrary, beat with a symmetrical undulation in which a wave is propagated along the flagellum. A flagellated cell usually carries only one or a few flagella, while a ciliated cell such as a paramecium may have several thousand cilia distributed on the

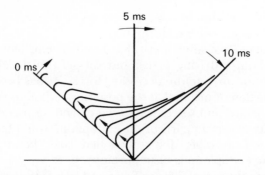

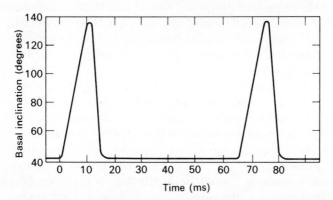

Fig. 11.1. The beat of the cilia of the ctenophore *Pleurobrachia*, drawn from a film taken at 17 °C. The position of one cilium is shown at 5 ms intervals. The power-stroke takes 10 ms, and the complete return stroke takes 50 ms. The cilium then remains in the initial position for 20 ms, until the next stroke begins at 65 ms (Sleigh, 1968).

surface. However, it is difficult to maintain a distinction between flagella and cilia. Morphologically, the structures have major charac-teristics in common, and it is now accepted that the same contractile system is responsible for the motion of both.

Cilia are found in many protozoans, where they are of primary importance in locomotion. However, ciliated cells and epithelia are of great importance for a number of other animals, both in locomo-tion and in transport of the external medium over respiratory sur-faces, as well as in the transport of material on internal epithelial surfaces (e.g. the respiratory tract and the oviduct). Cilia are, in fact, found in all animal phyla and serve in a wide variety of roles.

The electron microscope has revealed that cilia and flagella have a common structure, with a central pair of filaments surrounded by an

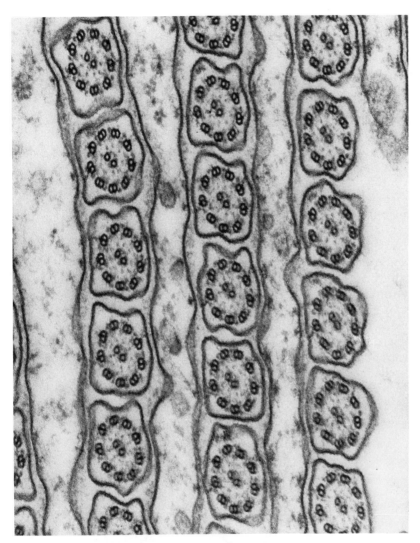

Plate 10. *Flagella.* This electron micrograph of the flagella of the protozoan *Trichonympha* shows very clearly the characteristic two-plus-nine arrangement of the filaments in each flagellum (× 150 000). (Courtesy of Dr. A. V. Grimstone, Zoology Department, Cambridge University.)

additional nine thin filaments. This arrangement of two plus nine filaments is very striking, it occurs from protozoans to vertebrates, and in the overwhelming majority of animal sperm. The particular number of nine plus two is almost universal, although not essential, for a few sperm are known which have three central filaments, one central filament, or none at all (Blum and Lubliner, 1973).

The characteristic nine-plus-two arrangement of the filaments has made it possible to recognize that cilia are part of the structure of several types of sensory organs, including hearing organs and visual receptors. It is on this discovery that we can base the statement that cilia occur in all animal phyla, for insects, in which cilia were previously believed to be absent, have certain sensory organs that contain this structure. It is correct, however, that the cilia of arthropods serve no motor function.

Cilia used for movement can function only in an aqueous medium, and are therefore found only on cell surfaces that are covered by a fluid or a film of an aqueous medium such as mucus.

We intuitively expect that an organism that swims with the aid of a single flagellum, such as a sperm, moves in the opposite direction to the wave that travels down the tail. This is correct for a smooth, undulating filament, which propels the organism away from the filament or 'tail'. Therefore it seems peculiar that some flagellates move with their long flagellum beating anteriorly, in the direction of movement, and that the waves move out from the body to the anterior tip of the flagellum (Jahn *et al.*, 1964). This seemingly paradoxical situation, that wave propagation and locomotion are in the same direction, may occur if the filament is rough or covered with projections. In the flagella of species which move in this way, the flagella have tiny appendages in the form of thin, lateral projections which are responsible for the unexpected direction of movement (fig. 11.2).

The mechanism of movement of cilia and flagella has long been a

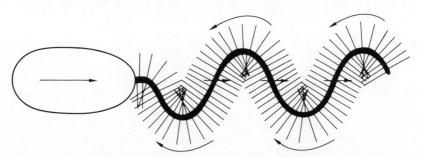

Fig. 11.2. The flagellate *Ochromonas* swims with the single flagellum extending in the direction of movement. The undular motion of the flagellum consists of a wave that travels from its base to its tip, i.e. in the same direction as the direction of swimming. This paradoxical situation is due to the fixed protrusions extending at a 90 ° angle from the flagellum, causing a movement of water as shown by the curved arrows, and the momentum imparted to the water gives the organism an equal momentum in the opposite direction (from Jahn, T. L. et al. 1964. *J. Protozool.*, **11**, 294).

matter of speculation. Three types of mechanism have been suggested: (1) that the flagellum is moved passively, much like a whip, by forces exerted at its base, (2) that elements along the inner curvature of a propagating wave contracts while the opposite side does not, and (3) that the thin filaments inside the cilium slide relative to each other due to forces between them of a nature similar to the sliding filaments of muscle contraction.

The idea that the flagellum is a passive element driven from its base has been shown to be incompatible with the form of the waves (Machin, 1958). In the sperm tail the bending waves pass along without decrease in amplitude, and in some cases with increasing amplitude (Rickmenspoel, 1965). It is therefore necessary to assume that there are active contractile elements in the flagellum and that the energy is generated locally. Present evidence indicates that the bending of flagella and cilia is accompanied by a longitudinal sliding of the nine peripheral filaments relative to one another, and that an active contraction or extension of the filaments is an unlikely mechanism. It is assumed that the active sliding process involves interaction between the filaments analogous to the active sliding process which is the basis for muscular contraction (Brokaw, 1971).

A protein that is believed to be associated with the bending has been extracted from flagella from a number of sources. This protein, known as *dynein,* shows little resemblance to any of the contractile proteins of muscle (see p. 507), except for its ability to split ATP enzymatically, a property that dynein and the important muscle protein myosin have in common (Gibbons and Rowe, 1965).

It is possible to prepare an experimental model of the ciliary apparatus of a paramecium by extracting the cell with a detergent which leaves the ciliary apparatus functional, but the cell membrane disrupted. If such an extracted paramecium is reactivated to swim in a solution of ATP and magnesium ions, the concentration of calcium ion determines the direction in which the cilia beat (Naitoh and Kaneko, 1972). When a living paramecium meets an obstacle, it retreats and swims backwards because of a reversal of the ciliary beat. It appears that the initial process in this reversal of the ciliary beat is an increase in the permeability of the cell to calcium ion (Eckert, 1972).

The over-all efficiency of conversion of metabolic energy into mechanical energy in the swimming of sperm has been estimated by Rikmenspoel to be at least 19%, and probably closer to 25% (Rikmenspoel, Sinton and Janick, 1969). This value is strikingly similar to the efficiency of muscles in performing external work.

MOVEMENT AND MUSCLE

Muscle, what it is – structure

As far as we know, the biochemical mechanism of muscle contraction is the same in all muscles. Two proteins, *actin* and *myosin,* are part of the machinery, and ATP (adenosine triphosphate) is used as the immediate energy source for the contraction. The detailed arrangement, however, varies a great deal, and it is convenient to classify various kinds of muscle accordingly. The classification is based primarily on vertebrate muscles for the simple reason that we know and understand more about these than about invertebrate muscle.

The broadest classification is based on the presence or absence in the muscle of regular cross-striations which can be seen in an ordinary light microscope. Vertebrate skeletal muscles and heart muscle are *striated,* and muscles of the internal organs, as found in the walls of the bladder, intestine, blood vessels, uterus, etc. are *unstriated* (also called *smooth*). The heart muscle, although striated, is often considered as a separate type because it differs from skeletal muscle in characteristic ways, the functionally most important being that a beginning contraction of the heart muscle spreads to the entire organ.

Skeletal muscles are usually called *voluntary,* for the muscles of limb and trunk are under control of the will. This does not mean that we are always aware of or decide about our movements; on the contrary locomotion, breathing, and so on take place without conscious knowledge of the muscles involved. Voluntary muscle is always striated, but the term is very poor, and, at best, useful only for man.

Vertebrate smooth muscle almost always occurs in the walls of hollow internal organs, it is not under control of the conscious mind, and is called *involuntary.* The contractions are usually much slower than those of striated muscle, and normally we are completely unaware of the state of contraction of the smooth muscle in our blood vessels, stomach, intestine, and so on.

Striated muscle

The organization of striated muscle is shown in diagram form in fig. 11.3. The muscle is made up of a large number of parallel *fibers* which are between 0.1 and 0.01 mm in diameter, but may be several cm long. These fibers in turn are made up of thinner *fibrils.* These fibrils have characteristic cross-striations, the so-called Z-bands, which are repeated at completely regular intervals of about 2.5 μm

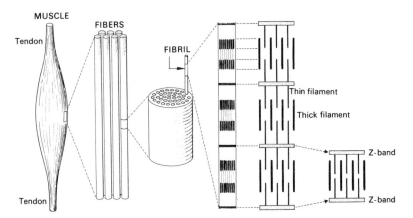

Fig. 11.3. Schematic diagram of a vertebrate striated muscle. The whole muscle is composed of fibers, which in the light microscope appear cross-striated. The fibers consist of fibrils, which have lighter and darker bands. The electron microscope reveals that these bands are due to a repeating pattern in the regular arrangement of thicker and thinner filaments.

in a relaxed muscle. From the narrow Z-band very thin *filaments* extend in both directions, and in the center these thin filaments are interspersed with somewhat thicker filaments. The result is a number of less conspicuous bands located between the Z-bands; the appearance of these bands changes with the state of contraction of the muscle. The thin filaments are about 0.005 μm in diameter, and the thick filaments are about twice the size, about 0.01 μm in diameter. The length of the thick filaments is about 1.5 μm. The thin filaments vary somewhat more in length, and are often between 2 and 2.6 μm from tip to tip (or about half this length measured from the Z-band to the tip). The arrangement of the filaments is extremely regular and well ordered. In striated muscle the thick filaments consist of myosin and the thin filaments of actin. The thick and thin filaments are linked together by a system of molecular cross-linkages, and when the muscle contracts and shortens, these cross-linkages are rearranged so that the thick filaments slide in between the thin filaments, thus reducing the distance between the Z-bands (Huxley, 1969).

This model of the structure of the muscle is well supported by electron micrographs, which show the filaments in the various states of contraction. Neither the thick filaments nor the thin filaments change in length during contraction, but it can clearly be seen that they move relative to each other. What is even more important in support of the sliding filament theory is that the force developed by

Plate 11. *Striated muscle.* Longitudinal section of the papillary muscle from the ventricle of a dog heart. The regularly interspersed thick and thin muscle filaments (M) form broad bands. The dark lines (Z) are the Z-bands, which in this preparation are spaced about 2.3 μm apart. Mitochondria (Mi) are found in large numbers in heart muscle. The small, dark granules (Gl) are glycogen (Joachim R. Sommer, Duke University).

a muscle is related to the amount of overlap of the filaments (Gordon, Huxley and Julian, 1966).

Cardiac and smooth muscle

Cardiac muscle is striated, like skeletal muscle, but its properties differ for reasons which are related to its structure. The muscle fibers, instead of being arranged like a bundle of parallel cylindrical fibers, are branched and connected somewhat like a meshwork. An important result of this structure is that, when a contraction starts in one place of the heart muscle, it rapidly spreads throughout the muscle mass. Another important property of cardiac muscle is that a contraction is immediately followed by a relaxation. As a result, a long-lasting contraction, like the sustained contraction of a skeletal muscle, does not occur. These two properties are essential for the normal rhythmic contraction of the heart.

Smooth muscle lacks the characteristic cross-striations of skeletal muscle, but the contraction depends on the same proteins as in striated muscle, actin and myosin, and a supply of energy from ATP. The lack of cross-striations seems to be due to an absence of a regular pattern in the arrangement of thick and thin filaments.

Smooth muscle has not been as extensively studied as striated muscle. There are several reasons, one is that smooth muscle is often interspersed with connective tissue fibers, another is that smooth muscle fibers do not form neat parallel bundles that can readily be isolated and studied. Smooth muscle also consists of much smaller cells, the fibers are often only a fraction of a millimeter long. Finally the rate of contraction is usually much, much slower than in striated muscle, in fact, virtually all smooth muscle is incapable of producing rapid contractions.

Substances for energy storage

The immediate source of energy for a muscle contraction is *adenosine triphosphate* or *ATP,* a compound which is the primary energy source for nearly every energy-requiring process in the body. ATP is the only substance that the muscle proteins can use directly. When the terminal phosphate group of this compound is split off, the high energy of the bond is available for the energy of muscle contraction.

In spite of its great importance, ATP is present in muscle in very small amounts. The total amount of ATP may be sufficient for no more than ten rapid contractions; it follows, therefore, that ATP must be restored again very rapidly, for otherwise the muscles would soon be exhausted. The source is another organic phosphate compound, creatine phosphate, which is present in larger amounts. Its phosphate group is transferred to adenosine diphosphate (ADP) and thus the supply of ATP is restored. The creatine phosphate must, however, eventually be replenished, and the ultimate energy source is the oxidation of carbohydrates or fatty acids. Carbohydrate is stored in the muscle in the form of glycogen, which is often present in an amount of between 0.5–2% of the weight of the muscle and provides an amount of energy perhaps 100 times as great as the total amount of creatine phosphate. In the absence of sufficient oxygen the glycogen can still yield energy by being split into lactic acid, but the amount of energy is then only a small fraction, about 7%, of that available from complete oxidation (see p. 218).

Creatine phosphate has been found in the muscles of all vertebrates that have been examined, but it is conspicuously absent in the muscles of some invertebrates. In these animals we find other organic phosphates, one such compound is arginine phosphate, which is never found in vertebrate muscle.

These various organic phosphate compounds all provide energy through the splitting-off of a phosphate group, the high bond energy being available for energy-requiring processes. The fact that creatine phosphate is found universally in vertebrates, but only in some invertebrate groups, has led to the suggestion that the biochemical similarity of these groups to the vertebrates indicate which invertebrate phylum has given rise to the vertebrates. Present knowledge indicates, however, that the distribution of arginine and creatine phosphate among invertebrates is very irregular and not related to the classification of these animals. In particular, there seems to be no pattern in the occurrence of creatine phosphate among invertebrates that can give a clue to the evolution of vertebrates.

Muscle, how it works – contraction

In muscle physiology the term *contraction* refers to a state of mechanical activity. It may involve a shortening of the muscle, but if the muscle is kept from shortening by having its ends solidly attached, we still use the word contraction to describe the active state. In the latter case, contraction of the muscle results in a force being exerted on the points of attachment, and since no change in length takes place, this is called an *isometric* contraction. If, on the other hand, we attached to one end of the muscle a weight which it can lift, the muscle will shorten during contraction; since the load remains the same throughout the contraction, this is called an *isotonic* contraction.

Virtually no movements of muscles in the body are purely isometric or purely isotonic, for usually both the length of a muscle and the load change during contraction. For example, when an arm lifts a certain weight, both the length of the contracting muscles and the load on them change continuously, for the leverage and the angle of attack change throughout the movement. Although pure isometric or pure isotonic contractions are unimportant for muscle function in the body, it is convenient in the study of isolated muscles and their function to distinguish between isometric and isotonic contractions. From isometric contractions, we obtain a great deal of information about the forces that a muscle can exert; in the study of isotonic contractions, on the other hand, the amount of work performed by the muscle can easily be calculated. In the laboratory we therefore usually chose one method or the other, according to the purpose of our study.

Force. A muscle can be stimulated to contract by applying a single brief electric pulse, and if the muscle is kept from shortening (isometric contraction) the force it produces gives a record as in fig. 11.4a. Immediately after the stimulus is applied, there is a very short period, the *latent period,* of a few milliseconds before contraction starts. Then the force exerted by the muscle rises rapidly to a maximum, and declines again somewhat more slowly. Such a single contraction in response to a single stimulus is called a *twitch.* The time required to reach maximum tension has been called the *contraction time,* which varies from muscle to muscle. The contraction time for the major muscles of locomotion of a cat may be as high as 100 ms, while the fastest muscles, such as the eye muscles, may have contraction times of less than 10 ms.

If a second stimulus is applied before a contraction is ended, a new contraction is superimposed on the first twitch (fig. 11.4b). The second contraction reaches a higher peak of force than the single twitch. If we continue to apply two stimuli spaced more closely, the two contractions become more and more fused (fig. 11.4c,d).

Repeated stimuli applied at suitably long intervals give a series of separate twitches, but if we space the stimuli more and more closely, the twitches become fused, and eventually yield a smooth sustained contraction as in fig. 11.5b. Such a smooth sustained contraction is called a *tetanus.*

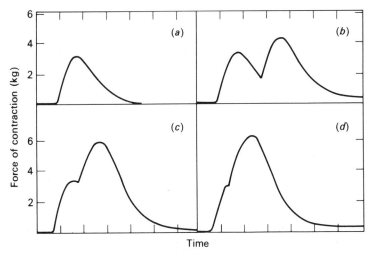

Fig. 11.4. (a) The force developed in isometric contraction by the gastrocnemius muscle from a cat's leg after a single stimulus.
(b), (c) and (d): Force curve after double stimuli applied at shorter and shorter intervals. Time marks are 20 ms apart. Temperature 34.5 °C (adapted from Cooper and Eccles, 1930).

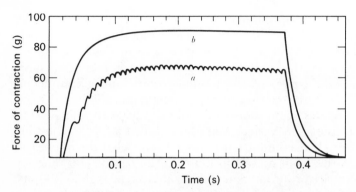

Fig. 11.5. Force developed in isometric contraction by an eye muscle of a cat in response to repeated stimuli. (*a*) Stimulation at a rate of 125 per second (8 ms interval). (*b*) Stimulation rate 210 per second. This gives almost complete fusion to a tetanus. Temperature 36 °C (adapted from Cooper and Eccles, 1930).

The maximum force developed in a tetanus depends on the initial length of the muscle. If a muscle is made to contract isometrically, i.e. with its ends fixed, the force is reduced if the distance between the ends is less than the resting length of the muscle, and the force also reduced if the length is higher than the resting length. The maximum force is obtained when the muscle is kept at its resting length.

This is closely related to the position of the filaments relative to each other (fig. 11.6). We can see that if the muscle is initially stretched so that there is no overlap between thick and thin filaments (fig. 11.6, position 1), there is no force developed at all. This is consistent with the complete inability of the filaments to form cross bridges in this position. If the muscle is permitted to be shorter, however, the force increases as the filament overlap is greater. The maximum force is developed when the spacing is such that all the cross bridges along the full length of the thick filaments are in contact with the thin filaments (position 2). If the muscle is even shorter before contraction begins, the force is again diminished and rapidly decreases to zero as the ends of the thick filaments make contact with the Z-bands (positions 3 and 4).

Shortening. If a muscle is free to contract when it is stimulated, it may shorten by, say, one-third. However, if a load is attached to the muscle, the shortening will be smaller. If the load is gradually increased, the muscle will be able to lift the load over a shorter and shorter distance until the load is so heavy that the muscle is unable to lift it at all. We thus obtain a curve for the shortening versus the

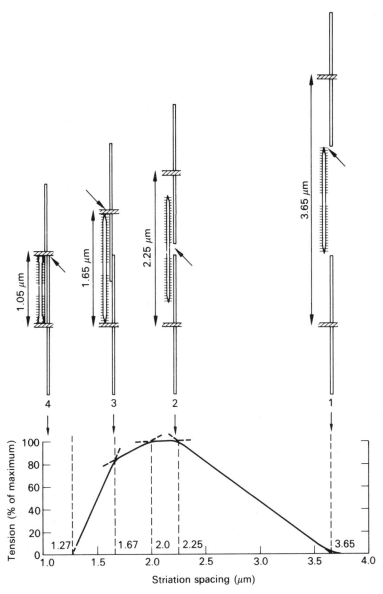

Fig. 11.6. The force developed by a contracting muscle (ordinate) in relation to its initial length (abscissa). The force is expressed in per cent of the maximal isometric force the muscle can develop. The amount of overlap between thick and thin filaments is shown at the top, each diagram placed above the corresponding place on the force curve (after Gordon *et al.*, 1966).

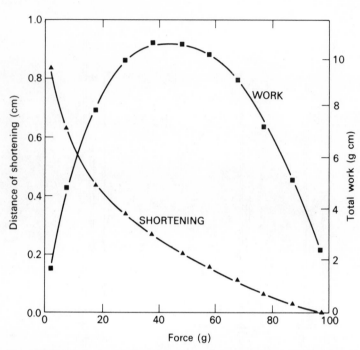

Fig. 11.7. Isotonic contraction of the sartorius muscle of the toad *Bufo marinus*. The shortening of the muscle decreases to zero as the load is increased to 97 gram. The work performed in the contraction increases to a maximum at a load of about 40% of the maximum, and decreases again as the load is increased (courtesy Frans Jöbsis).

load as shown in fig. 11.7. In this particular case, a load of just below 100 grams completely prevented shortening.

Do the muscles from various animals differ much in strength? When we see an ant carry in its jaws a seed that weighs more than the animal itself, we gain the impression that insect muscles must be inordinately strong. An objective comparison of muscle strength shows that this is not so. In order to compare different muscles we must, of course, take into consideration their different size. The force a muscle of a given kind can exert is directly related to its cross-sectional area, but not to its length. We can therefore compare widely different muscles if we relate the force they can exert to their cross-section, as table 11.1 shows for muscles from a variety of animals. Although many muscles are less powerful than those listed, it appears that the maximum limit is roughly the same for all muscle, irrespective of their origin, some 4 to 6 kg per cm² cross-sectional area.

The fact that force of contraction is similar in such a wide variety of muscle may at first seem surprising, but it is understandable in

Table 11.1. *The maximum force of contraction measured in muscles from a variety of animals*

		(kg cm^{-2})	
Annelid (*Arenicola*)	Body wall	3	(*a*)
Bivalve (*Anodonta*)	Adductor	5	(*b*)
Bivalve (*Mytilus*)	Ant. byssus retractor	4.5	(*c*)
Octopus (*Octopus*)	Funnel retractor	5.1	(*d*)
Insect (*Locusta*)	Hind legs	4.7	(*e*)
Insect (*Decticus*)	Flexor tibiae	5.9	(*e*)
Insect (*Drosophila*)	Wing muscles	5	(*f*)
Frog (*Rana*)	Anterior tibial (20 °C)	4.4	(*g*)
Rabbit	'Skeletal muscle' (37 °C)	5.0	(*h*)
Man	Ankle flexors	4.2 *	(*i*)

* Voluntary contraction.

(*a*) Trueman (1966).   (*f*) Roeder (1953).
(*b*) Weber and Portzehl (1954).   (*g*) Casella (1950).
(*c*) Abbott and Lowy (1953).   (*h*) Wilson (1972).
(*d*) Lowy and Millman (1962).   (*i*) Haxton (1944).
(*e*) Wigglesworth (1972).

view of the basic similarity in the contractile mechanism itself. All muscle contraction seems to be based on sliding filaments of actin and myosin, with the energy supplied by ATP. It would be unreasonable to expect that this mechanism could be improved to provide a greater force per cross-sectional area, for the maximal force should be related to the number of filaments that can be packed within that area, and this again depends on the size of the protein molecules that make up the filaments.*

Work. When a muscle lifts a weight, the external work performed is the product of the load and the distance over which it is moved. This is shown by the arched curve in fig. 11.7. When the load is zero, the external work performed is also zero, although the muscle shortens maximally. At the other extreme, when the load is too heavy to be lifted at all, the external work is also zero. Between these extremes,

---

* The diameter of the thick and thin filaments limits the number of filaments per cross-sectional area. However, the force exerted could conceivably be increased by lengthening the overlap between thin and thick filaments (i.e. increasing the distance between the Z-lines and having longer filaments). This would give a larger number of cross-bridges per filament and thus increase the force. This avenue for increasing the force is probably unavailable because of limitations on the structural strength of the filaments which must support the increased force.

the external work is at a maximum when the weight is about 40% of the maximal load the muscle can lift and the muscle shortens by about one-third of the maximal shortening.

The work that can be performed by a muscle is obviously related to its size. Consider two muscles that have the same dimensions and can develop the same maximum force per unit cross-sectional area. If they shorten to the same extent, the work they perform must be the same. Now consider two muscles of the same cross-section but different length, one twice as long as the other, and that they both can contract by the same fraction of their resting length, say, by one-third. The maximum work each can perform will be directly related to its initial length, i.e. twice as high in one as in the other, although the force of contraction will be the same (same cross-sectional area). We can extend this argument by saying that since work is the product of force and distance, and the volume of the muscle is proportional to the product of cross-section and length, the work a muscle can perform must be directly related to its volume. This, of course, is a generalization from which there are exceptions, but as a general rule it is a useful fact to keep in mind.

When it comes to the power output (work per unit time) the situation is quite different. Since a fast muscle contracts in a short period of time, the power produced will be greater than in a slow muscle. The consequence of all this is that, if as an example we compare the muscles of elephants and shrews, we can expect (a) that the muscles have approximately the same contractile force per cm^2 cross-sectional area, (b) that they can shorten to approximately the same fraction of their resting length, and (c) that the work performed during the contraction of one gram of muscle will be similar in the two animals. However, since the contraction of a shrew muscle takes place much faster, i.e. in a much shorter time, the power output per gram shrew muscle is correspondingly much higher than for the elephant muscle. This is precisely what is reflected in the metabolic rates of the animals (see p. 237), which are usually expressed in power units (work per unit time) and are much higher in the small animal.

Muscle, how it is used

The way muscle is used by various animals differs a great deal (depending primarily on the function of the particular muscle). The demands on the flight muscles of an insect which contract several hundred times per second, and on the muscle that closes the shells of a clam and remains contracted perhaps for several hours, are very different indeed. The best way to describe how muscle can serve different purposes is to examine some characteristic types of muscle and how they are modified to meet specific demands.

Vertebrate fast and slow muscles

Vertebrate striated muscle is composed of muscle fibers that fall into two (or more) distinct classes, which are frequently referred to as 'fast' and 'slow' fibers. This terminology can easily lead to confusion, and it is now common to designate the fast fibers as *twitch* fibers and the slower as *tonic* fibers. Usually there is a difference in the amount of myoglobin present in the two kinds of fiber; the twitch fibers which have a lower myoglobin content have been known as pale or white, the tonic fibers with a higher myoglobin content as red. Any one muscle may consist of only twitch fibers, only tonic fibers, or a mixture of both.

The distinction between twitch (fast) and tonic (slow) muscle fibers was first made for frog muscles. The main functional difference is that twitch fibers are used for rapid movements, while tonic fibers are used to maintain low-force prolonged contractions. The twitch system is associated with large nerve fibers (10–20 μm diameter) with conduction velocities about 8–40 meter s^{-1}, which lead to quick contractile responses. This twitch system is used e.g. in jumping. The tonic (slow) system has small-nerve fibers (about 5 μm diameter) which conduct at 2–8 meter s^{-1} and lead to slower, graded muscular contractions which are accompanied by non-propagated muscle potentials of small amplitude and long duration. This is the tonic system used e.g. for maintaining the posture of the body (Kuffler and Williams, 1953a).

The sartorius muscle of the frog, which runs along the thigh and is used primarily in jumping, consists entirely of twitch fibers. The nerve to the muscle contains large, fast conducting nerve fibers, and each makes contact with the muscle fibers through a single terminal (the end-plate). The twitch fibers respond according to the 'all-or-none' rule, which means that when a stimulus exceeds a certain minimum, the *threshold value,* the fiber responds with a complete contraction and a further increase in stimulus intensity gives no further increase in the response. Thus, the muscle fiber either doesn't respond, or responds maximally. This is known as an *all-or-none response.*

Tonic muscle is different. The muscle fibers, whose behavior in many respects bears little resemblance to that of the twitch fibers, are innervated by nerve fibers, all of small diameter, whose terminal contacts are distributed along the length of the muscle fiber. In contrast to the fast twitch-producing muscle fiber, the response of the tonic fibers does not necessarily spread throughout the fiber to give a complete all-or-none response, and there is a summation on repetitive stimulation. The response to a single stimulus is small, and the tension rises with repeated stimulus frequency. Thus, repeated stimulation of small nerves is necessary to cause a significant tension rise in tonic fibers. Increased frequencies speed up the rate of tension rise. Relaxation after slow muscle fiber contraction is at least 50 to 100 times slower than after twitch fiber action. This type of response plays an important role in the general postural activity of the frog (Kuffler and Williams, 1953*b*).

When we wish to study separately the function of the different types of muscle fiber, we do not search for muscles that contain only one type. The observed differences between a muscle consisting of only twitch fibers and one with only tonic fibers could just as well be due to the different location and use of the two muscles. It is therefore better to examine different fiber types as they are found in a single muscle.

The gastrocnemius muscle of the cat (which runs from the knee to the heel tendon and is used for extension of the foot) can be used for this purpose. It has not only two, but three distinct functional types of muscle fibers. This is shown in the following way. A single motor neuron in the appropriate region of the spinal cord is stimulated with the aid of an inserted microelectrode. This activates only one motor unit (which consists of many muscle fibers jointly innervated by the one single nerve axon). This may give one of three different types of contraction, depending on which particular neuron is stimulated. Fig. 11.8 shows the response of the three types of fiber, a fast contracting and fast-fatiguing type (*a*), a fast contracting and fatigue resistant type (*b*), and a slow contracting type which does not fatigue even during a prolonged stimulation (*c*). The slow contracting (tonic) fibers have contraction times more than twice as high as the contraction time for the fast contracting–fast fatigue (twitch) fibers (see table 11.2). The difference in the tension developed by the muscle is even more striking, the fast contracting fibers develop a tension more than ten times as high as the slow contracting fibers. This is evidently paid for in the much faster fatigue of these fibers.

Whether a muscle is twitch or tonic is not necessarily a constant inherent characteristic; somehow it seems to depend on how the muscle is used. By changing the position of insertion of the tendon of muscles in the hindlimb of rabbits, it has been possible to make tonic muscle acquire characteristics more nearly those of twitch muscle. In

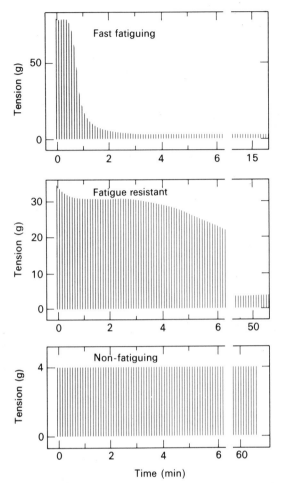

Fig. 11.8. The gastrocnemius muscle of a cat contains three types of fibers which respond to stimulation (40 per second) in three different ways. They may be fast contracting-fast fatiguing (top), fast contracting-fatigue resistant (center), or slow contracting-non fatiguing (bottom) (from Burke, R. E. 1971. Mammalian Motor Units: Physiological Histochemical Correlation in Three Types in Cat Gastrocemius. *Science,* **174**, 709–12, Fig. 1a. © American Association for the Advancement of Science).

the newborn cat all muscles seem to be equally slow, and the dif-ferentiation takes place as the muscles are used. By severing the nerves to two muscles and then re-connecting each nerve to the other muscle, it has been shown that when a nerve from a fast (or phasic) motor neuron is connected to a slow (tonic) muscle, the mus-cle is transformed into a fast (twitch) type muscle. Likewise, slow (tonic) motor neurons make twitch muscles change towards tonic

Table 11.2. *Contraction time and maximum tetanic tension developed by three different functional types of muscle fibers in the gastrocnemius muscle of a cat.* (Burke *et al.*, 1971)

	Fiber type		
	Fast contraction–fast fatigue	Fast contraction–fatigue resistant	Slow contracting
Twitch contraction time (ms)	34	40	73
Tetanic tension (gram force)	60	20	5

(Buller, Eccles and Eccles, 1960). It has become apparent, however, that cross-innervation does not completely convert one type of muscle to the other, but to some mixed form (Buller and Lewis, 1964).

The muscles of fishes also have twitch and tonic fibers. Pelagic fish such as mackerel and tuna, which swim continuously at relatively low speeds, have the two types of fibers separated in different muscle masses which have strikingly different appearance. The tonic muscle is deep red due to a high concentration of myoglobin, and is located along the sideline and stretches in towards the vertebral column. The basal swimming during cruising is entirely executed by this red muscle, while the large mass of white muscle (twitch type) represents a reserve of power for short bursts of high-speed activity (fig. 11.9).

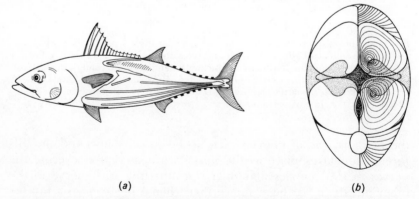

(a)   (b)

Fig. 11.9. In the skipjack tuna (*a*) the red (tonic) swimming muscle, which is used during steady cruising, is located along the sideline and in towards the vertebral column (stippled) (shown in (*b*)). The remaining large mass of white (twitch) muscle is used in bursts of high-speed swimming (Rayner and Kennan, 1967).

The swimming muscles of sharks are also divided into similar types, with less than 20% of the muscle mass made up of tonic fibers. Again, the tonic fibers are for cruising and the twitch fibers are used during vigorous swimming such as in the pursuit of prey. The twitch fibers are capable of a high level of anaerobic power output, as would be expected during maximal activity, and accordingly have a high glycogen content. After vigorous activity the glycogen level in the fast fibers falls markedly; during prolonged periods of slow swimming, however, there is no change in their glycogen level, which indicates that they are not used at all during cruising (Bone, 1966).

Smooth muscle

In contrast to striated muscle (in vertebrates the skeletal and heart muscle), the muscles of internal organs are smooth (also called un-striated, non-striated, plain, or involuntary muscle). In vertebrates this type of muscle occurs in the stomach, intestine, bladder, ureter, uterus, bronchi, blood vessels, etc. Smooth muscle also occurs in a vast number of invertebrates, but it does not form coherent functional groups in these.

Vertebrate smooth muscle is innervated from the autonomic nervous system by two sets of nerves. One set is stimulatory and the other is inhibitory, i.e. the two sets act in an antagonistic manner. Smooth muscle differs from voluntary muscle in that the muscle layers, mostly located in the walls of hollow organs, contain numerous nerve fibers and nerve cells. This has impeded experimentation with isolated smooth muscle, for when a stimulus is applied it is difficult to separate effects on the muscle cells themselves from effects on the nerve cells. A further difficulty in working with smooth muscle is that the fibers do not form neat bundles; the cells are smaller and shorter than in striated muscle and frequently run in many different directions.

Smooth muscle need not be stimulated through the nerves in order to contract. It shows spontaneous rhythmic contractions which can vary greatly both in frequency and intensity. The internal neurons and nerve fibers in the muscle may have considerable influence on this activity, but the details of such interaction remain to be clarified.

Sudden stretching of smooth muscle causes an immediate contraction. Distention of a hollow organ is therefore frequently followed by contraction. An isolated piece of smooth muscle behaves somewhat differently, however. If a piece is loaded slightly, it stretches slowly, and may, without change in the load, adopt different lengths at different times. Thus, smooth muscle has *no particular resting length.*

A distinctive feature of smooth vertebrate muscle is the slowness of response. Its most characteristic property is that it can maintain contraction for prolonged periods of time with very little energy expenditure. The ability to maintain a slight contraction or 'tone' at various different lengths is most important to the economy of maintaining such tension, for in general the cost of maintaining tension is inversely related to the speed of contraction (Ruegg, 1971).

The reason that no striations are visible in smooth muscle is that the filaments, which presumably are present, are not regularly aligned to form visible bands. Vertebrate smooth muscle contains thin (actin) filaments, but, although the muscles do contain myosin, it has so far been impossible to demonstrate thick filaments of the kind found in striated muscle.

Many invertebrate muscles are non-striated or smooth. One kind that will be discussed later is the slow closing muscle of bivalves, which not only pulls the shells together, but can maintain the contraction for hours or days. Not all smooth mollusc muscle is slow, however; the muscle of the octopus and squid mantle, which is responsible for the jet propulsion these animals use in swimming, is a fast contracting muscle. The duration of a full contraction in these muscles in the living animal may be one- or two-tenths of a second, which for a smooth muscle is very fast indeed (Trueman and Packard, 1968).

In structure and composition invertebrate smooth muscle differs somewhat from vertebrate smooth muscle. For example, mollusc smooth muscle contains a protein, tropomyosin A, which is not found elsewhere. Nevertheless, it appears that the basic mechanism of contraction always depends on a sliding filament displacement, which is similar in both smooth and striated muscle throughout the animal kingdom.

Molluscan catch muscle

Clams and mussels protect themselves by closing the shells. One or more adductor muscles pull the shells together, and keep them closed against the springy action of the elastic hinge. Species that live in the intertidal zone must keep their shells closed when the water recedes, and the muscle is kept contracted for hours with the shells tightly closed under conditions that are probably completely anaerobic. If a starfish attacks a mussel, it attaches its tube feet to the two shells and tries to pull them apart; the mussel must be able to withstand the strong pull in this endurance test between predator and prey.

In some species, but not all, the closing muscles are divided into two portions, one of smooth and one of striated fibers. Functionally, the striated portion contracts quickly and is referred to as the 'fast' or phasic (twitch) portion, while the non-striated is much slower and is known as the 'slow' or tonic portion. The long-maintained contractions which keep the shells closed for extended periods of time are produced primarily by the slow portion of the closing muscle (Hoyle, 1964).

It has been suggested that the closing muscles, once they have shortened, enter into a state of 'catch'. This means that the contraction is supposedly maintained without further expenditure of energy, in other words, that the relaxation phase is immensely prolonged and lasts for periods of up to many hours. The problem of how prolonged contractions are maintained in 'catch' muscle has only been partly resolved, and there is still much uncertainty about the mechanism.

The hypotheses to explain 'catch' are basically of two kinds. One postulates a change in the muscle at the molecular level which prevents the breakage of established cross-bridges between the filaments. The other postulates the need for repeated reactivation of the contractile system through the nervous system or through internal action potentials in the muscle.

The most studied muscle of the catch type is the anterior byssus retractor muscle (ABRM) of the blue mussel (*Mytilus edulis*), which contains only smooth fibers and can be made to contract by electric stimuli, but depending on the nature of the stimulus, relaxation may be complete within a matter of seconds or contraction may be maintained for minutes or hours (Johnson and Twarog, 1960).

If the state of tension can be maintained without the expenditure of energy, it would be particularly advantageous to the animal under anaerobic conditions. The oxygen consumption of the ABRM is too small to be discovered as an increase in the oxygen consumption of the whole animal, in part because it is difficult to exclude as an energy source the energy-rich phosphate compounds and the contribution of anaerobic processes.

The oxygen consumption of the isolated ABRM slowly diminishes with a time-course roughly similar to that of the decrease in tension in the muscle. During contraction of this muscle, the oxygen consumption is reduced if the external tension on the muscle is released, and increases again when the tension is restored by stretch-

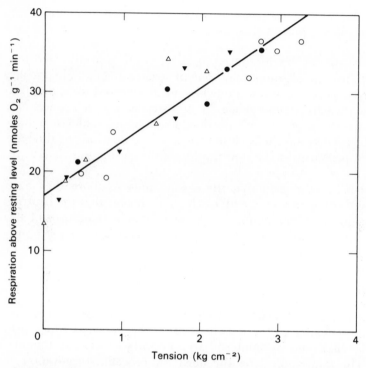

Fig. 11.10. The excess oxygen consumption (above the resting level) during contraction of the closing muscle of the blue mussel, *Mytilus*. The excess respiration is linearly related to the tension in the muscle and seems to be independent of the length of time the contraction has been maintained (o: 20 to 35 min; •: 35 to 50 min, ▼ 50 to 65 min; △ : 65 to 80 min) (Baguet and Gillis, 1968).

ing the muscle back to its original length. This confirms that the tension in the muscle determines the intensity of oxygen consumption during the 'catch'. If the excess oxygen consumption (that above the resting level) is studied, its relation to the tension in the muscle is linear, as shown in fig. 11.10.

These results are compatible with the hypothesis that the tonic tension in the catch muscle is maintained as an 'active' state which has a metabolic energy requirement. Due to the inherent slowness of the muscle, however, the energy requirement for maintained contraction is in the magnitude of more than 1000 times lower than for the maintenance of tetanic contractions of vertebrate fast muscle.

Crustacean muscle

Crustacean muscles show in pure form some characteristics which to a varying degree are present in the muscles of a number of other animals. We shall discuss two points, (1) the gross arrangement of the muscle fibers into parallel-fibered or pinnate muscles, (2) the multiple innervation of a single muscle fiber and its responses to the different nerves.

Nearly everybody who has handled live crabs or lobsters has made contact with the impressive force that these animals can exert with their claws. This is not because the strength of the muscle is extraordinarily high, but rather because of the anatomical arrangement of the fibers. In the claw of the crab the muscle is *pinnate,* which means that the fibers, instead of being parallel to the direction of the pull, are arranged at an angle which greatly increases their mechanical advantage (fig. 11.11). In the enclosed space of the claw, a pinnate muscle of a given volume can have more and shorter fibers than a parallel-fibered muscle. The increased force, due to this mechanical arrangement, is gained at the expense of the distance over which the attachment can be moved. In the crab claw, the closing muscle can exert, because of the pinnate arrangement, about twice as great a force as it could if it were parallel-fibered (Alexander, 1968).

Another advantage of the pinnate muscle in the crab claw is related to the confined space in which it works. As a parallel-fibered muscle contracts, its cross-section increases, and in the rigid enclosure of the claw this would cause difficulties. Due to the arrangement of the pinnate muscle, however, the cavity of the crab claw can be almost completely filled with muscle, for as the fibers shorten,

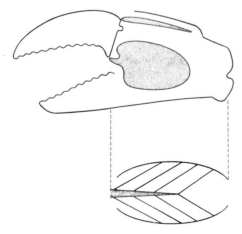

Fig. 11.11. The muscles in the claw of a crab or lobster are arranged so that a mechanical advantage is gained. This so-called pinnate arrangement also permits the muscle to thicken during contraction, which otherwise would not be possible within the confined space of the claw (from Alexander, R. 1968. *Animal Mechanics,* London: Sidgwick & Jackson).

their angle is changed so that there is space for the thickening. This is impossible with a parallel-fibered muscle in a limited space.

Pinnate muscles can perform the same amount of work as a parallel-fibered muscle of equal volume because the increased force of contraction is offset by a corresponding decrease in shortening distance. It should be noted that pinnate muscles are not unique to crustaceans, they also occur quite commonly in vertebrates.

The most interesting aspect of crustacean muscle is the multiple innervation, which means that individual muscle fibers may be innervated by two or more nerve fibers. In addition to multiple nerves that stimulate the muscle to contract, many arthropod muscles also have inhibitory nerves whose stimulation causes the muscle to relax if it is already in a state of contraction. Another crustacean characteristic is that a whole muscle is often innervated by only a few or a single axon. This makes the whole muscle act much as a single unit, and gradation of response is due to variations in the nerve impulses in combination with the balance between excitatory and inhibitory impulses. One muscle may, therefore, depending on the impulses it receives, act as a fast or as a slow muscle.

The difference in the innervation of vertebrate and crustacean muscle is shown in fig. 11.12. As stated above, the most striking characteristic of crustacean muscle is that it is supplied by two or more different nerve fibers. Often, there is a 'fast' excitatory fiber, a 'slow' excitatory fiber, and finally an inhibitory nerve fiber. In many

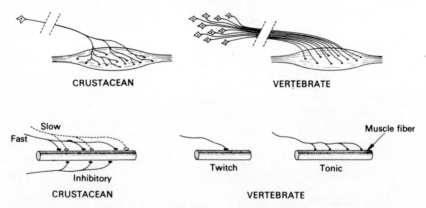

Fig. 11.12. Top: Crustacean muscle is innervated from a few neurons and the nerve fibers branch extensively and supply the entire muscle. Vertebrate muscle is innervated from a large number of neurons in the spinal cord, each one supplying a relatively small number of muscle fibers. Bottom: A crustacean muscle fiber is often supplied by three different nerve fibers, one for fast contraction, one for slow, and one for inhibition of contraction.

muscles the majority of the fibers are innervated by the 'fast' nerve, while a smaller number is innervated by the 'slow' nerve. In combination with the inhibitory fibers this permits a wide range in the gradation of the contractions.

The other characteristic of crustacean muscle that differs from vertebrate muscle is that a whole muscle is usually innervated from a very few neurons in the central nervous system, and is reached by a correspondingly small number of axons which branch and connect to the fibers of the entire muscle. Vertebrate muscle, in contrast, is innervated from a large number of neurons in the central nervous system, and each nerve fiber reaches only a small number of the fibers in the entire muscle. In the fast (twitch) vertebrate muscle a gradation in contraction of a muscle depends on the number of axons that are carrying excitatory impulses to their respective small groups of fibers. Inhibition does not occur in vertebrate muscle; in these, inhibition is achieved at the level of the neurons of the central nervous system and not at the level of the muscle itself (Wiersma, 1961; Hoyle, 1957).

Insect flight muscle

Many insects, such as dragonflies, moths, butterflies, and grasshoppers have relatively low wingbeat frequencies, and each muscle contraction occurs in response to a nerve impulse. Since the muscle contraction is synchronized with the nerve impulse, this type is known as *synchronous muscle.*

Many small insects such as bees, flies, and mosquitoes beat their wings at frequencies from 100 to more than 1000 beats per second (Sotavalta, 1953). This is much too fast for one nerve impulse to arrive at the muscle for each contraction, exert its effect, and decay before the next impulse arrives. It follows that for the muscles to contract at the high rates, they must be stimulated by some other mechanism. These fast muscles do have nerves, but the nerve impulses arrive at a lower frequency than the contractions, and they are therefore known as *asynchronous muscles.* This type of muscle is found in four insect orders, flies and mosquitoes (Diptera), wasps and bees (Hymenoptera), beetles (Coleoptera), and in some true bugs (Hemiptera).

The first surprise is that the muscles inside the thorax of these fast-beating insects are not attached to the wings at all, but rather to the wall of the thorax. There are two sets of muscles, which for convenience we can call vertical and horizontal. The primary effect of muscle contraction is to distort the thorax, and since the amount

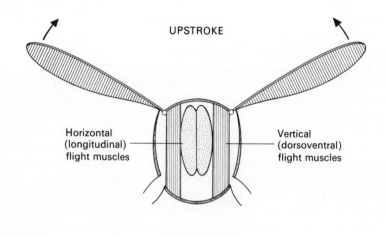

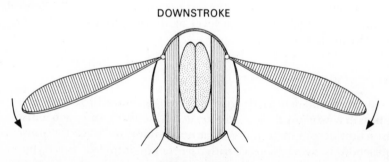

Fig. 11.13. In some insects the flight muscles together with the elastic thorax forms an oscillating system which permits the wing beat to occur at a much higher frequency than the frequency of nerve impulses. For details, see text.

of shortening is very small, only a few per cent of the muscle length, it can take place very fast. The most important property of asynchronous muscle is, however, that it contracts in response to being stretched.

When the vertical muscles contract and distort the thorax, it springs into a new position with a click, so that the tension is suddenly removed from the contracting vertical muscles (see fig. 11.13). However, the sudden change in shape of the thorax causes the horizontal set of flight muscles to be stretched, and this in turn acts as a stimulus for these to contract. This contraction now distorts the thorax in the opposite direction, and it suddenly clicks back to the initial position, thus releasing the tension on the horizontal muscles. This stretches the vertical muscles, which respond with a new contraction, and so on.

The elastic thorax with its muscles constitutes a mechanical oscillator, which by the energy input from the muscles is kept oscillating without being damped out by frictional forces. The most important damping is, of course, the drag on the wings as they propel the insect, and most of the muscle work is used to overcome this drag. The flight continues although the nerve impulses arrive at intervals of several muscle contractions. In some cases as many as 40 wingbeats have been recorded for each nerve impulse. Nerve impulses are needed, however, both for the initiation of flight and for maintenance of continued flight activity (Pringle, 1949; 1957).

The jump of the flea

Fleas and grasshoppers can make jumps that are perhaps fifty times the length of their bodies, and this has given rise to the popular pastime of calculating how far a man should be able to jump, could he jump in the same proportion. In North America he would jump to the top of the Empire State Building, while in France he would clear the top of the Eiffel Tower. Let us, instead, be a bit realistic when we examine this question.

Assume that a grasshopper weighs one gram and can make a standing jump of 50 cm, both figures in fact being quite realistic. Assume a larger animal of isometric build with all linear dimensions increased exactly tenfold. What is the effect of scaling on the expected height of a jump? The large animal will have a mass 1000 times that of the small one, but the cross-section of his jumping muscles (which determines the force the muscle can exert) will be increased only 100-fold. Relative to its larger mass, the large animal therefore has only one-tenth of the force available to accelerate the body. Since acceleration equals force/mass, the acceleration during take-off will also be one-tenth of that in the small animal. However, since all linear dimensions are increased ten times, acceleration continues throughout a take-off distance which is ten times the take-off distance of the small animal. The net result is therefore that the take-off speed will be the same for the two animals. After take-off both animals are slowed down due to the effect of the force of gravity. Since both the kinetic energy at take-off and the retarding force of gravity are proportional to the body mass, the deceleration due to gravity will permit the two animals to rise to equal heights. (Here we have disregarded the additional loss of speed due to air resistance, which for a very small animal such as a flea, becomes important.)

We can now conclude that if the muscles of the small and large animal exert the same force per cross sectional area, a small and a large animal of isometric build should be able to jump to the same height. Consider now the actual jumping performance of a variety of animals (table 11.3). (We should consider only standing jumps, for a running jump utilizes the kinetic energy of the animal to increase the height of the jump. We will therefore disregard the world record high jump for man, or a running horse, which both are about two meters.) In a standing jump a man can clear about 1.6 meters. His center of mass, however, is not lifted over this distance, but rather less, for it is not at ground level when he jumps but rather at about one meter (see footnote of table 11.3). Thus, man and grasshopper raise their center of mass by roughly the same amount, although the difference in their body mass is between 10 000 and 100 000-fold.

Of course, in reality animals are not isometric. Nevertheless, it is amazing how similar are the jumping records for a variety of animals. It is a matter of simple physics that man and other mammals do not jump in some proportion to their body length, and that this indeed would be impossible.

The conclusion that similar animals irrespective of body size should be able to jump to equal height, can also be stated as follows. Assume that the jumping muscles make up the same fraction of body mass. The muscle force is proportional to the cross-sectional area, and the shortening is proportional to the initial length of the

Table 11.3. *Comparison of the performance in a standing jump for a flea, a locust, and man*

	Flea (*Spilopsylla*)	Locust (*Schistocerca*)	Man (*Homo*)
Body mass, m	0.45 mg	3 g	70 kg
Height of jump, $h = v^2/2g$	5 cm	59 cm	60 cm *
Acceleration distance, s	0.04 cm **	4 cm	40 cm *
Take-off speed, v	100 cm s^{-1} **	340 cm s^{-1} ***	343 cm s^{-1}
Time of acceleration, $t = 2s/v$ (take-off time)	0.0008 s	0.00235 s	0.233 s
Acceleration, $a = v/t$	125 000 cm s^{-2}	14 500 cm s^{-2}	1471 cm s^{-2}
Acceleration relative to gravitational accel. (**g**)	127 **g**	15 **g**	1.5 **g**

* Estimate based on center of mass lowered to 60 cm above ground at beginning of jump, accelerated over 40 cm to 100 cm at take-off, and lifted to 160 cm. World record for standing jump is about 165 cm above ground.
** Bennet-Clark and Lucey (1967).
*** Alexander (1968).

muscle. The cross-section times the length is the volume of the muscle, and the energy output of a single contraction is the product of force and distance. Energy output is therefore proportional to the muscle mass, and in turn to body mass. Since a jump uses only a single contraction of the jumping muscles, the work performed during take-off and used for acceleration will be the same relative to body mass. Thus the conclusion is that isometrically built animals of different mass should all jump to the same height, provided that their muscles contract with the same force.

How is it possible, then, for some animals to jump much higher? The record standing jump is probably performed by the lesser galago, a small tropical primate that weighs about 250 gram. Under well-controlled conditions a galago has jumped 2.25 meter (Hall-Craggs, 1965). This is more than three times the height of a standing jump for man. Unless the muscles can produce more force per cm^2 cross-section (which is unlikely), the galago can improve its performance by having a larger muscle mass (more energy from the take-off contraction), and possibly by having a more favorable mechanical structure of the limbs. The galago indeed has large jumping muscles, nearly 10% of the body mass (Alexander, 1968) or about twice as much as in man. If we assume that all combined mechanical advantages of a highly specialized jumping animal could account for a 50% increase in performance, this in combination with a doubling in the mass of the jumping muscles, could account for a three-fold increase in jumping performance, by no means an unreasonable approximation. As a standing jump this is more impressive than the jump of the locust, but we must remember that the galago is a warm-blooded animal highly adapted to making long jumps in its natural jungle habitat.

Aside from the fact that animals are not completely isometric, we meet another difficulty. The smaller the animal, the shorter the take-off distance, and since the take-off speed should be the same for all, the smaller animal must accelerate its body mass much faster. Since much less time is available for the take-off, the power output of the muscle must be increased accordingly, i.e. the muscle must contract very fast.

For an animal the size of a flea, the take-off time is less than 1 ms, and the distance over which acceleration takes place is only 0.5 mm. The average acceleration during take-off is about 1000 meter s^{-2}, or roughly the equivalent of 100 **g** (Rothschild, Schlein, Parker and Sternberg, 1972).

Muscles just cannot contract this fast, so how can the flea jump at all? Fleas use the principle of the catapult and store energy in a piece of elastic material, *resilin,* at the base of the hind legs. Resilin is a protein which has properties very similar to rubber (Weis-Fogh, 1960; Andersen and Weis-Fogh, 1964). The relatively slow muscles are used to compress the resilin, which returns the total energy again with close to 100% efficiency when a release mechanism is tripped. In this way the elastic recoil works much the same way as a slingshot and imparts the necessary high acceleration to the flea.

Skeletons

To exert their force muscles must be connected to some mechanical structure; they obviously cannot do much for an animal unless they have something to pull on. The appropriate mechanical structure is the skeleton, which usually consists of a rigid structure, although a fluid can also be used to transmit force. We have the following major categories to consider:

		Examples
Rigid Skeleton	{ endoskeleton	Vertebrates
	{ exoskeleton	Insects, crustaceans
Hydraulic Skeleton	{ fluid + soft walls	Worms, octopus, starfish
	{ fluid + rigid elements	Spider legs

Thus, vertebrates have an *internal rigid skeleton,* the bones; insects and crustaceans have an *external cuticle* as a skeleton; and some animals, e.g. the earthworm, use their internal fluids for transmission of force, they have a hydraulic skeleton.

Rigid skeletons

Vertebrates have most of the soft tissues of the body draped around the skeleton. Is this the best arrangement? Arthropods have their skeleton on the outside, and all the soft parts are well protected. Is this a better arrangement?

Let us compare the two kinds of structures. For reasons of simple mechanics, a hollow cylindrical tube can support a much greater weight without buckling than a solid cylindrical rod made from the same amount of material. Therefore, if an animal can afford only a certain weight of skeleton, it is advantageous to use the material for a tube. This is precisely what arthropods have done, and they seem to have an advantage over vertebrates in this regard.

Why, then, do not vertebrates utilize a similar arrangement? First of all, for a constant weight of material, the rigidity of a hollow cylinder decreases very rapidly as its radius is increased, and eventually it will be so thin that the entire structure collapses. To prevent buckling of the hypothetical exoskeleton of a very large animal, it would be necessary to increase the thickness, and thus the weight. An exoskeleton has another mechanical disadvantage, even if it is strong enough to provide the necessary support, it will be very sensitive to impact with the risk of buckling as well as being punctured. This danger, which is of little significance for small animals such as insects, increases rapidly with increasing body size.

One difficulty with an exoskeleton is the question of growth. Most adult insects never grow, they retain the same rigid shape and size throughout their lifetime, and the larval growing phase is equipped with a softer cuticle that can be stretched. Even so, the larval skin is shed at intervals to permit adequate size increase. Aquatic arthropods which molt periodically go through a period of increase in size when the old exoskeleton has been shed and the new exoskeleton is still soft and can be stretched. Aside from the lack of protection and the inability to move during this 'soft' period, the mechanical disadvantage of a periodic loss of rigidity is not too inconvenient for an aquatic animal, for its body weight is supported by the water. For a land animal this is different, any sizeable animal would collapse under its own weight if it were to go through a molting process.

Nevertheless, some of the protective advantages of an exoskeleton are, in fact, utilized by vertebrates, for they enclose their brain, certainly a vital organ, in a rigid bone structure. The braincase is not a part of the locomotory skeleton, however.

One advantage of an exoskeleton is due to its elastic properties. Elastic deformation of skeletal elements constitutes a storage of energy which is utilized for purposes of locomotion; some well-known examples that were discussed above are insect flight and the jump of fleas.

Hydraulic skeletons

There is one important difference between rigid and hydraulic skeletons. A rigid skeleton consists of elements that usually act around a pivot point (a joint). The muscles are attached at both ends, and the force they exert is transmitted through the rigid element and acts somewhere else. For the hydraulic skeleton to work, a fluid must be enclosed in a limited space, and the muscle force must be used to produce pressure in this fluid, i.e. the muscle must usually enclose the fluid space in circular layers. In this case the muscle does not have a point of attachment, a ring of muscle pulls only on itself.

The role of hydraulic skeletons in the locomotion of worms has been studied in detail. The common earthworm is a familiar example. Its body wall has layers of muscles that run in two distinct directions, either in circular sheets around the body, or in the longitudinal direction.

The movement of the earthworm when it crawls forward is approximately as follows. A wave of contraction of the circular muscles begins at the anterior end. This causes the body to become thinner and elongated, pushing the anterior end forward. The circular contraction passes down the body as a wave, followed by a wave of contraction of the longitudinal muscles. Tiny bristles on the side of the body prevent back-sliding, and the anterior end therefore pulls the adjacent part of the worm in its direction. In this way a portion of the worm is pulled forward, the motion passing as a wave backwards on the heels of the wave of circular contraction. The cycle is repeated with a new elongation at the anterior end, the worm moving forward as the waves of contraction move backward (fig. 11.14).

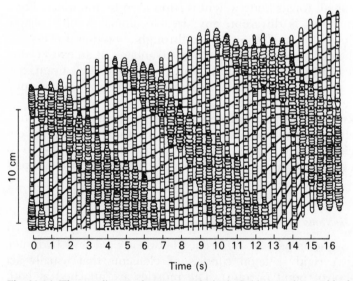

Time (s)

Fig. 11.14. The crawling earthworm uses its body fluids as an internal hydraulic skeleton. Contraction of circular muscles pushes the front end forwards, and is followed by a contraction in longitudinal muscles that thickens the body. The thickened segments remain in place relative to the ground as the other parts move forward. The track of individual points on the worm's body and their movements relative to each other are shown by the lines running obliquely forwards from left to right of the diagram. Diagram was prepared from a movie film (from Gray and Lissmann, 1938).

When the earthworm moves, the muscles exert their force on the fluid contents of the animal; the circular muscles build up a pressure which causes the longitudinal muscles to be stretched, and when the longitudinal muscles in turn contract, the pressure stretches the circular muscles and the worm thickens.

The body of the earthworm is separated into compartments by septa that separate each segment from the next. This helps a great deal in locomotion, for it makes the various regions of the worm relatively independent of the rest of the worm. Many worms, the lugworm (*Arenicola*) for example, lack such septa (Seymour, 1971). One disadvantage is that if the worm is wounded and body fluid is lost, it is virtually unable to move; as nearly everybody knows, if an earthworm is cut in two, each part can move about more or less like an intact worm.

Spider legs. Spiders cannot extend their legs with the aid of muscles for the simple reason that they have no extensor muscles. To flex the legs they use muscles, but how are the legs extended? There are two possibilities, the use of (1) elastic forces, or (2) hydraulic forces.

Elastic hinge joints in the legs could be used if they were arranged to extend the legs whenever the flexing muscles relax. This would be similar to the hinge of bivalves, in which the shells gape apart when the closing muscle relaxes. We can exclude this possibility by a very simple observation, the joints of a detached spider leg remain neutral over a wide angle.

The other possibility, the use of hydraulic pressure, turns out to be what spiders do. This can be shown in various ways, most simply by detaching a leg and pinching the open end with a pair of forceps. This closes the opening and raises the fluid pressure inside, and as a result the leg extends. Furthermore, there is a direct relationship between the internal pressure applied to the leg and the joint angles of the leg. Also, if a spider is wounded so that blood fluid is lost, it is completely unable to extend the legs, which remain flexed in a disorderly fashion up against the body. The blood pressure in spiders is surprisingly high, up to 400 mm Hg, or about 0.5 atm. This differs very much from the blood pressure of other arthropods, which is usually only a few mm Hg (Parry and Brown, 1959a).

Some small spiders, known as jumping spiders, can leap on their prey at distances of more than 10 cm. For jumping they use a sudden extension of one pair of legs, and it is now the question of whether the spiders can use hydraulic pressure for this very rapid motion.

The jump itself is due entirely to the sudden straightening of the last (fourth) pair of legs. The extension torque necessary for

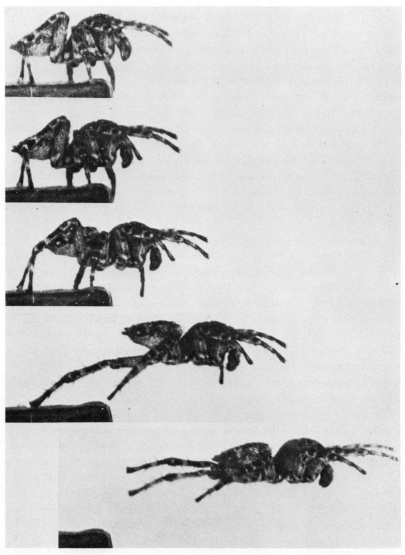

Plate 12. *Jumping spider.* The jumping spider (*Sitticus pubescens*), like other spiders, lacks muscles for extension of its legs. When this spider leaps on its prey, it uses the hind pair of legs which are extended hydraulically by blood pressure. (G. A. Parry, Cambridge University; from *Journal of Experimental Biology*, 1959, **36,** 654).

the jump can be estimated from the length of the jump, and the required hydraulic forces, which are difficult to measure directly in the jumping spider (they weigh only about 10 mg), fall within the range of blood pressures commonly measured in spiders. High-speed photographs of jumping spiders support the view that blood pressure is involved, for at the moment of the jump the minute spines on the legs become erect precisely in the way they do when there is an increase in fluid pressure in the leg. The exact release mechanism is not fully understood, however (Parry and Brown, 1959*b*).

BUOYANCY

If an animal that swims is heavier than the water, part of its energy expenditure goes to keep from sinking, and only part is available for forward locomotion. If the animal could have the same density as the water, i.e. have neutral buoyancy, it would have a great advantage in making more energy available for moving.

The problem of buoyancy is more important for large than for very small animals. Small organisms have a large relative surface and often have extensions of various sorts that reduce the sinking rate in water; the smallest can remain suspended much like dust particles in air and live through their entire life cycle in this way. Surface extensions large enough to act as a brake on sinking become mechanically impossible for large organisms, which must therefore use other means.

A reduction in the density or specific gravity of an organism is the only solution that does not involve some amount of swimming action to stay up, i.e. the expenditure of energy. There are five avenues available to reduce the tendency to sink and to shift the organism towards neutral buoyancy:

(1) Reduction in the amount of heavy substances, e.g. calcium carbonate or calcium phosphate.

(2) Replacement of heavy ions, e.g. Mg^{++} and SO_4^{--}, etc. by Na^+, Cl^-, or even lighter ions, H^+ and NH_4^+.

(3) Removal of ions without replacement, i.e. making the organism hypotonic.

(4) Increase in amount of light substances, primarily fats and oils.

(5) Use of gas floats, such as the swimbladder of fish.

We shall discuss these five possibilities, what advantages and disadvantages they may have, and examine the limitations of each. Since sea water is heavier (sp. gr. = 1.026) than fresh water (sp. gr. = 1.000), it is easier for an organism to achieve neutral buoyancy in the sea; in fact, some solutions that produce neutral buoyancy in the sea can never attain this goal in fresh water.

Reduction in heavy substances

Many skeletons consist of very heavy substances, often calcium salts, either calcium carbonate or calcium phosphate. Calcium carbonate is common among invertebrates (the shells of clams and snails, corals, etc.), and calcium phosphate is the major component of vertebrate bones where it is found in the form of the mineral apatite. Some organisms, e.g. some sponges, have skeletons made of silica, others use chitinous substances, which is common in arthropods (crabs, lobsters) either alone or together with calcium carbonate. The relative contribution of these substances to the weight of an animal in water is evident from a list of their specific gravities (table 11.4).

Table 11.4. *Specific gravity of substances commonly found as skeletal material in aquatic organisms.* (From Denton, 1961)

SUBSTANCE	Sp. gr.
Sea water	1.026
Fresh water	1.000
Calcium carbonate	
aragonite	2.9
calcite	2.7
Calcium phosphate,	
apatite	3.2
Chitin	1.2

It is easy to see that reducing the amount of calcium salts can go far towards reducing the weight of an organism. Accordingly, pelagic nudibranch snails lack shells in contrast to opistobranch snails that are bottom organisms and have well-developed shells. Swimming coelenterates (e.g. jellyfish) completely lack the heavy calcium carbonate skeletons that are characteristic of many sessile coelenterates such as corals.

It is interesting to note that one group of pelagic organisms, the swimming crabs, have not gone very far towards reducing their heavy exoskeleton (mostly calcium carbonate). This holds for the common blue crab (*Callinectes*), which swims continuously to keep from sinking.

There is, of course, a considerable structural disadvantage in the removal of mechanical support. Therefore we find this solution more common in relatively small organisms. There is also another way out, a heavy skeleton can be replaced by a skeleton built of lighter substances. Thus, in the ordinary squid the calcium carbonate skeleton that is found in its relative, the cuttlefish, has been replaced by a lighter structure made of chitin, the so-called 'pen' of the squid.

Replacement of dissolved ions by lighter ions

The best examples of this solution are perhaps found among plants. Two multicellular algae, *Valonia* and *Halicystis,* show what can be achieved by manipulating the ionic composition of the organism. Both are almost neutrally buoyant, the former barely sinks in sea water, and the latter has a slight positive buoyancy. Both have abundant cell sap which is isotonic with sea water but differs in composition (see table 11.5). Both exclude the heavy sulfate ion almost completely. *Valonia* likewise excludes magnesium, which is present in *Halicystis* in a concentration less than one-third that in sea water. Calcium, another important divalent ion, is also excluded by both, relative to its concentration in sea water.

Table 11.5. *Ionic composition of the sap of two multinuclear algae which both have nearly neutral buoyancy. Ionic molecular concentrations are expressed relative to chloride ($Cl^- = 100$).* (From Gross and Zeuthen, 1948)

	Sea water	Sap of *Valonia*	Sap of *Halicystis*
Cl^-	100	100	100
Na^+	85.87	15.08	92.4
K^+	2.15	86.24	1.01
Ca^{++}	2.05	0.285	1.33
Mg^{++}	9.74	trace	2.77
SO_4^{--}	6.26	trace	trace
Sp. gravity of sap	1.0277	1.0290	1.0250
Artificial solution *	1.0285	1.0285	1.0252

* Containing the ion concentrations listed in column above.

Of the monovalent ions K^+ is heavier than Na^+. *Valonia,* which is slightly heavier than sea water, accumulates primarily K^+ in the cell sap, as plants in general do. *Halicystis,* however, which is slightly lighter than sea water, has excluded most of the potassium and uses the lighter sodium as its most important cation.

A very common organism that causes luminescence in the sea is a dinoflagellate, *Noctiluca miliaris.* It is a little less than 1 mm in diameter and tends to accumulate at the surface because it is definitely lighter than sea water. When it is mechanically stimulated, it emits light flashes which are visible as phosphorescence at night. Its light emission has been the subject of extensive studies, but we shall discuss only its buoyancy problems.

The composition of the cell sap of *Noctiluca* is given in table 11.6. It is isotonic with sea water, and it achieves its positive buoyancy in part by reducing Ca^{++}, Mg^{++}, and especially the heavy SO_4^{--} ion. Furthermore, it accumulates ammonium ion, which is lighter than the two commonest monovalent cations, Na^+ and K^+.

Table 11.6. *The dinoflagellate* Noctiluca, *a common source of phosphorescence in the sea, achieves positive buoyancy by excluding heavy ions and by replacing them primarily with* NH_4^+. (From Kesseler, 1966)

	Sap meq liter^{-1}	Sea water meq liter^{-1}
Na^+	414	418
K^+	34	8.8
NH_4^+	58.5	0
Ca^{++}	9.5	18.5
Mg^{++}	15	95
Cl^-	496.5	498.9
SO_4^{--}	trace	46.6
$H_2PO_4^-$	13	trace
pH	4.35	8.2

An ion which is much lighter than any of the other cations is the hydrogen ion, H^+. It has been suggested that positive buoyancy could be achieved by increasing the hydrogen ion concentration, but this is not feasible. Although the cell sap of *Noctiluca* is rather acid, i.e. has a high hydrogen ion concentration, it is easy to see that the effect on the specific gravity must be utterly negligible. If we take the pH of the *Noctiluca* sap to be 4.0, the hydrogen ion concentration is 10^{-4} mol liter^{-1}. In other words, the hydrogen ion concentration is 0.1 meq liter^{-1}, and thus can replace no more than 1/5000 of the total cation content in *Noctiluca*. To replace a substantial amount of cation we would need a pH substantially less than 2, which corresponds to a hydrogen ion concentration in excess of 10 meq liter^{-1}. Such high acidity is not known to exist in any cells, it occurs only in a few acid secretions such as the gastric juice of vertebrates and the saliva of some marine snails which use acid to drill into the shell of their prey.

The principle of replacing heavier ions by light ones can be used just as effectively by large organisms. A striking example is found in the group of deep-sea squids known as cranchid squids. In these animals the fluid-filled coelomic cavity is very large and makes up about two-thirds of the entire animal. If it is opened and the fluid drained out, the animal loses its neutral buoyancy and sinks. The fluid has a specific gravity of about 1.010 to 1.012 (sea water is 1.026), it is osmotically isotonic with sea water, and the pH is 5.2. The concentration of NH_4^+ is about 480 meq liter^{-1}, and sodium is 80 meq liter^{-1}. The anion is almost exclusively chloride, in other words, the heavy SO_4^{--} ion is excluded (Denton, 1960; Denton, Gilpin-Brown and Shaw, 1969).

The explanation of the high ammonium concentration is really very simple. The coelomic fluid is slightly acid, and ammonia, which is the normal metabolic end product of protein metabolism in most aquatic animals, diffuses into the fluid and becomes trapped there. A mechanism that could produce the low pH would be a sodium pump which removes Na^+ from the coelomic fluid, for this would leave an excess of anions and thereby increase the hydrogen ion concentration, thus lowering the pH.

The method used by the cranchid squids is quite effective in achieving neutral buoyancy. It has some structural disadvantages, however, for two-thirds of the bulk of the animal is fluid, i.e. the volume of fluid carried around amounts to twice the volume of the living animal itself.

Hypotonicity

Removal of some ions without replacing them by other ions would seem a possible avenue for reducing weight. There would, of course, be osmotic problems, for such a dilute solution would be hypotonic to the sea water. In general, invertebrate organisms are in osmotic equilibrium with sea water, and none seems to have used hypotonicity as a buoyancy mechanism. Teleost fish, however, have ion concentrations much lower than those of sea water, and this contributes towards reducing their weight in water. This is only of minor importance, however, for teleost fish have skeletons and muscles which contribute sizeable amounts of relatively heavy substances. In many deep sea forms, which have greatly reduced skeleton and muscle masses, however, hypotonicity is of considerable help in balancing the relatively small amounts of heavy substances.

Fats and oils

Many planktonic organisms contain substantial amounts of fats. Fat, of course, is a common form of energy storage, both in animals and in many plants, but nevertheless the contribution of fats and oils towards buoyancy is also important. Planktonic plants frequently deposit fat, rather than the heavier starch which is a common storage compound in plants in general. All diatoms, for example, deposit only fat.

Elasmobranch fishes, as opposed to teleosts, never have swimbladders. On the other hand, their livers are large, especially in those sharks which are good swimmers. The liver may constitute about one-fifth of the body weight, while in teleost fish it usually is about 1 to 2% of the body weight.

A survey of the black spiny shark (*Etmopterus spinax*) showed that they had, on the average 17% of the body weight as liver, and 75% of the liver was oil (S. Schmidt-Nielsen, Flood and Stene, 1934). (In contrast, mammalian liver contains some 5% fat.) Half of the oil in the shark liver was *squalene,* an interesting unsaturated hydrocarbon ($C_{30}H_{50}$), which derives its name from *Squalus,* or shark. Most fats and oils have specific gravities of about 0.90–0.92, but squalene has a lower specific gravity, 0.86. The difference may not seem very great, but compared to sea water the buoyancy effect of a given volume of squalene will be about 50% higher than the effect of a similar volume of fat or oil. Whether this is a reason for the tremendous accumulation of squalene in elasmobranch livers is difficult to say, nevertheless, its presence definitely aids in bringing about near-neutral buoyancy for the fish. This is particularly effective because elasmobranchs, as opposed to teleosts, have a relatively light cartilaginous skeleton which is not weighted down by heavy calcium phosphates.

Interestingly, rays and skates, which are bottom-living animals, have smaller livers with a lower fat content. Five species of rays (*Raja*) averaged a liver size of 7.53% with less than 50% fat in the liver. This does indeed make sense, for rays and skates live at the bottom and are not very good swimmers.

It is worth noting that the relatively low ion concentrations in the body fluids of elasmobranchs, relative to sea water, do not contribute substantially to their buoyancy. The elasmobranchs are isosmotic with sea water, the difference in salt concentration is made up primarily by urea, and urea solutions give only very slight lift over sea water. Trimethylamine oxide (TMAO) which also is found in their body fluids is appreciably less dense than urea and contributes a minor amount towards giving lift.

Several deep-sea teleosts have swimbladders filled with fat, rather than with air. This is true of some species of *Gonostoma* and *Cyclothone* (Marshall, 1950), but since the fatty swimbladder of other fish have a substantial fraction of the fat as cholesterol (up to 49%), which has a higher specific gravity (1.067) than sea water (1.026), the role of the fat-invested swimbladder in buoyancy control is dubious (Phleger, 1971).

Gas floats

Compared to an equal volume of water, gas has a very low density, and even a relatively small volume is an excellent solution to the problem of buoyancy. This solution, however, has some limitations and disadvantages. The situation is quite different whether the gas space is surrounded by soft tissue or by rigid walls. Let us first examine what happens when the walls are soft.

Assume that a gas-filled rubber balloon is submerged – the further down it is brought, the greater is the pressure of water on its outside, and as the pressure of the gas inside matches the water pressure outside, it becomes increasingly compressed. The water pressure increases by about one atmosphere for each 10 meters' depth and for an organism living at considerable depth this causes problems. Say that an organism lives at 1000 meters' depth, where the water pressure is about 100 atm. To produce and maintain a volume of gas at that pressure entails three difficulties. First of all, the gas must be secreted into the space against a pressure of 100 atm, second, the *amount* of gas needed to fill the same volume at that pressure is 100 times as great as it would be at the surface, and thirdly, it is a problem to keep any gas at such a high pressure from diffusing out. The gas tensions in the water at any depth in the ocean are not usually very different from those at the surface, where there is equilibrium with 0.2 atm oxygen and 0.8 atm nitrogen. The gases in the float must be obtained from the water, secreted into the high pressure space, and must then be kept from diffusing out.

A gas float with soft walls leads to another difficulty, its volume changes if the organism moves up or down in the water. Say that an organism has enough gas in its float to be in neutral buoyancy at a given depth. If it moves down, the gas is compressed, it gets heavier and loses buoyancy; if it moves up, the gas expands and positive buoyancy increases. The latter can be dangerous, for unless the animal actively swims down to the neutral level, the gas will expand faster and faster as the animal rises at increasing speed. Unless it has a mechanism for releasing gas, an organism which has lost control of its rise may even rupture as the gas expands.

If the gas space has rigid walls, the limitations on vertical movements are less strict, but the walls must have sufficient mechanical strength to support changes in pressure. If the walls are made stronger, they also become heavier, and this defeats the purpose of having gas chambers to make the organism lighter. As we shall see later, the mechanical strength of the float is a limitation on the vertical distribution of the cuttlefish (*Sepia officinalis*), which uses rigid gas spaces for flotation.

The swimbladder of fish is undoubtedly the best known type of gas float, but the principle is also used by quite a few other organisms, in particular among the colonial siphonophores (Coelenterata). One of these is the conspicuous Portuguese Man-of-War (*Physalia*), which is commonly found along the Atlantic coast of North America. The colony is carried by a 'sail', a balloon-like gas chamber, which often is more than 10 cm long. It floats on the surface of tropical and semi-tropical oceans and carries the colony along as it drifts in the wind.

Since the gas chamber of *Physalia* floats on the surface rather than being submerged, the inside pressure is virtually equal to the atmospheric pressure. This simplifies the problem of filling it with gas. However, the composition of the gas is quite surprising, for it contains up to some 15% CO. The oxygen concentration is less than in the atmosphere, the carbon dioxide concentration is negligible, and nitrogen makes up the balance. The gas is produced by a 'gas gland', which uses the amino acid L-serine ($CH_2OHCHNH_2COOH$) as a substrate for the formation of carbon monoxide (Wittenberg, 1960). Other colonial siphonophores also carry floats which contain carbon monoxide. One, *Nanomia bijuga,* has very small floats with a volume of 0.14–0.46 μl; in these the carbon monoxide concentration is from 77 to 93% (Pickwell, Barham and Wilson, 1964). Carbon monoxide is, of course, highly toxic to man and many other higher animals, and usually occurs in nature only in trace amounts. It is found in the hollow stems of some marine algae, where it may comprise several per cent of the total gas. In minute amounts carbon monoxide is found in the exhaled air of man, even in non-smokers. It is not formed by bacteria in the intestine, but probably from metabolic breakdown of hemoglobin in the liver (Ludwig, Blakemore and Drabkin, 1957). The exhaled carbon monoxide does not originate externally from air pollution, for the antarctic Weddell seal (which

Plate 13. *Portuguese Man-of-War* (*Physalia pelagica*). The gas float of this colonial coelenterate is filled with a gas that contains a large proportion of carbon monoxide. The length of the float may exceed 20 cm, and the tentacles, which give extremely painful stings, may extend for more than 10 meters (Charles E. Lane, University of Miami).

presumably breathes entirely unpolluted air) also exhales a measurable amount of carbon monoxide (Pugh, 1959).

The presence of gas floats in plankton organisms may explain a rather mysterious phenomenon known as the 'deep scattering layer'. Ships that use echo-sounding for depth determination often find that sound waves are reflected not only by the bottom, but also by a layer in mid-water which thus gives the impression of a 'false bottom'. This extra layer, which often changes in depth from day to night, is the deep scattering layer. For many years what caused this sound reflection was unknown, but it is now almost certain that it is schools of fish with swimbladders or pelagic siphonophores with gas floats.

Siphonophores have gas floats with an opening so that gas can escape when they rise in the water. A very simple experiment has shown that gas can both be given off and secreted by these organisms. Pickwell *et al.* (1964) put them in a 10 ml syringe and exposed them to vacuum by pulling at the plunger. After gas had been pulled out, the floats were heavier than water, but within one hour they were again afloat. They were now filled primarily with carbon monoxide, presumably formed in a manner analogous to that in *Physalia*.

At night the deep scattering layer frequently migrates up towards the surface of the ocean. Since gas can escape from the floats of siphonophores, they presumably release bubbles as they ascend, and this can explain the observation that the intensity of the sound reflection of the layer frequently increases as it rises.

If organisms release gas on the way up, we would expect that a similar amount of gas is secreted as they descend again. However, they could avoid the need for repeatedly secreting gas and giving it up again by keeping a constant amount of gas and putting up with the volume changes during vertical migration. This might be feasible if the animal is in neutral buoyancy at the higher level, but it would hardly be possible if the neutral level is at depth. Let us look at an example. Say that a fish has a swimbladder with a volume (at neutral buoyancy) of 5% of the body volume. Descending from a night level of 50 meters' depth to 300 meters would reduce the gas volume to one-fifth, or 1% of the body volume. The fish would now be negatively buoyant, but could keep at the new level by a moderate amount of swimming. If, on the other hand, the fish were in neutral buoyancy at 300 meters' depth with the swimbladder constituting 5% of the body volume, rising to 50 meters would increase the swimbladder volume five-fold (unless gas were released). A fivefold expansion of the swimbladder volume would now involve a serious danger of losing control and rising all the way to the surface where, at one atmosphere pressure, the swimbladder would be a catastrophic 30 times its original volume. If organisms of the deep scattering layer move vertically without releasing gas, we must therefore expect that they are in neutral buoyancy at the higher, rather than the lower level.

Information about volume changes can be obtained to some extent without direct observation. The deep scattering layer reflects sound particularly well at frequencies which correspond to the resonant frequency of the gas bubbles. This depends in turn on the volume of the gas bubbles as well as on their pressure. If the pressure is known (the depth can be read directly from the echo-sounding record) and the resonant frequency is assumed to be that one which is best reflected, the size of the bubbles can be calculated. Now, when the layer moves up, one of two situations may prevail, either gas is given off and the bubbles maintain constant volume, or the gas is retained and the bubbles expand with the pressure change. If a bubble is kept to constant volume, their resonant frequency should vary with $\mathbf{P}^{1/2}$ ($\mathbf{P}$ is hydrostatic pressure); if the volume changes with pressure while keeping a constant mass of gas, the resonant frequency should vary with $\mathbf{P}^{5/6}$. The sound records are complex and often difficult to interpret, but it appears that organisms of the deep scattering layer may conform to either of these patterns (Hersey, Backus and Hellwig, 1962).

The problems connected with changes in volume, and thus in buoyancy with changing depth, are eliminated if the gas is enclosed in a rigid space. An example is the cuttlefish (*Sepia officinalis*), a cephalopod common in the upper reaches of coastal water masses. Nearly one-tenth of its body volume is made up of a 'cuttlebone', a calcified internal structure well known by bird-fanciers who use it to supply calcium to their birds (see fig. 11.15). The role of the cuttlebone has been studied by the British investigator Denton (Denton and Gilpin-Brown, 1959; 1961).

The volume of the cuttlebone is 9.3% of the body volume, and its specific gravity when freshly removed is in the range of 0.57 to 0.64. This provides sufficient lift to give the animal neutral buoyancy in sea water.

The cuttlebone has a peculiar laminar structure, it is made up of thin layers of calcium carbonate, reinforced by chitin and held about $2/3$ mm apart by a large number of 'pillars' of the same material. In a fully grown animal the number of parallel layers may be about 100, and most of the space between the layers is filled with gas.

If a cuttlebone is removed from an animal and punctured under

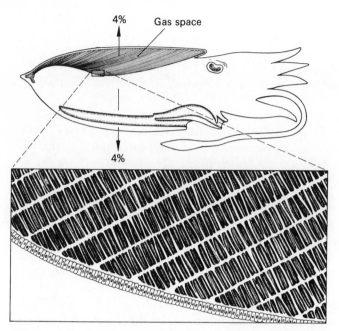

Fig. 11.15. Diagram of a 'cuttlefish', a marine cephalopod which uses a gas-filled rigid structure called the 'cuttlebone' to achieve neutral buoyancy in water. The posterior part of the cuttlebone (shown in black above) is filled with liquid. In sea water the cuttlebone gives a net lift of 4% of the animal's weight in air and thus balances the excess weight of the rest of the animal (after Denton and Gilpin-Brown, 1961).

water, no gas escapes. This shows that the gas it contains is not under pressure. If the bone is crushed, gas will escape and can be collected for analysis. The gas is mostly nitrogen under 0.8 atm pressure, plus a small amount of oxygen. Along the caudal surface of the cuttlebone, some liquid is present between the calcified layers. This liquid can be collected under paraffin oil by exposing the cuttlebone to a vacuum. It contains mostly sodium chloride, and is hypo-osmotic to sea water and the body fluids (which are isosmotic with sea water). The higher osmotic pressure outside the cuttlebone, there-fore, will tend to withdraw water from the cuttlebone. The osmotic withdrawal of water is opposed by the hydrostatic pressure, which tends to force water back into the cuttlebone. The maximum force of water withdrawal is limited by the concentration difference. If all ions were removed from the fluid within the cuttlebone, the concen-tration difference between the cuttlebone fluid and blood would be about 1.1 osmolar. This corresponds to 24 atm pressure; i.e. if the outside hydrostatic pressure exceeds 24 atm, fluid will be pressed into the cuttlebone in spite of the osmotic forces acting in the op-

Plate 14. *Cuttlebone.* The buoyancy mechanism of the cuttlefish (a relative of the squid and octopus) depends on a gas-filled structure, the cuttlebone, made of calcium carbonate. The cuttlebone consists of thin layers of calcium carbonate, spaced about 0.5 mm apart, and supported by 'pillars' of the same material (M. L. Blankenship, Duke University).

posite direction. At a depth of about 240 meters, the pressure is 24 atm and below this depth it is impossible to withdraw fluid osmotically from the cuttlebone. However, the cuttlefish live in the upper reaches of the water masses and probably do not descend below about 200 meters, i.e. the osmotic forces suffice to keep water from being forced into the cuttlebone.

The gas pressure inside the rigid walls of the cuttlebone is about 0.8 atm, and the cuttlebone must be sufficiently strong to support the hydrostatic pressure of the surrounding water. This is indeed the case, for if a cuttlebone is covered with a thin plastic bag (so that sea water cannot penetrate into it) and is exposed to high pressures, it withstands more than 25 atm before it is crushed by the pressure.

The advantage of a rigid system is, of course, that the buoyancy is virtually unaffected by changes in depth. This gives the cuttlefish a great deal of freedom in moving up and down without making ad-

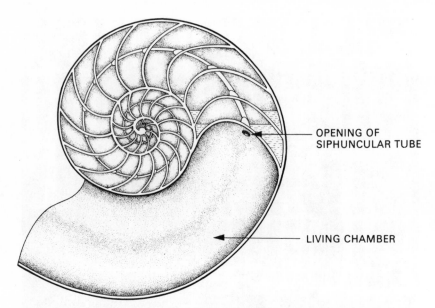

OPENING OF
SIPHUNCULAR TUBE

LIVING CHAMBER

Fig. 11.16. The shell of the Chambered Nautilus. Section of the shell of *Nautilus macromphalus*, oriented in its natural position. The chambers are gas filled, and new chambers are added as the animal grows. A new chamber is initially filled with fluid, which is withdrawn osmotically and replaced by gases. A small amount of liquid remains in the most recently formed chambers (from Denton and Gilpin-Brown, 1966.)

justments in the buoyancy mechanism. In addition to making the cuttlefish relatively independent of depth, the cuttlebone has the further advantage that it serves as a skeleton.

The mechanism responsible for filling the shell of the Pearly nautilus, or Chambered nautilus, is in principle the same as in the cuttlefish (see fig. 11.16). As the nautilus grows, it adds new air chambers to its shell, one by one. A newly formed chamber (next to the animal) is initially filled with liquid in which sodium chloride is the major solute. Sodium is removed from this fluid by active transport, water is therefore withdrawn osmotically in the same way as in the cuttlefish, and gas diffuses into the chamber to replace the water. Nitrogen will reach 0.8 atm, the same tension as in the water and the animal's body fluids. Oxygen, however, will be lower and not reach the concentration in the atmosphere, for the animal uses oxygen and its body fluids therefore have lower than atmospheric oxygen tension. The gas pressure in the shell is always about 0.9 atm, irrespective of the depth at which the animal is caught. Like the cuttlefish, nautilus has a freedom of vertical movement that is limited only by the pressure which its rigid shell can support.

Table 11.7. *Comparison of advantages and disadvantages of various mechanisms used by aquatic organisms to improve their buoyancy*

	Approx. volume needed, % of remaining body	Buoyancy effectiveness		Pressure independence	Structural qualities	Energy requirement
		Sea water	Fresh water			
Reduction of heavy structures	0	Fair	Poor	Excellent	Poor	No maintenance
Replacement of heavy ions	200	Good	Poor	Excellent	Fair	Needs continuous maintenance
Fat	50	Good	Good	Excellent	Good	High initial cost, no maintenance
Squalene	35	Good	Good	Excellent	Good	High initial cost, no maintenance
Air, soft walls	5	Excellent	Excellent	Poor	Excellent	Needs continuous maintenance
Air, rigid walls	10	Excellent	Excellent	Good	Excellent	Probably needs maintenance

We have now discussed examples of the different mechanisms that animals can use to improve their buoyancy. Each has advantages and disadvantages, which are outlined for comparison in table 11.7. It is evident that gas is one of the most favorable solutions to the problem of providing lift, and it has other advantages as well, but if contained in a soft structure it has one major disadvantage, it is extremely sensitive to pressure changes. Teleost fish use this mechanism in the form of an air-filled, soft-walled swimbladder, which is the subject we shall discuss next.

The swimbladder of fish

Many teleost fish have a *swimbladder* or *air-bladder* which provides the necessary lift to give them neutral buoyancy. Whether a particular fish has a swimbladder is related more to its life habits than to its systematic position. Some bottom fish lack a swimbladder, and it seems reasonable that neutral buoyancy has no particular advantage for a fish that is bound to the bottom. Many pelagic and surface forms do have a swimbladder, but others lack one. The Atlantic mackerel, for example, has no swimbladder. For a fast and powerful predator this is a definite advantage, for as we have seen, a soft-walled swimbladder places severe constraints on the freedom of moving up and down in the water masses. No elasmobranch (sharks and rays) has a swimbladder.

The swimbladder is a more or less oval, soft-walled sac, located in the abdominal cavity, just below the spinal column (see fig. 11.17). The shape varies a great deal, but the volume is rather constant from species to species, mostly about 5% of the body volume in sea

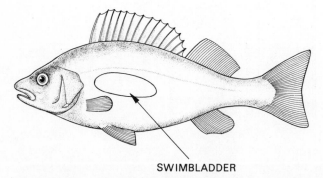

SWIMBLADDER

Fig. 11.17. The swimbladder of fish is located in the abdominal cavity, just below the vertebral column. The shape of the swimbladder can vary a great deal although its volume is normally similar in different species of fish.

water and 7% in fresh water species. The reason is that the specific gravity of a fish without the swimbladder is about 1.07. The volume of gas needed to give it neutral buoyancy in fresh water is therefore about 7% of the body volume, and in sea water (sp. gr. = 1.02) about 5%.

A swimbladder can provide a fish with perfect neutral buoyancy, and the fish can thus avoid the necessity of expending energy to keep from sinking. There is, however, a serious disadvantage, the fish is in equilibrium or neutral buoyancy at only one particular depth. If it swims down below this depth, the swimbladder is compressed by the increased water pressure, the buoyancy is decreased, and it must swim actively to keep from sinking further. This is not a very serious problem, for even if the swimbladder is almost completely compressed, the fish will not be much heavier than the water. If it rises above the level of neutral buoyancy, however, the situation can become catastrophic. The pressure is reduced, the swimbladder expands, and the fish gains increased lift. It must now swim actively down, or it will continue to rise, and as the swimbladder expands more and more it will eventually lose control. Some fish have a connection from the swimbladder to the esophagus which permits gas to escape, but others have a completely closed swimbladder. If such a fish rises from a depth of, say, 300 meters to 100 meters, the volume of its swimbladder increases threefold. This gives greatly increased lift, and as the swimbladder expands, it will also hinder the swimming movements. As the fish gets closer to the surface, expansion is more and more rapid, and the rise increasingly difficult to control. When a fish from 300 meters' depth (where the gas pressure in the swimbladder is 30 atm) arrives at the surface, the gas has expanded 30-fold and the swimbladder may have burst. This we often see when fish are brought up from deep waters, they arrive at the surface with the expanded swimbladder everted through the mouth or ruptured.

The swimbladder is embryologically formed as an evagination from the digestive tract. Those fish which maintain a connection between the swimbladder and the esophagus are known as *physostome* fish (Greek *physa* = bladder, *stoma* = mouth). These fish can fill their swimbladders by gulping air at the surface. For fish living at depth this would be impractical or impossible, for very large volumes of air would have to be filled into the bladder at the surface to achieve neutral buoyancy at depth where the pressure is many times as great.

In many fish the duct degenerates, and there is no connection from the swimbladder to the outside. In these fish, called *physoclist* fish (Greek *kleistos* = closed), the gases in the swimbladder must originate in the blood and be secreted into the swimbladder at a pressure equal to the depth at which the fish live. Fishes with swimbladders have been caught below 4000 meters' depth. Since the swimbladder wall is a soft structure, the swimbladder gas at that depth must be under a pressure in excess of 400 atm. This poses the problems already mentioned on p. 543.

The gases found in the swimbladder are the same as those in the atmosphere, oxygen, nitrogen, and carbon dioxide, but in different proportions. If we withdraw some gas from the swimbladder by puncturing it with a hypodermic needle, the fish is induced to secrete new gas to replace what was lost. In most fish the newly secreted gas will be rich in oxygen (see table 11.8). This is not true for all fish, the whitefish (*Coregonus albus*), for example, fills its swimbladder with virtually pure nitrogen. Since it lives in lakes at depths of 100 meters or so, nitrogen must be secreted at a pressure of some 10 atm. Nitrogen gas is highly inert, and it has been difficult to understand how an inert gas can be secreted. Also, the truly inert gases (such as argon) are likewise enriched in the swimbladder gas in about the same proportion to nitrogen as they occur in the atmosphere (Wittenberg, 1958). This indicates that the secretion mechanism cannot involve chemical reactions, it must be of a physical nature.

Even when the swimbladder gas is rich in oxygen, the nitrogen may still be more concentrated than in the surrounding water. Say,

Table 11.8. *Composition of gas secreted by the gas gland of codfish. The gland was covered by a thin plastic film and stimulated to secrete gas, whereupon bubbles of c. 0.5 µl size were collected and analyzed.* (Scholander, 1956)

	O_2 (%)	CO_2 (%)	N_2 (%)
	81.1	8.6	10.3
	70.3	5.0	24.7
	76.6	15.8	7.6
	49.5	6.9	43.6
	61.1	6.8	32.1
	66.5	5.9	27.6
	30.1	13.9	56.0
	69.2	7.2	23.6
Mean	63.1	8.8	28.1

for example, that we obtain a fish from a depth of 500 meters, and that a sample of the swimbladder gas shows 80% O_2 and 20% N_2 (and a trace of carbon dioxide). At the depth where the fish was caught, the gases were under 50 atm total pressure, so that their partial pressures were 40 atm for oxygen and 10 atm for nitrogen. Relative to the gas tensions in the surrounding water, 0.2 atm O_2 and 0.8 atm N_2, oxygen in the swimbladder gas was thus enriched 200-fold and nitrogen 12.5-fold. Before we discuss how gases can be secreted at these high pressures, we will examine the simpler question of how the gases are retained and kept from being lost from the swimbladder.

How to keep the gas in. As always, to understand function we must know a bit about structure. In the wall of the swimbladder of most fish we will find a gas gland, which stands out because of its bright red color. If the gas gland were supplied with ordinary arterial blood, which comes from the gills and therefore has gas tensions in equilibrium with the water, the blood would tend to dissolve the swimbladder gases and carry them away. How can this be prevented?

The gas gland is supplied with blood through a peculiar structure known as the *rete mirabile* (plural, *retia mirabilia*). The form of the rete and the gas gland varies a great deal from species to species, but in principle it is as follows: before the artery reaches the gas gland, it splits up into an enormous number of straight, parallel capillaries, which may unite again into a single vessel before reaching the gas gland (fig. 11.18). The vein from the gland likewise splits up into a similar number of parallel capillaries which run interspersed between the arterial capillaries, whereupon they unite and form a single vein.

Let us see how this structure can help to keep gases inside the swimbladder, once they have been deposited there. The system is analogous to the countercurrent heat exchanger in the flipper of the whale, only that we now deal with dissolved gases. Assume that the swimbladder contains gas under a pressure of 100 atm. As venous blood leaves the swimbladder, it will contain dissolved gases at this high pressure. When this blood enters the venous capillary, only the thin capillary wall separates it from blood in the arterial capillaries. Gases therefore diffuse across the wall to the arterial blood. As the blood runs along the venous capillaries, it loses more and more gas by diffusion to the arterial capillaries. At the end of the venous capillaries, as the blood is about to leave the rete, the surrounding capillaries contain arterial blood which comes directly from the

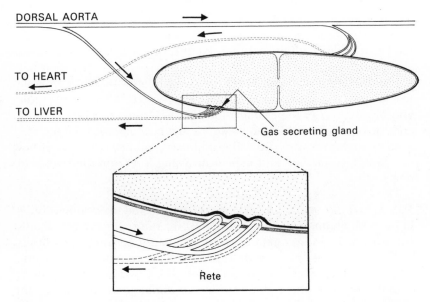

Fig. 11.18. Diagram of the circulation to the swimbladder of fish. Blood may reach the swimbladder either (*a*) through a set of parallel capillaries, the rete, which supply the gas gland, or (*b*) a vessel to the posterior portion where fine blood vessels that spread over the wall can serve to absorb gases rapidly. When gas is being secreted, the gas absorption vessels are closed and carry no blood.

gills and is not yet enriched with additional gas. The venous blood can therefore lose gases until it is in diffusion equilibrium with the incoming arterial blood, i.e. the venous blood that leaves the rete contains no more gas than the incoming arterial blood. In this way the rete serves as a trap that is used to retain the gases in the swimbladder and avoid loss of gas to the circulating blood. The rete is a typical countercurrent exchange system which depends on passive diffusion between two liquid streams which run in opposite directions. The exchange of gas is aided by a large surface (i.e. the large number of capillaries), by a short diffusion distance (i.e. the thin capillary wall which is only a fraction of 1 μm thick), and by the length of the capillaries (which are exceptionally long).

Some measurements are informative in this regard. The rete from a common eel weighed about 65 mg, and had some 100 000 arterial capillaries and about the same number of venous capillaries. The capillaries were about 4 mm long, and this gives a total length of 400 meters of each kind of capillary. Their diameter was, as most capillaries, some 7 to 10 μm, and simple arithmetic then gives their total wall surface area as being in excess of 100 cm^2 for each kind, all within the volume of a water drop (Krogh, 1929). This large area is thus available for gas exchange between the arterial and venous blood.

The capillaries in the swimbladder rete are exceptionally long, often several millimeters. Muscle capillaries, which are otherwise among the longest anywhere, are only about 0.5 mm long. There is a clear correlation between the length of the rete capillaries and the gas pressure in the swimbladder. Those fish which live at the greatest depth and secrete gases under the highest pressures have the longest capillaries, the record being held by a deep-sea fish, *Basso-zetus taenia,* which has rete capillaries with the enormous length of 25 mm (Marshall, 1960).

The effectiveness of the rete has been estimated by Scholander, who found that its dimensions would suffice for maintaining pressures in excess of 4000 atm in the swimbladder (Scholander, 1954). Fish that have swimbladders almost certainly do not occur below 5000 meters (500 atm), and we must therefore conclude that the rete is more than adequate as a device to keep the swimbladder gas from being carried away, dissolved in the blood. Since the rete depends entirely on diffusion and requires no energy, it provides a rather advantageous solution to the problem.

How is gas secreted into the swimbladder? The question of depositing gas into the swimbladder against a high pressure is a more formidable problem than keeping the gas in, once it is there. We will first discuss oxygen, for this is the gas most commonly secreted into the swimbladder. Let us base this discussion on a swimbladder that already contains oxygen at a high pressure, say 100 atm (corresponding to 1000 meters' depth), and that additional gas is to be secreted against this enormous pressure. Since the oxygen tension in the arterial blood which comes from the gills will at best equal that of the surrounding water, the arterial oxygen tension will be no more than 0.2 atm.

Let us consider the oxygen *content* of the blood, rather than the gas tension. If oxygen is being removed from the blood and secreted into the swimbladder, it follows that the venous blood leaving the rete must contain less oxygen than the entering arterial blood. As an arbitrary example, say that the arterial blood contains 10 ml O_2 per 100 ml blood and that the venous blood contains 9 ml O_2 per 100 ml when leaving the rete. In fig. 11.19, which for simplicity has the entire rete structure represented by a single loop, these two numbers are placed at the far left of the diagram. The difference between the arterial and the venous blood, 1 ml O_2 per 100 ml blood, is the oxygen that is being deposited into the swimbladder.

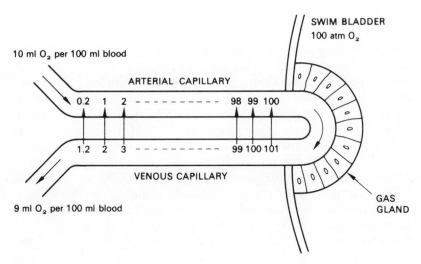

Fig. 11.19. Diagram of the countercurrent multiplier system of the fish swimbladder. The numerous capillaries of the rete are represented by a single loop. As the gas gland produces lactic acid, the oxygen tension increases in the venous capillary, and gas diffuses across to the arterial capillary, thus remaining within the loop. The venous blood, as it exits, contains less oxygen than the incoming arterial blood.

Let us now assume that the gas gland begins to produce lactic acid (which in reality it does when gas secretion takes place). The lactic acid enters the blood and reduces its affinity for oxygen. This effect of acid is much more pronounced in fish blood than in mammalian blood and is known as the *Root effect* (Scholander and Van Dam, 1954). The lactic acid drives oxygen off the hemoglobin, and the oxygen tension in the blood which is leaving the swimbladder is therefore increased (see fig. 11.20). The oxygen tension now is higher in the venous than in the arterial capillary, and oxygen diffuses across to the arterial capillary. This continues as long as lactic acid is added to the venous blood, which thus leaves the rete with less oxygen than is contained in the incoming arterial blood. Owing to the countercurrent flow in the rete, lactic acid also tends to remain in the loop, thus increasing the effect.

In this system it is important to distinguish clearly between *amount* and *tension*. The amount of oxygen in the venous blood which leaves the rete must be less than in the incoming arterial blood, but for oxygen to diffuse across from the venous to the arterial capillary, the oxygen tension must be higher in the venous than in the arterial capillary. This higher tension is caused by the acid which is produced by the gas gland. In this way oxygen is constantly returned to the arterial capillary and gradually accumulates in the loop, where it

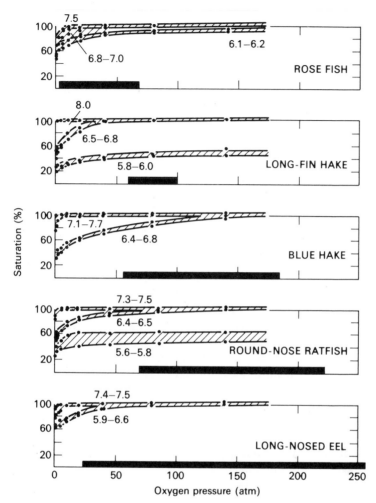

Fig. 11.20. Oxygen dissociation curves of fish blood demonstrate the pronounced effect of acid, known as the Root effect. The upper curve in each diagram represents untreated blood at 4 °C; in the lower curve of each diagram lactic acid has been added to give the pH stated for each curve. The solid bar on the abscissa gives the depth range at which the fish lives. Note that the abscissa gives the oxygen pressure in atmospheres (from Scholander and Van Dam. 1954. *Biol. Bull.*, **107**, 247).

can reach very high concentrations. How the gas gland itself functions is not known in detail, but the system we have described explains how the blood that enters the gland can have very high oxygen concentrations.

The mechanism which we have described has received experimental verification. The Norwegian investigator, John Steen, has succeeded in taking minute blood samples from the artery and the vein of the rete in eels which were in the process of secreting oxygen. He did indeed find that the venous blood had a lowered oxygen *content*, but an increased oxygen *tension* which was due to the production of lactic acid and carbon dioxide in the gas gland (Steen, 1963).

We can now see that the rete, when it secretes, is a countercurrent *multiplier* system. A small difference in oxygen tension between venous and arterial blood is multiplied along the length of its capillaries. Thus, the longer the capillaries, the higher the multiplication effect. Again, this is in accord with the much longer rete capillaries found in fish from greater depths. The multiplication requires energy, as any process that transports a substance against a gradient; in this case the lactic acid released by the gas gland provides the necessary driving force.

The system we have described depends on the use of acid to decrease the affinity of hemoglobin for oxygen. However, acid is not the only substance that has an effect on the blood gases, any solute added to the blood decreases the solubility for gases. This we could call a *salting-out effect*. Thus, lactic acid, in addition to its Root effect, also has a salting-out effect on the blood gases, not only on oxygen, but also on the other gases, including nitrogen. Since the addition of solute decreases the solubility for nitrogen, the rete will act as a multiplier for this gas as well as for oxygen, and we now can explain how inert gases can accumulate in the swimbladder. Until the salting-out effect was understood, it was very difficult to explain how any inert gas could be secreted. The possibility was pointed out as early as 1934 by Henri Koch, but received little attention until the late 1960s when sufficient experimental work had accumulated to yield a clear understanding of how the rete acts as a countercurrent multiplier system.

The salting-out effect, however, does not seem to be as effective as the Root effect. Careful estimates indicate that the salting-out effect of an increase in the blood salts of 0.02 mol liter^{-1} could result in a nitrogen concentration of 25 atm, but due to the Root effect a lactic acid concentration of only 0.005 mol liter^{-1} (as observed in actively secreting eel) could potentially produce some 3000 atm of oxygen (Kuhn, Ramel, Kuhn and Marti, 1963).

One further finding seems to put the final piece of the puzzle in place. As acid is added to fish blood and oxygen is forced off the hemoglobin, the effect is spoken of as the 'Root-off' reaction. The reverse, that oxygen is again bound to the hemoglobin when the pH is increased is known as the 'Root-on' effect. The interesting finding is that the reaction rates for the Root-off and Root-on effects are quite different. The half-time for the off reaction is 50 ms (at 23 °C); for the on reaction is in the order of 10–20 s. Blood in the arterial capillary of the rete, as acid diffuses into it, therefore rapidly gives off its oxygen, but as the same blood runs in the venous capillary and the lactic acid diffuses out and into the arterial capillary, the oxygen does not immediately enter into combination with hemoglobin, it remains as uncombined oxygen, in other words, a high oxygen tension persists. Thus, the slow Root-on effect permits a large fraction of the hemoglobin to leave the venous capillary in the deoxygenated state, although the tension of oxygen may still be quite high (Berg and Steen, 1968).

Although the function of the rete as a countercurrent multiplier system is clear, the function of the secretory epithelium of the gas gland itself has not been well studied. One of the unsolved problems is how gas is moved from the capillary into the gas phase of the bladder itself. The role of the secretory epithelium in this process is not understood.

REFERENCES

ABBOTT, B. C. and LOWY, J. (1953). Mechanical properties of *Mytilus* muscle. *J. Physiol.*, **120**, 50P.

ALEXANDER, R. MCN. (1968). *Animal Mechanics,* London: Sidgwick & Jackson, 346 pp.

ALLEN, R. D., FRANCIS, D. and ZEH, R. (1971). Direct test of the positive pressure gradient theory of pseudopod extension and retraction in amoebae. *Science,* **174**, 1237–40.

ANDERSEN, S. O. and WEIS-FOGH, T. (1964). Resilin. A rubber-like protein in arthropod cuticle. *Adv. Insect. Physiol.*, **2**, 1–65.

BAGUET, F. and GILLIS, J. M. (1968). Energy cost of tonic contraction of a lamellibranch catch muscle. *J. Physiol.*, **198**, 127–43.

BENNET-CLARK, H. C. and LUCEY, E.C.A. (1967). The jump of the flea: a study of the energetics and a model of the mechanism. *J. Exp. Biot.*, **47**, 59–76.

BERG, T. and STEEN, J. B. (1968). The mechanism of oxygen concentration in the swim-bladder of the eel. *J. Physiol.*, **195**, 631–8.

BLUM, J. J. and LUBLINER, J. (1973). Biophysics of flagellar motility. *Ann. Rev. Biophys. Biomed. Engin.*, **2**, 181–219.

BONE, Q. (1966). On the function of the two types of myotomal muscle fibre in elasmobranch fish. *J. mar. Biol. Ass. UK,* **46**, 321–49.

BROKAW, C. J. (1971). Bend propagation by a sliding filament model for flagella. *J. Exp. Biol.*, **55**, 289–304.

BULLER, A. J., ECCLES, J. C. and ECCLES, R. M. (1960). Interactions between motoneurons and muscles in respect to the characteristic speeds of their responses. *J. Physiol.*, **150**, 417–39.

BULLER, A. J. and LEWIS, D. M. (1964). The rate of rise of tension in isometric tetani of cross-innervated mammalian skeletal muscles. *J. Physiol.*, **170**, 67P–68P.

BURKE, R. E., LEVINE, D. N., ZAJACK, F. E., III, TASIRIS, P. and ENGEL, W. K. (1971). Mammalian motor units: Physiological-histochemical correlation in three types in cat gastrocnemius. *Science*, **174**, 709–12.

CASELLA, C. (1950). Tensile force in total striated muscle, isolated fibre and sarcolemma. *Acta. Physiol. Scand.*, **21**, 380–401.

COOPER, S. and ECCLES, J. C. (1953). The isometric responses of mammalian muscles. *J. Physiol.*, **69**, 377–85.

DENTON, E. (1960). The buoyancy of marine animals. *Sci. Am.*, **203**, 119–28.

DENTON, E. J. (1961). The buoyancy of fish and cephalopods. *Progr. Biophys. biophys. Chem.*, **11**, 178–234.

DENTON, E. J. and GILPIN-BROWN, J. B. (1959). On the buoyancy of the cuttlefish. *Nature, Lond.*, **184**, 1330–32.

DENTON, E. J. and GILPIN-BROWN, J. B. (1961). The distribution of gas and liquid within the cuttlebone. *J. mar. Biol. Ass. UK*, **41**, 365–81.

DENTON, E. J. and GILPIN-BROWN, J. B. (1966). On the buoyancy of the pearly Nautilus. *J. Mar. Biol. Ass., UK*, **46**, 723–59.

DENTON, E. J., GILPIN-BROWN, J. B. and SHAW, T. L. (1969). A buoyancy mechanism found in cranchid squid. *Proc. Roy. Soc. Lond. B*, **174**, 271–9.

ECKERT, R. (1972). Bioelectric control of ciliary activity. *Science*, **176**, 473–81.

GIBBONS, I. R. and ROWE, A. J. (1965). Dynein: a protein with adenosine triphosphatase activity from cilia. *Science*, **149**, 424–6.

GORDON, A. M., HUXLEY, A. F. and JULIAN, F. J. (1966). The variation in isometric tension with sarcomere length in vertebrate muscle fibres. *J. Physiol.*, **184**, 170–92.

GRAY, J. and LISSMANN, H. W. (1938). Studies in animal locomotion. VII. Locomotory reflexes in the earthworm. *J. Exp. Biol.*, **15**, 506–17.

GROSS, F. and ZEUTHEN, E. (1948). The buoyancy of plankton diatoms: a problem of cell physiology. *Proc. Roy. Soc. Lond. B.*, **135**, 382–9.

HALL-CRAGGS, E.C.B. (1965). An analysis of the jump of the Lesser Galago (*Galago senegalensis*). *J. Zool.*, **147**, 20–9.

HAXTON, H. A. (1944). Absolute muscle force in the ankle flexors of man. *J. Physiol.*, **103**, 267–73.

HERSEY, J. B., BACKUS, R. H. and HELLWIG, J. (1962). Sound-scattering spectra of deep scattering layers in the western North Atlantic Ocean. *Deep Sea Res.*, **8**, 196–210.

HOYLE, G. (1957). *Comparative Physiology of the Nervous Control of Muscular Contraction*, Cambridge, England: Cambridge University Press, 147 pp.

HOYLE, G. (1964). Muscle and neuromuscular physiology. In *Physiology of Mollusca*, vol. I (K. M. Wilbur and C. M. Yonge, eds), pp. 313–51, New York: Academic Press.

HUXLEY, H. E. (1969). The mechanism of muscular contraction. *Science*, **164**, 1356–66.

JAHN, T. L. and BOVEE, E. C. (1971). Effects of environmental conditions on the motile behavior of amebas. *Adv. Comp. Physiol. Biochem.*, **4**, 1–35.

JAHN, T. L., LANDMAN, M. D. and FONSECA, J. R. (1964). The mechanism of locomotion of flagellates. II. Function of the mastigonemes of *Ochromonas*. *J. Protozool.*, **11**, 291–6.

JAHN, T. L. and VOTTA, J. J. (1972). Locomotion of Protozoa. *Ann. Rev. Fluid Mech.* **4**, 93–116.

JOHNSON, W. H. and TWAROG, B. M. (1960). The basis for prolonged contractions in molluscan muscles. *J. Gen. Physiol.*, **43**, 941–60.

KESSELER, H. (1966). Beitrag zur Kenntnis der chemischen und physikalischen Eigenschaften des Zellsaftes von *Noctiluca miliaris*. *Veröffentlichungen des Instituts für Meeresforschung in Bremerhaven*, Sonderband **2**, 357–68.

KOCH, H. (1934). L'émission de gas dans la vésicule gazeuse des poissons. *Rev. Quest. sci.*, **26**, 385–409.

KROGH, A. (1929). *The Anatomy and Physiology of Capillaries*, 2nd edn, New Haven, Conn.: Yale University Press, 422 pp.

KUFFLER, S. W. and WILLIAMS, E.M.V. (1953a). Small-nerve-junctional potentials. The distribution of small motor nerves to frog skeletal muscle, and the membrane characteristics of the fibres they innervate. *J. Physiol.*, **121**, 289–317.

KUFFLER, S. W. and WILLIAMS, E.M.V. (1953b). Properties of the 'slow' skeletal muscle fibres of the frog. *J. Physiol.*, **121**, 318–40.

KUHN, W., RAMEL, A., KUHN, H. J. and MARTI, E. (1963). The filling mechanism of the swimbladder. Generation of high gas pressures through hairpin countercurrent multiplication. *Experientia*, **19**, 497–511.

LOWY, J. and MILLMAN, B. M. (1962). Mechanical properties of smooth muscles of cephalopod molluscs. *J. Physiol.*, **160**, 353–63.

LUDWIG, G. D., BLAKEMORE, W. S. and DRABKIN, D. L. (1957). Production of carbon monoxide and bile pigment by haemin oxidation. *Biochem. J.*, **66**, 38P.

MACHIN, K. E. (1958). Wave propagation along flagella. *J. Exp. Biol.*, **35**, 796–806.

MARSHALL, N. B. (1960). Swimbladder structure of deep-sea fishes in relation to their systematics and biology. *Discovery Reports*, **31**, 1–122.

NAITOH, Y. and KANEKO, H. (1972). Reactivated triton-extracted models of paramecium: modification of ciliary movement by calcium ion. *Science*, **176**, 523–4.

PARRY, D. A. and BROWN, R.H.J. (1959a). The hydraulic mechanism of the spider leg. *J. Exp. Biol.*, **36**, 423–33.

PARRY, D. A. and BROWN, R.H.J. (1959b). The jumping mechanism of salticid spiders. *J. Exp. Biol.*, **36**, 654–64.

PHLEGER, C. F. (1971). Pressure effects on cholesterol and lipid synthesis by the swimbladder of an abyssal *Coryphaenoides* species. *Am. Zool.*, **11**, 559–70.

PICKWELL, G. V., BARHAM, E. G. and WILSON, J. W. (1964). Carbon monoxide production by a bathypelagic siphonophore. *Science*, **144**, 860–2.

PRINGLE, J.W.S. (1949). The excitation and contraction of the flight muscles of insects. *J. Physiol.*, **108**, 226–32.

PRINGLE, J.W.S. (1957). *Insect Flight,* Cambridge, England: Cambridge University Press, 132 pp.

PUGH, L.G.C.E. (1959). Carbon monoxide content of the blood and other observations on Weddell seals. *Nature, Lond.,* **183,** 74–6.

RAYNER, M. D. and KENNAN, M. J. (1967). Role of red and white muscles in the swimming of the skipjack tuna. *Nature, Lond.,* **214,** 392–3.

RIKMENSPOEL, R. (1965). The tail movement of bull spermatozoa. Observations and model calculations. *Biophys. J.,* **5,** 365–92.

RIKMENSPOEL, R., SINTON, S. and JANICK, J. J. (1969). Energy conversion in bull sperm flagella. *J. Gen. Physiol.,* **54,** 782–805.

ROEDER, K. D. (1953). *Insect Physiology,* New York: John Wiley & Sons, 1100 pp.

ROOT, W. S. (1965). Carbon monoxide. *Handbook of Physiology,* Sec. 3: *Respiration,* Vol. 2: 1087–1098. Washington, D.C.: American Physiological Society.

ROTHSCHILD, M., SCHLEIN, Y., PARKER, K. and STERNBERG, S. (1972). Jump of the oriental rat flea *Xenopsylla cheopis* (Roths.). *Nature, Lond.,* **239,** 45–8.

RÜEGG, J. C. (1971). Smooth muscle tone. *Physiol. Rev.,* **51,** 201–48.

SCHMIDT-NIELSEN, S., FLOOD, A. and STENE, J. (1934). On the size of the liver of some gristly fishes, their content of fat and vitamin A. *Det Kongelige Norske Videnskabers Selskab Forhandlinger,* **7,** 47–50.

SCHOLANDER, P. F. (1954). Secretion of gases against high pressures in the swimbladder of deep sea fishes. II. The rete mirabile. *Biol. Bull.,* **107,** 260–77.

SCHOLANDER, P. F. (1956). Observations on the gas gland in living fish. *J. Cell. Comp. Physiol.,* **48,** 523–8.

SCHOLANDER, P. F. and VAN DAM, L. (1954). Secretion of gases against high pressures in the swimbladder of deep sea fishes. I. Oxygen dissociation in blood. *Biol. Bull.,* **107,** 247–59.

SEYMOUR, M. K. (1971). Burrowing behaviour in the European lugworm *Arenicola marina* (Polychaeta: Arenicolidae). *J. Zool.,* **164,** 93–132.

SLEIGH, M. A. (1968). Patterns of ciliary beating. *Symp. Soc. Exp. Biol.,* **22,** 131–50.

SMITH, D. S. (1965). The flight muscles of insects. *Sci. Am.,* **212,** 76–88.

SOTAVALTA, O. (1953). Recordings of high wing-stroke and thoracic vibration frequency in some midges. *Biol. Bull.,* **104,** 439–44.

STEEN, J. B. (1963). The physiology of the swimbladder in the eel *Anguilla vulgaris.* III. The mechanism of gas secretion. *Acta Physiol. Scand.,* **59,** 221–41.

TRUEMAN, E. R. (1966). Observations on the burrowing of *Arenicola marina* (L.). *J. Exp. Biol.,* **44,** 93–118.

TRUEMAN, E. R. and PACKARD, A. (1968). Motor performances of some cephalopods. *J. Exp. Biol.,* **49,** 495–507.

WEBER, H. H. and PORTZEHL, H. (1954). The transference of the muscle energy in the contraction cycle. *Prog. Biophys. biophys. Chem.,* **4,** 60–111.

WEIS-FOGH, T. (1960). A rubber-like protein in insect cuticle. *J. Exp. Biol.,* **37,** 889–907.

WIERSMA, C.A.G. (1961). The neuromuscular system. In *The Physiology of Crustacea*, vol. II (T. H. Waterman, ed.), pp. 191–240, New York: Academic Press.

WIGGLESWORTH, V. B. (1972). *The Principles of Insect Physiology*, 7th edn, London: Chapman and Hall, 827 pp.

WILSON, J. A. (1972). *Principles of Animal Physiology*, New York: Macmillan Co.

WITTENBERG, J. B. (1958). The secretion of inert gas into the swim-bladder of fish. *J. Gen. Physiol.*, **41**, 783–804.

WITTENBERG, J. B. (1960). The source of carbon monoxide in the float of the Portuguese Man-of-War, *Physalia physalis* L. *J. Exp. Biol.*, **37**, 698–705.

12

Information and senses

In the preceding chapters we have discussed some of the most important aspects of the environment in which animals live – oxygen, food, temperature, and water. In this chapter we shall turn to the question of how animals obtain detailed information about their environment and how the information is utilized. We shall be concerned with the kinds of information that are available and how information is processed and passed on to the central nervous system. The next and final chapter will deal with the integration of information from outside as well as from within the body, and how the various functions are controlled and integrated.

Virtually every animal depends on information about its surroundings, whether it is for finding food and mates or escaping from predators. They must find their way about, and also assess the quality of the environment, such as its temperature, for example.

Most information about the environment is obtained through specialized sensory organs. Traditionally, sense organs are separated into *exteroceptors*, which respond to stimuli coming from outside, such as light, sound, etc., and *proprioceptors*, which refer to internal information such as the position of the limbs, for example. This separation does not have much inherent meaning, and, at best, is a matter of convenience. Another traditional classification of senses is based on the five most obvious senses of humans – vision, hearing, taste, smell, and touch. In reality our sensory equipment is not nearly so limited.

A list of external stimuli to which at least some animals respond is quite extensive (see table 12.1). The listed categories are not discrete and the separation is somewhat arbitrary. However, as a matter of convenience we shall follow approximately this outline in the discussion of possibilities and limitations that apply to the various kinds of available information. The information naturally falls into three major categories, the first dealing with electromagnetic and thermal

Table 12.1. *Environmental stimuli significant in sensory perception*

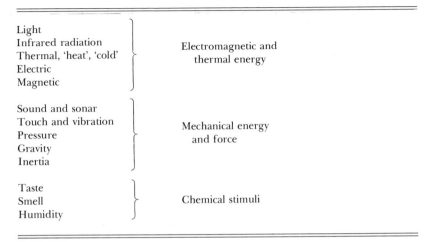

energy, the second with mechanical energy and mechanical force, and the third with chemical information.

In this context the question of whether information can reach the central nervous system via other avenues, outside the sensory organs (extrasensory perception, or ESP) is irrelevant. In this chapter we are dealing with measurable physical quantities which can be recognized, described, and manipulated in controlled ways. Although some sensory mechanisms may still be unknown (this is true of the response to magnetic fields), the word extrasensory by definition means that no sensory structure is involved.

The wide variety of stimuli and the striking structural differences in the sense organs may seem confusing until we realize that the general sequence of events is much the same for all of them (fig. 12.1). The external stimulus impinges on an accessory structure, which may be highly complex such as the eye or the ear, or much simpler such as the touch receptors of the skin. Through these accessory structures the external stimulus reaches one or more sensory

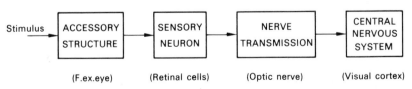

Fig. 12.1. Simplified diagram of the sequence of events in sensory perception. Light perception is used as an example, although in higher animals vision is far more complex than indicated here.

neurons. These have many properties in common, but they differ from organ to organ in that each kind is particularly sensitive to one specific kind of stimulus. The sensory neurons of the retina are highly sensitive to light, those of the ear to vibrations at frequencies in the range that we identify as 'sound,' and so on. However, all sensory neurons respond in the same way, they translate the stimulus into nerve impulses which, via the appropriate sensory nerves, are transmitted to the central nervous system.

The fundamental nature of the nerve impulses is the same in all sensory nerves, and the sorting out of what the original stimulus was is carried out in the central nervous system. Thus, if the auditory nerve is stimulated artificially, the induced nerve impulses will be perceived by the central nervous system as 'sound', and artificial stimulation of the optic nerve will be perceived as 'light'. Most of us already know an example of this kind of interpretation, we know that mechanical pressure on the eyeball is perceived as light, for pressure or a sharp blow on the eyeball makes us 'see stars before the eyes' although no light is involved.

Returning to the sequence of fig. 12.1, we say that the sensory neurons are the *transducers* which receive the external information and encode it as impulses in the sensory nerves.* In the central nervous system the nerve signals are again decoded and the pertinent information is integrated and utilized.

In sensory physiology the most interesting events are the transducing and encoding of information on the one hand, and the decoding and utilization on the other. The sensory nerves are transmission lines and the least interesting part of the system. In this chapter we shall first consider what kinds of information can be obtained and the possibilities and limitations inherent in the nature of the stimulus. Afterwards we shall discuss the encoding, processing, and decoding of pertinent information.

Sensory information – possibilities and limitations

Determination of direction and distance

The direction from which a stimulus comes is of utmost importance, for this permits an animal to move towards the source of the

---

* A transducer, in a general sense, is a device that transforms or 'translates' energy supplied in one form into another form. For example, a microphone is a transducer which translates sound energy into electric energy. A transducer often requires an auxiliary source of energy and thus, strictly, is an energy controller. Furthermore, the output is often at a higher energy level than the input, i.e. the transducer is also an amplifier.

stimulus or to evade it, as need may be. Without knowledge of direction this is impossible. Distance is also important, although usually not by itself but in combination with information about direction. There are strict limitations, based on the nature of the stimulus, to what is possible and what the accessory structure can help to achieve in this regard.

The most directional sense of all is vision, to a majority of animals probably the most important single sense. The reason for the directionality of vision is, of course, that light rays travel in straight lines. In addition to information about the location, the sense of vision also gives much information about the nature of various objects. Although vision is the most highly directional sense, information about direction can also be obtained with other senses, but there are certain inevitable limitations.

In principle, there are only three ways in which directional information can be obtained.

(1) The sense organ itself can be directional, which means that it must give different signals if turned towards or away from the source of the stimulus (e.g. the eye).

(2) Signals obtained from a pair of similar sense organs can be compared. Directional information about sound is obtained in this way by the use of two ears, although the hearing organ itself is not very directional.

(3) Signals can be compared in space and time; this is achieved by successively assessing the condition at a series of points in space.

Information about distance is more difficult to obtain. Here again vision is superior, and with the aid of two eyes and the differences in the images they receive, distance can be evaluated. Most other sense organs can only achieve a rough estimate of distance, based on the attenuation of the signals. Certain auditory mechanisms can, however, be used for extremely precise information about distance as well as direction, as we shall see later (p. 585–8).

Light and vision

Undoubtedly vision can give more detailed information than any other sense. With the aid of suitable accessory structures (eyes) light can form a detailed image of the environment.

Information obtained with the aid of light depends on differences in *intensity*, for a uniformly luminous environment conveys no information of interest. Further information is contained in differences in *wave length* (color). For animals that have appropriate accessory organs the *plane of polarization* of the light can also be highly informative. As far as an animal is concerned, the transmission of a light signal is instantaneous, which means that changes in the environment are perceived without delay, except for the time used in transducing the signal and transmitting it to the central nervous system. Compare this with, for example, the slowness with which a change in an olfactory signal is perceived.

Image-forming eyes. Light sensitivity is extremely widespread, and even many unicellular organisms are sensitive to light. Multicellular animals usually have this sensitivity concentrated in certain spots, eye-spots. If these are shielded by a pigment on one side, they become direction-sensitive as well.

More highly organized animals have increasingly complex light-sensitive structures which, in their ultimate form, are excellent image-forming eyes. Well developed image-forming eyes are found in four different animal phyla, worms, molluscs, arthropods, and vertebrates. The various eyes differ in design and development, and it must be assumed that they have evolved independently in these animals.

The image-forming eyes can be based on two different principles, either a multi-faceted eye as in insects, or a single-lens, camera-like structure as in vertebrates (see fig. 12.2). The multi-faceted eye has the same angle of resolution, whether an object is distant or close by, but the lens-type eye must have some device for focusing if it shall be able to perform equally well at different distances. It is interesting that among the few worms that have well-developed, image-forming eyes, both single-lens and multi-faceted eyes occur.

Wavelength sensitivity. What man perceives as light lies within a very narrow wavelength band, 380 to 760 nm, out of the wide spectrum of electromagnetic radiation which ranges from the extremely short gamma rays to long wavelength radio waves (see fig. 12.3).

All other animals have their visual sensitivity within the same range of wavelengths as man, or very close to it. The remarkable fact is that not only animals, but also plants respond to light within this same range. This includes both photosynthesis and phototropic growth of plants. The reason for this universal importance of a very narrow band of electromagnetic radiation is simple enough. The energy carried by each quantum of radiation is inversely related to the wavelength. Therefore, the longer wavelengths do not carry sufficient energy in each quantum to have any appreciable photochemical effect, and the shorter wavelengths (ultraviolet and shorter) carry so much energy that they are destructive to organic

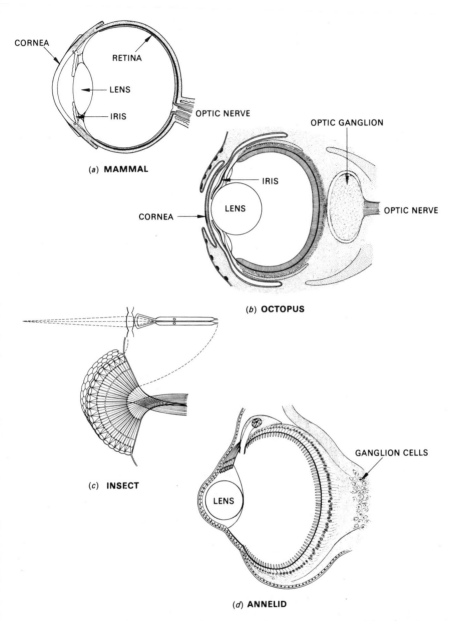

CORNEA

RETINA

LENS

IRIS

OPTIC NERVE

OPTIC GANGLION

(*a*) **MAMMAL**

IRIS

CORNEA

LENS

OPTIC NERVE

(*b*) **OCTOPUS**

(*c*) **INSECT**

GANGLION CELLS

LENS

(*d*) **ANNELID**

Fig. 12.2. The eyes of vertebrates and some invertebrates have a lens and function on the same principle as a camera. Insects have a compound eye in which the single elements combine to form the image. (*a*) Mammal; (*b*) octopus; (*c*) insect eye with one single ommatidium shown in detail; (*d*) annelid.

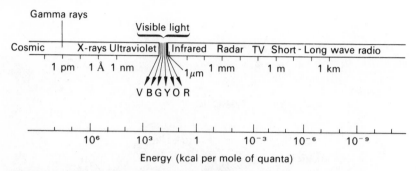

Fig. 12.3. The range of electromagnetic radiation ranges from the shortest cosmic rays to long-wave radio waves. Only a narrow band between about 380 and 760 nm is perceived as light. The energy carried by each quantum of electromagnetic radiation increases tenfold for each tenfold decrease in wavelength. V, violet; B, blue; G, green; Y, yellow, O, orange; R, red.

materials. The universal biological use of what we know as 'light' is therefore a result of the unique suitability of these particular wavelengths.

The range is not exactly the same for all animals. The vision of insects, for example, extends into the nearest ultraviolet range, to slightly shorter wavelengths than the vertebrate eye. The mammalian retina would be sensitive to ultraviolet light, but these wavelengths do not penetrate to the retina, primarily because of a slight yellowness of the lens which acts as a filter. This narrowing of the band of wavelengths that can be perceived is probably advantageous. A lens, if made of a uniform material, refracts short wavelength radiation more strongly than longer wavelengths. This means that the various wavelengths cannot be brought into focus simultaneously. In man-made lenses this difficulty is known as chromatic aberrration, and is corrected for by the use of composite lenses consisting of several elements with different refractive indices. For an eye that does not have color correction, the simplest way of reducing the difficulty of simultaneously focusing different wavelengths is to narrow the wavelength band that is permitted to enter. It is then best to eliminate the shortest wavelengths for which the refraction error is the greatest. In insects, which have a non-focusing composite eye, this is irrelevant, for the resolution is determined by the angular distance between the single elements of the eye.

Light absorption – retinal pigments. For the sensory neurons of the retina to be stimulated by light and respond, a sufficient number of light quanta must be absorbed. This is achieved with the aid of a light-absorbing pigment. The best known such pigment is *rhodopsin,* which is found in the rods of the vertebrate eye. (The rods are the retinal elements responsible for vision in dim light, while the cones, which will be discussed below, are involved in vision in bright light and the perception of color.)

Rhodopsin can be isolated from retinas of animals whose eyes are fully dark-adapted. It is light sensitive, and when exposed to light breaks down to *retinene,* a molecule closely related to vitamin A, and a protein called *opsin.* This process is the initial step in the perception of light by the rods of the vertebrate eye.

The strongest argument for the role of rhodopsin in the visual process is that, if the rhodopsin in the retina is depleted by exposing the eye to strong light, the eye is insensitive to dim light until the rhodopsin has regenerated. Another strong argument for the role of rhodopsin comes from its absorption spectrum, which has its maximum at 497 nm. This maximum and the detailed absorption spectrum coincide with the maximum sensitivity and the sensitivity spectrum of the human eye in dim light (see fig. 12.4).

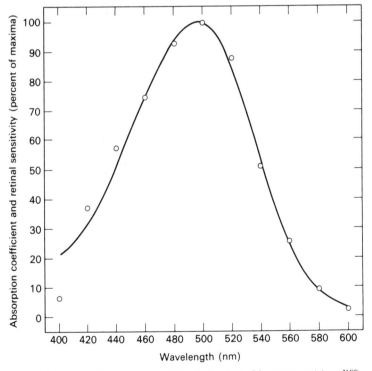

Fig. 12.4. The relative sensitivity of the dark-adapted human eye (o) at different wavelengths, compared to the absorption spectrum of the visual pigment (fully-drawn curve) (Crescitelli and Dartnall, 1953).

It has been more difficult to establish the role of pigments in the perception of light by the cones, primarily because of difficulties in extracting appropriate pigments. It has long been understood that the perception of color would require more than one pigment, and also that three pigments with absorption peaks in blue, green, and red would be sufficient to explain the perception of all colors that we see (the tri-chromatic vision theory).

The difficulties have been overcome by making measurements of the absorption spectra directly in the human eye. By using an ingenious instrumentation, measurements on single cones have revealed that there are three kinds of cones, blue-sensitive cones with an absorption maximum about 450 nm, green-sensitive cones with an absorption maximum about 525 nm, and red-sensitive cones with an absorption maximum at 555 nm. Presumably, these three types of cones are those that are responsible for human color vision in bright light (Brown and Wald, 1964).

The spectral sensitivity of the human eye is in accord with the three-pigment theory and the measured absorption spectra. Furthermore, what is known about deficiencies in the color vision of man (often called color blindness) is satisfactorily explained on the basis of the existence of these same three pigments (Wald, 1964).

Polarized light. One quality of light that man is unaware of is its plane of polarization.* For this reason, the information carried by the polarization of light and its importance for animals remained unnoticed until the German zoologist Karl von Frisch discovered that honey bees use the plane of polarization as a directional cue.

Von Frisch studied the communication between bees and how information is relayed from a bee that has found a good food source to other bees. Usually the bees indicate the direction of the food source relative to the direction of the sun, i.e. they use the sun as a compass, and they have an internal clock which compensates for the movements of the sun across the sky.

---

* In polarized light the vibrations of the propagated wave (ray) are all in one plane, perpendicular to the direction of the ray. In ordinary unpolarized light the vibrations are in all directions (yet perpendicular to the ray). Unpolarized light may become polarized by reflection from water and other non-metallic surfaces, and by transmission through certain materials. The plane of polarization can carry information not contained in unpolarized light. It appears that, in nature, the most important information animals can obtain in this way is due to the fact that the light from the blue sky is polarized, the polarization at any point in the sky depending on its position relative to the sun. Indirectly, this tells the location of the sun, information that is most important in animal orientation and navigation.

It was found, however, that the bees could orient correctly, even in the absence of a visible sun, provided that a small piece of the blue sky could be seen. The light from the blue sky is polarized to a degree that gives information about the actual position of the sun, and this is used by the bees. One way to show the dependence on the plane of polarization of the light is to reflect the light by a mirror. This makes the bees orient in the opposite of the correct direction, as expected from the reversal of the plane of polarization caused by the mirror. Also, by imposing filters which change the direction of polarization, any degree of deviation from the correct orientation can be imposed on the bees (von Frisch, 1948).

After the importance of polarized light had been demonstrated, it was found that insects in general are sensitive to polarized light. It was a surprise, however, that several aquatic animals are also sensitive to the polarization of light. This was first observed in the horseshoe crab, *Limulus* (Waterman, 1950). Many other aquatic animals, including octopus and fish, are also sensitive to polarized light (Waterman and Forward, 1970).

A substantial amount of evidence indicates that the ability to perceive the plane of polarization of light is important in orientation and navigation, not only for terrestrial insects, but for a large number of aquatic animals as well (Waterman and Horch, 1966).

Temperature

We may think that we are able to sense the thermal condition of the environment, and our behavior often seems to indicate that this is so. What is sensed, however, is not the environmental temperature, but the temperature of the skin at the depth of the appropriate receptors. A good example of our inability to judge air temperature is provided by the kind of electric hand-drier that is installed in some public washrooms. These blow a stream of hot air (as high as 100 °C) over the hands, and as long as the skin is wet, the air feels moderate, but as soon as the skin is dry, the air stream becomes painfully hot.

Heat and cold are perceived by different sense organs in the skin. This can be shown by an examination of the activity in the individual nerve fibers from the temperature-sensitive skin receptors. The fibers from a heat-sensitive receptor show no activity until the temperature increases above a certain point; from then on the activity increases, roughly in proportion to the temperature rise, until a certain upper limit is reached, above which the nerve activity again decreases.

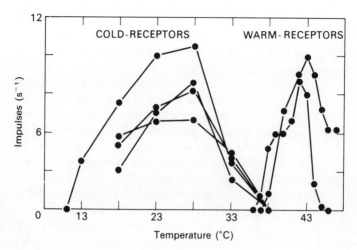

Fig. 12.5. Response curves obtained from single 'warm-receptors' and 'cold-receptors' in the scrotal skin of rats (Iggo, 1969).

Records obtained from the scrotal skin of the rat show exactly this (fig. 12.5). The warm-receptors are 'silent' below approximately 37 °C. Above this temperature their output increases rapidly with increasing temperature, but above 43 °C their output falls off. Below 37 °C the warm-receptors remain silent, while cold-receptors show an increasing response, again roughly proportional to the decrease in temperature. Below a certain level, the cold-receptors show a decreased activity which eventually falls to zero.

Within the narrow 'neutral' range, which corresponds to a normal scrotal skin temperature for a rat which is in heat balance, there is a low-level activity from both warm- and cold-receptors. A minor change in either direction can therefore be perceived with rapidity and precision. Thus we have the makings for a sensitive control system which is based on obtaining accurate information about slight changes in skin temperature, and hence on changes in heat transfer between the skin and the environment.

Infrared radiation. Cutaneous temperature receptors of the kind we have just discussed are sensitive to the local skin temperature. They give information about the environment only indirectly, for skin temperature is a result of heat flow from the core of the animal as well as the conduction and radiation to the environment. They discover radiation only through its effect on skin temperature, as we know when we can feel the radiation from a hot stove or other hot object. To some extent we can also sense a loss of heat by radiation. For example, a cold brick wall in a house may give a definite feeling of 'cold', we can feel that it comes from the wall and we often interpret it as a 'draft' although obviously no air flows through the wall.

Infrared radiation can, however, be perceived directly by a few animals that have specialized sense organs which respond to this type of radiation. The so-called *facial pits* or *pit organs* on the head of some snakes are such specialized infrared receptors.

When a rattlesnake strikes, the direction of the strike seems to be guided by the infrared radiation from its prey. A rattlesnake strikes only at warm-blooded prey, and when the prey is dead and at room temperature, the snake will not strike. However, a blind-folded snake strikes correctly at a moving dead rat, provided that the rat is warmer than the surroundings. Blind-folded, it cannot be guided by vision, and yet, it will strike correctly even at a moving, cloth-wrapped electric bulb, that is, the sense of smell is definitely not what guides the snake. The pit organ is evidently involved in sensing the location of warm objects. All those snakes which have pit organs feed preferentially on warm-blooded prey, and this further supports the view that these organs are infrared sensors. In rattlesnakes the pit organ is located, one on each side, between the nostril and the eye, it is richly innervated, and this in itself suggests a sensory role of the organ.

The sensitivity of the facial pit has been examined by recording the activity in the nerve leading from the organ. A variety of stimuli, such as sound, vibration, or light of moderate intensity (with the infrared part of the spectrum filtered out) has no detectable effect on the activity in the nerve. However, if objects of a temperature different from the surroundings are brought into the receptive field around the head, there is a striking change in nerve activity, regardless of the temperature of the intervening air (Bullock and Cowles, 1952).

The question now arises as to how the infrared radiation is sensed. The pit is covered by a thin transparent membrane, and it has been suggested that a rise in temperature in the chamber behind the membrane could cause an expansion of the gas with a consequent deformation of the membrane, which in turn could be sensed by a suitable receptor. This hypothesis is highly improbable, for a cut in the membrane which opens the chamber to the outside air causes no loss in responsiveness. This result is incompatible with the hypothesis of a pressure change being sensed.

We are now left with two other possibilities to consider, either that infrared radiation is absorbed by a specific compound, analogous to the light-sensitive pigments in the eye, or that the pit organ is sensitive to the slight temperature rise caused when infrared radiation reaches it. The infrared radiation from a mammalian body has its peak around 10 000 nm, and because of the low quantum energy in this wavelength, it is extremely unlikely that it can have any photochemical effect on a pigment. Pure infrared radiation can be produced by a laser, and experiments with such radiation of known wavelength provide strong evidence that the mode of reception in the facial pit organ is entirely thermal (Harris and Gamow, 1971).

It is unlikely that specialized infrared sensitive organs are of widespread importance among animals. In particular, it is almost unthinkable that any similar organ could be of importance for aquatic animals, partly because of the very low penetration of infrared radiation in water, and also because the direct contact of the body surface with a medium of high thermal conductivity and thermal capacity would make it impossible to perceive the small amounts of heat involved.

Animal electricity

It is well known that some fish can produce strong electric shocks. Ancient people such as the Greeks and Egyptians knew and wrote about the powerful shocks delivered by the electric ray (*Torpedo*) and the electric catfish (*Malapterurus*), but the phenomenon must have appeared utterly mysterious until the nature of electricity became known.

Powerful electric discharges serve obvious offensive and defensive purposes. The strongest shocks are produced by the South American electric eel (*Electrophorus*). It can deliver discharges of between 500 and 600 volts, which are powerful enough to kill other fish and possibly animals the size of a man. Such strong electric discharges can be delivered by only a few species of fish, which are not particularly closely related. Some are teleosts while others are elasmobranchs, and they occur in both fresh water and in the sea. Freshwater electric fish often live in stagnant and murky waters. Their electric discharges seemed more meaningful after it was understood that they can be used, not only to stun and catch prey, but also to obtain information about the environment, including the location of prey.

As electric fish have been studied more closely, it has become apparent that even weak discharges, too weak to have any effect on other fish, serve a variety of useful purposes. Obviously, very weak discharges cannot stun the prey, but it is now clear that they can be used to obtain information about the environment in general. Electric discharges are particularly suitable for distinguishing between non-conducting objects and good conductors, such as another animal. A fresh-water animal, say a fish, has a much higher electric conductivity (because of the salt content of its body fluids) than the poorly conducting water. Weak electric discharges can be used for navigation in murky water, as well as for the location of predators and prey. Furthermore, electric discharges are used in communication between individuals, and in some cases a fish will respond with sexual behavior only to the electric signals from the opposite sex (Black-Cleworth, 1970; Hopkins, 1972).

Finally, it has recently become known that many fish which cannot produce electric signals themselves are quite sensitive to the weak electric activity which results from the ordinary muscle function of other organisms. In this way sharks and rays are able to locate another fish, even if it is at rest. A hungry dogfish (a shark, *Scyliorhinus*) will even respond to a flounder that is resting at the bottom of an aquarium, completely covered by sand. If the shark passes within less than 15 cm of the flounder, it makes a sudden turn towards the hidden prey, removes the sand by sucking it up and expelling it, and seizes the flounder. Careful experiments have shown that this ability to locate the flounder is not associated with the sense of smell. Apparently the dogfish responds to the minute electric potentials produced by the breathing movements of the hidden flounder, for it will react to an artificial electric discharge of four microamperes (the same order of magnitude as that produced by a living flounder) and try to dig out this imaginary prey (Kalmijn, 1971).

To summarize, electric fish can use their discharges to stun prey, to obtain information about their environment (much more common), and for communication. Furthermore, non-electric fish can use their extremely sensitive electro-receptors to locate prey by means of the action potentials caused by normal muscle activity.

Production of the electric discharges. In nearly all electric fish the discharges are produced by discrete electric organs, which consist of modified muscle. These organs have been most carefully studied in the electric eel, which can produce discharges of over 500 volts. The large electric organs of the electric eel run along most of the body, one large mass on each side, and make up about 40% of the animal's volume.

These organs consist of thin, wafer-like cells, known as *electroplaques* or *electroplates*. The electroplaques are stacked in columns that may contain between 5000 and 10 000 plaques, and there are some 70 such columns on each side of the body. The two faces of each electroplaque are markedly different, one face is innervated by a dense network of nerve terminals and the other is deeply folded and convoluted.

When the organ is at rest, the two surfaces of the electroplaque are positively charged on the outside, each being +84 mV relative to the inside (see fig. 12.6). The over-all potential from the outside of one membrane to the outside of the other therefore is zero. During a discharge, however, the potential on the innervated surface is reversed, and the total voltage across the single electroplaque now becomes about 150 mV. With the serial arrangement of the electroplaques, the voltage adds up as when we connect a number of batteries in series. With several thousand electroplaques arranged in this way, the electric eel can reach several hundred volts. (Due to internal losses the voltage does not reach the full magnitude of an ideal serial connection of the electroplaques.)

The arrangement of electroplaques in parallel columns increases the current flow. In the electric eel the current may be about one amp. The giant electric ray (*Torpedo*) lives in sea water, which has a much lower resistance than fresh water. Its voltage is lower (some 50 volts), but since it has some 2000 columns in parallel in each electric organ, the total current is extraordinarily high, up to 50 amperes. A current of 50 amperes at 50 volts corresponds to a power of 2500 watts during the discharge (Keynes and Martins-Ferreira, 1953).

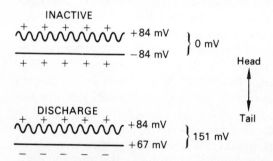

Fig. 12.6. The electric organs of fish consist of thin, wafer-like cells (electroplaques) which are stacked in columns of several thousand. When a plaque is inactive and at rest (top), the outside of both faces is positively charged relative to the inside, the outsides being at +84 mV. Therefore no potential difference exists between the two outside faces. During discharge (bottom) the potential on the posterior face of the plaque is reversed and becomes −67 mV on the outside. The potential difference between the two outside faces therefore is 84 + 67 mV = 151 mV (after Keynes and Martins-Ferreira, 1953).

Electroreceptors. Most fish have a large number of sensory organs in the skin, mostly around the head and along the lateral line. One kind is known as the ampullae of Lorenzini, which have a characteristic small flask-shaped appearance. They are found on the head of all marine elasmobranchs (see fig. 12.7), and similar organs are found in a variety of teleosts. Appropriate nerve sections and behavior experiments have shown that the ampullae of Lorenzini of the dogfish (*Scyliorhinus*) are the electroreceptors (Dijkgraaf and Kalmijn, 1963). The electroreceptors of electric fish are of several different types, the commonest has a duct leading from the surface of the skin to an enlarged ampulla situated deeper in the skin (fig. 12.8).

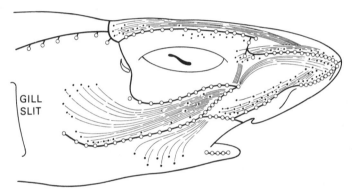

Fig. 12.7. Location of sense organs on the head of the dogfish *Scyliorhinus*. Openings of the ampullae of Lorenzini shown by black dots. The open circles show the pores of the lateral line system and the black lines its location (Dijkgraff and Kalmijn, 1963).

The electric reception has been studied in detail in the African electric fish *Gymnarchus niloticus* (Machin and Lissmann, 1960). This fish, which lacks a common name, discharges a continuous stream of pulses at a frequency of about 300 to 400 per second. During each discharge the tip of the tail is momentarily negative with reference to the head, so that an electric current flows into the surrounding water. The configuration of the electric field depends on the conductivity of the surroundings, and is distorted if an object with higher or lower conductivity than the water is introduced into the field. The lines of current flow will converge on an object of higher conductivity than water, and diverge around a poorer conductor.

The sense organs are sensitive to such changes in the field. However, they do not respond to single pulses, but to the average current over a time period of about 25 ms, a period sufficient for the emis-

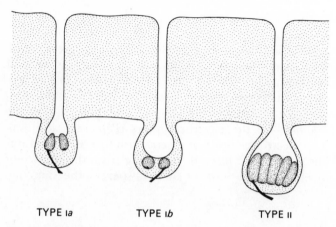

TYPE ia   TYPE ib   TYPE ii

Fig. 12.8. Diagram of the different types of cutaneous ampullary receptors of an electric fish (*Eigenmannia virescens*). The receptor cells (shown stippled, with their nerves in black) are located at the bottom of a long canal connecting to the surface of the skin (Lissmann and Mullinger, 1968).

sion of some seven to ten discharges. The total current needed to excite an individual electric receptor is extremely small, about 3×10^{-15} amperes (Machin and Lissmann, 1960).

Many electric fish live in murky water where visibility is poor, and they often have poorly developed eyes. The value of an electric sense that permits scanning of the environment and location of prey is therefore obvious, and it has the advantage above vision that the system is independent of the day-and-night light cycle. A disadvantage, compared to light, is the limited range. Another question is that of interference, for if many individual electric fish live in the same area, how can they avoid confusing their own signals with those of others? If an electric fish is subjected to an artificial electric pulsing of the same frequency as its own, it responds by shifting its frequency away from the artificial signal. This presumably can prevent the fish from confusing its signals with those originating from other fish. This conclusion is supported by the tendency of electric fish in nature to shift and find their own individual private frequencies when approached by other fish with the same frequency as their own (Bullock, Hamstra and Scheich, 1972).

Magnetic sense

It can no longer be doubted that many animals respond to magnetic fields, although innumerable experiments with artificial magnets have failed to yield clear results. However, the fact that a magnetic field has some effect on an animal does not mean that, in nature, the animal senses a magnetic field or makes any use of it.

Apparently one reason that many experiments have failed to show an effect is that experimenters have used very strong magnetic fields. Fields of the same magnitude as the natural magnetic field of the earth systematically influence the direction in which mud snails (*Nassarius*) and planarians (*Dugesia*) move. When the strength of the artificial magnetic field is decreased, response drops off, but more importantly, if it is increased substantially, response also becomes minimal. It seems that the magnetic response (compass orientation) is exquisitely adjusted to the strength of the natural geomagnetism, and this may explain the lack of success in so many earlier experiments (Brown, Barnwell and Webb, 1964).

Many other invertebrates also react to weak magnetic fields, but the meaning of their responses is often difficult to establish. We can assume that, if the response is more than incidental, an influence on direction and orientation is important, as has been established for honey bees. When bees communicate to other worker bees the direction of a food source, the 'waggle dance' they perform on a vertical honeycomb is affected by the earth's magnetic field (Lindauer and Martin, 1968).

It seems that in the orientation of birds, the role of a magnetic sense is beyond doubt. When pigeons are homing from unknown sites, they use a number of different cues to obtain navigational information, one of them being the sun which is used to establish compass direction. Even on overcast days, however, pigeons can carry out successful homing, but if they carry small magnets, they become disoriented, whereas control birds which carry a similar but non-magnetic piece of metal show no disorienting effects. It seems that magnetic cues are only one of several kinds of information used by homing pigeons, and therefore do not fully explain the bird's excellent orientation abilities (Keeton, 1971).

In laboratory experiments, where single environmental cues can be better isolated and controlled, the effect of magnetic fields can be studied more accurately. The European robin (*Erithacus rubecula*) displays a period of migratory restlessness in the spring and again in the fall. If it is placed in a circular cage it will tend to move towards the direction which would be its natural migratory direction for that time of the year. This preferred direction is maintained in completely closed rooms without any optical reference points or other known cues that would make the birds orient correctly. However, if the test cage is moved to an all-steel chamber, which provides a complete shield to magnetic fields, the birds can no longer find the natural migratory direction (Fromme, 1961). Furthermore, artificially generated magnetic fields influence the directional choice made by the birds. It appears that the magnetic 'compass' of these birds is not sensitive to the polarity of the field, but that the inclination of the axial direction of the field lines is used as a cue for deriving information about north–south polarity (Wiltschko and Wiltschko, 1972).

The question of what organ or structure responds to magnetic fields is completely obscure. No information is available about any sensory organs that could be responsible for a magnetic sense, and there are no convincing hypotheses about mechanisms that could be involved.

Sound and hearing

In trying to define sound, we easily end up with a circular definition. We know that sound consists of regular compression waves which we perceive with the ears, and in turn, an ear can be defined as an organ sensitive to sound. Since compression waves can be transmitted in air, in water, and in solids, we obviously cannot restrict the word 'hearing' to perception in air only.

For human hearing, we meet no great conceptual difficulties. Our ears are sensitive to regular compression waves in the air in a range from approximately 40 to 20 000 hertz (cycles per second). However, a dog can perceive higher frequencies which are completely inaudible to humans, up to 30 000 or 40 000 Hz. Bats can perceive frequencies as high as 100 000 Hz, and we still consider it hearing. There are two good reasons, one is that dogs and bats have ears that are very similar to ours, and another is that the frequency range that is perceived is continuous; our own ear is just designed so that it is insensitive to the very high frequencies that many other animals can hear.

On the other hand, we know that compression waves at frequencies below those which we perceive as sound can often be perceived as a vibration, i.e. we still sense frequencies lower than those which we consider 'hearing'.

When a fish senses footsteps at the edge of the water, is this hearing? The vibration from the steps is transmitted through ground and water, and these low-frequency vibrations are probably perceived by sensory organs in the lateral line, rather than in the ear. Is it then correct to say that the fish 'hears' or 'feels' the approaching steps? In general, we tend to speak about 'hearing' when animals have specially developed organs which are sensitive to what we consider to be sound, but this is obviously not a good way to define hearing.

The question of how to define hearing becomes much more difficult for invertebrate animals, in which the organs responsible for perception of vibration are very different from the vertebrate ear, and in many cases have not been identified at all.

What kinds of information can be obtained from sound waves? Man can perceive sound waves and their temporal pattern, determine intensity, distinguish different frequencies (pitch), carry out very complex frequency analysis, and he is also able to determine the direction from which sound comes. In addition, some animals can use self-generated sound for obtaining detailed information about the physical structure of the environment.

The complexity of frequency analysis of which man is capable is astonishing if we consider the fact that what he hears basically is nothing more than a minute in-and-out motion of the ear drum. In spite of this fundamental physical simplicity, the average musical ear can distinguish individual instruments in an orchestra with a large number of simultaneously performing instruments.

Insects can produce and perceive a large variety of sounds, which are used extensively for communication, often with the opposite sex. The hearing organs of insects vary greatly in structure and may be located in various parts of the body. Apparently, insect hearing organs are not sensitive to pitch the way our ear is; instead information is transmitted mainly as changes in intensity, duration, and pattern of the sound. Records of sound produced by male insects to attract females can be played back with so much frequency distortion that they are unrecognizable to the human ear, and still attract the females. The meaningful information seems to be carried in the pattern of pulses and not in pitch and tone quality. The pattern of insect sound is relatively fixed and is highly species specific. The human ear is inherently unsuited for perceiving the important features of insect sound, but as electronic equipment for recording and analysis has become available, the pattern of insect song can be transcribed to visual patterns which we understand better and can evaluate more intelligently.

Sound location and sonar. Determining the direction of sound depends mainly on having two ears, separated in space. Some information is derived from the fact that the signals arriving at the two ears often differ in intensity, but more important is the difference in time of arrival of the sound from a given source.

For a complex signal the interpretation of direction may be

brought about by a simultaneous comparison of intensity, frequency spectrum, phase, and time of arrival of the signals at the two ears. The comparisons must be made by the central nervous system, which must also be capable of separating the important signal from other noise in the environment (Erulkar, 1972). For man the ability to locate sound is much less precise than for predators that normally work in the dark, such as cats or owls. The barn owl, for example, can locate prey in total darkness, using only the sense of hearing, with an error of less than 1° in both the vertical and the horizontal planes. For this accurate location of the source of a sound, they depend on frequencies above 5000 Hz, which are more directional than lower frequencies (Payne, 1971).

The owls use sounds coming from the prey to determine its location, but some vertebrates obtain information from the faint reflections or echoes of sound they themselves produce. This is known as *echo-location* or *animal sonar,* and is particularly well developed in bats, but is found also in whales and dolphins, in shrews, and a few birds.

It has long been known that bats can fly unhindered around in a completely darkened room, and it makes no difference whether or not their poorly developed eyes are covered.

One of the best studied bats is the Big Brown bat (*Eptesicus fuscus*). As this bat cruises around in the dark, it emits pulses of sound which are inaudible to man, for the frequency is between some 25 000 and 50 000 Hz. Each pulse or click lasts about 10 to 15 ms, and may be given off at a rate of some 5 single clicks per second. The bat obtains information about the environment from the echoes, and is able to avoid all sorts of obstacles and can even fly between wires strung across the room. If the bat approaches a particularly 'interesting' object such as a flying insect, the number of clicks increases and the duration of the single click is shortened, so that the rate may be as high as 200 clicks per second, each click lasting less than one millisecond.

One advantage of using a high frequency sound (short wavelength) is that the directional precision is much better than for low frequency sound, which spreads too widely and gives diffuse reflections unsuited for pinpointing accurately the location of objects. Furthermore, in order to reflect sound waves an object must be above a certain size, relative to the wavelength. The shorter the wavelength, the smaller the size of object that gives a distinct reflection, and the highest frequencies therefore permit the detection of the smallest objects.

The impressive skill at avoiding fine wires can be tested in a long room where the bats can fly back and forth. If the center of the

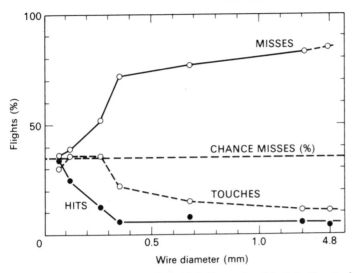

Fig. 12.9. Obstacle avoidance of the Little Brown bat (*Myotis lucifugus*) when flying in darkness through a row of vertical wires spaced 30 cm apart. The statistical chance of missing the wires was 35%. The ability to avoid the wires deteriorated when the diameter of the wires was decreased to less than 0.3 mm (from Griffin, D R. 1958. *Listening in the Dark:* Yale University Press, Fig. 6a).

room is 'divided' by vertical wires, spaced 30 cm apart, a bat flying without information about the wires, has a 35% random chance of missing a wire. Wires of several millimeters diameter are successfully avoided, but if the diameter of the wire is reduced to less than 0.3 mm, there are more hits, and with 0.07 mm wire the average number of hits approaches the chance value of 35%, the same as in experiments with bats with plugged ears (fig. 12.9).

The shortest wavelength used by the Little Brown bat, which was used in these experiments, is about 3 mm (100 000 Hz), which means that it detects objects that are no more than about 0.1 wavelength in diameter.

Acoustic orientation is, of course, particularly useful for animals that are active in the dark, and it is also of importance in deep or murky waters. Both dolphins and whales use acoustic echoes to avoid colliding with objects and with the ocean bottom, as well as for finding food. When they swim in open water, there are few obstacles, but at night and at depths where little or no light penetrates, their sonar helps them avoid collision with the bottom. It is, of course, very valuable to be able to locate food both in the dark and when the visibility is low.

How dolphins use acoustic information has been tested in captive animals. A trained dolphin can find a dead fish thrown into the water and reach it within seconds from the opposite end of the tank, even if the water is so murky that a man cannot see the fish at all. If a panel of clear plastic is used to block their path, they can immediately find an opening which they swim through without hesitation (Kellogg, 1958; Norris, Prescott, Asa-Dorian and Perkins, 1961).

Other mammals that use acoustic orientation or echo-location include the shrews, which employ sound pulses of short duration and high frequency (up to over 50 000 Hz) (Gould, Negus and Novick, 1964).

At least two species of birds use echo-location, the oilbirds (*Steatornis*) of South America and the cave swiftlets (*Colocalia*) of South-East Asia. They are not closely related, but both live and nest in deep caves. The best known is the oil bird or guacharo, which can fly around freely in dark caves without hitting the walls or other obstacles. They use a sonar system, much like bats, except that their sounds have a frequency of only about 7000 Hz. Therefore they are fully audible to man, and their sound has been described as something like the ticking of a typewriter. If the bird's ears are plugged, they lose their sense of orientation in the dark, but they can still fly about in a lighted room and evidently use their eyes for orientation (Griffin, 1953; 1954).

Other mechanical senses

We have seen that vision and hearing can provide a great deal of detailed information about the environment, hearing in fact far more than we initially would think. There is no doubt that those animals that use echo-location obtain a three-dimensional impression of their immediate environment, somewhat analogous to that which humans obtain through vision, except that the range is more limited.

A number of other senses provide additional information, but mostly of a less comprehensive nature. This includes a variety of mechanical senses about which we, in many cases, have only the most rudimentary information.

Sensitivity to *touch,* which actually discovers a deformation of the skin surface, is widespread and gives information about objects in immediate contact with the body. The sensation of vibrations is of a similar nature. The lateral line of fish also belongs in the same category. The lateral line is, in fact, a highly complex organ that probably is used for a multiplicity of purposes, among others to sense the rate of flow of water over the fish, thus giving information about speed (Cahn, 1967).

Sensitivity to *hydrostatic pressure* has been demonstrated in a variety of aquatic invertebrates. Some are quite sensitive to small pressure changes, but the mechanism that is involved remains obscure. It is easy to understand how an animal that carries a volume of gas can be very sensitive to pressure changes, for the resulting distortion can easily be sensed with the aid of mechanoreceptors. However, the majority of plankton organisms that are pressure sensitive carry no bodies of air that can explain their high sensitivity to small pressure changes (Enright, 1963).

Sensitivity to the force of *gravity* is extremely widespread, and gives information about what is up and down. More complex animals have gravity receptors, known as *statocysts*, which are uniformly built on the same principle (see fig. 12.10). A heavy body, the statolith, often of calcium carbonate, rests on sensory hairs. Any displacement relative to the force of gravity causes the statolith to stimulate the ciliary bundles of other sensory cells, which then send information to the central nervous system.

Angular acceleration is sensed in a similar way; fluid within a suitable structure is set in motion and deflects the ciliary cells in a direction that depends on the direction of acceleration. In vertebrates the organ responsible for sensation of angular acceleration is located in the inner ear, and consists of the three semi-circular canals, oriented

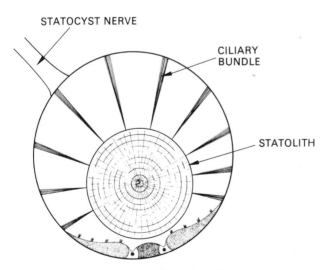

STATOCYST NERVE

CILIARY BUNDLE

STATOLITH

Fig. 12.10. The statocyst of a pelagic snail. Within the statocyst, a heavy body of calcium carbonate, the *statolith,* rests on sensitive sensory cells. These respond to changes in the position of the statolith as the animal changes its position in the field of gravity (Tschachotin, 1908).

in three planes at right angles to each other. They are filled with fluid, and any angular movement causes the inertia of the fluid to stimulate the sensory cells in the plane of the movement. With the canals oriented in three different planes, any angular acceleration can be perceived.

Chemical senses – taste and smell

To a human the sense of taste refers to materials in contact with the mouth, while smell refers to gaseous substances that reach the nose via the air. The separation is highly subjective, for the sense of taste, located in the mouth, separates only the four basic qualities of salt, sweet, sour, and bitter. When we speak about the 'taste' of food, most of the information is obtained through the nose.* When we extend the use of the terms taste and smell to other air-breathing vertebrates, the analogy is justified, for their sense organs have a great deal of structural similarity to ours.

For aquatic animals a distinction between taste and smell rapidly becomes meaningless. For example, some fish have chemical senses of unbelievable acuity, but should we say that they 'smell' or 'taste' the water? In common language it is often convenient to use the term smell for objects at a distance, and taste for materials in direct contact with the animal, in particular with the mouth. However, a catfish has sensitive chemoreceptors located all over the body, and we find it natural to call these receptors tastebuds (see fig. 12.11).

The chemical senses cannot be used for scanning the environment, the way that light and sound are used, for the chemical compounds to be sensed must first arrive at the animal's sensory organ, either by diffusion or by mass movement of the medium. Light has, for physiological purposes, the quality of instantaneous arrival, and sound arrives very rapidly, but with such a delay that the delay itself carries valuable information (as in directional hearing and in echolocation). The limitations on the use of the chemical senses are much more severe, and detailed information is usually obtained only over extended periods of time, often in combination with the animal moving about.

The most useful information obtained with the chemical senses concerns the finding of food, or mates, and in locating enemies. They must be considered primarily as useful alerting or warning devices, so that the efforts of other senses can be concentrated.

---

* The texture of the food is also an important quality in 'taste', and if familiar foods have been homogenized they often are very difficult to recognize.

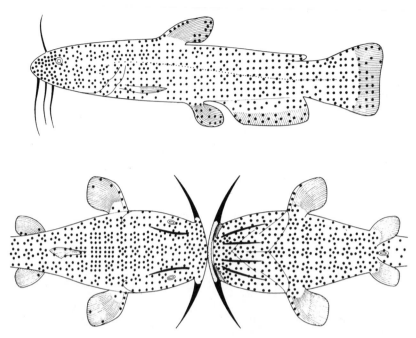

Fig. 12.11. The catfish has 'tastebuds' distributed all over the body. Each black dot represents 100 tastebuds. Higher concentrations on lips and barbels are indicated by small dots and solid black respectively. Top: side view; bottom: dorsal (left) and ventral (right) view (Atema, 1971).

Some kind of chemical sense seems to be universally present among animals, even very simple organisms and protozoans respond to chemical stimuli. Obviously, it is of interest to an amoeba whether or not a particle is digestible and thus should be ingested. However, discrete chemical sensory organs do not occur in the simplest animals. Their highest development is reached in vertebrates and in arthropods.

The presence of a chemical sense in less highly organized animals is easily demonstrated by observation of their behavior, but when it comes to an exact study of the physiological events in the chemoreceptor organs, virtually all our information has been obtained from arthropods and vertebrates. The reason is that the chemosensitive organs in these animals form discrete structures which, especially in insects, are easily accessible and eminently well suited for electrophysiological studies.

Insect taste and olfaction. Let us separate taste and olfaction according to the medium that carries the chemical stimulus, and say that the transfer medium for taste is water and for olfaction is air. We can then say that taste and olfactory receptors in the insect are distinctly separate structures. They can be located almost anywhere, on the mouth parts, the antennae, the feet, etc. The taste organs are usually bristles or hairs, a fraction of a millimeter long, open at the tip, and with one or more sensory neurons at the base. The olfactory organs are of a variety of types, they may be hair- or bristle-like, but other types consist of thin-walled pegs or pits in a wide variety of forms and shapes.

The taste organs usually require the stimulating molecules to be in much higher concentrations than the olfactory receptors. The latter in some cases are unbelievably sensitive to specific odoriferous molecules which are important for locating food, prey, and mates.

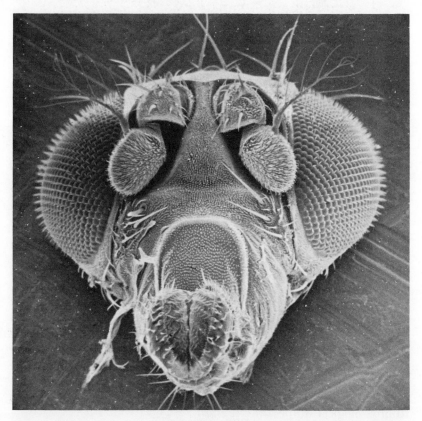

Plate 15. The head of a fruit-fly (*Drosophila*) carries numerous sensory hairs. The large bulging areas on each side of the head are the two compound eyes. The mouth parts, with a number of sensory hairs attached, are seen at the bottom of the photograph (R. Falk, Hebrew University, Jerusalem).

Substances used to find other individuals of the same species, as well as for a variety of social communication activities in colonial insects, are called *pheromones* (Greek *pherein* = to carry). Such pheromones constitute a chemical language which is of great importance in laying trails, recognizing individuals from the same nest, or marking the location of a food source, as alarm substances, and so on.

The sensory hairs on the proboscis of the ordinary blowfly (*Phormia regina*) have turned out to be an excellent experimental model. Severing the head from the fly and attaching it to a fine micropipette filled with saline provides a way of mounting it, as well as an electrical connection to the inside. By penetrating the wall of a sensory hair, at some distance from the tip, with a microelectrode, the electrical activity in the hair can now be recorded. A minute glass tube, filled with a solution to be tested, can then with the aid of a micromanipulator be brought into contact with the open tip of the hair, and if the solution contains a suitable stimulant, it causes electric activity which the microelectrode records (Morita and Yamashita, 1959).

A typical taste hair (see fig. 12.12) may have five sensory neurons at its base. One of these is always a mechanoreceptor, sensitive to deflection of the hair, and of the others one is usually sensitive to sugar, one to water, and one or two to salts and various other com-

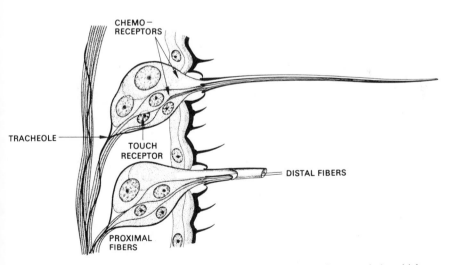

Fig. 12.12. The proboscis of the blowfly carries a large number of sensory hairs which can sense various chemicals and thus are taste receptors. At the base of the hollow hair is a number of sensory neurons, each more or less specialized in the range of compounds it responds to, and one of them always being a touch receptor.

pounds (Hodgson, 1961). By direct recording from the taste hair, it is possible to determine which substances stimulate the receptor, and what concentration is necessary.

In contrast to the taste receptors, which are sensitive to a very small range of compounds, the olfactory receptors of insects are highly sensitive and respond to a wide variety of stimuli. They are more complex than the taste receptors and often contain a large number of sensory neurons.

The olfactory receptors are often, but not always, located on the antennae. The ability of certain male moths to locate females of the same species, even at distances of several kilometers, is well known. The female gives off a sex attractant, and the male is equipped with giant antennae which have a primary role in sensing the sex attractant in immensely dilute concentrations (see fig. 12.13). In the male polyphemus moth (*Telea polyphemus*) each antenna carries about 70 000 sensory organs with about 150 000 sensory cells. The female, in contrast, has much smaller antennae with only about 14 000 sensory organs and 35 000 sensory cells (Boeckh, Kaissling and Schneider, 1960). Approximately two-thirds of the receptor cells of

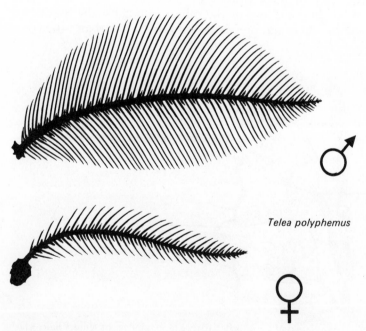

Telea polyphemus

Fig. 12.13. The large antennae of the male polyphemus moth are sensory organs which are highly sensitive to the sex attractant of the female moth. Their shape serves to attain contact with the largest possible volume of air.

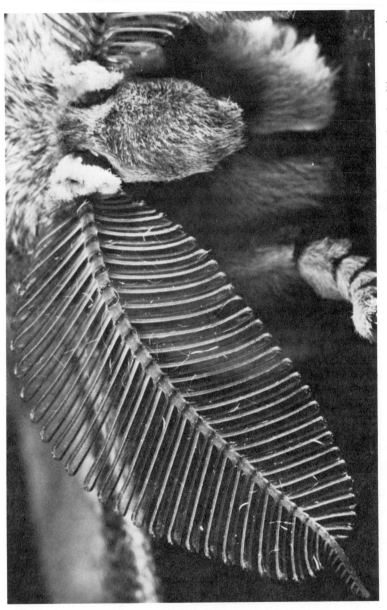

Plate 16.   The large antennae of the male Antherae moth (*Antheraea pernyi*) are used in locating the female of the same species by means of the sex attractant she gives off (Muriel V. Williams, Harvard University).

the male are specialized for sensing the female sex attractant. The shape of the antennae makes them serve as a sieve, almost combing through the air for the appropriate molecules. With the aid of radioactively labelled sex attractant it has been demonstrated that more than 25% of the sex attractant molecules from the air streaming through the antennae are absorbed (Schneider, Kasang and Kaissling, 1968). Experiments with male silk moths (*Bombyx*) have provided an estimate of how many molecules are necessary to cause a reaction in the moth. If the air contains 10 000 molecules of the sex attractant per cm^3 and flows at a speed of 60 cm s^{-1} for a total time of two seconds, the male responds. The number of molecules that reaches the antennae corresponds approximately to a level where each of the 10 000 receptors have received a maximum of one molecule of the attractant during the two seconds (Schneider, 1969).

A male moth that senses a female from a distance of several kilometers cannot possibly use concentration gradients in the air to discover the direction in which he can find the female. The typical behavior of the male is to take off up-wind, and this gets him started in the right direction. He then follows the aerial trail of the attractant in a characteristic zig-zag flight across it. Animals that follow an odoriferous trail commonly do this; when they pass out of the trail and lose it, they tend to turn in the reverse direction, so that they enter the trail again. Leaving the trail thus automatically brings them back in again, so that their path becomes a sinuous crossing in and out of the trail. When they have reached the immediate vicinity of their goal, the final localization can take place with the aid of concentration gradients.

The localization of the source of a chemical signal at a long distance (such as the sex attractant) requires a long-lasting relatively stable molecule. For alarm substances, which should act briefly and in a localized area, the molecule should be relatively unstable to avoid long-lasting residual effects that would be confusing. The steep gradients in the vicinity of the source which result when the odoriferous substance is unstable thus are more easily pinpointed.

Transduction and transmission of information

Sensory information, such as light, sound, etc., must be transformed into nerve impulses, and these must be conveyed to the central nervous system which coordinates and uses the information in directing appropriate responses. In the preceding section we were concerned with the nature of the information and how it is received by the sensory organs. We shall now examine the effect of the stimulus on the receptor organs, how the stimulus is transduced into nerve signals, and in what form the information is transmitted to the central nervous system.

To understand this, it is necessary to build on an elementary knowledge of the nature of nerve impulses and how nerve cells function. In the next chapter we shall return to this subject and examine in detail how impulses are communicated from one nerve cell to another.

Nerve cells, membrane potential, and nerve impulses.

All nervous systems consist of a large number of single nerve cells, *neurons*. Neurons can have a wide variety of shapes and sizes, but they have certain important features in common (fig. 12.14). There

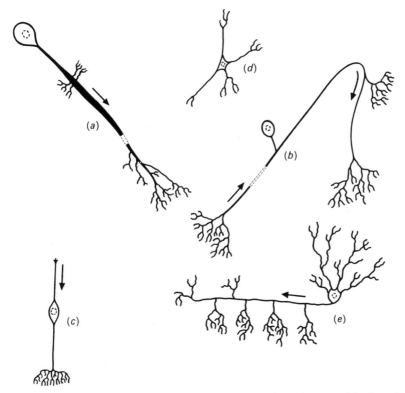

Fig. 12.14. Neurons can have various forms, but they always have a cell body and a number of long extensions. These schematic drawings represent (*a*) an arthropod motorneuron; (*b*) a mammalian spinal sensory neuron; (*c*) a bipolar neuron from the vertebrate retina; (*d*) a neuron from the nerve net of a coelenterate; and (*e*) a basket cell from the mammalian cerebellum (Aidley, 1971).

is a cell body which contains the nucleus, and a large number of thin fibers extending from it. Each neuron usually has a single long fiber, the *axon,* which in a large animal may be several meters long, and a large number of shorter fibers, *dendrites,* which are heavily branched and for the most part are less than 1 mm long. In vertebrates the cell body of the neuron is usually quite small, often less than 0.1 mm in diameter and the fibers less than 0.01 mm thick. The entire cell, including all the fibers, is surrounded by a thin membrane, the *nerve membrane.* A complex nervous system contains an immense number of neurons. In man, for example, the brain alone contains around 10 000 000 000 cells.

The long fibers, the axons, are the main conductance lines in the body. What is commonly known as a nerve, or nerve trunk, consists of hundreds or thousands of axons, each originating from a different neuron. A nerve has no cell bodies in it; these are found in the central nervous system, in special aggregations known as ganglia, and in the sensory organs.

The points where nerve cells and their extensions make contact with other nerve cells are called *synapses.* A single nerve cell may, through the synapses, be connected to hundreds and hundreds of other neurons. The most important feature of the synapse is that it functions as a one-way valve. Transmission of an impulse can take place in one direction only, from the axon to the next cell, and not in the opposite direction (for details, see p. 629). As a result, conduction in an axon anywhere in the nervous system takes place in one direction only. An axon in itself is perfectly capable of conducting impulses in either direction, but in the integrated nervous system all conductance in any given axon is always in the same direction. In a nerve trunk, however, some axons may conduct in one direction, and others in the opposite direction. For example, in a nerve to a given muscle, the impulses stimulating the muscle to contract are transmitted from the central nervous system to the muscle, and at the same time sensory information from the muscle is carried in other axons back to the central nervous system.

The nature of nerve impulses. The normal neuron, including its axon, shows a potential difference between the inside and the outside of the nerve membrane. In the inactive or resting neuron this membrane potential is known as the *resting potential.* The potential is due to an uneven distribution of ions on the two sides, caused primarily by the active extrusion (pumping) of sodium ions out of the cell, making the outside positively charged. In nearly all neurons from all sorts of animals the difference between the outside and the inside of the membrane is of the same magnitude, some 60 to 90 mV, with the inside negative relative to the outside.

Activity in an axon is most easily discovered as a change in the membrane potential. Such activity can be induced by applying a suitable *stimulus,* and in experimental work we almost universally use a brief pulse of electric current for this purpose. Electricity is particularly suitable because it can be applied at any desired moment in time, in precisely the desired amount, and for extremely short periods of time. Many other stimuli could be used to induce nerve activity, various chemicals, for example, or just pinching the nerve, cause activity. However, such stimuli are more damaging than electricity and are less suitable because the moment of application, strength, duration, and instantaneous removal are not easily controlled.

If we apply a suitable stimulus to an axon, we obtain a record of the change in the membrane potential similar to that shown in fig. 12.15. An electrode placed inside an axon at rest may show a potential of, say, −70 mV relative to the outside. The stimulus causes a near-instantaneous change in the permeability of the membrane.

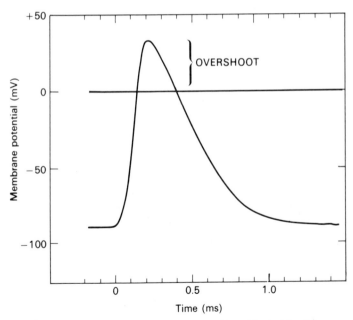

Fig. 12.15. Action potential in a single nerve fiber. The inside of the nerve membrane at rest has a negative potential, relative to the outside, of some − 70 to − 90 mV (the resting membrane potential). During the passage of an impulse, the membrane potential is momentarily abolished, then for a moment reversed (the overshoot), whereupon it rapidly returns to the initial resting potential.

The membrane is suddenly permeable to sodium ions, and because of the high sodium concentration outside, sodium rapidly diffuses in. As a result of the sudden flow of ions, there is a momentary abolishment of the membrane potential (*depolarization*), or even a reversal (*overshoot*) which for an instant makes the inside slightly positive. The membrane also becomes permeable to potassium, which diffuses out of the cell, but at a slightly delayed rate. Then the membrane again becomes impermeable, and the potential rapidly returns to the normal resting potential. The entire event, which is known as an *action potential*, takes place very rapidly, in some cases in a fraction of a millisecond.

A very weak stimulus does not result in an action potential. For this to happen, the stimulus must have a certain strength, referred to as the *threshold* strength. Below this threshold there is no action potential, but if the stimulus exceeds the threshold value, an action potential results. The size of the action potential, however, is not influenced by the magnitude of the simulus. Let us assume that we double, or triple, the strength of the stimulus above the threshold; the resulting action potential remains the same. The reason for this is that the action potential is due to the concentration of ions on the two sides of the membrane. Therefore, the stimulus either causes a full-strength action potential, or none at all. In physiology, this type of response is called an *all-or-none response*.

Once the axon membrane has been depolarized and an action potential generated, this acts as a stimulus for depolarization of the adjacent portions of the membrane. Depolarization therefore spreads and travels as a rapidly propagated action potential along the axon. Since the action potential is due to the local concentrations of ions, it spreads along the axon without change in magnitude. This is called *conduction without decrement,* and is a fundamental characteristic of the axon, different from conduction in, for example, an electric conductor.

If an action potential is always of the same magnitude and uninfluenced by the strength of the stimulus, how can an axon then convey information about the strength of a stimulus, information that is obviously of the greatest interest?

The answer to this question is simple, a change in frequency of the action potentials in the axon can be used as an indicator of stimulus strength. This is illustrated in fig. 12.16, which shows that a pressure sensitive skin receptor cell in the finger of a man responds to an increased stimulus strength with increasing frequency of the action potentials in its axon. This particular receptor had a threshold of about 0.5 g, and therefore showed no response to a force 0.2 g. A

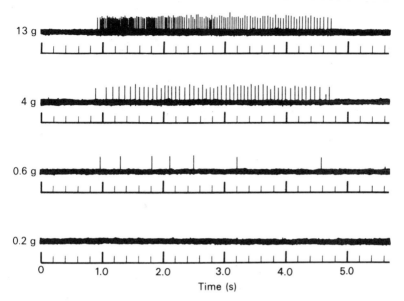

Fig. 12.16. Impulses recorded from a single pressure-sensitive receptor on the human hand. A plastic rod of 1 mm diameter was pressed against the skin with the force indicated on each record. This receptor did not respond to temperature changes, i.e. it was a specific mechanoreceptor (from Hensel and Boman. 1960. *J. Neurophysiol.*, **23**, 564–68, Fig. 2).

force of 0.6 g, however, gave a clear response, which increased in frequency with the application of increasing forces. Thus, we can say that the magnitude of a sensory stimulus is coded and transmitted as a frequency-modulated signal.

We have now established some basic principles in sensory physiology. Information is conveyed in the sensory nerves as action potentials, the action potentials in all sensory nerves are of the same nature, the magnitude of the action potentials is constant and uninfluenced by stimulus intensity, and finally, information about stimulus intensity is coded as a frequency modulation of the action potentials. We can now proceed to the question of how sensory information is further processed.

Sorting and processing of sensory information

If all information available to the sensory organs were transmitted to the central nervous system, the mass of signals would be formidable and probably utterly unmanageable. However, a great deal of screening, filtering, and processing takes place before the signals are passed on, beginning at the sensory neuron, and continuing at several levels on the way to the brain. The filtering networks pass on only selected portions of the information they receive, and furthermore, they carry out certain steps of processing which improve on the information that is transmitted to higher levels. We shall discuss these principles using the processing of visual information as an example.

In spite of the complexity of the visual system, the processing of visual signals is better understood than that of other complex senses. There are several reasons, one being that artificial stimuli (light) can be directed, timed, and quantified with great accuracy. Another reason is the use of the arthropod eye as experimental material, because the structure of the compound eye permits ready access to single sensory receptor units. However, even the complex signal processing in the vertebrate eye is amazingly well understood.

Lateral inhibition

The horseshoe crab, *Limulus,* has a compound eye in which individual receptor units can readily be stimulated by a fine light-beam.

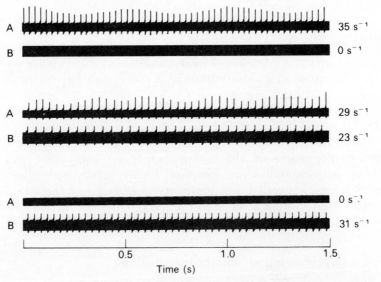

Fig. 12.17. Action potentials recorded simultaneously from two adjacent ommatidia, A and B, in the compound eye of the horseshoe crab *Limulus.* The top record, illumination of ommatidium A alone. Discharge rate indicated at the right of record. Center record, illumination of A and B, resulting in lateral inhibition (see text). Bottom record, illumination of B alone (from Hartline, H. K. and Ratliff, F. 1957. *J. Gen. Physiol.,* **40,** 357–76, Fig. 1).

It is also possible to record the impulses from that fiber of the optic nerve which connects to the particular receptor in question. It turns out that the signals in the axon do not completely represent the stimulus, but that the pattern also depends on the amount of light falling on other receptor units. This is because each visual receptor is connected to its neighbors and inhibits their activity. This characteristic, known as *lateral inhibition* has the effect of enhancing the contrast between the amounts of light falling on two adjacent receptors.

This may need some further explanation. Fig. 12.17 shows the records from two adjacent receptor units of the *Limulus* eye. The top record shows the even discharges when a single receptor, A, is stimulated. The rate of discharge was 35 impulses per second. When the adjacent receptor, B, was also stimulated, the discharge rate in unit A decreased to 29. Unit B was likewise influenced by its neighbor, for when the stimulus was removed from receptor A, the discharge rate from receptor B alone increased from 23 to 31 impulses per second. Thus, there is a mutual lateral inhibition between these two units (Hartline and Ratliff, 1957).

The effect of lateral inhibition can be understood from the diagram in fig. 12.18. Assume that we consider 12 units which are stimulated by uniform light of two intensities, there being a sharp transition between units 6 and 7. In this system, units 2, 3, 4, and 5 are all bordered by units receiving bright illumination, and are therefore subject to lateral inhibition. Unit 6, however, is not subjected to the same degree of lateral inhibition from unit 7, which is within the dimly stimulated zone. As a result, its discharge frequency will be higher. Next, consider the dimly lit units, of which receptors 8, 9, 10, and 11 receive the same small amount of lateral inhibition. Unit 7, however, is inhibited by its brightly lit neighbor, receptor 6, and therefore discharges at a lower frequency than the other dimly lit receptors. The over-all effect of this is that the transition between bright and dimly lit receptors is emphasized. As a result, the messages in the optic nerve give a correct picture of the edge, but with an emphasized contrast between the two zones.

Another aspect of visual reception should also be emphasized before we move on to the vertebrate eye. It has been observed in the scallop (*Pecten*), which has well developed image-forming eyes along the edge of the mantle, that the retina has two layers; one of which, as expected, responds to light as a normal stimulus. The other layer of the retina, however, behaves in a very different way, it does not respond to increased light but is sensitive only to a decrease in the intensity of illumination. This phenomenon is common in the eye of vertebrates, and we shall return to it shortly. The importance of this phenomenon can easily be imagined. When a shadow suddenly falls

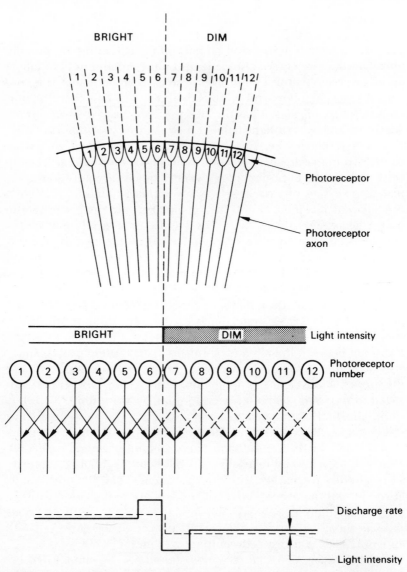

Fig. 12.18. Diagram showing that lateral inhibition causes an enhanced edge effect. For details, see text.

on an animal, it often means the approach of a predator, and in this case, a decrease in stimulus intensity is far more important to the animal than any other information.

We are now familiar with two relatively simple but important aspects of visual systems, the amplification of edge effects and the importance of dark areas in the visual field.

Information processing

The frog's eye. The frog's retina has no central fovea,* and rods and cones are uniformly distributed so that the structure of the retina is much the same from place to place. The photoreceptors connect to several kinds of neurons which are located within the retina at various levels. We shall be concerned only with those neurons which are known as *ganglion cells,* and whose axons make up the optic nerve. There are about half a million of these ganglion cells, a number which corresponds approximately to the number of fibers in the optic nerve (Maturana, 1959). However, there are more than a million receptor cells (rods and cones) in the retina. This makes it clear that the optic nerve cannot carry a complete point-by-point picture of the image that falls on the retina; the analogy between the retina and the photographic film in a camera is not fully valid, for there must be some discrimination or processing taking place before the signals are sent on to the brain.

In the optic nerve there are five different kinds of fibers which originate from the ganglion cells. They have been clearly identified because they respond differently to specific kinds of stimulation of the retina. Some fibers respond only to the onset of illumination; these are therefore called '*on fibers*'. Other fibers respond only to the termination of a light stimulus, and are therefore called '*off fibers*'. A third kind of fiber responds to either onset of a light stimulus or its termination; these are known as '*on–off fibers*'. Their response to the movement over the retina of a linear shape is marked, and they could therefore also be called 'moving edge detectors'. There are also fibers which respond to the presence of a sharp edge in the visual field, whether it is stationary or moving, and these '*edge receptors*' differ distinctly from the on–off fibers. Finally, there is a fifth category which is particularly interesting because they could be called '*bug detectors*'. These fibers respond to small, dark, moving objects, but not to large dark objects or to stationary objects. The two last categories do not react to changes in the general intensity of the light, even switching a light on or off does not affect them.

We can now see that the frog's retina can carry out a great deal of analysis of the visual signals before information is transmitted to the optic nerve, and if we examine this information in view of what may be important to the frog, the system makes a great deal of sense. From the frog's viewpoint the insects on which it feeds are some of the most relevant objects in its life. Therefore, a small dark object is

---

* In the mammalian eye a small, central area of the retina, the *fovea centralis*, contains only cones. It is the area which provides the most acute vision in strong light, but due to the absence of rods the fovea is insensitive to dim light.

important, in particular if it moves, and the frog is equipped to respond specifically to this stimulus through the 'bug-detector' fibers. Small stationary objects, such as spots, shadows, pebbles, etc. are of no special interest, and information about these does not reach that part of the system which is designed for a rapid response to live prey. The most meaningful processing of signals related to food has therefore already taken place in the retina, and is transmitted as specialized information to the central nervous system (Lettvin, Maturana, Pitts and McCulloch, 1961). We shall not discuss the details of the other types of retinal detection systems but instead move on to the mammalian visual system.

The mammalian eye. The eye of a cat, and of many other mammals, has about a hundred million receptor cells in the retina. The optic nerve carries about one million axons. This immediately tells us that the brain cannot receive separate information from each individual receptor cell; the simple fact is that a great deal of sorting and processing takes place already before the information is sent to the central nervous system.

In spite of the complexity, what takes place in the mammalian optical system is reasonably well known. In a simplified way we can regard the transmission as taking place over six levels, three of which are in the retina of the eye, one in the lateral geniculate body of the brain, and two levels in the visual cortex of the brain (see fig. 12.19).

Let us now follow these levels and examine the processing that takes place before information reaches the visual cortex, which in turn is connected with other parts of the central nervous system.

The light *receptor neurons* in the retina (see fig. 12.20) connect to a layer of nerve cells known as the *bipolar cells.** These in turn are connected with a layer of retinal *ganglion cells* whose axons make up the optic nerve. The connections between these three types of cells are very complex. A receptor cell may connect to more than one bipolar cell, and several receptors may be connected to one and the same bipolar cell. The same holds for the connections between the bipolar cells and the ganglion cells. Since there are more than 100 receptor cells for each ganglion cell, the ganglion cell must evidently receive information from a large number of receptor cells. Together such a group of receptor cells make up what is called a *receptive field,* which designates that area of the retina with which the particular ganglion cell is connected.

---

* For simplicity the discussion of amacrine and horizontal cells is omitted.

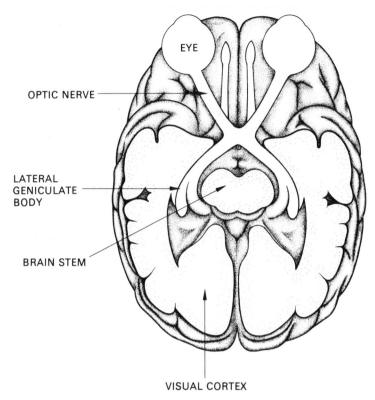

EYE

OPTIC NERVE

LATERAL
GENICULATE
BODY

BRAIN STEM

VISUAL CORTEX

Fig. 12.19. Diagram of human brain, seen from below, to indicate main pathway between the eye and the visual cortex.

Without going into detail, let us remember that retina cells may have both 'on' and 'off' characteristics, and that the synaptic connections may be excitatory as well as inhibitory (for a further discussion of excitation and inhibition, see next chapter, p. 629 to 642).

One of the most striking characteristics of the ganglion cells is that, in the absence of any stimulation, they pour impulses at a steady rate of about 20 to 30 per second into the optic nerve. Even more surprising is the fact that subjecting the entire retina to illumination does not have any prominent effect on the number of impulses from the ganglion cells. However, if a small spot of light falls on one receptive field, it has a very marked although highly complex effect on the corresponding ganglion cell.

It was discovered by Kuffler (1953) that the receptive fields in fact consist of two concentric areas, one with 'on' receptors, and the other with 'off' receptors. The receptive fields are of two kinds, the *'on-center'* fields, in which the central area consists of 'on' receptors,

Direction of light

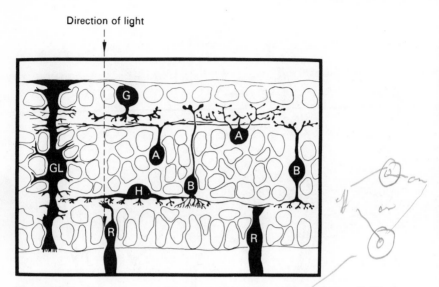

Fig. 12.20. Simplified diagram of the connections between the receptor cells (R), the bipolar cells (B), and the ganglion cells (G) in the retina of the eye. The diagram also shows horizontal cells (H), amacrine cells (A) and a glial or supporting cell (GI). (Drawing supplied by John E. Dowling, Harvard University.)

surrounded by a ring of 'off' receptors. The other kind, the *'off-center'* areas have a central 'off' region, surrounded by a circle of 'on' receptors. If a small spot of light shines on the center of an 'on-center' field, the corresponding ganglion cell shows increased activity. Shining two spots at adjacent points within the 'on' region increases the response, but if one of the spots falls on the 'on' region, and the other on the surrounding 'off' region, the response is greatly reduced. This explains why uniform illumination of the entire retina gives rise to very little response, for the various 'on' and 'off' regions to a great extent cancel each other.

The meaning of this signal evaluation or processing is obvious. It is of no particular interest to record and inform the brain about the general intensity of uniform light; the detail of light and dark contrast is of much greater importance. What we have is in fact a case of selective destruction of information, carried out in the eye. The eye selects what kind of information is to be transmitted in the optic nerve and thus reduces the information load on the brain.

The fourth level in the processing of visual stimuli is in the brain, in the *geniculate body*. The cells of the geniculate body correspond to the receptive fields of the retina, they have some characteristics in common with the retinal ganglion cells, and one of their character-

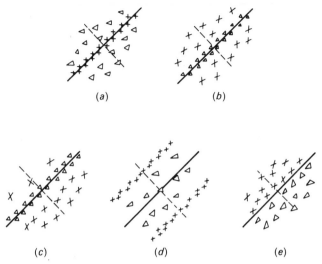

Fig. 12.21. Various arrangements of receptive fields in the visual cortex. Areas giving 'on' response indicated by crosses, those giving 'off' responses by triangles. The orientation of the receptive fields, whether of the 'on' or 'off' type, is indicated by straight lines through the center of the field (from Hubel and Wiesel 1962. *J. Physiol.*, **160**, 106–54, Fig. 2).

istics is that they function to increase the contrast between small spots of light and changes in diffuse lighting. For our purposes, however, we shall just regard the geniculate body as a way-station, consisting of synaptic connections between the optic nerve and the visual cortex of the brain.

Of the many different kinds of cells in the *visual cortex,* we shall discuss only two. The first of these is known as *'simple'* cells. They respond primarily to contrasting lines, such as light bars on a dark background or dark bars on a light background, and to sharp light–dark edges. Whether or not a given 'simple' cell responds to such a bar depends on the orientation and position of the bar within the receptive field of the retina. For example, if a vertical bar gives a response in a given 'simple' cell of the visual cortex, the cell will not respond if the bar is rotated some 10 or 15°, or if the bar is displaced slightly to one side or the other. A careful testing of the retina with minute light spots has shown that the simple cells respond to arrangements such as shown in fig. 12.21.

Since the response of a given 'simple' cell depends on the orientation as well as the position on the retina of a straight line, we have a built-in filter which removes or destroys a great deal of information and transmits only specialized information about certain kinds of contrasting lines. Each simple cell thus has a very specific function, it corresponds to one restricted part of the retina, and responds only to one particular kind of information falling on this spot. There is no evidence that in the cat's retina there is any preferred orientation, the receptive fields may occur in all possible orientations with no preference for vertical, horizontal, or oblique orientations.

The next step, the sixth in our sequence, is the 'complex' cells of the visual cortex. These cells respond, like 'simple' cells, to the orientation of a given stimulus. If a 'complex' cell responds to a certain orientation of a light bar on the retina, rotation of the bar causes the cell to cease its activity. If the bar is moved over the retina with unchanged orientation, however, the complex cell continues to be active. This can most readily be explained by assuming that the 'complex' cell receives information from a large number of 'simple' cells, all of which are of the same general type and have the same orientation of their fields. This requires an enormous complexity of the connections in the visual cortex, but all the evidence that has been obtained indicates that the cells are indeed located in such anatomically well defined groups.

We can now summarize the main steps in the processing of visual information in the mammalian eye. The receptors can be 'on'- or 'off'-type receptors. They do not individually send information to the central nervous system, but are arranged in fields that readily process information about contrast between light and dark, rather than about the general level of illumination. This takes place with the aid of two levels of transmitting neurons within the retina, which select the most pertinent information and reject less meaningful information. The selected information is transmitted to the geniculate body, which accentuates the information obtained from the 'on-center' and 'off-center' receptive fields. The visual cortex, finally, rearranges the information from the geniculate body so that lines and contours are selected for further processing. Here the first step (in the 'simple' cells) is concerned with the orientation of lines, whereupon the information converges on the 'complex' cells, from which the information is transmitted to other parts of the brain for integration with the general functioning of the body.

The processing and sorting of sensory signals which is carried out already before they are passed on to the central nervous system removes much irrelevant information and greatly reduces the load on the control systems that are responsible for adequate and proper reactions. We shall now pass on to the question of how these control systems work and coordinate the many functions and activities of the animal body.

REFERENCES

AIDLEY, D. J. (1971). *The Physiology of Excitable Cells,* Cambridge, England: Cambridge University Press, 468 pp.

ATEMA, J. (1971). Structures and functions of the sense of taste in the catfish (*Ictalurus natalis*). *Brain, Behav. Evol.,* **4,** 273–94.

BARRETT, R., MADERSON. P.F.A., and MESZLER, R. M. (1970). The pit organs of snakes. In *Biology of the Reptilia,* vol. II (C. Gans and T. S. Parsons, eds), pp. 277–300, London: Academic Press.

BLACK-CLEWORTH, P. (1970). The role of electrical discharges in the non-reproductive social behaviour of *Gymnotus carapo* (Gymnotidae, Pisces). *Animal Behav. Monogr.,* **3,** 1–77.

BOECKH, J., KAISSLING, K.-E. and SCHNEIDER, D. (1960). Sensillen und Bau der Antennengeissel von *Telea polyphemus* (Vergleiche mit weiteren Saturniden: *Antheraea, Platysamia* und *Philosamia*). *Zool. Jahrb. Abt. Anat. Ontog. Tiere,* **78,** 559–84.

BROWN, F. A., JR., BARNWELL, F. H. and WEBB, H. M. (1964). Adaptation of the magnetoreceptive mechanism of mud-snails to geomagnetic strength. *Biol. Bull.,* **127,** 221–31.

BROWN, P. K. and WALD, G. (1964). Visual pigments in single rods and cones of the human retina. *Science,* **144,** 45–52.

BULLOCK, T. H. and COWLES, R. B. (1952). Physiology of an infrared receptor: the facial pit of pit vipers. *Science,* **115,** 541–3.

BULLOCK, T. H., HAMSTRA, R. H., JR. and SCHEICH, H. (1972). The jamming avoidance response of high frequency electric fish. I. General features. *J. Comp. Physiol.,* **77,** 1–22.

CAHN, P. H. (ed.) (1967). *Lateral Line Detectors,* Bloomington, Inc.: Indiana University Press, 496 pp.

CRESCITELLI, P. and DARTNALL, H.J.A. (1953). Human visual purple. *Nature, Lond.,* **172,** 195–7.

DIJKGRAAF, S. and KALMIJN, A. J. (1963). Untersuchungen über die Funktion der Lorenzinischen Ampullen an Haifischen. *Zeits. vergl. Physiol.,* **47,** 438–56.

ENRIGHT, J. T. (1963). Estimates of the compressibility of some marine crustaceans. *Limnol. Oceangr.,* **8,** 382–7.

ERULKAR, S. D. (1972). Comparative aspects of spatial localization of sound. *Physiol. Rev.,* **52,** 237–360.

FRISCH, K. VON (1948). Gelöste und ungelöste Rätsel der Bienensprache. *Die Naturwissenschaften,* **35,** 38–43.

FROMME, H. G. (1961). Untersuchungen über das Orientierungsvermögen nächtlich ziehender Kleinvögel (*Erithacus rubecula, Sylvia communis*). *Zeits. Tierpsychol.,* **18,** 205–20.

GOULD, E., NEGUS, N. C. and NOVICK, A. (1964). Evidence for echolocation in shrews. *J. Exp. Zool.,* **156,** 19–39.

GRIFFIN, D. R. (1953). Acoustic orientation in the oil bird, *Steatornis. Proc. Nat. Acad. Sci.,* **39,** 884–93.

GRIFFIN, D. R. (1954). Bird sonar. *Sci. Am.,* **190,** 78–83.

GRIFFIN, D. R. (1958). *Listening in the Dark: The Acoustic Orientation of Bats and Men,* New Haven, Conn.: Yale University Press, 413 pp.

HARRIS, J. F. and GAMOW, R. I. (1971). Snake infrared receptors: thermal or photochemical mechanism? *Science,* **172,** 1252–3.

HARTLINE, H. K. and RATLIFF, F. (1957). Inhibitory interaction of receptor units in the eye of *Limulus. J. Gen. Physiol.,* **40,** 357–76.

HENSEL, H. and BOMAN, K.K.A. (1960). Afferent impulses in cutaneous sensory nerves in human subjects. *J. Neurophysiol.,* **23,** 564–78.

HODGSON, E. S. (1961). Taste receptors. *Sci. Am.,* **204,** 125–44.

HOPKINS, C. D. (1972). Sex differences in electric signaling in an electric fish. *Science,* **176,** 1035–7.

HUBEL, D. H. and WIESEL, T. N. (1962). Receptive fields, binocular interaction and functional architecture in the cat's visual cortex. *J. Physiol.,* **160,** 106–54.

IGGO, A. (1969). Cutaneous themoreceptors in primates and sub-primates. *J. Physiol.,* **200,** 403–30.

KALMIJN, A. J. (1971). The electric sense of sharks and rays. *J. Exp. Biol.,* **55,** 371–83.

KEETON, W. T. (1971). Magnets interfere with pigeon homing. *Proc. Nat. Acad. Sci.,* **68,** 102–6.

KELLOGG, W. N. (1958). Echo ranging in the porpoise. *Science,* **128,** 982–8.

KEYNES, R. D. and MARTINS-FERREIRA, H. (1953). Membrane potentials in the electroplates of the electric eel. *J. Physiol.,* **119,** 315–51.

KUFFLER, S. W. (1953). Discharge patterns and functional organization of mammalian retina. *J. Neurophysiol.,* **16,** 37–68.

LETTVIN, J. Y., MATURANA, H. R., PITTS, W. H. and MCCULLOCH, W. S. (1961). Two remarks on the visual system of the frog. In *Sensory Communication* (W. A. Rosenblith, ed.), pp. 757–76, New York: M.I.T. Press and John Wiley and Sons.

LINDAUER, M. and MARTIN, H. (1968). Die Schwereorientierung der Bienen unter dem Einfluss des Erdmagnetfeldes. *Zeits. vergl. Physiol.,* **60,** 219–43.

LISSMANN, H. W. and MULLINGER, A. M. (1968). Organization of ampullary electric receptors in Gymnotidae (Pisces). *Proc. Roy. Soc. Lond. B,* **169,** 345–78.

MACHIN, K. E. and LISSMANN, H. W. (1960). The mode of operation of the electric receptors in *Gymnarchus niloticus. J. Exp. Biol.,* **37,** 801–11.

MATURANA, H. R. (1959). Number of fibers in the optic nerve and the number of ganglion cells in the retina of anurans. *Nature, Lond.,* **183,** 1406–7.

MORITA, H. and YAMASHITA, S. (1959). Generator potential of insect chemoreceptors. *Science,* **130,** 922.

NORRIS, K. S., PRESCOTT, J. H., ASA-DORIAN, P. V. and PERKINS, P. (1961). An experimental demonstration of echo-location behavior in the porpoise, *Tursiops truncatus* (Montagu). *Biol. Bull.,* **120,** 163–76.

PAYNE, R. S. (1971). Acoustic location of prey by barn owls. *J. Exp. Biol.,* **54,** 535–73.

SCHNEIDER, D. (1969). Insect olfaction: deciphering system for chemical messages. *Science,* **163,** 1031–7.

SCHNEIDER, D., KASANG, G. und KAISSLING, K.-E. (1968). Bestimmung der Riechschwelle von *Bombyx mori* mit Tritium-markiertem Bombykol. *Die Naturwissenschaften,* **55,** 395.

TSCHACHOTIN, S. (1908). Die Statocyste der Heteropoden. *Zeits. Wissensch. Zool.,* **90,** 343–422.

WALD, G. (1964). The receptors of human color vision. *Science,* **145,** 1007–16.

WATERMAN, T. H. (1950). A light polarization analyzer in the compound eye of *Limulus. Science,* **111,** 252–4.

WATERMAN, T. H. and FORWARD, R. B., JR. (1970). Field evidence for polarized light sensitivity in the fish *Zenarchopterus. Nature, Lond.,* **228,** 85–7.

WATERMAN, T. H. and HORCH, K. W. (1966). Mechanism of polarized light perception. *Science,* **154,** 467–75.

WILTSCHKO, W. and WILTSCHKO, R. (1972). Magnetic compass of European robins. *Science,* **176,** 62–4.

13
Control and integration

In this book we have often mentioned regulation and control, but without discussing any details of the process or mechanism of regulation. To regulate means to adjust an amount, a concentration, a rate, or some other variable, usually in order to attain or keep it at some desired level. An example is the regulation of respiration. We take it for granted that respiration should be adjusted to provide oxygen at the rate it is used by the organism, i.e. the rate of oxygen uptake in the lung should be coordinated with its use in the tissues. Similarly, all the various physiological processes should be controlled, regulated, and integrated.

Integration means to put parts together. In physiology we use the word to describe the process of controlling all the functional components so that they merge into a smoothly operating organism in which no single process or function is permitted to run wild or proceed at its own independent rate.

There are several ways to control physiological function. We have seen that carbon dioxide is an important element in the control of respiration in air-breathing animals, but carbon dioxide does not directly influence the breathing muscles. In a mammal the contraction of the diaphragm during inspiration is controlled by a nerve from the respiratory center in the brain, this center in turn is sensitive to the carbon dioxide level in the blood. The respiratory movements are thus under nervous control. We have learned that the secretion of digestive juice from the pancreas is stimulated by a hormone (secretin), but in addition the pancreas is supplied with nerves that can also stimulate secretion. Some processes, such as the release of stored sugar from the liver, seem to be primarily under endocrine control (in this case noradrenalin). Evidently both nerves and hormones can control physiological processes, the two can work together in a well-coordinated fashion, and it may be difficult to draw the line as to where one influence terminates and the other takes over.

Although there is much overlap between nervous and hormonal control there are two important differences that stand out, one relates to the speed of action, the other to the size of the target.

When a quick response is required, such as the contraction of skeletal muscles, the rapid conduction in a nerve is necessary for fast action. Nerve impulses move with speeds up to about 100 meters per second, and the delay in transmission of a message therefore need not be more than milliseconds. A process regulated by a hormone, on the other hand, requires that the hormone reach the target organ before there is any effect, i.e. the speed of transmission is limited by the transport of the hormone via the blood. The minimum response time will therefore be in the magnitude of seconds or more.

The second important difference between nervous and hormonal control is the high precision that can be obtained in the spatial distribution of nervous control, while hormonal control is usually more diffuse. The single axon in a motor nerve connects only to a limited number of muscle fibers, and this permits the separate stimulation (or inhibition) of a single muscle without affecting other muscles, and it even permits the stimulation of a small fraction of a single muscle. Hormones, in contrast, affect all sensitive cells that they reach via the circulation, in other words, whole organs or organ systems are affected. For example, noradrenalin affects the sugar release from the entire liver. Control of the degree of effect is still possible, however, through the amount of the hormone released into the blood. Hormones usually control processes in which the response is slow, such as the secretion of digestive juices, the control of urine concentration and volume, the excretion of sodium, as well as some very slow processes such as the development of the gonads and the growth of the body. However, as already mentioned, we shall find no sharp separation between nervous and hormonal control; the nervous system not only contributes to the regulation of endocrine function but is important in the production of hormones as well.

CONTROL AND CONTROL THEORY

The control mechanisms responsible for the maintenance of steady states in living organisms constitute one of the major chapters in physiology and it is helpful to examine their function in the light of simple control theory. For example, in birds and mammals the body core temperature is maintained nearly constant in spite of wide variations in external temperature and in internal heat production.

The regulation is carried out with the aid of a complex control system.

In engineering, control systems are so important that control theory and design are an independent branch of engineering with its own terminology and theoretical approach. For physiology this is important in two respects, (*a*) many ill-defined and hazy old terms have been replaced by exact and well-defined concepts, and (*b*) the theoretical approach to control theory has led to a more precise definition of the components of physiological mechanisms and a better understanding of the relation between the component parts.

Let us examine a familiar control system, the thermostat that controls the temperature in a house or a waterbath. In this and all other control systems we operate with a *controlled variable* (in this case, temperature) which is kept within a more or less narrow range around a desired value. A measurement of the value of the controlled variable is compared with the desired value, the *set point*. This is done by an *error detector,* which delivers a signal which in turn activates a control mechanism that results in the necessary correction.

If we want to keep a water bath at a 'constant' temperature, we could use the arrangement shown in fig. 13.1, which is nearly self-explanatory. Person A can decide to add heat to the water bath by throwing an electric switch, thus causing the temperature to rise. A thermometer tells him the temperature of the water. Person B, who

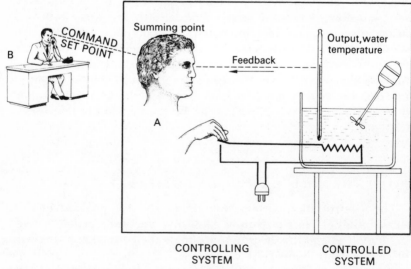

Fig. 13.1 Diagram of a temperature-controlled waterbath which includes a controlling system and a command. Details in text.

is in charge, has already told A what temperature he wants maintained in the water bath. If the water temperature falls below the set point, A can throw the switch to raise the water temperature. By looking at the thermometer, he can see when the temperature has reached the desired level, and then break the contact again.

We could, of course, replace A by an automatic mechanism, a thermostat, similar to that which controls the central heating system in houses. We can then represent the entire system by the block diagram shown in fig. 13.2*a*. Information about the output of the system, in this case water temperature, is fed back into the thermostat so that appropriate action is taken to correct any deviation in water temperature and keep it at the desired level. This is known as *feedback,* a term used when we compare the condition of the output of the control system with the set point. In this case, a rise in water temperature is corrected by a decrease in the heat input. This is called *negative feedback,* a term used when a deviation leads to a corrective action in the opposite direction.

For the further discussion it will be useful to expand the diagram slightly and introduce standardized terms that are used by control system engineers (fig. 13.2*b*).

The typical negative feedback system is known as a *closed-loop control system,* a term which is obvious from a glance at the diagram. The signal from the controlled variable is fed back into the system, thus forming a closed loop. For completeness we should also briefly discuss *open-loop systems,* although they are less important in physiological regulation. Assume that a house furnace has a variable fuel supply which we arrange so that a drop in outside temperature increases the flow of fuel to the furnace. We could then carefully adjust the system so that a decrease in the outside temperature gives precisely the amount of fuel needed to keep the room temperature constant. In this example the input is the outside temperature, and the output is the heat supplied by the furnace. However, if a disturbance enters into the carefully calibrated system, for example, a strong wind, more heat is carried away from the house, but the furnace supplies no more heat and the room temperature falls. In this system there is no feedback, and we therefore have an *open-loop control system.*

We discussed negative feedback, which serves to reduce the difference between the output and the desired value, and a thermostat is a good example. What about *positive feedback,* does it exist, and does it have any importance in biology?

The following situation has been suggested by my colleague, Steven Vogel, as an illustration of positive feedback. Assume that a hus-

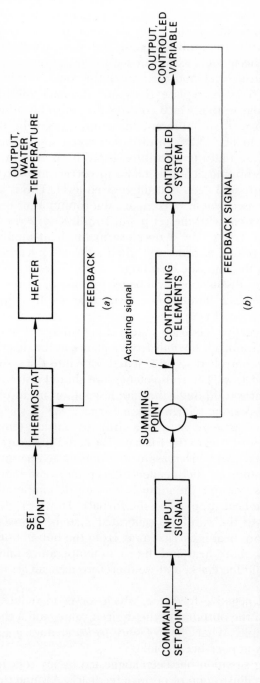

Fig. 13.2 (a) Diagram of the control system represented in fig. 13.1. (b) General diagram of a control system which includes the terms most commonly used in control theory.

band and wife have separate electric blankets, each thermostatically
controlled through negative feedback. Say that the husband prefers
a rather cool blanket, and the wife a higher temperature. Let us now
assume that the thermostats inadvertently get interchanged. The
husband will set 'his' thermostat at his preferred low temperature,
and his wife who now finds her blanket cooler than she likes will
turn up 'her' thermostat. The husband soon finds his blanket much
too warm, and turns the thermostat further down, whereupon the
wife turns 'hers' up even more. This is *positive feedback,* in which a
deviation leads to an ever increasing augmentation of the deviation.

Obviously, positive feedback is no good at all for control purposes,
for the system will proceed to some extreme condition. However,
positive feedback systems can be useful in some biological situations.
While negative feedback is used to achieve a steady state control,
positive feedback can be used where we need either a rapidly developing maximal response, or certain kinds of synchronized events. A
well-known example of useful positive feedback is found in mating
and reproduction. As the mating act proceeds, progress is reinforced by positive feedback, the partners get increasingly interested,
mutual response reinforces the progress, and continued positive
feedback leads to copulation and completion of mating.

On–off or proportional control. The thermostatic control of a house furnace is an *on–off system.* The furnace is either on or off, and therefore gives a discontinuous supply of heat. This inevitably leads to an
oscillation in temperature. In order to start the on cycle, a certain
deviation from the set point is needed. Similarly, when the heater is
on, there is usually an over-shoot in temperature before the heater is
cut off. The oscillations can be reduced by making the system more
sensitive, but there is no way to eliminate them entirely from an
on–off system.

There are, however, other types of control systems that give a
more constant output of the controlled variable. One important system is known as *proportional control,* which is best illustrated perhaps
by a mechanical analog (see fig. 13.3). The water level in the tank is
controlled with the aid of a float which gives continuous instead of
an on–off control. If the outflow from the tank for some reason
increases, the water level sinks, thus opening for increased inflow.
The more the level sinks, the greater the increase in the inflow.
Should the outflow be impeded, the water level rises and causes the
inflow valve to close. This system has continuous control, and
the degree of control action is directly related to the deviation
from the set point. (The term 'proportional' implies a continuous

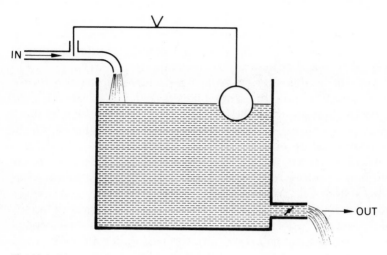

Fig. 13.3. Diagram of a simple proportional control system that can adjust the water inflow into a tank to match the outflow.

linear relation between output and input with the origin at zero, but the term is commonly used for systems that do not meet this strict definition of proportionality.)

A proportional controller has an interesting characteristic which is important. Assume that we have a system which controls exactly at the set point. If a disturbance is now introduced (e.g. a change in outflow), the system readily reaches a new steady state, but it is not possible to attain the original set point. This may need explanation. Assume a given steady state, which is the original set point of the water level. Let us now decrease the outflow, and examine the corrective action. The increased water level lifts the float and decreases the inflow. The new steady state involves a higher water level and the float must remain at this higher level to keep the inflow equal to the reduced outflow, i.e. the new level represents an error from the original set point. The corrective action cannot return the fluid level exactly to the original set point (zero error), for the float would then be in its original position and inflow would not match the decreased outflow. Now assume that a different kind of disturbance is introduced by adding a second inflow. The fluid level again rises and reduces the regulated inflow, but the new steady state will again be above the set point.

A physiological system which represents proportional control, is the control of mammalian respiration (ventilation of the lungs) by the carbon dioxide level in the blood. When a mammal at rest has an arterial P_{CO_2} of 40 mm Hg, we can consider this as the set point for the arterial P_{CO_2}. If we increase the carbon dioxide content of the inspired air, there will be an increase in arterial P_{CO_2} which results in an increase in ventilation, and the amount of increase is directly related to the increase in P_{CO_2} (see p. 40).*

As we have seen, a steady state with zero error cannot be achieved through simple proportional control. This is possible, however, by using an *integral control system*. In this system the output is proportional to the time integral of the input (i.e. the rate of change of output is proportional to the input). If a disturbance is introduced into such a system, and the disturbance remains constant, the error in the output will tend towards zero with time. Whatever the value of the disturbance, if it remains constant the integral action can achieve a steady state with zero error.

An additional type of control action is known as *derivative control action,* also known as *rate control,* for it responds to the time derivative of a signal, or the rate of change in the signal. Since this control action responds only to a change, and not to the signal itself, it is usually useful only in combination with other control action. In physiology such combination is particularly valuable, for it can be used to take appropriate action when there are temporary transients in a controlled variable.

Let us now return to the regulation of body temperature. We used as an example the analogy with the thermostatic control of a water bath through a negative feedback loop. Central heating in a house can be regulated in the same way, and if in summer the house tends to get too warm, we can add an air conditioning (cooling) system. If the house temperature rises unduly, the thermostat starts the cooling cycle, an action which serves to decrease the deviation from the desired set point, i.e. there is negative feedback. Thus regulation of both heating and cooling depends on negative feedback control.

The analogy to body temperature control is obvious. If body temperature tends to fall, heat production is increased, mainly through involuntary muscle contractions (shivering). If body temperature increases due to a heat load, whether external or internal, there is increased cooling achieved through sweating or panting. Thus both

---

* The respiratory control is in reality more complex. We know that in exercise carbon dioxide production is increased; if there were a simple proportional control system, the arterial P_{CO_2} should be slightly increased in exercise to provide the actuating signal for the increased ventilation. Actually, the situation is the opposite, during exercise arterial P_{CO_2} is slightly decreased. This is often 'explained' as a result of a 'resetting' of the set point in exercise. The real reason is that the respiratory center receives and integrates several different control inputs, including nerve impulses from the working muscles.

heating and cooling of the body depend on negative feedback control systems.

The temperature regulation center is located in the hypothalamus of the brain. This can be demonstrated in various ways, for example, if the blood in the carotid artery of a dog is heated, it causes the dog to pant. This shows that the regulation takes place in the head, because as the experimental animal is caused to pant, it loses too much heat and its core temperature drops. Conversely, if the carotid blood is cooled, the dog begins to shiver, and the body core temperature rises. The exact location of the heat regulation center can be pinpointed by heating or cooling of small areas in the hypothalamus.

The heat regulation center can be regarded as equivalent to the thermostat, and the 'normal' body temperature as the set point. The system is not so simple, however, for there is no constant set point and there are multiple inputs. To begin with, the body temperature fluctuates with a daily cycle, even if the external temperature and internal heat production remain constant. This means that the 'set point' undergoes a diurnal cycle. Also, we learned that during exercise the core temperature is 'reset' and regulated at a higher level than at rest. Of the many inputs to the heat regulation center, the warm- and cold-receptors in the skin provide important information, but many other inputs provide additional information, one of them being the temperature of the arterial blood which reaches the center.

We mentioned that regulation of body temperature depends on both heating and cooling, i.e. control action takes place in either direction. This is an important principle in physiological regulation, and much more common than a superficial examination might indicate. The heart is another example. Its rhythmic contraction is an inherent characteristic of the heart muscle, it starts at the sinus node, and spreads throughout the heart muscle, followed by a relaxation, whereupon a new contraction begins. The rate of contraction is under the control of two nerves. One, the *accelerator nerve,* speeds up the heart beat. The other, the *decelerator nerve,* is a branch of the vagus nerve and slows down the rate. The heart rate is thus determined by the balance between two antagonistic nerves, one stimulating and the other inhibiting. We shall see that a combination of stimulation and inhibition is very common in physiological control systems. An understanding of nerve cells and how they transmit excitatory and inhibitory impulses will help to clarify this common principle of physiological regulation, and this is what we will turn to next.

PHYSIOLOGICAL CONTROL AND INTEGRATION

Nervous control systems

Nerve cells or neurons are the basic component of all nervous systems. We shall now focus our attention on two of their most important functional parts, the long fibrous extensions, the axons, and the connections between cells, the synapses. The single neuron has only one axon but may have hundreds or even thousands of synaptic connections, and their role in integration is extremely important. The axons function as cables, and the synapses are highly complex contact or switching devices.

We shall first describe some important characteristics of axon physiology, and then discuss how synapses handle the information which impinges on the neuron. As an example of integration we shall then describe how simultaneous stimulatory and inhibitory control by the central nervous system is used to control the muscles of a limb used in locomotion.

Axon characteristics

The nature of the nerve impulse, the action potential, was discussed on p. 599. It consists of a rapid transient increase in membrane permeability, followed by ion movement and a change in membrane potential. Although there is some minor variation between different animal species, the magnitude of both resting and action potential is similar throughout the animal kingdom. The resting membrane potential of both axon and cell body is mostly of the order of 60 to 90 mV (outside positive). The action potential, which consists of a localized drop in membrane potential, is mostly about -80 to -120 mV relative to the resting potential. There are no regular variations, either with the size of the animal, with the size of the axon, or whether the animal is an invertebrate or a vertebrate. The time course of the action potential, on the other hand, is much more variable (see table 13.1).

Speed of conduction is an entirely different matter; it varies tremendously from nerve to nerve and from animal to animal. A few typical values are listed in table 13.2. We can see that vertebrate motor nerves have a much higher speed of conduction than the regular motor nerves of invertebrates. This is not because the invertebrates live at lower temperatures, for the cold-blooded vertebrates also have relatively high conduction speeds. If the values for the cold-blooded vertebrates were recalculated to the body temperature of mammals, using a Q_{10} of 1.8 (which is characteristic of nerve fibers), they would all be within the mammalian range. For the invertebrates this would not be true. Some invertebrates, however, have a few fast-conducting axons known as *giant axons*. The giant axons are much larger than ordinary axons, up to nearly 1 mm in diameter. In these giant axons the speed of conduction is around ten times as fast as in ordinary axons from the same animal.

The biological role of high-speed conduction is obvious, it is always related to a quick response mechanism which the animal uses in locomotion, mostly to avoid predators. One of the fastest such responses is that of the cockroach, which within 25 ms reacts to an air puff on the tip of the abdomen (which is equipped with hair receptors). Another example is the squid, which has giant fibers that run the length of the mantle. Due to the rapid conduction in these giant axons the entire mantle musculature can contract almost simultaneously, which is needed when the squid swim by jet propulsion. If conduction to the more distant parts of the mantle were to take appreciably longer time, contraction would spread slowly and not give the necessary sudden forceful jet of water. In an earthworm, which on a moist morning reaches half-way out of its burrow, giant fibers permit an almost instantaneous withdrawal in response to a

Table 13.1. *Resting and action potentials in neurons from a variety of animals.* (Data from Bullock and Horridge, 1965)

ANIMAL	FIBERS	Resting potential (mV)	Action potential peak (mV)	Spike duration (ms)
Squid (*Loligo*)	Giant axon	60	120	0.75
Earthworm (*Lumbricus*)	Median giant	70	100	1.0
Crayfish (*Cambarus*)	Median giant	90	145	2
Cockroach (*Periplaneta*)	Giant fibers	70	80–104	0.4
Shore crab (*Carcinus*)	30 μm leg axon	71–94	116–153	1.0
Frog (*Rana*)	Sciatic nerve axon	60–80	100–130	1.0
	CELLS			
Sea slug (*Aplysia*)	Visceral ganglion	40–60	80–120	10
Land snail (*Onchidium*)	Visceral ganglion	60–70	80–100	9
Crayfish (*Cambarus*)	Stretch receptor	70–80	80–90	2.5
Puffer fish (*Sphaeroides*)	Supramedullary	50–80	80–110	3
Toad (*Bufo*)	Dorsal root ganglion	50–80	80–125	2.8
Toad (*Bufo*)	Spinal motoneuron	40–60	40–84	2
Rabbit (*Oryctolagus*)	Sympathetic	65–82	75–103	4–7
Cat (*Felis*)	Spinal motoneuron	55–80	80–110	1–1.5

Table 13.2. *Conduction velocities in nerves from various animals.* (Data from Bullock and Horridge, 1965)

VERTEBRATES	Motor nerves (meter s^{-1})	
Cat	30–120	
Snake	10–35	
Frog	7–30	
Fish	3–36	

INVERTEBRATES	Regular motor nerves (meter s^{-1})	Giant axons (meter s^{-1})
Cockroach	2	10
Squid	4	35
Earthworm	0.6	30
Crab	4	—
Snail	0.8	—
Sea anemone	0.1	—

mechanical disturbance, in sharp contrast to the relatively slow locomotion which is otherwise characteristic of this animal.

In general, there is a direct relationship between conduction velocity and axon diameter. It can be derived on theoretical grounds, based on cable theory, that the speed of conduction (u) should be approximately proportional to the square root of the fiber diameter (d), or $u = k\sqrt{d}$ (Hodgkin, 1954).

A great deal of observational material indicates that this relationship is approximately correct, but the value of the constant k varies from animal to animal. The equation therefore tells us that, for a given type of fiber, we can expect a tenfold increase in conduction speed for a 100-fold increase in fiber diameter. The giant axons mentioned above may be 50 or even 100 times thicker than the ordinary axons in the same animals, and their high conduction velocities therefore fall within the expected range.

Myelinated fibers. Vertebrates do not have giant axons, and yet, speed of conduction in their motor nerves is very high, the highest found anywhere in the animal kingdom. The reason that mammalian axons conduct rapidly, although they are very thin, is their peculiar structure. The entire axon is covered with a thin sheath of a fat-like substance, *myelin,* which is interrupted at short intervals to expose the nerve membrane. The exposed sites are known as the nodes, and the distance between the nodes is from a fraction of a millimeter up to a few millimeters (see fig. 13.4). The myelin sheath is formed from glial (or supporting) cells, which grow into a many-layered wrapping from which the protoplasm disappears so that multiple layers of the glial cell membrane remain.

Let us examine the conduction of an impulse in a myelinated axon. At a node the action potential is exactly like that of any other action potential, i.e. there is a local depolarization of the membrane. This means that this node, relative to the neighboring node, will appear to be negative (the usual positive membrane charge at our node has disappeared). This instantaneously sets up an electric current between this and the neighboring node, sufficient to trigger an action potential at the neighbor. This is a virtually instantaneous process, so that the action potential appears to have jumped from one node to the next. The depolarization at the node itself is less rapid, but as it develops, it in turn causes depolarization at the next node, and so

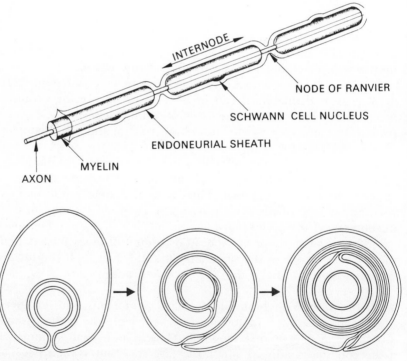

Fig. 13.4. Diagram of a myelinated nerve fiber (top). The myelin sheath is developed from a supporting cell which grows around the nerve fiber until it is covered by a multilayered wrapping of the membrane of the supporting cell (bottom) (Aidley, 1971).

on. The rapid transmission from node to node is known as *saltatory conduction* (from Latin *saltare,* to dance or jump), and since it takes place almost without delay, the result is a rapid transmission of action potentials in fibers whose small diameter would otherwise give a very slow conduction.

The conduction velocity in a myelinated frog axon has been accounted for as follows. An axon of 10 μm diameter had a conduction velocity of 20 meter s^{-1} and the delay at each node was approximately 0.06 ms. Since the nodes in a frog axon of this size are spaced about 1.6 mm apart, it can be seen that most of the conduction time is accounted for by the delay at the nodes, and transmission from node to node occupies only a small fraction of the time (Tasaki, 1959).

This conclusion can be further supported by a rather elegant method which involves the cooling of the nerve. This slows down the rate of propagation of an impulse. However, if the cooling is restricted to the area between two nodes, the duration of the transmission between the nodes is nearly unchanged. This result is again consistent with the hypothesis that the conduction between the nodes is an electric phenomenon, and that the potential change at the node itself is of the same nature as the action potential in a non-myelinated axon (Hodler, Stämpfli and Tasaki, 1951).

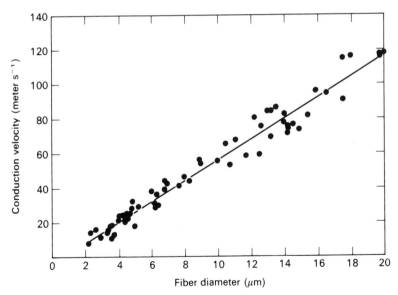

Fig. 13.5. The speed of conduction in myelinated fibers of the cat, plotted against fiber diameter (from Hursh, J. B. 1939. *Am. J. Physiol.,* **127,** 131–39, Fig. 2).

The conduction velocity in myelinated axons still depends on the fiber diameter, but in these fibers the conduction velocity is directly proportional to fiber diameter (fig. 13.5). The reason that the conduction velocity is higher in the larger-diameter myelinated axon seems to be that the distance between the nodes increases with the diameter of the axon. Since virtually all delay is at the nodes, a smaller number of nodes permits a faster over-all propagation of the action potential (Rushton, 1951).

The fact that, in myelinated fibers, velocity is linearly related to the fiber diameter, while in non-myelinated fibers it is proportional to the square root of the diameter, has the interesting consequence that for very small diameters, less than about 1 μm, myelinated axons will conduct more slowly than non-myelinated axons (fig. 13.6). This coincides with the lower limit for the actual size of myelinated fibers in the organism, which never seem to be smaller than 1 μm in diameter. The so-called C fibers of the sympathetic nervous system are non-myelinated, and the fastest and largest C fibers have a diameter of 1.1 μm and conduct at 2.3 meter s^{-1}. This point is marked with a circle on fig. 13.6.

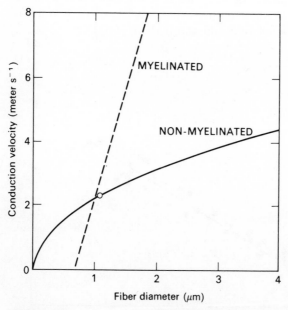

Fig. 13.6. The conduction velocities in myelinated and non-myelinated fibers are different functions of fiber diameter. Below a certain diameter the myelinated fiber will therefore conduct more slowly than a non-myelinated fiber of the same diameter. Dotted line, extrapolation from data in fig. 13.5; curved line, theoretical curve based on the fastest observed non-myelinated C-fibers from cat (Rushton, 1951).

The greatest advantage of myelinated axons comes from their small size, which permits a highly complex nervous system with high conduction velocities without undue space occupied by the conduits. Let us say that we wish to increase the conduction velocity tenfold in a given non-myelinated fiber. This would require a 100-fold increase in its diameter, and the volume per unit length would in turn be increased 10 000-fold. Obviously, if this avenue were used for increasing conduction velocity, a nerve trunk which should contain hundreds or thousands of non-myelinated axons would be unreasonably voluminous. Imagine the size of the optic nerve in humans if a high conduction velocity were to be achieved without myelination; this nerve has a diameter of 3 mm, and if it were to contain the same number of fibers without myelination it would require a diameter of 300 mm.

The synapse. Excitation, inhibition, and computation

The information which travels in an axon is transferred to another neuron at the synapses. The structure of the synapse is important for the understanding of its function (see fig. 13.7). The terminal end of the axon spreads out and forms the *axon knob*, which in turn makes contact with a dendrite or a cell body of another neuron. There is no fusion of the axon and the other cell; there remains a narrow space or gap, the *synaptic cleft*, which has a width of about 20 nm. The knob is referred to as *pre-synaptic*, and the dendrite or

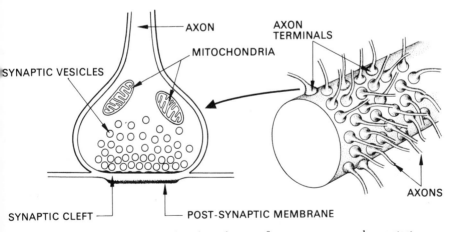

Fig. 13.7. At the synapse, the point where the axon from one neuron makes contact with another neuron, the end of the axon is swollen into a synaptic knob. Impulses are transmitted to the receiving neuron by release of a chemical transmitter substance contained in the synaptic vesicles.

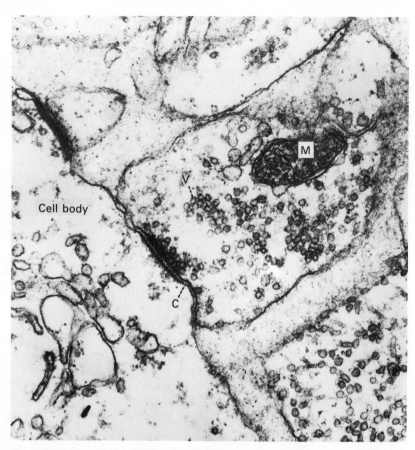

Plate 17. Synapse in the medulla of a goldfish. The nerve ending (the large rectangular area stretching from the center of the photograph towards the upper right) contains numerous synaptic vesicles (V) and a mitochondrion (M). The synaptic cleft (C) separates the pre-synaptic membrane of the nerve ending from the post-synaptic cell body (which fills the entire lower left area of the photograph). The synaptic vesicles are 40 to 50 nm in diameter (J. David Robertson, Duke University).

neuron as the *post-synaptic* structure. Both the appearance of the cleft and its width is surprisingly similar throughout the animal kingdom, but it is so small that information about its contents has remained virtually unknown. It has been suggested that the relatively constant width of the cleft can best be explained by assuming that it is not merely a fluid-filled space, but that it contains a structure of oriented molecules arranged between the two nerve membranes.

The synaptic knob has an important structural characteristic, it contains a large number of small vesicles which are usually about

20–100 nm in diameter. These vesicles are closely packed near the pre-synaptic membrane. It is well established that the transmission of an impulse from the pre-synaptic knob to the post-synaptic neuron takes place through the release of a *chemical transmitter substance* from the synaptic vesicles, which diffuses across the synaptic cleft and affects the post-synaptic membrane. The occurrence of vesicles is so typical of all synapses that their presence is used by electron microscopists as a criterion for the presence of a synapse.*

We see now why synapses can transmit impulses in only one direction. Since transmission across the synaptic cleft depends on the release of a transmitter substance which is present on the pre-synaptic side only, impulses could not be transferred in the opposite direction.

The diagram in fig. 13.7 may leave the impression that the transfer of information to a neuron is a simple matter. However, axons are highly branched and connect to a large number of other neurons, as we saw in the discussion of the structure of the retina (see p. 606). Although the single neuron gives off only one axon, this axon may branch widely and connect to a large number of other neurons. As a consequence, each single neuron must also receive a large number of axon branches which terminate on the neuron or its dendrites (see fig. 13.8). There are often hundreds of synapses on a single neuron in the central nervous system, and motor neurons in the vertebrate spinal cord may have over 1000 synapses. Some specialized neurons have such high density of synaptic knobs on their surface that it has been estimated that a single cell may have about 10 000 synaptic connections. This should give an idea of the complexity of the central nervous system, when viewed in combination with the enormous number of neurons (estimated at 10 000 000 000 in the central nervous system of man). It should be emphasized, however, that the connections are by no means random; they are highly specific and form precisely functioning tracts within the central nervous system. These have been studied and described in great detail by neuroanatomists as well as neurophysiologists.

Post-synaptic potentials. The understanding of synaptic transmission was revolutionized in the 1950s because methods were developed to record electrically from single neurons. Extremely fine glass pipettes with tip diameters of less than 1 μm are prepared. These are filled with a salt solution (usually concentrated potassium chloride), thus

---

* It is probable that synaptic transmission in some types of specialized neurons may be of an electrical nature. This appears to be the case with certain giant neurons (Mauthner cells) in the central nervous system of fish (Furshpan, 1964).

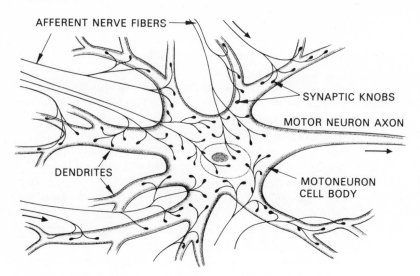

Fig. 13.8. Diagram showing a motor neuron cell body and dendrites branching from it. The synaptic knobs are the terminals of axons, or impulse carrying fibers, from other nerve cells. The motor neuron axon leads to a muscle fiber (from Eccles, J. © *Sci. Am.,* January 1965, p. 58. All rights reserved).

making microelectrodes which permit the recording of any potential change between the tip and a common 'ground'. As such a micro-electrode is slowly moved towards and inserted into a neuron, it will first show the same potential as the 'ground', but the moment it penetrates the nerve membrane it will show a negative potential relative to the 'ground'. The membrane seems to seal around the glass, and it is now possible to record from a neuron for several hours while it continues to function apparently normally. By using pipettes with two or more channels, instead of a single one, it is even possible to apply minute amounts of various chemicals at the site from which the recording is being made.

With this technique we can study the events at the synapse. As an impulse arrives at the pre-synaptic membrane, there is a slight delay of a fraction of a millisecond before there is a change in the potential at the post-synaptic membrane. This latter potential is the *post-synaptic potential* or *PSP* (see fig. 13.9). The PSP initially rises rapidly, and then decays again, but at a much slower rate. The PSPs differ from action potentials in two important respects, they are usually much smaller in amplitude, and they have a much longer duration, sometimes as much as ten or a hundred times as long. These characteristics have two important consequences, a single PSP is rarely (if ever) sufficient to cause an action potential in the post-synaptic

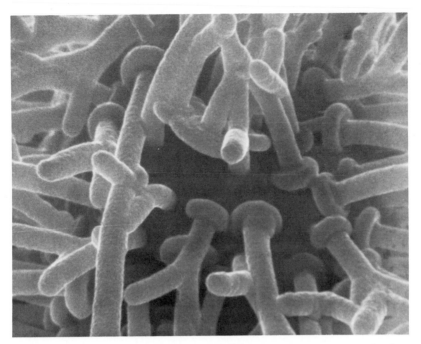

Plate 18. This scanning electron micrograph shows the multitude of synaptic knobs that may be located on the surface of a single neuron in the nervous system of *Aplysia*, a marine snail. The diameter of each disc-like terminal knob is in the order of 1 μm (Edwin R. Lewis, University of California, Berkeley).

neuron, and the long duration permits a great deal of interaction with other PSPs in the same neuron, both at the same synapse and at other neighboring synapses on the same neuron. This interaction is very important and requires a more detailed discussion of the post-synaptic events.

What causes the PSP? We will assume that the PSP is caused by a transmitter substance released from the pre-synaptic membrane, and that the delay corresponds to the diffusion time across the synaptic cleft. The best known and understood such transmitter substance is *acetylcholine*, whose function in synapses and at neuromuscular junctions is beyond doubt. In the synapse, the magnitude of the PSP seems to be directly related to the amount of acetylcholine released. The system could not function, however, unless the acetylcholine were rapidly removed again, for otherwise it would gradually accumulate and maintain a continuous PSP. The enzyme *acetylcholinesterase* is always present at the synapse and serves to hydrolyze and thus remove the transmitter. As the transmitter substance disappears, the PSP gradually decays.

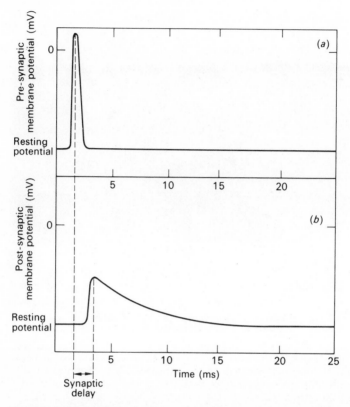

Fig. 13.9. The arrival of an action potential at the terminal knob (*a*) gives rise to a potential at the post-synaptic membrane, the *post-synaptic potential* (*b*) which is smaller in magnitude but has a much longer duration (from Stevens, C. F. 1966. *Neurophysiology, A Primer:* John Wiley, Fig. 3-2).

If a second impulse arrives at the same axon terminal before the first PSP has fully decayed, the amount of transmitter substance is increased, and the PSP will be greater. We therefore observe a summation of the two impulses, and since this is a summation in time, it is referred to as *temporal summation*.

Let us assume that a long train of impulses arrive at an axon terminal at a constant rate. The PSPs will be summed, giving an over-all PSP which is directly related to the frequency of impulse arrival. If the frequency is increased, the PSP will increase, and a new steady state PSP is established at a level where the decay rate of the transmitter equals the release rate from the axon terminal. The magnitude of the PSP therefore is a direct expression of the impulse frequency in the axon; in other words, the PSP is a *frequency modulated* potential. We can now see how a series of action potentials, each being an all-or-none event of constant magnitude, can be used to transmit information about a changing signal. For example, an increase in the magnitude of a signal from a sensory neuron is encoded and transmitted in its axon as an increase in the frequency of action potentials; on arrival at the synapse it is then decoded as a PSP of a magnitude directly related to the magnitude of the original signal.

Let us now return to the post-synaptic neuron and remember that there are numerous synapses on its surface. The PSP at any one synapse is not completely restricted to the area under the axon knob; it spreads to the immediate neighborhood with a decreasing magnitude as the distance increases. This spatial spread means that there is a slight change in membrane potential in the neighborhood of the synapse, and should an impulse arrive at another synapse within this area, the new PSP will be added to the existing potential. Such summation in space is known as *spatial summation.*

We have now seen that various inputs to a neuron, arriving at different synapses, will influence each other and can be summated through both temporal and spatial summation. These two types of summation form the basis for computations in each neuron, and thus in the entire nervous system.

Before we discuss these integration processes further, we must know what effect post-synaptic potentials have on the neuron. We said that the PSPs are usually insufficient to cause an action potential in the post-synaptic neuron. There is, however, an area of the neuron, called the *hillock,* near the origin of its axon, which fires action potentials if it is sufficiently stimulated by depolarization of the membrane. If a large number of impulses arrive at various synapses on a neuron, a combination of temporal and spatial summation may lead to a PSP sufficiently great to cause the passive spread to reach and depolarize the *axon hillock,* thus setting up an action potential in the axon.

Excitation and inhibition. Until now we have not been concerned with the polarity of the PSPs. A PSP can, in fact, change the normal membrane potential in either direction. If the post-synaptic membrane becomes slightly depolarized by the PSP, the direction of potential change is the same as that of an ordinary action potential. This is referred to as an *excitatory post-synaptic potential,* or *EPSP.* If the PSP has the opposite sign, however, it causes an increase in the normal membrane potential, or a hyperpolarization. The hyperpolarized PSP is the opposite of that which would normally lead to an action potential, and is known as an *inhibitory post-synaptic potential* or

IPSP, because it tends to inhibit the generation of an action potential.

Current evidence indicates that any one synapse is always of one kind or the other, and that a given axon causes the same type of PSP at all its synaptic terminals. One particular synapse therefore is always either inhibitory or excitatory and does not change.

Except for the polarity, the two kinds of post-synaptic potentials are of the same nature. The EPSPs transform an arriving train of impulses into a sustained depolarization, while the IPSPs convert a similar train of impulses into a sustained hyperpolarization. A given neuron therefore responds differently to arriving information, depending on what information it has recently received or is receiving from other sources. Under circumstances when a neuron is already partly depolarized, an additional excitatory potential may be sufficient to fire an action potential from the hillock, although at other times the same EPSP would be far from sufficient.

We now begin to see how a single neuron can carry out extensive integration of information received from various sources. Although the synapse is a one-way valve, the amount of integration and computation that results from excitatory and inhibitory synapses, in combination with spatial and temporal summation, makes the single neuron a formidable device in the computation processes in the nervous system.

We now will consider one further characteristic of neurons. Without going into details, we should mention that there is an important process known as *pre-synaptic inhibition.* An axon terminal may, for a length of time, remain slightly hyperpolarized, or slightly depolarized (without any action potential). An impulse which arrives during a period of slight depolarization will cause the release of a smaller than usual amount of transmitter, and therefore cause a reduced PSP. Since the reason for the smaller PSP is located in the axon terminal, this type of inhibition is pre-synaptic.

At first glance it may seem trivial whether inhibition is pre- or post-synaptic, but in fact it is important. Post-synaptic inhibition works by subtracting from excitatory PSPs which arrive at the neuron, and it therefore is a non-selective inhibition. Pre-synaptic inhibition, on the other hand, is highly selective, for it affects only signals which arrive at that particular synapse. It also has another characteristic, it is not a simple additive inhibition (as in post-synaptic summation), for it influences the transmission of each impulse in a larger train of an arriving signal. Thus, pre-synaptic inhibition increases both the specificity and the complexity of the integration that can take place at the neuronal level.

Control of muscle function

In the chapter on muscle we saw that crustacean muscles are controlled by a small number of axons (often three to five). Each axon branches to most or all of the muscle fibers within a muscle. Vertebrate muscles, in contrast, are innervated by a large number of axons, each branching to a small number of muscle fibers (*a motor unit*). This difference in innervation causes much of the control of crustacean muscle to take place peripherally, while vertebrate muscle is essentially under central control, directed by a highly complex central nervous system.

We can now see why this must be so. Myelinated axons permit the fast transmission of impulses without undue increase in axon diameter, thus, the necessary large number of connections can be carried in a nerve of moderate size. For the crustacean muscle, in contrast, rapid conduction requires a substantial increase in fiber size, and a finely graded central control which depends on a large number of fibers in the motor nerve is therefore not feasible. We have already described how crustacean muscles are nevertheless under precise control, although they may receive no more than two or three stimulatory and one or two inhibitory axons. We shall now examine the control of vertebrate muscle.

The contraction of vertebrate muscle is under precise central control, and there is no need for inhibitory axons to reach the muscle and modify the degree of contraction.* The vertebrate muscle system acts under the direction of a central computer which receives a great deal of information from various sources, including a very important part which comes as feedback directly from the muscles themselves. We shall use the admirable performance of this system as an example of what can be achieved through central control and integration.

The nerve to a vertebrate muscle has fibers which carry the impulses for stimulation of contraction, but in addition, it contains other fibers that carry sensory information from the muscle to the central nervous system. There are two major kinds of such sensory fibers (see fig. 13.10). Some come from small sensory units located in

---

* While vertebrate skeletal muscle apparently has no inhibitory innervation, this is not true for other kinds of vertebrate muscle. The heart, which is a striated muscle, has double innervation, both stimulatory and inhibitory. All smooth muscle receives both adrenergic and cholinergic fibers from the autonomic nervous system. For some smooth muscle the adrenergic fibers are inhibitory (bronchial muscle for example), for others they are stimulatory (vascular smooth muscle). Since cholinergic fibers are always antagonistic to the adrenergic fibers, they may likewise be either inhibitory or stimulatory.

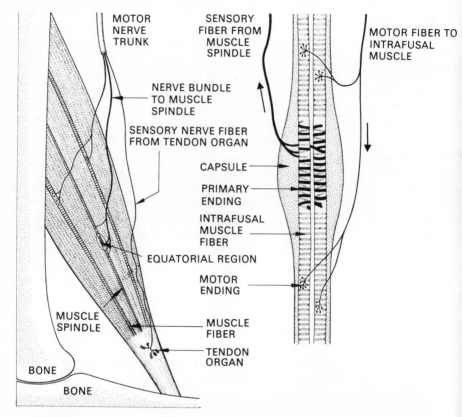

Fig. 13.10. A vertebrate muscle is equipped with sensory organs which transmit information to the central nervous system. The tendon organs provide information about the force of contraction and the muscle spindles primarily about the length of the muscle. This information is used as important feedback signals in the precise control of the muscle and its contraction (from Merton, P. A. May 1972. How We Control the Contraction of Our Muscles. © *Sci. Am.,* p. 35.).

the tendons, the *tendon organs,* others come from a specialized type of muscle fiber, known as *muscle spindles.* The tendon organs seem to be used for sensing the deformation produced by tension in the tendon, and thus give information about the force of muscle contraction. The muscle spindles, which we shall discuss in greater detail, are used for obtaining information about length, but as we shall see, they are even more useful because of the way their information is used in the feedback control of muscle contraction. In fact, it has been stated that the muscle spindles, next to the eye and the ear, are the most complex sensory organs in the body.

Let us use a highly simplified diagram (fig. 13.11). A person can easily, in response to a constant load, maintain a constant force in a

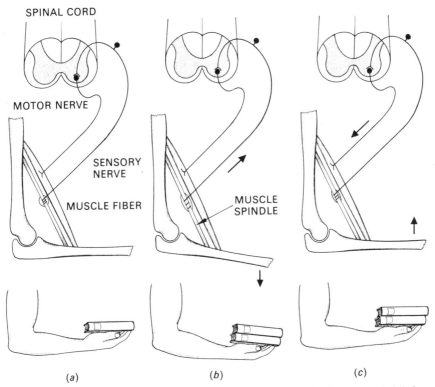

SPINAL CORD

MOTOR NERVE

SENSORY NERVE

MUSCLE FIBER

MUSCLE SPINDLE

(a)   (b)   (c)

Fig. 13.11. Feedback control of muscle contraction. A constant load on a muscle is balanced by a constant force (*a*). A sudden increase in the load (*b*) stretches the muscle, the muscle spindles send signals back to the spinal cord, where additional motor neurons are excited and send impulses back in the motor nerve, causing increased contraction to balance the load (*c*) (from Merton, P. A. May 1972. How We Control the Contraction of Our Muscles. © *Sci. Am.,* p. 35. All rights reserved).

muscle. If an additional load is unexpectedly added, it causes a stretching of the muscle which almost instantly is compensated for by an increased force of contraction in the muscle. This is due to a reflex which originates in the muscle spindles. When stretched, the spindles send impulses to the spinal cord where the sensory fibers make synaptic contact with motor neurons, which in turn cause sufficient contraction in the muscle to compensate for the initial stretching.

In reality, more than one synapse is usually involved, and there are connections from the spinal cord which carry information to different levels of the central nervous system, including the brain, but we need not be concerned with this complexity. The important matter is that the muscle spindles can serve to maintain a constant length of the muscle in the face of variations in the load, an obvious advantage in coordinating movements of all sorts.

Another important characteristic of the muscle spindles is that these are themselves contractile and innervated by motor axons which are separate from the axons that cause contraction of the regular muscle fibers. We now need to return to the detailed structure of the muscle spindles (see fig. 13.10). The spindle consists of a bundle of a few modified muscle fibers, called the *intrafusal muscle fibers* (Latin *fusus* means spindle). These fibers are contractile, but in the middle of the fiber the contractile apparatus is absent and the fiber has sensory nerve endings wrapped around it. Let us say that an intrafusal fiber is stimulated to contract; this stretches its central region which sends to the spinal cord a volley of impulses. These in turn stimulate the motor neurons, which respond by sending action potentials in the motor nerve, causing contraction in the muscle. When the contraction in the muscle has reached the same level as the contraction in the intrafusal fiber, the equatorial region is no more stretched and is therefore silenced. This is a very elegant system, which in a way is analogous to a mechanical servosystem. For example, in the power steering of a heavy automobile a small movement of the steering wheel is used to direct a motor which performs the heavy work of turning the wheels on the road. Servomechanisms are really no more than automatic feedback control mechanisms in which the controlled variable is the position of a mechanical device. Such systems are used widely in the design of all sorts of mechanical and industrial control processes.

The characteristic of intrafusal fibers which we have just described provides a very elegant mechanism for maintaining a graded speed of contraction against a variable load. If the contraction of a given muscle is directed by a certain slow contraction of the intrafusal fibers, an increase in the load on the muscle will only slow down its rate of contraction momentarily. The intrafusal fibers continue to contract at a set rate, but if the muscle follows too slowly because of the increased load, the equatorial region is stretched and immediately sends additional pulses to the spinal cord, which in turn increases the contraction of the muscle. The result is an automatic compensation for increases or decreases in the load which occur during muscle contraction.

When we turn to the question of how the muscles are used in the movements of the limbs, we find a need for additional control. We know that the muscles which serve in maintaining posture normally maintain some degree of light contraction; if they were not maintaining such tonus, the body would collapse, as we see in a person who faints. Therefore, to move a limb, it is not sufficient that a given set of muscles contract, it is also necessary that the antagonistic muscles relax. This is achieved through a relatively simple reflex mechanism which is shown in schematic form in fig. 13.12. Suppose that stimulation of the muscle spindles in muscle A evokes contraction of this muscle. Information about this contraction is received by certain neurons in the spinal cord, which are called *interneurons* * and act by inhibiting other neurons. The pathways are arranged in such a way that these *inhibitory interneurons* cause the inhibition of the motor neurons of the antagonistic muscle, B. The reciprocal action of antagonistic muscles, with one relaxing as the other contracts, is functionally very useful and could not take place without the aid of inhibitory neurons or other inhibition.

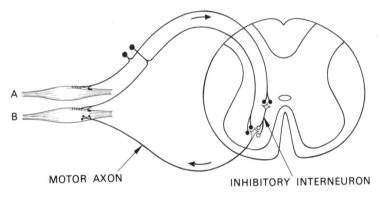

A

B

MOTOR AXON   INHIBITORY INTERNEURON

Fig. 13.12. When two antagonistic muscles (A and B) participate in the movement of a limb, the contraction of one muscle (A) causes impulses to be sent to the spinal cord. Here an inhibitory interneuron, by inhibiting the appropriate motorneuron, causes relaxation of the antagonistic muscle (B).

In addition to the control of antagonistic muscles, inhibition serves other important purposes as well. For example, if a contracted muscle is stretched very forcefully, the tendon organs are activated and initiate impulses which reach inhibitory interneurons in the spinal cord. This causes inhibition of contraction in the muscle that is being stretched, and the active resistance to stretching is decreased. This reflex contributes to the protection of the muscles from rupture when they are suddenly loaded. Without this protection there would be far more 'pulled muscles' than we now see.

The integration of muscular movement in locomotion is more

---

* Neurons whose axons do not extend outside the central nervous system are known as *interneurons*. They may be excitatory or inhibitory.

complex than described in the examples mentioned here. It depends on various commands from the brain, and also on information received from sensory organs such as eyes, ears, skin, etc. The central nervous system is a highly complex control system that acts on information and feedback signals from a variety of sources.

We shall now turn to the close connection between the central nervous system and the endocrine system, and in fact see that the central nervous system is not only a control system but is also a major producer of hormones.

Integration of endocrine and nervous control

A great number of physiological functions are under hormonal control. Oftentimes this is referred to as chemical control, in contrast to nervous control, but the term is both unfortunate and misleading. We have already seen that synaptic transmission of nerve impulses is of a chemical nature, and we shall soon see that the nervous system is a major factor in hormone production and control. Furthermore, some chemical substances that are normally formed in the body, carbon dioxide for example, have pronounced physiological effects and are important in regulation, but they are not hormones and have nothing to do with endocrine function.

The term hormone can best be defined as a substance which is released from a well-defined organ or structure and has a specific effect on some other discrete structure or function. The most important vertebrate hormones and their major functions are listed in table 13.3.

Until recently, endocrinology and hormone research was an empirical science which centered around a fairly uniform approach. An organ suspected of producing a hormone would be removed, thus depriving the organism of the normal source of this hormone. The results were observed and described, and if the symptoms could be relieved by injections of extracts of the organ, further work consisted of purification of the extract, chemical isolation of the active component, and eventually its chemical synthesis. Another important method of replacement therapy consisted of transplantation of hormone-producing tissue. These general methods sound simple, and are quite successful in research on organs that can be removed without hazard to life, such as the gonads. However, many hormone-producing organs have functions which are essential in other respects, and their removal leads to serious difficulties. The liver or kidneys, for example, produce hormones, but their removal rapidly leads to fatal effects unconnected with their role as a source of hormones.

Table 13.3. *Important vertebrate hormones*

Source	Hormone	Major functions
Hypothalamus	Releasing and release-inhibiting hormones acting on adenohypophysis	Hormones delivered via portal circulation to adenohypophysis. Functions, see table 13.4
Hypothalamus (via neurohypophysis)	Oxytocin	Stimulates contraction of uterine muscle; releases milk
	Antidiuretic hormone (ADH) *	Stimulates water reabsorption in kidney
(Neurohypophysis)	See above	
Adenohypophysis	Adrenocorticotropic hormone (ACTH)	Stimulates adrenal cortex
	Thyrotropic hormone (TSH)	Stimulates thyroid
	Follicle-stimulating hormone (FSH)	Stimulates ovarian follicle development; seminiferous tubule development in testes
	Luteinizing hormone (LH)	Stimulates conversion of ovarian follicle to corpus luteum; stimulates progesterone and testosterone production
	Prolactin	Stimulates milk production
	Melanocyte-stimulating hormone (MSH)	Stimulates dispersion of melanin in amphibian skin pigment cells
	Growth-stimulating hormone (GSH)	Stimulates growth (acts via liver, see below)
Liver	Somatomedin	Stimulates growth (is effector for hypophyseal GSH)
Adrenal cortex	Glucocorticoids (corticosterone, cortisone, hydrocortisone, etc.)	Regulate carbohydrate metabolism
	Mineralocorticoids (aldosterone, deoxycorticosterone, etc.)	Regulate sodium metabolism and excretion
	Cortical androgens, progesterone	Stimulate secondary sexual characteristics, predominantly male
Ovary	Estrogens	Initiate and maintain female secondary sexual characteristics; initiate periodic thickening of uterine mucosa; inhibit FSH.
	Progesterone	Cooperates with estrogens in stimulating female secondary characteristics; supports and glandularizes uterine mucosa; inhibits LH and FSH
Testis	Testosterone	Initiates and maintains male secondary sexual characteristics
Thyroid	Thyroxin and trriodothyronin	Stimulates oxidative metabolism; stimulates amphibian metamorphosis; inhibits TSH
	Calcitonin **	Inhibits excessive rise in blood calcium

Table 13.3. *Important vertebrate hormones* (*Continued*)

Source	Hormone	Major functions
Parathyroid	Parathormone	Increases blood calcium
Stomach	Gastrin	Stimulates secretion of gastric juice
Duodenum	Secretin	Stimulates secretion of pancreatic juice
	Pancreozymin ***	Stimulates secretion of pancreatic enzymes
	Cholecystokinin ***	Stimulates release of bile by gall bladder
	Enterogastrone	Inhibits gastric secretion
Pancreas	Insulin	Reduces blood glucose, stimulates formation and storage of carbohydrates
	Glucagon	Increases blood glucose by mobilization of glycogen from liver
Adrenal medulla	Adrenalin, noradrenalin	Augment sympathetic function; vasodilation in muscle, liver, lungs, vasoconstriction in many visceral organs. Increases blood sugar.

NOTE: The terms 'tropic hormone' and 'tropin' are equivalent; e.g. 'thyrotropic hormone' is also known as thyrotropin. In analogy with other use of the suffix -tropin (from tropein = to turn or change), the preferred spelling is tropin, rather than trophin (from trophos = feeder) (cf. phototropic and hypertrophy) (Stewart and Li, 1962).

  * ADH is also known as vasopressin, a name it received because in high concentrations it stimulates contraction of vascular smooth muscle and causes increased blood pressure.
 ** In mammals from the C-cells of the thyroid, in lower vertebrates from the ultimobranchial bodies.
*** Pancreozymin and cholecystokinin are believed to be identical.

The hypothalamic control system

The *hypothalamus* is located at the base of the brain, immediately above the hypophysis (or pituitary gland).* It is located just posterior to the optic chiasma and forms the floor of the third ventricle.

The hypothalamus is the seat of several nervous regulatory functions, notably temperature regulation and the regulation of intake of

* The terms *hypophysis* and *pituitary* are equivalent. The organ was known to the ancient Greeks as the hypophysis, which refers to its location under the brain. The name pituitary came into use later because it was mistakenly believed that this gland secretes the mucus of the nose (Latin *pituita* = mucus or slime). The use of the terms

water and of food. Regulation of body temperature has already been discussed as a feedback system. The role in regulation of food intake can be demonstrated by destruction of certain parts of the hypothalamus, and by electric stimulation, which, if properly located, make animals ingest huge quantities of food and grow abnormally obese. Regulation of water intake is shown similarly; electric stimulation or the injection of small amounts of hypertonic salt solution into certain areas of the hypothalamus cause animals to drink excessively. Goats, for example, have in this way been induced to drink, within minutes, 40% of their body weight in water.

The hypothalamus is of major importance in the endocrine system, for it controls the function of the hypophysis, which in turn has been called the master gland of the endocrine system. This control is mediated to the neurohypophysis via neural connections and to the adenohypophysis via special blood vessels, known as the portal circulation. It is now well established that there are at least nine and probably ten hypothalamic regulating hormones involved in the system which controls the hypophysis (see table 13.4).

The neurohypophysis contains two well-known hormones, antidiuretic hormone and oxytocin. It has been known for more than half a century that the neurohypophysis contains substances which affect the reabsorption of water in the kidney and are necessary for the formation of a concentrated urine. It is now known that the mammalian *antidiuretic hormone,* ADH, is identical with *vasopressin,* a hormone that, if injected in large and unphysiological amounts, causes a marked rise in blood pressure due to the constriction of arterioles. Another substance in the neurohypophysis, *oxytocin,* causes contraction of the smooth muscle of the uterus in pregnant females at term. These two substances are octapeptides; both are formed in the cells of the hypothalamus, and are transported along the axons to the nerve endings in the neurohypophysis from where they are released into the blood. The neurohypophysis therefore serves as a storage and release organ, or a *neurohemal organ,* for these two hormones.

Let us next examine the hormones from the adenohypophysis.

'anterior' and 'posterior' in connection with the hypophysis is unfortunate, for the relative position of these two major parts of the gland differs in different vertebrates. The *adenohypophysis,* also called the anterior lobe, is that part which is of glandular origin, formed by an invagination from the digestive tract (Greek *aden* = gland). The *neurohypophysis,* also called the posterior lobe, is that part of the hypophysis which is of neural origin. Further confusion is caused by the term 'intermediate lobe', a part which arises with the adenohypophysis but is included with the neurohypophysis in the 'posterior lobe'.

Table 13.4. *Hypothalamic hormones which control the release of hormones from the adenohypophysis.* (Schally, Arimura and Kastin, 1973)

	Abbreviation
Growth hormone releasing hormone	GH-RH
Growth hormone release-inhibiting hormone	GH-RIH
Prolactin releasing hormone	P-RH
Prolactin release-inhibiting hormone	P-RIH
Melanocyte-stimulating hormone releasing hormone	MSH-RH
Melanocyte-stimulating hormone release-inhibiting hormone	MSH-RIH
Cortiocotropin (ACTH) releasing hormone	C-RH
Thyrotropin releasing hormone	T-RH
Luteinizing hormone releasing hormone	LH-RH
Follicle-stimulating hormone releasing hormone	FSH-RH

For three of these, *growth hormone* (GH), *prolactin* (P), and *melanocyte-stimulating hormone* (MSH), the hypothalamic control is dual, one is inhibitory and one is stimulatory. The release of GH, prolactin, and MSH thus is not regulated by simple feedback systems, although feedback signals are undoubtedly involved in their control.

For the four other hormones, the regulation of their release seems to depend on a regular negative feedback system. *Corticotropin* (ACTH), *thyrotropin* (TSH), *luteinizing hormone* (LH), and *follicle stimulating hormone* (FSH) have as their target the adrenal cortex, the thyroid, and the gonads, respectively. These glands when stimulated release the appropriate hormones into the blood. These hormones (cortiocosteroids, thyroxin, and sex steroids) in turn inhibit, by negative feedback, the secretion from the adenohypophysis of the tropic hormones. There is now strong evidence that this inhibition acts via the hypothalamus (except that thyroxin may have a shorter feedback loop and act via the adenohypophysis).

The hypothalamic control system is summarized in fig. 13.13, which shows the central role of the hypothalamus in endocrine function, controlling the adenohypophysis through a series of regulating hormones. The adenohypophysis in turn controls the secretion from the thyroid, gonads, and adrenal cortex through a negative feedback system, while three other functions are regulated by dual sets of hormones of which one is stimulating and the other inhibiting.

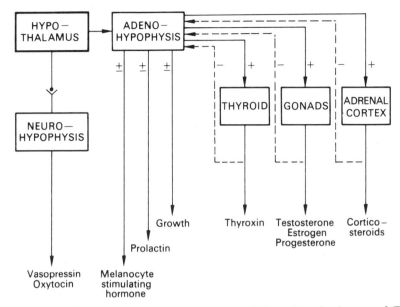

Fig. 13.13. Diagram of the central role of the hypothalamus in endocrine control. The neurohypophysis is controlled via neural pathways. The release of some hormones from the adenohypophysis is under dual (stimulatory and inhibitory) control (indicated by ±); the release of other hormones is controlled by negative feedback from the target organ (indicated by dashed lines and minus signs).

With this central role of the hypothalamus, the question arises of how this important organ in turn is controlled. We know that there are extensive nerve connections to the hypothalamus from other parts of the brain and that these are influenced by many environmental as well as emotional factors, by light cycles, seasons, etc. We can therefore conclude that the major portion of the entire endocrine system is under nervous control, acting through the central role of the hypothalamus in the control system.

Endocrines outside direct hypothalamic control

The parathyroid glands, which are located near to or imbedded in the thyroid gland, produce a hormone, *parathormone,* which causes the calcium ion concentration in the blood plasma to increase. An antagonistic hormone, *calcitonin,* which lowers the plasma calcium ion concentration, is released from the C-cells of the thyroid in mammals. An excess of parathormone causes the plasma calcium level to increase to an abnormal level, the calcium being taken primarily from the bones. This in turn results in an increased renal excretion of calcium, and because of the excessive loss of calcium the bones become decalcified. Removal of the parathyroids has the opposite effect, leading to a fall in blood calcium concentration, followed by tetanic muscle cramps caused by the low blood calcium level. The normal secretion from the parathyroid is apparently controlled by a simple feedback mechanism, because a high plasma level of calcium inhibits release of the hormone from the parathyroids, and a low level stimulates release. The effect of the calcium level on the release of calcitonin is the opposite; a high calcium level stimulates hormone release and a low level inhibits release. The calcium ion concentration in the blood plasma is thus controlled by the balance between two negative feedback systems.

Various organs of the digestive tract secrete hormones which control the secretory activity of the digestive glands. Their release is to a great extent under nervous control. In addition, the pancreas (specifically the small groups of endocrine cells known as the *islets of Langerhans*) produces two important hormones which have no direct role in digestion but are extremely important in carbohydrate metabolism. These are *insulin* and *glucagon*. The most obvious effect of insulin is that it decreases the glucose concentration in the blood by stimulating the formation and deposition of glycogen in the cells from the blood glucose. In the disease diabetes mellitus, the production or release of insulin is inadequate (there may be an insulin deficiency or a tissue resistance to the effects of insulin). As a result the blood glucose concentration remains too high, and glucose is excreted in the urine, often in vast quantities. However, insulin has other and more complex roles in intermediary metabolism, and there are several additional serious effects of insulin deficiency.

Insulin is produced by the so-called beta-cells of the islets of Langerhans, while glucagon is produced by the alpha cells. If the glucose concentration in the blood rises above the normal level (for example after ingestion of carbohydrates), insulin is released and stimulates glucose uptake in the muscles and the formation of glycogen. If the blood glucose level falls too low, release of glucagon causes mobilization of glucose from the liver and an increase in blood glucose concentration. The glucagon effect is opposite to the effect of insulin; the normal blood glucose level thus is regulated by two antagonistic hormones, one inhibitory and one stimulating, each acting through a negative feedback loop.

The last endocrine organ that we shall discuss is the *adrenal medulla*. The cells of the adrenal medulla are of neural origin, and belong to the sympathetic nervous system. The adrenal medulla secretes two hormones, *noradrenalin* and *adrenalin*. Adrenalin causes acceleration of the heart beat, increased blood pressure, increase in blood sugar through conversion of glycogen to glucose, vasodilation and increased blood flow in heart muscle, lungs, and skeletal muscle, but causes vasoconstriction and decrease in blood flow in smooth muscle, digestive tract, and skin. These effects of adrenalin are well known as the 'fight or flight syndrome', which occurs in response to fear, pain, and anger. These reactions help mobilize the physical resources of the body in response to an emergency situation.

Noradrenalin has very similar effects, and the differences between the two hormones are mostly quantitative. Adrenalin, for example, has a more potent effect on the heart rate, while in some organs noradrenalin gives more vasoconstriction. The two hormones act through receptor sites known as alpha-receptors (most affected by noradrenalin) and beta-receptors (most affected by adrenalin); many organs have both alpha- and beta-receptors, and in addition each receptor type is to some degree sensitive to both hormones. This explains why it is difficult to give a definite listing of similarities and differences between the two hormones – there is too much overlap in their effects.*

Chemical transmission and transmitter substances

The effects of adrenalin and noradrenalin released into the blood from the adrenals resemble the effects of a general stimulation of the sympathetic nervous system.† This similarity is easy to understand when we realize that the cells of the adrenal medulla are of neural origin and analogous to the large sympathetic ganglia located along the spinal column. The similarity is even more clearly understood when we learn that noradrenalin is the terminal transmitter substance of the sympathetic nervous system as a whole.

We have already seen that most synaptic transmission is accomplished through the release of a transmitter substance, acetylcholine, and also that acetylcholine is released from the terminal

---

* It should be mentioned that noradrenalin has been found in the nervous system of insects and annelids, but it seems to be absent in crustaceans, mollusca, echinoderms, and many other phyla.
† The major components of the peripheral vertebrate nervous system are sensory nerves, motor nerves, and the autonomic nervous system. The autonomic nervous system in turn is divided into the *sympathetic* and the *parasympathetic* nervous systems, which jointly control the function of internal organs such as heart, stomach, glands, intestine, kidney, and so on, i.e. functions which are outside voluntary control. These organs are innervated by both sympathetic and parasympathetic nerves which act antagonistically, one being stimulatory and the other inhibitory.

end of the motor nerves at the neuromuscular junction. The autonomic nervous system also exerts its effects through transmitter substances. The sympathetic nerves release noradrenalin from their nerve terminals, and the parasympathetic nerves release acetylcholine. These two substances have antagonistic effects; when one stimulates, the other inhibits. One of the best known examples of such antagonistic action is found in the control of the heart beat. The heart is innervated by two nerves. One is a branch of the splanchnic nerve (sympathetic) which when stimulated releases noradrenalin and speeds up the heart beat, the other is a branch of the vagus nerve, which is parasympathetic and when stimulated releases acetylcholine from its nerve endings and causes the heart to slow down.

The effect of acetylcholine in slowing down the heart is a striking example of the fact that the response to a given substance depends on the target organ rather than on the nature of the transmitter substance. Ordinary striated muscle is stimulated to contract through the release of acetylcholine, while the effect of acetylcholine on the heart is to slow down the rate of contraction.

The major aspects of transmitter action in the peripheral nervous system of vertebrates are summarized in fig. 13.14. We can now see that acetylcholine is the common transmitter substance, not only for the terminals of the motor nerves and synapses within the parasympathetic nervous system, but also in sympathetic ganglia and for the nerves that innervate the adrenal medulla. Thus, in the nerve-to-

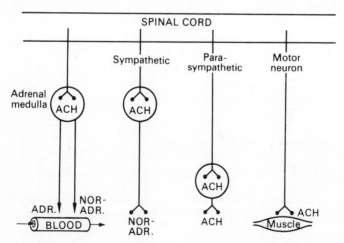

Fig. 13.14. Diagram of the peripheral transmitter substances in vertebrates. For details, see text. ACH, acetylcholine; ADR., adrenaline; NORADR., noradrenaline.

nerve transmission outside the brain and spinal cord, acetylcholine seems to be the universal transmitter substance in vertebrates. Within the central nervous system the situation is less clear.

Although acetylcholine is so widespread in peripheral synaptic transmission, there is strong evidence that other substances function in transmission, particularly within the vertebrate central nervous system. For invertebrates a number of other substances, most of them amines, seem to be important transmitter substances. The structure of a few of these substances is shown in fig. 13.15. The reason that the information about transmitter substances is so uncertain is that it is extremely difficult to establish with certainty that a given substance has a role as a transmitter.

The greatest difficulty is that many of the substances which occur in the organism in small amounts have rather profound physio-

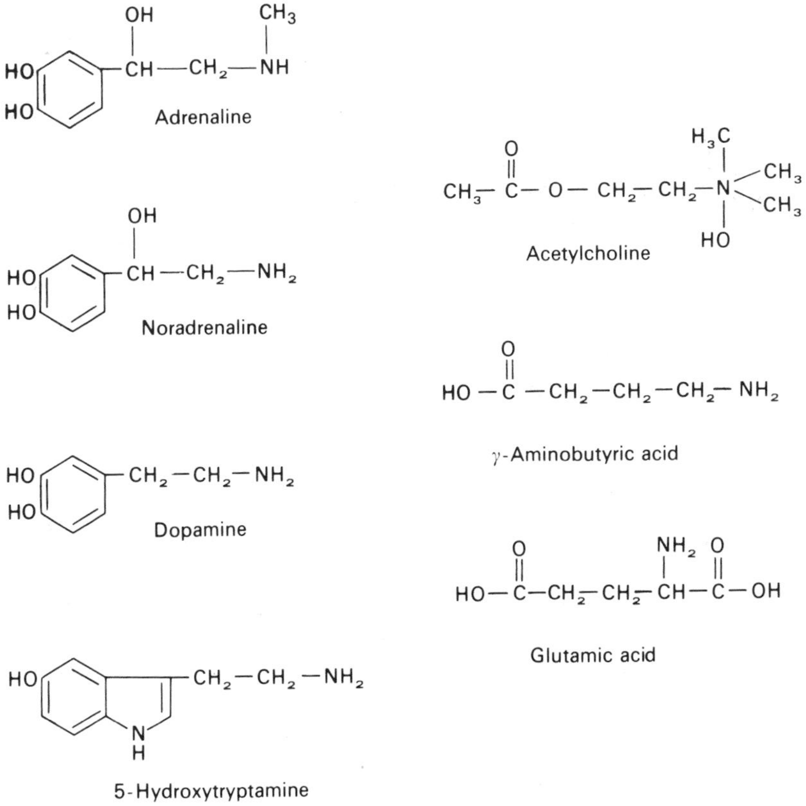

Fig. 13.15. Structural formulas of some compounds that are or are believed to be transmitter substances.

logical effects, in particular on the nervous system. The fact that a substance causes a reaction which is similar to some normally occurring nervous phenomenon does not establish this substance as a transmitter substance. A number of criteria should be fulfilled before this can be established, among others the following:

(1) The substance or a precursor should be present in the neuron from which the suspected transmitter is released.

(2) The substance should be present in the extra-cellular fluid in the region of the activated synapse.

(3) When applied to the post-synaptic structure, the substance should mimic the action of the transmitter.

(4) A mechanism for removal or inactivation of the substance should be present, either through enzymatic inactivation or through specific uptake or reabsorption.

In addition, pharmacological agents have helped to clarify mechanisms by interfering with synthesis, release, removal, inactivation, or target, but they do not provide direct evidence for transmitter function.

The crucial demonstration of a transmitter substance consists in proving that, at the arrival of an action potential at the pre-synaptic ending, the substance is released in sufficient quantity to produce the observed physiological effect on the post-synaptic structure.

It is indeed difficult to meet all these criteria, and very often the technical obstacles have been insurmountable. As a result much work on suspected transmitter substances, especially in invertebrate animals, has been unsatisfactory because it has not progressed much beyond the demonstration of an effect of some suspected agent. Therefore, present information about transmitter substances is not very adequate and is insufficient for generalizations. Nevertheless, two substances should be mentioned here because they are invariably recognized as being highly active in many invertebrates.

5-Hydroxytryptamine, often written 5HT, is found in many invertebrate neurons, notably of molluscs and arthropods. 5HT is not restricted to nervous structures, however, but the concentration in other tissues is usually lower. Large amounts of it are found in certain animal toxins, such as those from octopus, the snail *Murex,* and the venom of wasps. The heart of bivalve molluscs is quite sensitive to 5HT, and there is good evidence that this substance is a normal transmitter for acceleration of the mollusc heart. The effect of 5HT on vertebrate blood pressure (which it increases) was known before the substance was chemically identified. It was then given the name *serotonin,* a name which is still in use. 5HT is also suspected of being a transmitter within the vertebrate central nervous system, but whether it has a normal role in vertebrates still remains uncertain.

Gamma-aminobutyric acid, often written GABA, is another substance suspected of being a transmitter. It occurs in certain regions of the mammalian central nervous system, both in the brain and spinal cord. When applied to crustacean synapses, it mimics the action of inhibitor neurons, and it may well be a normal transmitter. Although it is found in the vertebrate nervous system, it is again impossible to say with certainty whether it has a normal function as a transmitter substance.

Some amino acids, notably *glutamic acid* and *aspartic acid,* have also been suggested as transmitter substances. They occur both in vertebrate and in invertebrate nervous tissue, but substantial evidence for their normal role is still lacking. One major difficulty in establishing amino acids as transmitter substances is that they also occur normally in the animals, and their mere presence therefore carries no particular significance. Another problem is that it is extremely difficult to establish an inactivation mechanism specifically for amino acids as related to their suspected role as transmitter substances, for all the amino acids are regular participants in normal metabolic pathways.

Control and integration in invertebrates

In the more highly organized invertebrates the control of physiological function is under both nervous and endocrine control. As in vertebrates, the nervous system serves for rapid communication, which is essential for actions that are related to escape, feeding, mating, and so on. As in vertebrates, an endocrine system produces hormones which control slower processes, such as growth, maturation, and many other metabolic functions. The nervous system has a direct and primary role in hormone production and the connection between the nervous and endocrine systems seems to be even closer than in vertebrates.

The function and role of nervous systems.

We have seen that in vertebrates the central nervous system coordinates a large number of separately controlled functions, both nervous and endocrine. The apparent contrast between nervous control (usually rapid) and hormonal control (usually slow to very slow) breaks down under closer scrutiny. Not only is the transmission from neuron to neuron, and from neuron to effector, usually of a chemical nature, but the central nervous system itself is a major producer of hormones and exerts much of its over-all regulatory function in this way.

When we examine the role of nervous systems in the less highly organized invertebrates, we will need a more precise definition of the term *nervous system* than we used for vertebrates, where the meaning of the term is self evident. A nervous system is an assembly of neurons which is specialized for repeated and organized transmission of information from sensory receptor sites to neurons, or between neurons, or from neurons to effectors (e.g. muscles, glands, etc.). This definition precludes the existence of a nervous system in unicellular animals (protozoans), although functions within such cells may be coordinated with the aid of conducting organelles.

It is, on the whole, reasonably easy to recognize a neuron or nerve cell, mostly based on its membrane potential and the ability to produce repeated action potentials. However, it may be difficult to recognize whether a 'neurosecretory' cell is truly a modified nerve cell or not. Its location in association with nerve cells is helpful, and further information can be gained from recognition of action potentials, structural characteristics, and the function of similar structures in related species.

Invertebrate neurons have the same characteristics as vertebrate neurons. They are able to produce *action potentials* which consist of changes in membrane permeability and potential. Such action potentials are of an *all-or-none* nature and can be propagated in a nerve fiber, or axon. Typical neurons are able to produce repeated brief action potentials which last from a fraction of a millisecond to several milliseconds, and such neurons occur in all major phyla except protozoa and sponges. The transfer of information from one neuron to another takes place at *synapses,* which usually show evidence of *chemical transmission.* As in vertebrates, the nerve membranes of the two neurons most often remain spatially separated, and transmission is achieved through the release of an intermediary *chemical transmitter substance.* It appears, however, that there may be direct electrical transmission at certain synapses where the nerve membranes of the two cells become contiguous, a situation that also may occur, although rarely, at some vertebrate synapses. Typical *post-synaptic potentials* seem to be universally present in at least four major phyla, molluscs, annelids, arthropods, and chordates. These may be *excitatory post-synaptic potentials (EPSP)* or *inhibitory post-synaptic potentials (IPSP).* In general, the excitatory post-synaptic potentials consist in a decrease in membrane polarization (in the direction of complete depolarization), and inhibitory potentials in an increase in membrane potential. These general characteristics, then, are in principle similar in invertebrate and vertebrate neurons.

Neurosecretion. The importance of specialized neurosecretory cells and organs has been recognized in invertebrates long before their importance was generally established in vertebrates. Neurosecretory cells are difficult to define and characterize so that they are easily and unequivocally recognized; the major approach for locating and pinpointing such cells still depends on the use of certain stains that have a special affinity for neurosecretory cells.

Neurosecretory organs consist of a group of neurons which are the source of secretion. These cells have nerve fibers, in which the secreted agent is transported, and which usually terminate in close association with a vascular structure. Here they form a *neurohemal organ,* where the secreted product is stored and released. Neurohemal organs are very widespread, but it is not certain that they are universally present in connection with all neurosecretory organs.

In all the more highly organized animal phyla neurosecretory systems are central to the control and operation of endocrine mechanisms. We have already seen the importance of neurosecretion in the control of vertebrate endocrine organs; among invertebrates neurosecretion seems to be even more important. We shall discuss later, examples of endocrine function in some insects, where neurosecretion and neurohemal organs have a central role in the control systems. Neurosecretory systems are also important in annelids and crustaceans, in which they are involved in control of reproduction, metabolism, molting, pigmentation, etc. Although some of the endocrine glands of vertebrates as well as invertebrates are outside direct neuroendocrine control, the normal function of most or all glands ultimately depends on regulation from the nervous system, either directly through nerve pathways or through hormonal mechanisms. It seems that control by neurosecretory mechanisms is the general rule in all animals in which endocrine mechanisms play a major role.

Hormones and endocrine function

Endocrine function is best understood and has been most extensively studied in those invertebrates which are morphologically highly organized. There are two reasons for this situation; one is that a highly organized animal such as an insect needs more detailed control and integration than a less organized animal such as a sea anemone. The other reason that the more highly organized animals are better understood is that many functions are delegated to specialized organs. This permits experiments such as removal of organs, reimplantation, extraction, and other procedures that are difficult or impossible in less highly organized animals.

Most invertebrate hormones are different from vertebrate hormones, both in chemical constitution and in their effects. Many vertebrate hormones which have profound effects on vertebrate growth, development of gonads, metabolic processes, and so on, have no effect whatsoever if introduced into invertebrate animals.

Endocrine function has been demonstrated in many invertebrates, but the mere fact that some extract has a physiological effect does not demonstrate that the extract contains a substance that has a normal role as a hormone in this animal. More careful study is required and the necessary criteria include the demonstration of facts such as a normal occurrence of the substance in question, demonstration of where it originates, examination of the effects of removal of its source, or of the blocking of its secretion, the purification and isolation of the pure substance, and eventually its synthesis, including the demonstration that the synthetic product is effective in similar concentrations to the natural product. As a result, detailed knowledge of invertebrate endocrinology is restricted to a relatively few forms.

The groups in which organs of internal secretion have been clearly demonstrated and are reasonably well understood are molluscs (notably cephalapods), annelids, crustaceans, insects, and tunicates. All these are highly organized animals where the complexity of function and the specialization of organs facilitate endocrine research. In particular, insects have been studied in great detail for reasons which include their easy accessibility, the ease of collecting and breeding them, their tolerance to drastic surgical procedures, and their economic importance.

Insect endocrinology. Hormones play a major role in the physiology of insects, especially in growth, molting, pupation, and metamorphosis into the mature adult form. These phenomena have been studied in a number of different species, but considering the fact that there may be around one million different insect species, the fraction is very small indeed. However, there is a striking similarity in the endocrine function of different insects. For example, the *hemimetabolic insects,* which are those which go through a number of molts as they change from the newly hatched form and gradually develop into the adult, employ the same hormones as the *holometabolic insects,* which remain in the completely larval form through a number of molts, whereupon they transform into a pupa from which the fully formed adult emerges. We shall discuss one of each type.

As an example of a hemimetabolic insect we shall use the South American blood-sucking bug, *Rhodnius,* which is a relative of the bed bug. *Rhodnius* hatches from the egg as a tiny blood-sucking bug, a nymph, which through five stages (*instars*) gradually develops into the adult form. In each stage the nymph must obtain a meal of blood before it can develop into the next stage by shedding its old cuticle. After it has sucked blood, about four weeks pass by, the nymph then sheds the old cuticle and increases in size by filling the tracheal system with air before the new cuticle hardens. It is then ready to suck blood again, but if the opportunity does not present itself, it can survive for many months. However, if it obtains blood, another molt occurs about four weeks later.

The molt of *Rhodnius* is stimulated by a hormone, *ecdysone* (also known as molting hormone), which is secreted after the blood meal has been ingested. This hormone is secreted by the two prothoracic glands, located in the thorax of the insect. The prothoracic glands, in turn, are stimulated by a hormone secreted by specialized neurosecretory cells in the brain.

The adult *Rhodnius* has wings and mature gonads, and differs from the larval forms in other respects as well. What directs this development? It turns out that the reason for a lack of adult characters in the earlier nymphal stages is due to another hormone, the *juvenile hormone,* which is secreted by the corpus allatum, a tiny cluster of cells just behind the brain. The juvenile hormone determines that the new cuticle will have nymphal characteristics, and its presence therefore prevents the formation of adult characters. Thus, in all the early nymphal stages, when juvenile hormone is present, ecdysone causes molting which results in a larger nymph. In the fourth nymphal stage the amount of juvenile hormone declines, and the fifth nymphal stage shows the beginning of the development of wings. In the very last molt the juvenile hormone is virtually absent, and an adult emerges (see fig. 13.16).

Let us examine the evidence for endocrine control of these events. If a fifth-stage *Rhodnius* is decapitated shortly after a blood meal, the final molt does not take place, although the headless animal may live for more than a year. This makes it appear that the molting hormone, ecdysone, is produced in the head. However, if the decapitation takes place after a certain 'critical period' which occurs around the seventh day after the meal, the animal will molt and develop into a headless adult. Thus, the head is needed only early in the period and not for the molt itself. The difference can be shown to be due to

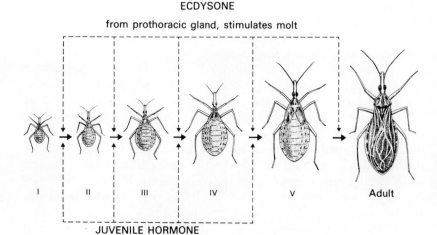

ECDYSONE

from prothoracic gland, stimulates molt

I   II   III   IV   V   Adult

JUVENILE HORMONE

from corpora allata, causes formation of juvenile cuticle

Fig. 13.16. The bloodsucking bug *Rhodnius* develops into an adult through five molts. Each molt is caused by the release of the hormone ecdysone from the prothoracic glands. During the first four molts the presence of juvenile hormone (from the corpora allata) causes the new cuticle to be of the larval type, thus preventing the formation of an adult. During the last molt the absence of juvenile hormone results in a fully developed adult.

endocrine factors, for if blood from an insect decapitated after the 'critical period' is transferred to an insect decapitated before the 'critical period', the latter is induced to molt. The explanation is that during the critical period the brain releases a hormone, the *brain hormone,** which stimulates the prothoracic gland. If the decapitation occurs early, the prothoracic gland is never stimulated, if decapitation is later, the brain hormone has taken effect and is not needed, and ecdysone is released and induces molting in due time.

The simplest way of performing a blood transfusion in these insects is to connect them with a short glass capillary that can be fixed into the neck of the decapitated insect with a droplet of wax (see fig. 13.17). If we join together a fifth-stage nymph that has had a blood meal and has passed through the critical period, and a small first-stage nymph recently emerged from the egg and not yet fed, the small nymph will develop adult characteristics such as wings and genital organs and appear as a midget adult. If the experiment is

* The brain is the source of several hormones, but it is conventional among insect endocrinologists to use the term *brain hormone* for the agent which specifically activates the prothoracic gland and leads to release of ecdysone. It is also called the *prothoracicotropic hormone.*

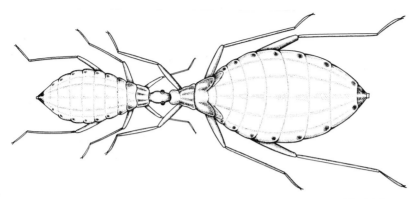

Fig. 13.17. The transfer of hormones via the blood can be demonstrated by joining together two animals. In this case a fourth stage of a *Rhodnius* nymph (to the left) with only the tip of the head removed (leaving the brain intact) has been joined to a head-less fifth-stage nymph (to the right) (from Wigglesworth, V. B. 1959. *The Control of Growth and Form.* © Cornell University. By permission of Cornell University Press).

repeated with an older first-stage nymph, this animal is again in-duced to molt, but does not acquire adult characteristics because by this time it has produced and released enough juvenile hormone to molt into a juvenile nymph.

Ecdysone is obviously needed for each molt, and its effects are merely modified by the juvenile hormone which is secreted later. Therefore, if decapitation is carried out after the brain hormone has been released and has stimulated the prothoracic gland, molting will be induced; the type of molt depends on whether or not the decapi-tation occurred before or after juvenile hormone was secreted. By accurately timing the decapitation in relation to the blood meal and the 'critical period', it has been possible to obtain small, headless *Rhodnius* with adult morphological characteristics.

The same endocrine functions have been extensively studied in a number of moths, including the silk moth (*Bombyx mori*) and the cecropia moth (*Hyalophora cecropia*). The larva of the cecropia silk-moth goes through four molts (five stages), all initiated by the brain hormone and its action on the prothoracic gland, releasing ec-dysone. In the younger larval stages the *corpora allata* release juvenile hormone, so that the larva molts into another larva, which because of its loose and wrinkled skin can grow and increase in size as it feeds. At the end of the fifth larval stage, the corpora allata cease to secrete juvenile hormone, and at the next molt, a harder cuticle is formed. The larval tissues break down and are transformed into pupal structures, and after a final molt, when no juvenile hormone remains, a fully developed adult moth emerges (see fig. 13.18).

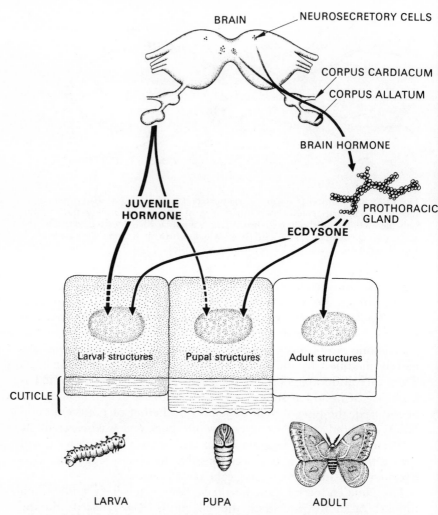

Fig. 13.18. Endocrine organs which control development of the cecropia moth. Each molt is caused by the hormone ecdysone, which is released from the prothoracic gland. This gland in turn is controlled by the brain hormone, produced by neurosecretory cells in the brain. The juvenile hormone, secreted from the corpora allata, causes the new cuticle to be of the larval type; in the last larval molt less juvenile hormone is present, and a pupa with a heavier pupal cuticle is formed; in the final molt juvenile hormone is absent and an adult moth emerges (from Schneiderman, H. A. and Gilbert, L. I. 1964. Control of Growth and Development in Insects. *Science*, **143**, 325–33, Fig. 1. © American Association for the Advancement of Science).

The simplest way to confirm the function of the corpus allatum is to remove this organ in the young silkworm during one of the early instars. If this is done, molting takes place as usual (the brain–prothoracic system is untouched), but instead of developing another

Plate 19. *Silk worm.* This photo shows the larva of the silk moth, *Hyalophora cecropia,* in the first, third, and fifth instars. After the fifth instar the larva pupates and goes through a complete metamorphosis whereupon the adult moth emerges (Charles Walcott, State University of New York, Stony Brook).

larva, the molt results in a diminutive pupa, from which, in turn, a tiny adult emerges (see fig. 13.19). There is no other effect of the removal of the corpus allatum (and thus the juvenile hormone) so that, except for size, a normal-looking adult develops from the pupa.

The role of the corpora allata can be confirmed by transplantation. If a number of corpora allata from early larvae are implanted into the last larval instar, the next molt does not produce a pupa, but instead an over-sized larva. This larva may continue to grow, and by introducing additional corpora allata it is possible to produce a giant larva which can molt into a giant adult. There is a limit to this development, however, it is not likely that the animal will survive beyond an artificial seventh instar.

The chemical structure of the hormones we have discussed should be mentioned for comparison with vertebrate hormones. Ecdysone is a steroid (Karlson and Sekeris, 1966) which was originally isolated and crystallized by extraction of 500 kg silkworm pupae, which yielded 25 mg of the crystalline hormone. The full chemical structure of the hormone was finally established on material extracted from a batch of 1000 kg dried silkworm pupae (corresponding to

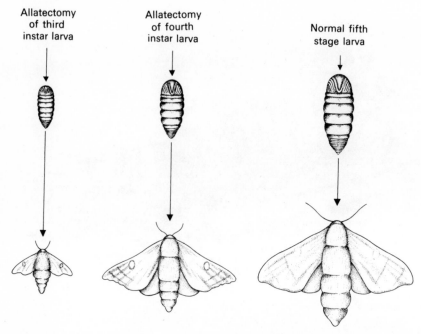

Fig. 13.19. If the corpora allata are removed from a young moth larva, the larva is deprived of juvenile hormone, and instead of going through a larval molt, the animal pupates and a small adult emerges (Highnam and Hill, 1969).

about four tons fresh weight). It is interesting that ecdysone is a steroid because of the widespread occurrence of hormones of steroid nature in vertebrates.

The juvenile hormone, on the other hand, is entirely different and is similar in structure to terpenes. A large number of compounds with juvenile hormone activity have been isolated, and many have been synthesized, some apparently far more potent than the naturally occurring ones. The major natural hormone from cecropia has the following structure (Röller, Dahm, Sweely and Trost, 1967):

$$H_3C\overset{\displaystyle CH_3}{\underset{\displaystyle CH_2}{\underset{\displaystyle O-CH}{\overset{|}{\underset{|}{C}}}}} CH_2 \underset{CH_2}{\overset{CH_3}{\underset{|}{\overset{|}{\underset{CH_2}{\overset{|}{C}}}}}} CH_2 \underset{CH_2}{\overset{CH_3}{\underset{|}{\overset{|}{C}}}} \overset{O}{\overset{\|}{C}}-OCH_3$$

The synthesis of compounds with juvenile hormone activity has some challenging perspectives since they may be potential insecticides. These compounds are effective in extremely small amounts;

and if they are applied at suitable periods during the life history of insects, they can prevent normal adult development, and thus reproduction. The advantage of juvenile hormones as insecticides is that insects are less likely to develop immunity to substances on which they normally depend, than to the various poisonous substances now in use. Since some of the artificial 'juvenile hormones' act more specifically on some insects than on others, it may be feasible to develop insecticides which are more specific for one insect species than for another, a far more attractive prospect than the use of highly toxic substances such as DDT, which kill desirable and undesirable insects alike. Also, since juvenile hormone has little or no effect on many other animals, at least not on vertebrate predators, such substances probably will have less disastrous ecological effects.

An accidental discovery of a naturally occurring juvenile hormone substance suggests that plants in fact already use such substances for their own protection. This was discovered when a Czech investigator, Dr. Karel Sláma, spent a year at Harvard University and brought with him his stock of the bug *Pyrrhocoris*. In the laboratory these bugs are reared on linden seeds with paper towels as a climbing surface. At Harvard they would not develop into adults after the fifth instar, but developed into oversized sixth or even seventh instar immature forms. It appeared as if the bugs had unintentionally been exposed to contamination with juvenile hormone, and a systematic comparison of all possible differences between the conditions at Harvard and in Europe revealed that the paper towels used at Harvard contained a substance with juvenile hormone activity (Sláma and Williams, 1965; 1966). It turned out that all American-made paper, including newspapers, contained the substance. It was not added during manufacture, but originated in the wood from fir trees used for paper manufacture in the United States. Paper of Japanese or European origin had no such effect. The substance, called the *paper factor,* has been isolated and has some chemical resemblance to the juvenile hormone. This suggests that many terpenes, which occur commonly in plant material, especially in evergreen trees, may be natural protective substances that have evolved as a counter-measure to insect pests.

The insect hormones we have discussed so far are all concerned with the control of growth and development, but many other physiological processes are also under endocrine control. As an example, we shall briefly describe a hormone that is concerned with the regulation of water excretion in a blood-sucking insect, which, because of its feeding habits, is periodically subjected to a heavy water load.

The blood-sucking bug *Rhodnius* is faced with this problem each time it takes a blood meal. Immediately after feeding it excretes a large volume of dilute urine, thus eliminating most of the excess water. This sudden surge in the excretion of water is regulated by a hormone which stimulates water excretion.

The action of this hormone has been studied on isolated Malpighian tubules of *Rhodnius*. These organs can continue to secrete urine if they are placed in blood from *Rhodnius* or in a suitable saline solution. If they are placed in the blood from a recently fed *Rhodnius*, the tubules secrete a copious urine at a much higher rate than if they are placed in blood from an unfed animal. This suggests that the blood contains a hormone which is directly responsible for stimulating the secretion of urine, in other words, a *diuretic hormone*. Extracts of various known hormone-producing organs of *Rhodnius* have no effect on urine secretion, but extracts of the ganglia from the first abdominal segment are effective. These ganglia contain several groups of large neurosecretory cells, and 97% of the diuretic activity is contained in the most posterior group of these.

After a *Rhodnius* has sucked blood, urine production increases within less than a minute, showing that the hormone must be released very rapidly, undoubtedly stimulated by nervous reflexes from the feeding process or from the abdominal expansion caused by the feeding (Maddrell, 1963).

The few examples of insect hormones we have described here give no more than an indication of the multitude and complexity of endocrine regulation, and the field of invertebrate endocrinology will continue to be an area of much fruitful exploration and research into control mechanisms.

REFERENCES

AIDLEY, D. J. (1971). *The Physiology of Excitable Cells,* Cambridge, England: Cambridge University Press, 468 pp.

BULLOCK, T. H. and HORRIDGE, G. A. (1965). *Structure and Function in the Nervous Systems of Invertebrates,* vols. I and II. San Francisco: W. H. Freeman & Co., 1719 pp.

ECCLES, J. (1961). The nature of central inhibition. *Proc. Roy. Soc. Lond. B,* **153,** 445–476.

ECCLES, J. (1965). The synapse. *Sci. Am.* **212,** 56–66.

FURSHPAN, E. J. (1964). 'Electrical transmission' at an excitatory synapse in a vertebrate brain. *Science,* **144,** 878–80.

HIGHNAM, K. C. and HILL, L. (1969). *The Comparative Endocrinology of the Invertebrates,* London: Edward Arnold Publ. Ltd, 270 pp.

HODGKIN, A. L. (1954). A note on conduction velocity. *J. Physiol.,* **125,** 221–4.

HODLER, J., STÄMPFLI, R. and TASAKI, I. (1951). Über die Wirkung internodaler Abkühlung auf die Erregungsleitung in der isolierten markhaltigen Nervenfaser des Frosches. *Pflügers Archiv.* **253**, 380–385.

HURSH, J. B. (1939). Conduction velocity and diameter of nerve fibers. *Am. J. Physiol.*, **127**, 131–9.

KARLSON, P. and SEKERIS, C. E. (1966). Ecdysone, an insect steroid hormone, and its mode of action. *Recent Progress in Hormone Research*, **22**, 473–502.

MADDRELL, S.H.P. (1963). Excretion in the blood-sucking bug, *Rhodnius prolixus* Stål. *J. Exp. Biol.*, **40**, 247–56.

MERTON, P. A. (1972). How we control the contraction of our muscles. *Sci. Amer.* **226**, 30–7.

PHILLIS, J. W. (1970). *The Pharmacology of Synapses.* Oxford, England: Pergamon Press. 358 pp.

RÖLLER, H., DAHM, K. H., SWEELY, C. C. and TROST, B. M. (1967). The structure of the juvenile hormone. *Angew. Chem., Internat. Edit.*, **6**, 179–80.

RUSHTON, W.A.H. (1951). A theory of the effects of fibre size in the medulated nerve. *J. Physiol.*, **115**, 101–22.

SCHALLY, A., ARIMURA, A. and KASTIN, A. J. (1973). Hypothalamic regulatory hormones. *Science*, **179**, 341–9.

SCHNEIDERMAN, H. A. and GILBERT, L. I. (1964). Control of growth and development in insects. *Science*, **143**, 325–33.

SLÁMA, K. and WILLIAMS, C. M. (1965). Juvenile hormone activity for the bug *Pyrrhocoris apterus*. *Proc. Nat. Acad. Sci.*, **54**, 411–14.

SLÁMA, K. and WILLIAMS, C. M. (1966). The juvenile hormone. V. The sensitivity of the bug, *Pyrrhocoris apterus*, to a hormonally active factor in American paper-pulp. *Biol. Bull.*, **130**, 235–46.

STEWART, J. and LI, C. H. (1962). On the use of -tropin or -trophin in connection with anterior pituitary hormones. *Science*, **137**, 336–7.

TASAKI, I. (1959). Conduction of the nerve impulse. *Handbook of Physiology*, sect. 1: *Neurophysiology*, vol. I, pp. 75–121, Washington, D.C.: American Physiological Society.

WIGGLESWORTH, V. B. (1959). *The Control of Growth and Form: A Study of the Epidermal Cell in an Insect*, Ithaca, N.Y.: Cornell University Press, 140 pp.

USEFUL REFERENCE MATERIAL FOR PART V

AIDLEY, D. J. (1971). *The Physiology of Excitable Cells,* Cambridge, England: Cambridge University Press, 468 pp.

ALEXANDER, R. MCN. (1968). *Animal Mechanics,* London: Sidgwick & Jackson, 346 pp.

AUTRUM, H. *et al.* (editorial Board). *Handbook of Sensory Physiology,* **1,** *Principles of Receptor Physiology;* **2,** *Somatosensory System;* **3,** *Enteroceptors;* **4,** *Chemical Senses;* **5,** *Auditory System;* **6,** *Vestibular System;* **7,** *Photochemistry of Vision;* **8,** *Perception,* New York: Springer-Verlag, 1971–74.

BENSON, G. K. and PHILLIPS, J. G. (eds) (1970). *Hormones and the Environment, Mem. Soc. Endocrinol.,* **18,** London: Cambridge University Press, 629 pp.

BULLOCK, T. H and HORRIDGE, G. A. (1965). *Structure and Function in the Nervous Systems of Invertebrates,* vols I and II, San Francisco: W. H. Freeman and Co., 1719 pp.

BÜLBRING, E., BRADING, A. F., JONES, A. W. and TOMITA, T. (1970). *Smooth Muscle,* London: Arnold. 676 pp.

CLOSE, R. I. (1972). Dynamic properties of mammalian skeletal muscles. *Physiol. Rev.,* **52,** 129–97.

DAW, N. W. (1973). Neurophysiology of color vision. *Physiol. Rev.,* **53,** 571–611.

DE VRIES, H. (1956). Physical aspects of the sense organs. *Progr. Biophys.,* **6,** 207–64.

ECCLES, J. C. (1973). *The Understanding of the Brain,* New York: McGraw-Hill, 328 pp.

ERULKAR, S. D. (1972). Comparative aspects of spatial localization of sound. *Physiol. Rev.,* **52,** 237–360.

FIELD, J., MAGOUN, H. W. and HALL, V. C. (eds) *Handbook of Physiology.* Sect. 1: *Neurophysiology,* vol. I, 1–780 (1959); vol. II, 781–1440 (1960); vol. III, 1441–966 (1960); Washington, D.C.: American Physiological Society.

GERSHENFELD, H. M. (1973). Chemical transmission in invertebrate central nervous systems and neuromuscular junctions. *Physiol. Rev.,* **53,** 1–119.

GRAY, J. (1968). *Animal Locomotion,* London: Weidenfeld and Nicolson.

HORRIDGE, G. A. (1968). *Interneurons. Their Origin, Action, Specificity, Growth, and Plasticity,* San Francisco: W. H. Freeman Co., 436 pp.

HUBBARD, J. I. (1973). Microphysiology of vertebrate neuromuscular transmission. *Physiol. Rev.,* **53,** 674–723.

KARPOVICH, P. V. and SINNING, W. E (1971). *Physiology of Muscular Activity,* 7th edn, Philadelphia, Pa.: W. B. Saunders Co., 374 pp.

KATZ, B. (1966). *Nerve, Muscle, and Synapse,* New York: McGraw-Hill Book Co., 193 pp.

ROEDER, K. D. (1967). *Nerve Cells and Insect Behavior* (revised edn), Cambridge, Mass.: Harvard University Press, 238 pp.

TURNER, C. D. and BAGNARA, J. T. (1971). *General Endocrinology* (5th edn), Philadelphia, Pa.: W. B. Saunders Co., 659 pp.

WEBER, A. and MURRAY, J. M. (1973). Molecular control mechanisms in muscle contraction. *Physiol. Rev.,* **53,** 612–73.

WESTFALL, J. A. *et al.* (1973). Symposium on invertebrate neuromuscular systems. *Am. Zool.,* **13,** 233–445.

APPENDIX 1

Measurements and units

The purpose of physiological measurement is to determine the magnitude of certain physical quantities. This appendix provides (1) a list of names and symbols used for some physical quantities, (2) names of units and their symbols, and (3) a brief list of conversion factors. The units, their symbols, and the signs connected with their use are standardized and internationally adopted through a long-standing permanent international body, *The General Conference on Weights and Measures* (Conférence Générale des Poids et Mésures). The system of units is known as *The International System of Units*, or commonly the *SI System* (=*Système Internationale*).

PHYSICAL QUANTITIES

The names of some physical quantities and *recommended* symbols are given below. Alternative symbols are separated by commas.

length	l	time	t
height	h	velocity	u, v
radius	r	frequency	f
area	A	force	F
volume	V	work	W
mass	m	energy	E, W
density (m/V)	ρ	power	P
pressure	p	temperature	T

The symbol for a physical quantity should be a single letter of the Latin or Greek alphabet, and when necessary, subscripts (and/or superscripts) are used to indicate a specific meaning of the symbol. (Example: volume of the lung, V_{lu}; blood volume, V_{bl}.)

Equations which represent physical or physiological relationships use these and similar symbols for the quantities. Symbols for units (m, kg, s, etc.) are *not* to be used in such equations.

UNITS

The SI System uses seven *base units* for seven dimensionally independent basic physical quantities; each base unit is represented by a *symbol*. These symbols are *mandatory*.

Physical quantity	SI base unit	Symbol
length	meter	m
mass	kilogram	kg
time	second	s
thermodynamic temperature	kelvin	K
amount of substance	mole	mol
electric current	ampere	A
luminous intensity	candela	cd

The SI System recognizes several *derived units* (as well as some supplementary units). Although some of these have special symbols, they are derived from and defined in terms of the base units. A few of particular interest in physiology are listed below:

Physical quantity	SI derived unit	Symbol for unit	Definition in terms of base units
force	newton	N	$m \, kg \, s^{-2}$
energy	joule	J	$m^2 \, kg \, s^{-2}$
power	watt	W	$m^2 \, kg \, s^{-3}$
pressure	pascal	Pa	$m^{-1} \, kg \, s^{-2}$
electric potential difference	volt	V	$m^2 \, kg \, s^{-3} \, A^{-1}$
electric resistance	ohm	Ω	$m^2 \, kg \, s^{-3} \, A^{-2}$
electric charge	coulomb	C	$s \, A$
frequency	hertz	Hz	s^{-1}

The following should be noted about the use of these units and symbols:

The symbols are printed in Roman (upright) letters. The symbols are not abbreviations and are not followed by a period sign, except at the end of a sentence.

The plural form of a symbol is unchanged, and the ending -s should not be used.

Certain units are derived from proper names (e.g. Volta), the symbols for these units are capital Roman (upright) letters (V), but the units are not capitalized (volt).

There is frequent need for smaller and larger decimal multiples of the units. These are obtained by the use of the following prefixes (six additional prefixes are approved but not listed here):

Prefix	Symbol	Equivalent
mega	M	10^6
kilo	k	10^3
deci	d	10^{-1}
centi	c	10^{-2}
milli	m	10^{-3}
micro	μ	10^{-6}
nano	n	10^{-9}
pico	p	10^{-12}

These prefixes may be attached to any SI base and derived unit. (Large multiples of seconds, however, are rarely expressed this way.) The combination of prefix and symbol is regarded as a single symbol, and compound prefixes are not allowed.

Products or quotients of two or more units are treated according to the rules of mathematics. Note that the use of more than one solidus (/) is ambiguous. Thus, a specific rate of oxygen consumption can be written:

$$\mu l/(kg\ s), \text{ or } \mu l\ (kg\ s)^{-1}, \text{ or } \mu l\ kg^{-1}\ s^{-1}, \text{ but } not\ \mu l/kg/s.$$

Several circumstances make it impractical, at the present time, to express all physiological measurements in SI base and derived units. One reason is the large body of existing information, including widely known and used tables of physical and chemical constants, which use other units (e.g. calorie). Another reason is that methods in common use make it simpler to use certain traditional units (e.g. gram force, rather than the SI unit, the newton). This practice will continue (and has been used in this book), but should progressively be abandoned in favor of SI units.

USEFUL CONVERSION FACTORS

Traditional units are commonly used in those fields of physiology which concern energy (in the area of metabolic rates, etc.) and pressure (in the area of gas exchange, etc.).

Force

1 kilogram force $= 9.807$ N
1 newton $= 0.102$ kilogram force

Energy, work (force × distance)

1 cal $= 4.184$ J
1 joule $= 0.239$ cal

One liter O_2 (at 0 °C, 760 mm Hg), when used in oxidative metabolism, is often equated with 4.8 kcal; thus: 1 liter $O_2 \simeq 20.112$ kJ.

Power (work per unit time)

$1 \text{ cal s}^{-1} = 4.184 \text{ J s}^{-1} = 4.184 \text{ W}$

$1 \text{ kcal h}^{-1} = 1.163 \text{ W}$

$1 \text{ watt} = 0.239 \text{ cal s}^{-1} = 14.34 \text{ cal min}^{-1} = 860.4 \text{ cal h}^{-1}$

Pressure (force per unit area)

$1 \text{ atm} = 1.013 \times 10^5 \text{ N m}^{-2}$

$1 \text{ mm Hg at } 0 \text{ °C (1 torr)} = 1.333 \times 10^2 \text{ N m}^{-2}$

$1 \text{ kN m}^{-2} = 7.50 \text{ mm Hg} = 9.869 \times 10^{-6} \text{ atm}$

Volume

1 liter is (since 1964) defined as exactly equal to one dm^3 (0.001 m^3), i.e. $\text{ml} = \text{cm}^3$, and $\mu\text{l} = \text{mm}^3$.

$1 \text{ U.S. gallon} = 3.785 \text{ liter}$

$1 \text{ U.S. quart } = 0.946 \text{ liter}$

Length

$1 \text{ inch} = 25.4 \text{ mm}$

$1 \text{ foot} = 304.8 \text{ mm} = 0.3048 \text{ m}$

$1 \text{ yard} = 0.914 \text{ m}$

$1 \text{ mile} = 1609.3 \text{ m}$

$1 \text{ ångström (Å)} = 10^{-10} \text{ m} = 0.1 \text{ nm}$

Mass

$1 \text{ pound (avoirdupois)} = 0.4536 \text{ kg}$

$1 \text{ ounce } = 0.0283 \text{ kg}$

Temperature

The temperature unit of one degree Celsius (°C) * equals the temperature unit of one kelvin (K).

The Celsius temperature is defined as the excess of the thermodynamic temperature over 273.15 K.

For practical purposes, the melting point of ice at 1 atm pressure is 0 °C, the triple point of ice, water, and water vapor is 0.01 °C, and the boiling point of water at 1 atm pressure is 100 °C.

GENERAL REFERENCES

MECHTLY, E. A. (1969). *The International System of Units. Physical Constants and Conversion Factors* (revised). NASA SP-9012. Washington, D.C.: National Aeronautics and Space Administration, 19 pp.

PAGE, C. H. and VIGOUREUX, P. (1972). *The International System of Units (SI)*. NBS Special Publ. No. 330, Washington, D.C.: National Bureau of Standards, 42 pp.

SYMBOLS COMMITTEE OF THE ROYAL SOCIETY (1971). *Quantities, Units, and Symbols,* London, England: The Royal Society, 48 pp.

* Also known as 'centigrade'.

APPENDIX 2

Diffusion

Diffusion is a fundamental process in the movement of materials. Those diffusion processes that are of physiological importance take place over short distances, a fraction of a millimeter, a few micrometers, or less. Examples are the diffusion of respiratory gases across the respiratory membranes, or the diffusion of ions across the nerve membrane during the action potential. Over longer distances, diffusion is physiologically unimportant, and transport takes place by mass movement (convection), e.g. protoplasmic streaming or circulation of blood. Why is diffusion limited to short-distance transport? The reason is that diffusion time increases with the square of the diffusion distance.

Let us consider the time it takes for particles to travel a given distance by diffusion:

The average distance, R, covered by a diffusing substance, depends on the average length, l, of each track in the random thermal motion of the particles (the mean free path), and on the number of individual tracks, n. The relationship is $R = l\sqrt{n}$. Since the number of individual tracks is proportional to time, t, we can write $R \propto l\sqrt{t}$ or $t \propto \frac{R^2}{l^2}$. This means that diffusion time, t, increases with the square of the average distance, R, covered by the diffusing substance.

What does this mean for diffusion in the tissues of an organism? Suppose that we impose a step-wise increase in oxygen concentration at a given point, A. Oxygen will diffuse into the surrounding tissues, and will attain an average diffusion distance of 1 μm in about 10^{-4} s. An average diffusion distance ten times as long will require 100 times the time, and so on. Thus, for an average diffusion distance of 10 μm the time required will be 10^{-2} s, for 1 mm it will be 100 s, for 10 mm, 10^4 s (nearly 3 h), and for 1 meter, the time will be 10 000 times as long again, or more than 3 years. The distance from the lungs of a man to his extremities is in the order of 1 meter, and this makes it clear that transport by diffusion alone is impossibly slow and that mass movement is necessary.

It is important to realize that we have discussed the time needed for diffusion over a certain distance when we have imposed a change in a concentration of the diffusing substance at one point and observe the effect of this change at another point at a certain distance.

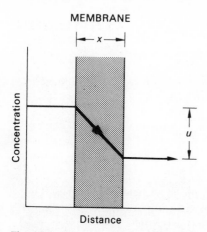

MEMBRANE

Fig. A2.1. Diagram of steady-state diffusion through a homogeneous membrane.

Let us now consider a steady-state diffusion over an unchanged gradient (as we find in many physiological processes).

The diffusion over a constant concentration gradient (see fig. A2.1) is expressed by Fick's equation, which is:

$$dQ = -D A \frac{du}{dx} dt$$

in which Q is the quantity of a substance diffused, A is the area through which diffusion takes place, u is the concentration at point x (du/dx thus being the concentration gradient), t is time, and D is the diffusion coefficient. Diffusion coefficients for various substances can be found in standard tabular works.*

From this equation we can see that the amount of a substance diffusing is proportional to elapsed time. This also makes sense, intuitively. To take a physiological example, alveolar air has a partial pressure of oxygen of 100 mm Hg, and mixed venous blood arriving at the lung may have an oxygen partial pressure (tension) of 40 mm Hg. Normally, these concentrations remain very constant, and oxygen therefore diffuses from the air through the alveolar wall at a constant rate. It is now obvious that in two minutes twice as much oxygen diffuses as in one minute, in ten minutes ten times as much, and so on. Thus, if a constant gradient is maintained, the amount of the substance diffusing is directly proportional to time.

Another consequence of Fick's equation is that if the diffusion distance, x is increased, the amount of substance diffusing, Q, will decrease. If x is doubled, Q is reduced to half, if x is ten times as large, Q is one-tenth, and so on. The amount diffusing is thus inversely proportional to the diffusing distance, but again, this applies only when a steady state is maintained. This requires that we must have, on one side of the membrane, a constant supply

---

* At 20 °C the diffusion coefficient, D, for oxygen in air is $D = 0.196$ cm^2 s^{-1}, and for oxygen in water $D = 0.183 \times 10^{-4}$ cm^2 s^{-1} (Altman, 1958).

or an infinite pool of the diffusing substance, and on the other side a constant lower concentration maintained by the continuous removal of the diffusing substance. This, for practical purposes, applies to our example of the lung.

The diffusion constant. The diffusion constant, K, which was used for the calculation on p. 15, differs from the diffusion coefficient, D, used in Fick's equation. The need for the use of the diffusion constant in physiology arose in connection with the study of diffusion of gases between air and tissues, notably in the mammalian lung. For this purpose it is practical to refer the diffusion rate of a gas, say oxygen, to the partial pressure of the gas in the alveolar air. The diffusion coefficient, D, cannot be used in this case, for it requires knowledge of the concentration gradient (du/dx) in the tissue. The concentration in the tissue is difficult to determine directly, and cannot be calculated unless the solubility for the particular gas in that particular tissue is known. The diffusion constant circumvents this problem by relating the rate of diffusion directly to the partial pressure of the gas in the gas phase.

The diffusion constant, K, is defined as the number of cm^3 of a gas which diffuses in 1 min through an area of 1 cm^2 when the pressure gradient is 1 atm per cm. These units can be written out and reduced to a simpler form as follows:

$$\frac{cm^3}{min \times cm^2 \times atm \times cm^{-1}} = cm^2\ atm^{-1}\ min^{-1}.$$

Diffusion constants have been determined for only a small number of animal tissues; some examples are listed in table A2.1. These constants are empirically determined and include the solubility of gas in the tissues in question. The figures show that the diffusion in those tissues that have a rel-

Table A2.1. *The diffusion constant (K) for oxygen at 20 °C in various materials. (Note that these figures are not diffusion coefficients.)* (From Krogh, 1919)

	(cm^2 atm^{-1} min^{-1})
Air	11
Water	34 $\times 10^{-6}$
Gelatine	28 $\times 10^{-6}$
Muscle	14 $\times 10^{-6}$
Connective tissue	11.5 $\times 10^{-6}$
Chitin *	1.3 $\times 10^{-6}$
Rubber	7.7 $\times 10^{-6}$

* The diffusion constant for chitin often is given incorrectly as ten times the value listed here. This is due to a misprint in a book by Krogh (1941) which has been copied by later authors.

atively high water content, muscle and connective tissue, is somewhat lower than for water but of the same order of magnitude. The diffusion through chitin (the hard cuticle of insects), on the other hand, is lower by an order of magnitude. It is worth noting that the diffusion of oxygen through rubber is considerable, about one-half of the rate for muscle. This is of importance in physiological experimentation, for if rubber tubing is used for connections, oxygen can diffuse through the walls of the tubing, and carbon dioxide can diffuse even faster because of its high solubility. The actual rate of diffusion in rubber depends on the kind of rubber used; some kinds may be more permeable than those listed and others may be much less permeable.

GENERAL REFERENCES

ALTMAN, P. L. (1958). *Handbook of Respiration,* Philadelphia: W. B. Saunders Co., 403 pp.

KROGH, A. (1919). The rate of diffusion of gases through animal tissues, with some remarks on the coefficient of invasion. *J. Physiol.,* **52,** 391–608.

KROGH, A. (1968). *The Comparative Physiology of Respiratory Mechanisms* (reprint edn), New York: Dover Publ. Inc., 172 pp.

APPENDIX 3

Logarithmic and exponential equations

Logarithmic equations are of the general form:

$$y = b \cdot x^a \tag{1}$$

The logarithmic form of this equation is:

$$\log y = \log b + a \log x \tag{2}$$

Equation (2) shows that $\log y$ is a linear function of $\log x$, i.e. by plotting $\log y$ against $\log x$ we obtain a straight line with the slope a.

Example: Rate of oxygen consumption ($\dot{V}_{O_2}$) plotted against body mass (M_b) (see e.g. fig. 6.12, p. 242).

$$\dot{V}_{O_2} = k \cdot M_b^a \tag{3}$$

The independent variable, body mass (M_b) is customarily plotted on the abscissa. The rate of oxygen consumption per unit mass is obtained by dividing both sides of equation (3) by M_b:

$$\frac{\dot{V}_{O_2}}{M_b} = k \frac{M_b^a}{M_b} = k \, M_b^{(a-1)} \tag{4}$$

Example: See fig. 6.10, p. 238.

Exponential equations are of the general form:

$$y = b \cdot a^x \tag{5}$$

The logarithmic form of this equation is:

$$\log y = \log b + x \log a \tag{6}$$

Equation (6) shows that $\log y$ is a linear function of x, and plotting $\log y$ against x gives a straight line. Use of semi-log graph paper (linear abscissa and logarithmic ordinate) for plotting y against x gives the same result.

Example: Rate of oxygen consumption plotted against temperature (see e.g. fig. 7.4, p. 268).

APPENDIX 4

Thermodynamic expression of temperature effects

The term Q_{10} (defined on p. 263) is widely used in biology, for it gives a simple and easily understood expression of the effect of temperature on rates (e.g. rate of oxygen consumption). In physical chemistry, however, temperature effects are expressed differently, based on thermodynamic considerations of the frequency of molecular collision as a function of temperature. The equation, as developed by Arrhenius, is usually given in the following form

$$K_{T_2} = K_{T_1} \cdot e^{-\mu/R[(T_2 - T_1)/T_2 T_1]} \tag{1}$$

in which K stands for rate constant, T for thermodynamic temperature (absolute temperature, in kelvin), μ for activation energy, R for the universal gas constant, and e is the base of the natural logarithm.

We can obtain a thermodynamically correct expression for Q_{10} (the ratio of two rates 10 degrees C apart) by solving equation (1) for $\dfrac{K_{T_2}}{K_{T_1}}$ when $T_2 - T_1 = 10$. We then find:

$$Q_{10} = \frac{K_T + 10}{K_T} = e^{-\mu/R\{10/[T(T + 10)]\}} \tag{2}$$

When we examine equation (2), we find that Q_{10} is a function of and changes with temperature (T), and thus cannot be a constant. R, e, and (for a given process) μ are constants, and Q_{10} therefore decreases with increasing temperature. An example of this change in Q_{10} with temperature is given in table A4.1 for $\mu = 12\,286$ (a value of μ chosen to correspond to $Q_{10} = 2.00$ for the temperature interval between 20 and 30 °C). If $\mu = 19\,506$, Q_{10} is 3.0 for the interval between 20 and 30 °C, but will be higher at low temperatures and lower at high temperatures. The use of Q_{10} to describe temperature effects therefore lacks a theoretical foundation in thermodynamics; however, Q_{10} is a highly descriptive term which is convenient and much used in biology.

Table A4.1. *The change in Q_{10} for temperature intervals between 0 °C and 50 °C, calculated for a constant activation energy, μ, of 12 286*

T_1 (°C)	T_2 (°C)	Q_{10}
0	10	2.22
10	20	2.10
20	30	2.00
30	40	1.91
40	50	1.84

APPENDIX 5

Solutions and osmosis

Water is the universal solvent for virtually all biological reactions and related phenomena. The following brief summary therefore is restricted to aqueous solutions. For a more extensive treatment, the reader is referred to an introductory chemistry or physical chemistry text.

SOLUTIONS AND CONCENTRATIONS

A one *molar* solution contains one mole of solute per liter of solution. Thus a one molar solution of sodium chloride contains 58.45 g NaCl per liter solution. In biological work it is common to refer to a one-thousandth of a one molar solution, a millimolar solution. This helps to eliminate small decimal fractions.

A one *molal* solution contains one mole of solute per kg water. For example, a one molal solution of sodium chloride contains 58.45 g NaCl and one kg water. Since the volume of this solution is more than one liter, its concentration is less than 1.0 molar.

The molarity designation is often used in biological work and is convenient because it is easy to determine the volume of a given sample of a biological fluid while it is cumbersome or impossible to determine its exact water content. For example, the volume of a blood sample is easily measured, and its content of sodium can readily be determined. Referring the sodium concentration to the amount of water in the sample is more difficult, for both plasma and red cells contain many other solutes, including substantial amounts of proteins. In physical chemistry, the molality concept is more useful; this includes the consideration of osmotic phenomena.

Concentrations are sometimes given as *per cent* (%). Unfortunately, this expression is ambiguous and there is no accepted convention for how it should be used. This often leads to confusion. For example, does a 1% solution of sodium chloride contain 1 g NaCl and 99 g water? Or does it contain 1 g NaCl in 100 ml solution? Or since the term *per cent* means *per hundred,* does it mean 1 g sodium chloride per 100 g water? Even more confusing is the expression *milligram per cent* (mg %), which should logically mean 'milligrams per hundred milligrams'. Usually it means the number of milligrams of a substance per hundred ml solution. Ambiguities can be avoided by using the appropriate units, such as '100 mg glucose per 100 ml blood', or often better, the molar concentration.

Addition of modifiers such as 'weight per cent' helps little. It commonly, but not always, refers to parts of solute by weight per 100 parts of solution by weight. Accordingly, a 1% solution of sucrose contains 1 g sucrose in 99 g water. However, if a salt which contains water of crystallization is weighed out, both the amount of the salt and the total amount of water may be uncertain.

Strictly, the term *per cent* should express a non-dimensional fraction. This means that the ratio of the two quantities should be expressed in the same units which cancel out in the division. It should be noted that gas concentrations can properly be expressed in per cent. For example, dry atmospheric air contains 20.95% oxygen, whether the quantity of oxygen relative to the total is given in number of molecules, in ml, or in pressure units.

OSMOSIS

Osmosis is the movement of water between two solutions with different solute concentrations when the solutions are separated by a semi-permeable membrane, i.e. a membrane which is permeable to water but impermeable to the solute.

Consider the system in fig. A5.1. Water will tend to move from compartment B into compartment A, causing a pressure, the osmotic pressure, to

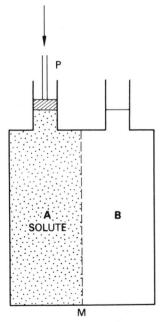

Fig. A5.1. When two solutions are separated by a semi-permeable membrane, M, water tends to move from the more dilute solution to the more concentrated solution. If the membrane is rigid, the water movement can be prevented by exerting a pressure on the piston P, equal to the difference in osmotic pressure between the two solutions.

develop in A. The movement of water can be counteracted by pressure exerted on the plunger, and when this pressure exactly balances the osmotic pressure there is no net movement of water.

The force which tends to move water, π, is defined by van't Hoff's equation:

$$\pi V = n R T$$

in which V = volume, n = number of moles of solute particles, R = the universal gas constant, and T = absolute temperature. This equation is similar to the gas equation $P V = n R T$ (in which P = pressure).

A one molal solution of an ideal non-electrolyte has an osmotic pressure of 22.4 atm at 0 °C (273 K). This equals the pressure exerted by one mole of an ideal gas when the volume is one liter.

Solutes depress the freezing point of water, and a one molal solution of an ideal non-electrolyte has a freezing point of −1.86 °C. Solutes also reduce the vapor pressure and increase the boiling point of a solution.

The osmotic pressure of a solution ideally depends only on the number of particles present, and not on their size or nature. An electrolyte such as sodium chloride, which in water ionizes into Na^+ and Cl^-, therefore has an osmotic concentration higher than expected from its molality.

Unfortunately, observations of actual osmotic pressures in solutions of electrolytes do not completely conform to the expected results as estimated from the expected number of ions. This is because of interaction between the positive and negative ions and also interaction between the ions and the water molecules. The osmotic pressure of a solution of an electrolyte therefore cannot be calculated directly from the molal concentration of the electrolyte, and it is necessary to introduce an empirical *osmotic coefficient, ϕ.*

If sodium chloride were completely ionized in solution, we would expect a 1.00 molal solution to have a freezing point depression of 2×1.86 °C, or 3.72 °C. In fact, a one molal solution of NaCl freezes at −3.38 °C. Its osmotic coefficient therefore is $3.38/3.72 = 0.91$. The osmotic coefficient varies with the concentration of the electrolyte and differs from one electrolyte to another.

THE DONNAN EQUILIBRIUM

Assume a system similar to that in fig. A5.1, but with the compartments separated by a rigid membrane permeable to water and electrolytes, but not to a large ionized molecule, say a negatively charged protein, Pr^- (fig. A5.2). Let the electrolyte be sodium chloride.

If the compartments have rigid walls, pressure on the plunger P prevents any net movement of water through the membrane. It can now be shown that at equilibrium:

$$[Na^+]_A \times [Cl^-]_A = [Na^+]_B \times [Cl^-]_B$$

Rearranging the terms, we find that the ratio of the diffusible cations on one side to those on the other is inversely proportional to the ratio of the anions on the two sides:

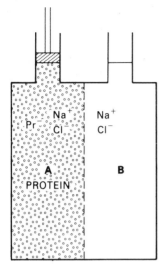

Fig. A5.2. An osmotic system consisting of two chambers, **A** and **B**, separated by a rigid membrane which is freely permeable to water and small ions but impermeable to large ions such as a negatively charged protein, Pr−. The distribution of ions and the osmotic pressure at equilibrium (the Donnan equilibrium) are described in the text.

$$\frac{[Na^+]_A}{[Na^+]_B} = \frac{[Cl^-]_B}{[Cl^-]_A} = r$$

These ratios are known as the *Donnan constant, r.*

Assume that the initial concentration of sodium chloride is the same on the two sides of the membrane, and that compartment A also contains sodium proteinate, NaPr, dissociated as Na^+ and Pr^-. Thus, the total concentration of Na^+ in compartment A is higher than in B, and sodium ions will diffuse into B. This will give an excess of Na^+ ions relative to Cl^- ions in compartment B, and an equal number of chloride ions will therefore cross the membrane into compartment B. However, this tendency is counteracted by the resulting higher chloride concentration in B than in A. At equilibrium there will be unequal concentrations of ions across the membrane and a net negative charge in chamber A relative to chamber B. This equilibrium situation with the unequal distribution of diffusible ions and the resulting electric charge is known as the *Donnan equilibrium.*

We should note that the potential difference is due to the presence of the non-diffusible anion in chamber A. Since the system is in equilibrium, the potential is maintained without expenditure of energy and remains indefinitely.

The potential difference (E) can be calculated from Nernst's equation:

$$E = \frac{RT}{zF} \ln \frac{[Na^+]_A}{[Na^+]_B}$$

in which $R =$ the universal gas constant, $T =$ absolute temperature, $z =$ valency of the ion (in this case unity), and $F =$ Faraday's constant (96 500 coulomb).

The pressure applied to the plunger in fig. A5.2 necessary to prevent osmotic flow of water into compartment A is known as the *colloidal osmotic pressure*. In the event that the pressure on the plunger exceeds the colloidal osmotic pressure, water and electrolytes are forced out through the membrane while the non-permeable protein is withheld in A. This process is called *ultrafiltration*. The osmotic pressure of the ultrafiltrate will equal the osmotic pressure of the solution in compartment B.

APPENDIX 6

The animal kingdom

An abbreviated classification of living animals

PHYLUM PROTOZOA
    CLASS FLAGELLATA. Flagellates
    CLASS SARCODINA. Protozoans with pseudopods. *Amoeba*
    CLASS CILIATA. Ciliates. *Paramecium*

PHYLUM PORIFERA. Sponges

PHYLUM COELENTERATA (or Cnidaria)
    CLASS HYDROZOA. Hydrozoans. *Hydra*
    CLASS SCYPHOZOA. Jellyfishes
    CLASS ANTHOZOA. Sea anemones and corals

PHYLUM CTENOPHORA. Comb jellies

PHYLUM PLATYHELMINTHES. Flatworms
    CLASS TURBELLARIA. Free-living flatworms
    CLASS TREMATODA. Flukes
    CLASS CESTODA. Tapeworms

PHYLUM NEMATODA. Round worms. *Ascaris*

PHYLUM MOLLUSCA. Molluscs
    CLASS AMPHINEURA. Chitons
    CLASS GASTROPODA. Snails
    CLASS PELECYPODA. Bivalve molluscs, clams, mussels, etc.
    CLASS CEPHALOPODA. Squids, octopuses, and *Nautilus*

PHYLUM ANNELIDA. Segmented worms
    CLASS POLYCHAETA. Marine worms, tubeworms, etc.
    CLASS OLIGOCHAETA. Earthworms and many fresh-water annelids
    CLASS HIRUDINEA. Leeches

PHYLUM ARTHROPODA. Arthropods
    CLASS XIPHOSURA. Horseshoe crabs. *Limulus*
    CLASS ARACHNIDA. Spiders, ticks, mites, scorpions, etc.
    CLASS CRUSTACEA. Crustaceans. Lobster, *Daphnia*, *Artemia*
    CLASS CHILOPODA. Centipedes

CLASS DIPLOPODA. Millipedes
CLASS INSECTA. Insects

PHYLUM ECHINODERMATA. Echinoderms
CLASS ASTEROIDEA. Sea stars
CLASS OPHIUROIDEA. Brittle stars
CLASS ECHINOIDEA. Sea urchins and sand dollars
CLASS HOLOTHUROIDEA. Sea cucumbers

PHYLUM CHORDATA. Chordates
SUBPHYLUM UROCHORDATA (or TUNICATA). Tunicates
SUBPHYLUM CEPHALOCHORDATA. Amphioxus
SUBPHYLUM VERTEBRATA. Vertebrates
CLASS AGNATHA. Jawless fishes. Lampreys and hagfish

CLASS CHONDRICHTHYES. Cartilaginous fishes. Elasmobranchs
(sharks, skates, rays)

CLASS OSTEICHTHYES. Bony fishes
Subclass Crossopterygii. *Latimeria*
Subclass Dipnoi. Lung fishes
Subclass Actinopterygii. Higher bony fishes

CLASS AMPHIBIA. Amphibians
Order Anura. Frogs and toads
Order Urodela. Salamanders

CLASS REPTILIA. Reptiles
Order Chelonia. Turtles
Order Crocodylia. Crocodiles and alligators
Order Squamata. Lacertilia, lizards. Ophidia, snakes

CLASS AVES. Birds

CLASS MAMMALIA. Mammals
Prototheria
Order Monotremata. Egg-laying mammals (echidna and platypus)
Metatheria
Order Marsupialia. Marsupials (kangaroos, opossum, etc.)
Eutheria. Eutherian placental mammals
Orders: Insectivores (moles, shrews, etc.), bats, primates (lemurs, monkeys, apes, man), sloths and armadillos, rabbits and hares, rodents, whales and dolphins, carnivores and seals, elephants, manatees and sea cows, ungulates.

Index